Biopharmaceutical Manufacturing, Volume 1

Regulatory processes

Biopharmaceutical Manufacturing, Volume 1

Regulatory processes

Sarfaraz Niazi
University of Illinois, Chicago, IL, USA

Sunitha Lokesh
Pharmaceutical Scientist LLC, Deerfield, IL, USA

IOP Publishing, Bristol, UK

ISBN 978-0-7503-3175-3 (ebook)
ISBN 978-0-7503-3173-9 (print)
ISBN 978-0-7503-3176-0 (myPrint)
ISBN 978-0-7503-3174-6 (mobi)

DOI 10.1088/978-0-7503-3175-3

Version: 20211201

IOP ebooks

British Library Cataloguing-in-Publication Data: A catalogue record for this book is available from the British Library.

Published by IOP Publishing, wholly owned by The Institute of Physics, London

IOP Publishing, Temple Circus, Temple Way, Bristol, BS1 6HG, UK

US Office: IOP Publishing, Inc., 190 North Independence Mall West, Suite 601, Philadelphia, PA 19106, USA

To Dr Sarah Yim, whose humility, scientific wisdom, and vision help many minds grow at the FDA and in the biosimilars community. —Sarfaraz K Niazi

I am thankful to my father for encouraging me to be a curious scientist and my husband for supporting me to practice my passion. —Sunitha Lokesh

Contents

Preface

Overview

The evolution of humanity owes much to the resilience of *Homo sapiens* against the challenges of Nature through changing lifestyles as knowledge became widespread. In prehistoric times, from about 3.5 million years ago until writing emerged around 5500 years ago, there was less spread of knowledge because there was little long-distance travel. Writing did not become a widespread tool for disseminating knowledge until the nineteenth century when education became more accessible.

The end of the era of prehistory occurred at very different dates in different places. It came earlier in regions such as Mesopotamia, the Indus River Valley, and Egypt, where the development of medicines was recorded. However, myths were also believed widely, such as osteoarthritis being caused by lifting heavy objects, which was routine work for most.

Throughout human history, it has proven to be advantageous to utilize other life forms as resources. The use of genetic engineering to transform yeast, bacteria, mammalian cells, plants, and viruses into recombinant therapeutic proteins showcases a continuous timeline of discovery and exploration, beginning with hunting and gathering, then the domestication of animals and crops, ultimately leading to the production of therapeutic proteins.

In prehistoric cultures, plant products (herbs and chemicals derived from natural sources) were among the oldest cures for illness. Earths and clays may also have offered some of the first medicines of prehistoric peoples. Moving from prehistoric times to ancient civilizations, the Egyptian Imhotep describes the diagnosis and treatment of 200 diseases in 2600 BC. The birth of Hippocrates in 460 BC created much of the concept of medicine, and Galen brought much to the science of medicine in 130 AD. Pedanius Dioscorides wrote *De Materia Medica* in around 60 AD. Vaccination against communicable diseases occupies a large portion of medicine's history, starting with the Persian physician Rhazes, who identified smallpox in 910 (although Pylarini gave the first smallpox inoculations in 1701). Avicenna wrote *The Book of Healing* and *The Canon of Medicine* in 1010. Anton van Leeuwenhoek observed bacteria under a microscope in 1683. However, it was not until 1857 when Louis Pasteur identified germs as the cause of disease and developed vaccines for anthrax, rabies, and tuberculosis. Many vaccines have followed.

The first chemically synthesized drug was chloral hydrate, in 1832, and aspirin was first produced in 1899. Since then millions of new molecules have been synthesized. A large number have ended up as effective drugs to treat just about every type of ailment, from modulating immune systems to inactivating viruses.

The discovery of insulin in 1922 revolutionized biological medicine. Insulin was also the first biopharmaceutical drug developed by recombinant technology. Penicillin became available in 1928 thanks to Fleming and streptomycin thanks to Waksman.

Individual enzymatic transformation phases with micro-organisms in chemical manufacturing pathways, such as the biotransformation of steroids in 1950, expanded the scope of biotechnological pharmaceutical manufacturing. Growing knowledge of the regulation of primary and secondary metabolite production, as well as expertise with microbes as biological agents brought significant advances in the field of biotechnology drugs.

In 2006 the first vaccine directed against a specific cause of cancer was approved. The first mRNA vaccine was approved in 2021 to treat the coronavirus COVID-19. This is a significant milestone that has opened the door to many new and novel biopharmaceutical products.

Definitions

Biopharmaceuticals

In writing this book, we need to define clearly what a 'biopharmaceutical product' is. This is necessary to bring a focus to the scope of the book. Chapter 1 describes how the definition of biopharmaceutical products varies between regulatory agencies and within the scientific literature, and why we created a new definition to encompass the concept that is described in this book:

> A biopharmaceutical drug product, a biologic(al) medical product, or a therapeutic biologic, is any nonendogenous macromolecule manufactured in, extracted from, or semi-synthesized from an engineered living entity.

Biotechnology

A similar dilemma arose in defining 'biotechnology', which is the critical technology for the manufacturing of biopharmaceuticals. The term 'biotechnology' has only been defined recently. It is important to look at how it is understood in different regions since this affects how regulatory controls on biotechnology-derived items are assessed. The definitions of the world's principal regulatory regions are listed in table P.1.

The use of biological entities to manufacture industrial products is known as biotechnology. The term 'biotechnology' defines numerous areas of application, including alcohol brewing, antibiotic production, and dairy processing.

For hundreds of years, brewers have used yeast to ferment grain into alcohol. Similarly, farmers and breeders use 'genetic engineering' to develop crops and livestock by selecting desirable traits in plants and animals. Scientists have only recently altered an organism's genetic material at the cellular or molecular level thanks to 'new' biotechnology techniques. Although these approaches are more accurate, the findings are comparable to those obtained using traditional genetic techniques that include whole species. The word 'biotechnology-derived products' (BDPs) refers to things produced using modern biotechnology techniques. The development of BDPs presents several obstacles, including the convergence of numerous sciences and an almost artful application of technology to produce consistent results.

However, the current interest in biotechnology is primarily the result of two major advances:

- Gene transplantation made possible by recombinant DNA (rDNA) technology (gene coding). For example, a desired protein could be inserted into a prokaryotic or eukaryotic cell, and that cell could express the desired protein.

Table P.1. Global definitions of biotechnology.

Jurisdiction	Definition	Comments
Britain	A strategy for applying biological systems, structures, or processes to both manufacturing and service industries.	Manufacturing and services industries are non-committal, broad, technology-driven, and bureaucratic.
European	Biotechnological applications applied to the manufacturing and service industries.	Changes in the definition of 'components and targets,' with the word 'integrated' playing a prominent role in defining biotechnology.
Japan	The integration of biochemistry, microbiology, and engineering sciences for the technological application of microbiological, biotechnological, and engineering sciences.	The complete utilization of technological means, such as copying and various other types, can be deemed significant because it is helping to produce 'useful' substances.
USA	Under the supervision of a licensed specialist, the use of biological agents such as bacteria or cellular components.	Harmful effects of using micro-organisms are evident, and it is also apparent that control and concerns about the technology being used unlawfully are also important.

- The spike in the production of monoclonal antibodies during times of increasing demand (i.e. antibodies arising from a single lymphocyte).

According to the Biotechnology Industry Organization, '[b]iotechnology [is] the combination of biology and technology, includ[ing] biologic applications, diagnostic tools, and businesses that improve everyday life by providing solutions to some of life's most vexing problems'.

We are adopting a simple definition of biotechnology as:

> Any nonendogenous macromolecule made in, extracted from, or semi-synthesized from an engineered living entity produced by any technology and labeled as a biopharmaceutical product.

Biopharmaceuticals are macromolecules produced in a genetically modified organism, excluding naturally derived macromolecules. There are several conflicting definitions of biopharmaceuticals; for example, the FDA considers proteins produced by a recombinant process as a biological product subject to a biological license application (BLA) filing. Protein molecules with less than 40 amino acids are not considered proteins, regardless of how they are produced. These proteins are called peptides, some of which are produced using the same technology as used for recombinant antibodies; in this context, biopharmaceuticals are products such as liraglutide with fewer than 40 amino acids produced by recombinant technology with multiple unit processes.

The impact of biopharmaceuticals

Approximately 400 biotechnology medications and vaccines have aided hundreds of millions of people since 1982. Even more are in development, treating illnesses that were once thought to be incurable, such as AIDS, Alzheimer's disease, and stroke prevention. Diseases such as cancer tend to be preventable in a limited period. For example, many enzymes used in the food industry are likely to cause cancer.

Most biopharmaceuticals are recombinant proteins manufactured by expression in genetically modified biological entities such as mammalian cell lines, bacteria, insects, fungi, and plants. The productivity of these entities continues to improve as we learn more about the ways to change their genomes. These technological advances have led to well-characterized genomes, plasmid vector versatility, the availability of different host strains, and cost-effectiveness compared to other expression systems. Biopharmaceuticals are expected to account for up to 50% of all drugs in production over the next five to ten years.

The total number of biopharmaceutical entities approved by the FDA is 218, with multiple BLAs for some. The highest number of BLAs assigned was for somatropin, followed by albumin, which was also the first BLA approved. Trend analysis shows that the trend for new biological entity approvals is becoming significantly higher, albeit slower (figure P.1).

Moreover, while injections have been the primary delivery method for the first biopharmaceuticals, future products will be supplied through, for example, oral and dermatological routes, with inhibited formulations using a range of encapsulation approaches to reduce the biological instability resulting from protein aggregation and denaturation, due to physicochemical changes such as deamination.

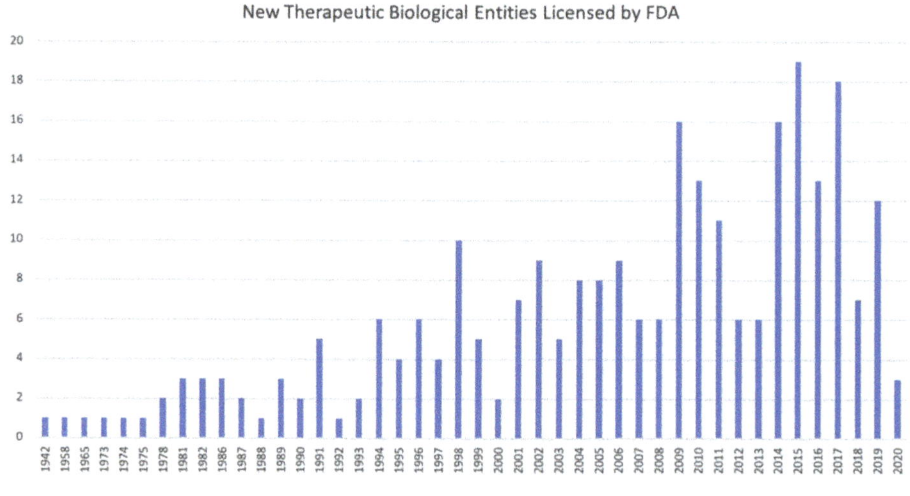

Figure P. 1. BLAs licensed by the FDA from 1942 to March 2020. (Source: *FDA Purple Book*, https://www.fda.gov/drugs/therapeutic-biologics-applications-bla/purple-book-lists-licensed-biological-products-reference-product-exclusivity-and-biosimilarity-or).

Another derivative of organic medicines has been specialist products, a new classification of medicinal products for advanced treatment based on 'genes, cells or tissue engineering', including medicines for gene therapy, somatic cell therapy drugs, and combinations thereof. Advanced therapy medical products (ATMPs) are the most recent classification of pharmaceuticals for advanced therapy products. In European Medicines Agency contexts the term 'advanced therapy' specifically refers to ATMPs, although outside these contexts this term is relatively unspecific.

For example, gene-based and cellular biologics are often at the forefront of biomedical science to treat several medical conditions for which no other therapies exist. The DNA and RNA vaccines will fall in the category of gene therapy, as described in this book.

The FDA has approved several cellular and gene therapy products as of early 2020. Cancer, genetic disorders, and infectious diseases are among the diseases for which gene therapy drugs are being researched. Plasmid DNA is one of the many forms of gene therapy products available. Therapeutic genes may be genetically modified into circular DNA molecules and delivered to human cells. In addition, several cord blood products are permitted for use only in the 'hematopoietic stem cell transplant' procedure in patients with blood-forming (hematopoietic) system disorders. Cord blood contains blood-forming blood cells, which can be used to treat blood cancer patients with conditions such as leukemia and lymphomas, and certain blood-forming and immune system disorders, such as sickle cells and Wiskott–Aldrich syndrome.

Biopharmaceutical development

Drug development is a time-consuming and costly procedure. The costs for a new medicine range from $314 million to $2.8 billion on average, with a 15 year regulatory clearance process. To support the return on investment, research companies have embarked on extensive intellectual property protection that goes well beyond protein identity and gene sequence expression; the patents now include protection of formulations, manufacturing process, cell lines, indications, delivery devices, and many more. Figure P.2 shows the top 20 biopharmaceuticals and their associated patents. A complete list is provided in chapter 8.

The technology for the development and manufacturing of biopharmaceuticals has transformed significantly over the past couple of decades, with the introduction of single-use systems and novel concepts such as continuous manufacturing. The drivers for new technology include the regulatory constraints for assuring product safety in addition to the cost of goods. For example, not long ago many companies were manufacturing blood products across the globe. However, when the FDA brought in new guidelines in the 1980s on the virus clearance of animal products, the entire industry almost collapsed, leaving only a few ventures into this complex product that requires extensive testing to assure safety.

Automation is becoming a norm for most industries, the biopharmaceutical industry being no exception, but the regulatory constraints make the adoption of such modifications cumbersome and expensive.

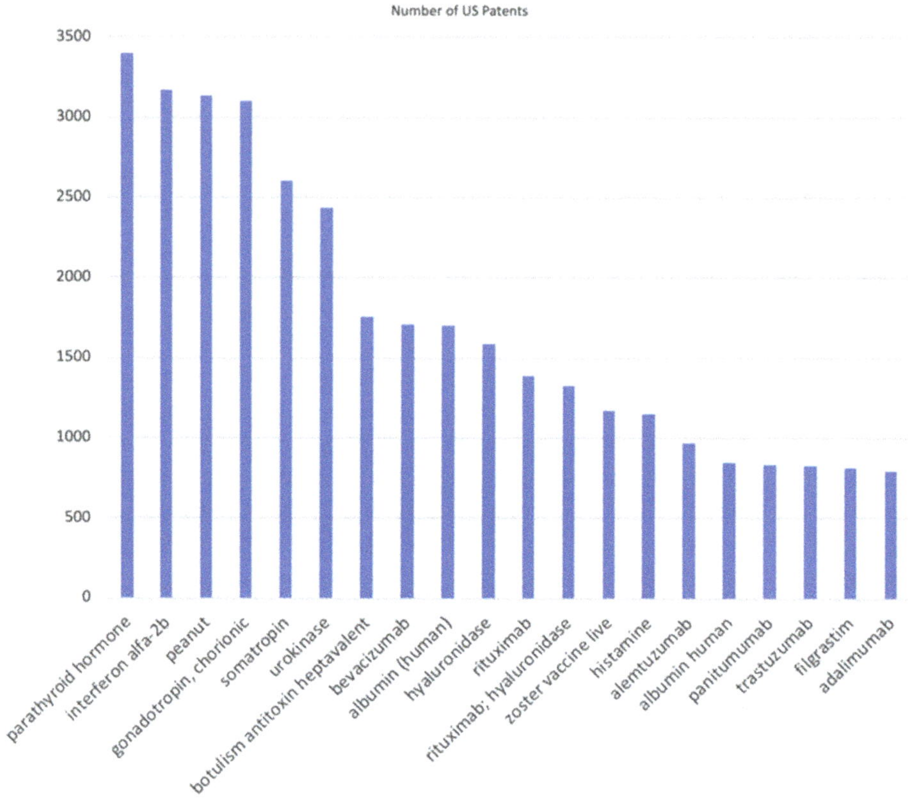

Figure P.2. The number of patents awarded to biological molecules—top candidates.

Global markets

The global pharmaceutical market amounted to $1.2 trillion in 2018, up from $100 billion in 2017, and will rise to $1.5 trillion (based on invoice pricing) in 2023 with a composite annual growth rate (CAGR) of 4%–5%. This is below 6.3% of the CAGR for 2014–2018. In 2018 expenditure in the United States amounted to $485 billion, 5.2% above the previous year, and spending in 2023 will rise to $625 billion.

The top-ten selling products in 2019 included seven biotherapeutic products (figure P.1). All except three, Xarelto, Revlimid, and Eliquis, are biopharmaceuticals. In 2019 biopharmaceutical products had a total market of about $300 billion, growing at a CAGR of 12%. With more than 50% of all new drug applications now constituting biological drugs, biological medicines will soon reach 50% of the global market. In addition, a significant push in the biopharmaceutical industry comes from biosimilars as more patents expire (figure P.3).

About this book

This book on biopharmaceutical manufacturing is divided into two volumes. The first volume, *Regulatory Aspects*, provides an overview of the science behind

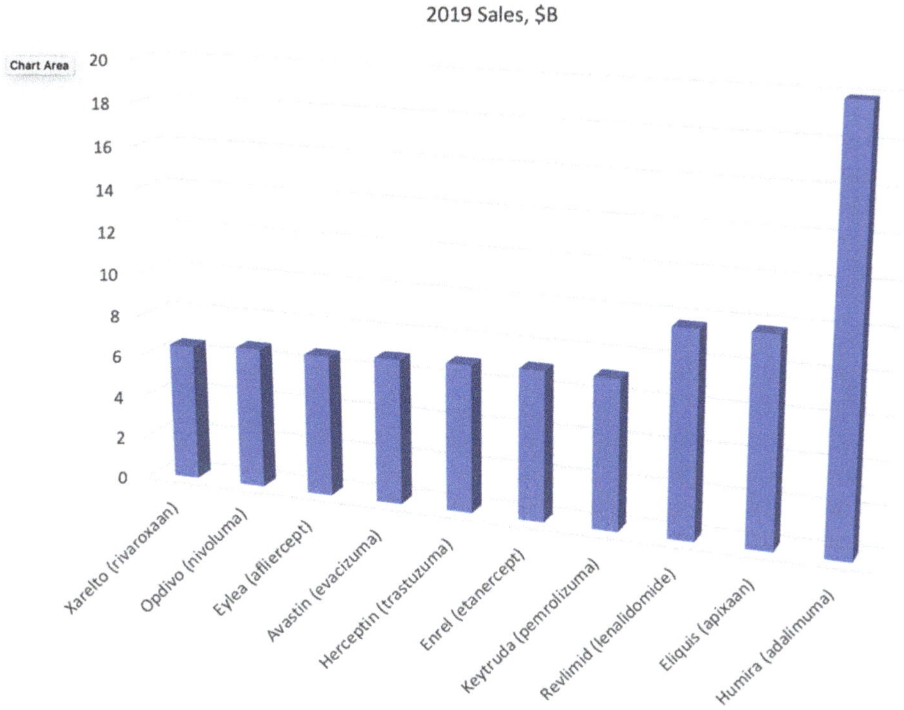

Figure P.3. Sales of the top-ten drugs in 2019 (in billions of dollars). All except three are biopharmaceuticals.

biopharmaceuticals pivotal to securing regulatory approval for commercial biopharmaceutical products. The second volume, *Unit Processes*, is a detailed review of the manufacturing steps, technology implementation, and systems analysis leading to the commercial production of biopharmaceuticals.

The leading technology used in the manufacture of biopharmaceuticals is relatively new, and we foresee many spin-offs of the current recombinant engineering to manufacture many products in the future. Therefore, to expand the utility of this book we have included descriptions of a few emerging technologies that have yet to be approved for human use.

In writing this book we focused on the emerging role of recombinant technology in the manufacture of biological products, evaluated the scientific changes that are arriving rapidly in biotechnology, examined the fast expansion of the industry, and realized that there is a need to combine multiple disciplines to achieve successful commercial manufacturing. The workforce needed for this fast-expanding industry is in extremely short supply, while higher qualification requirements continue to increase as newer concepts and applications come to the market. This treatise should serve well the students who intend to adopt this industry as their career, graduate students who are specializing in biotechnology, and above all those who are actively engaged in the development, regulatory filing, manufacture, and assuring regulatory compliance to secure the market authorization of BLA for these products.

One indication of how fast the field of biopharmaceutical manufacturing is expanding can be gauged from the publications listed under the heading of 'biopharmaceutical manufacturing', with over 5000 publications over the last five years and 10 000 publications over ten years[1]. A search for 'biopharmaceuticals' shows more than 1.4 million hits in ten years and a half as many over five years.

We have developed and manufactured biopharmaceutical products, and this experience is offered in writing these chapters. SKN has written several books on this subject, including the first book on the topic, *Handbook of Biogeneric Therapeutic Proteins*, followed by the two-volume *Handbook of Biosimilars and Interchangeable Products*, the book *Disposable Manufacturing*, a book on regulatory aspects, *FDA Perspective of Biosimilarity*, and the textbook *Modern Bioprocessing*. Each of these books presented a different view of the industry and technology, with a focus on biopharmaceuticals. While we have focused on the scientific principles, we have merged the learning with practical examples. The diversity of topics in this book is intentional. In our experience, the scientists and engineers involved in development and manufacturing need to be fully aware of the patent systems and regulatory compliance needs, and stay informed on the fast-emerging trends in the art and science of biopharmaceuticals. Given below is a summary of the chapters in each volume.

Volume 1. *Biopharmaceutical Manufacturing: Regulatory Processes*

The commercial manufacture of biopharmaceuticals is challenging because of the highly rigorous regulatory compliance requirements that may not be as restrictive for other types of drugs. The reason for this is that any change in the structure of biopharmaceuticals during the manufacturing process can result in severely altered responses and side effects. The impact of these changes cannot be readily established since many effects of these drugs appear in the long term, such as in the form of altered immune response that may create new diseases such as diabetes, multiple sclerosis, and other autoimmune disorders. Understanding the regulatory concerns requires a deep understanding of the nature of the products, the strategic and tactical challenges in their manufacture, and a keen sense of their development process.

Volume 1 encompasses all the necessary details intended to teach and train the scientists and technicians involved in commercial manufacturing. Their role extends to assuring the quality of the product and the manufacturing facility's compliance. In all types of pharmaceutical manufacturing it is important that all those involved in any stage of the manufacturing be fully engaged since it is impossible to maintain current good manufacturing practice (cGMP) compliance without the collaboration of all teams—this is the focus of this volume.

Chapter 1. Introduction to biopharmaceuticals

This introductory chapter provides a clear definition of biopharmaceuticals, including biosimilars, commercial production, and global regulatory compliance requirements. Detailed descriptions of the nature and properties of biopharmaceuticals and how these macromolecules have shifted the arena of drug therapy are provided. A classification of

[1] https://pubmed.ncbi.nlm.nih.gov/?term=biopharmaceutical+manufacturing&sort=pubdate&sort_order=asc.

the drug types that constitute biopharmaceuticals is given and examples are provided. There is extensive discussion of protein structure and the properties of hormones, peptides, enzymes, protein scaffolds, antibodies, and others, and the structural modifications that can change the disposition characteristics of biological molecules are described in detail. The critical concern in using biological drugs is their immunogenicity, which we cover in detail to understand the structural elements that can produce an immune response. The factors that affect the pharmacokinetics of therapeutic proteins are essential for understanding in developing a safe and effective product. The manufacturing of biopharmaceuticals involves recombinant technology, and detailed analyses of the expression systems and their relative advantages and disadvantages are provided. This chapter's main objective is to educate scientists and technicians engaged in the manufacture of biopharmaceuticals about the essential science elements required to develop and manufacture these products and secure their regulatory approval.

Chapter 2. Antibody biopharmaceuticals
Antibodies form the most significant commercial category of biopharmaceutical products; this chapter describes in detail their structure and mode of action, with a listing of 83 FDA approved monoclonal antibodies, types of antibodies, and their targets including bispecific, multi-specific, antibody fragments, single chain, humanized, chimeric, and fully human monoclonal antibodies. Related topics include affinity maturation, antigenized antibodies, IgG1 fusion proteins, and drug or toxin conjugates. The development technologies of mouse hybridoma, transgenic animals, phage display, and single B-cells are detailed and compared for clinical and cost advantages. The commercial production of monoclonal antibodies is summarized, and an extensive list of online databases to find antibody properties is provided.

Chapter 3. Gene and cell therapy biopharmaceuticals
The newer technologies of gene and cell therapy and gene editing are described in a summary form in chapter 3. Gene and cell therapy constitute recent advances in the field of biopharmaceutical products. An overview of the diseases, risks, development, and ethical issues is provided, along with a comprehensive list of currently approved products. Gene therapy and cell therapy, including DNA and mRNA vaccines and CAR-T (T-cell therapy using chimeric antigen receptor (CAR)) techniques, are introduced. Gene editing technologies define the methodologies and their relative advantages. Upstream and downstream technologies for gene and cell therapy products and allogenic products are described, including regulatory controls, characterization of the cell population, release testing, and radioisotope tagging. Issues related to vectors and vector preparation are discussed. Finally, preclinical evaluation methods of evaluation and challenges in commercializing gene and cell therapy products are described.

Chapter 4. Formulation of biopharmaceuticals
The delivery of biopharmaceuticals is highly complex due to their structural instability, quick physical degradation, and complex chemical interactions with excipients and

degradation that can lead to immunogenicity and other side effects. Extensive descriptions of chemical degradation and the methods to obtain stability in the formulated products are provided with examples of commercial formulations. While the current methodology for the delivery biopharmaceuticals remains parenteral, a large number of new dosage forms have emerged, including oral, nasal, transdermal, pulmonary, ocular polymer-based, hydrogels, lipid-based emulsions, liposomes, and nanoparticle systems. Also included is a description of nontraditional dosage forms.

Chapter 5. Drug development cycle
Regulatory approval of biopharmaceutical products is subject to strict compliance and the creation of a registration dossier. In this chapter we describe in sufficient detail all the steps related to the filing of a BLA in the US or a marketing authorization application (MAA) in the EU. The topics include early discovery documentation, pharmacopeia, preclinical research, and IND steps to phase 4. The expectations of regulatory inspections (audits) are presented based on the authors' long experience in both the US and EU systems. This chapter also details the stepwise approach taken in the US, EU, and Japan to evaluate a regulatory dossier. Biopharmaceutical manufacturing has changed significantly since the 1980s, primarily because of the regulatory authorities' stricter controls on therapeutic products. This chapter will provide a global view of how the agencies approve products and facilities to manufacture these products. A comprehensive approach requires a complete understanding of these steps to make the large-scale commercial manufacturing of biopharmaceuticals possible.

Chapter 6. Biosimilar biopharmaceuticals
Biosimilars represent the fastest growing category of biopharmaceuticals, where a copy of the originator's biological drug is developed as a low-cost competitor to the first BLA product. The regulatory agencies have placed extreme caution on safety, and the development process takes a stepwise approach, described in detail in this chapter. Regulations across the globe are compared, and advice is provided on cost optimization of development. The key elements to establish biosimilarity, including analytical assessment, nonclinical testing, clinical pharmacology testing, and clinical efficacy testing, are described to focus on expediting the development process. A summary of FDA licensed products with more information on the studies submitted and the status of biosimilars in the EU is also provided.

Chapter 7. Intellectual property considerations
The development of biopharmaceuticals is an expensive exercise and the burden of reducing the risk of litigation for the infringement of other patents and protecting one's own intellectual property requires close collaboration between scientists and legal teams. This chapter describes the protection of intellectual property and walks the reader through a process to avoid infringing on that of others. The finer points of the definition of the vocabulary used in a patent application and the legal language are described in simple language. All types of patents, differences in global patent laws, and a detailed description of patents related to biological drugs are provided. Details on writing freedom-to-operate documents are provided, along with a

comprehensive list of the patent expiry of biological drugs and the approved BLAs and their patent expiry.

Volume 2. *Biopharmaceutical Manufacturing: Unit Processes*
In recent years, most biopharmaceutical manufacturing operations have narrowed down to upstream processes. A product is produced (expressed) and separated from a culture medium and purified (downstream) to a quality suitable for human use. Commercial manufacturing differs from a laboratory-scale set-up where the cost and time for production are often not pivotal. Today antibodies are made in quantities of thousands of kilograms, and some cytokines only into the hundreds of grams; both scales require different scopes of the same unit operations and processes. Volume 2 is geared toward providing a practical plan for deploying recombinant manufacturing technology, starting with the creation of a productive cell line and finishing with packaging a product for human use. Economy and safety of the process may not always be mutually exclusive; the key is to assure a consistent high yield while maintaining the quality standards assured in the regulatory filing. This volume provides practical training for technicians and advises scientists on modifying the process and maintaining cGMP compliance.

Chapter 1. Understanding bioprocessing
Biopharmaceutical manufacturing is complex due to the multiple steps that must integrate well within meeting cGMP compliance requirements. This chapter summarizes the manufacturing process, which is discussed in greater detail in the following chapters of this volume. This chapter further includes sections that do not fit into other unit processes, such as cost containment, documentation, and cGMP compliance assurance. The personnel involved in biopharmaceutical manufacturing must understand the entire process regardless of their specific role, which may be limited to a specific unit process, to enable smooth integration. A biopharmaceutical manufacturing chain's strength is only as strong as its weakest link, making it essential that the entire team understands each step. This chapter provides a stepwise description of clearly distinct unit processes or actions, a discussion of how they are connected, and the routine practices and development choices that make a manufacturing process commercially feasible. We strongly urge the reader to read this chapter with great scrutiny to make best use of the rest of the book.

Chapter 2. Recombinant manufacturing system
The manufacturing engine for biopharmaceutical products is a living entity—a bacterium, a yeast, a mammalian cell, or transgenic cells and species. The DNA is modified to combine with a foreign gene of expression entity. The genetically modified entity can express proteins of interest. This chapter discusses recombinant DNA technology, the types of entities available for recombination of their DNA, and a detailed comparison of each of these expression systems' advantages and disadvantages. Understanding the fundamentals governing these processes is extremely important to process design. This chapter focuses on the fundamentals of protein expression, the availability of various host systems, and the selection basis for the intended application.

Chapter 3. Cell line development

The cell line expressing the desired target protein is the core of the recombinant therapeutic process. Given the various choices of expression system available, the simplicity of microbial systems (*Escherichia coli*) and the popular mammalian systems (Chinese hamster ovary cells) are preferred. The advantages of microbial systems lie in their fast growth and easy manipulation, but the lack of post-translational modifications (glycosylation) is a major drawback. In contrast, mammalian cell cultures are more complex than microbial cultures. They have a longer doubling time, but this is compensated for in their mechanism to allow posttranslational modifications essential for protein functionality. Stable cell line development, a time-consuming process, entails incorporating linearized plasmid DNA encoding the therapeutic protein into the host genome. When the cells divide the transgenes are transferred to the subsequent generations. Screening for recombinants is performed by including a selectable marker on the plasmid. This can be an antibiotic resistance gene or a metabolic gene. The recombinants are heterogeneous concerning their growth characteristics, expression (productivity), and product quality. Therefore it is essential to screen the potential clones to establish the right production cell line based on these features. A necessary consideration for mammalian cell lines is to ensure that the cell line is clonally derived from a single cell, confirming the homogeneity (genotypic and phenotypic). The chosen cell line (production) must demonstrate proper growth, productivity, and product quality over the generations required to run the manufacturing process.

Chapter 4. Upstream equipment and systems

The technology for the upstream unit process has evolved significantly over the past few decades through new uses of systems initially developed for fermentation to allow the growth of many different cell cultures to yield a target protein molecule. The biological process of cell growth and expression when a genetically modified organism is involved is more sensitive to bioprocess conditions. To overcome these shortcomings, bioreactors are now designed with high-level technology to enable reproducible batch yields. This chapter describes many types of equipment and operational processes that now find their place at the core of biopharmaceutical production. While the science of bioprocessing requires a greater depth of learning about the physics and chemistry of bioprocess, in this chapter the most commonly used equipment and how it is made operational in a commercial setting is the focus.

Chapter 5. Upstream process

The upstream unit process yields a crude product either as an inclusion body in bacteria or a secreted product from a bacterial or a mammalian cell. Optimization of the upstream process is pivotal in producing a cost-effective product at a commercial scale. This requires making an optimal choice of the cell line, culture media, and types of operation, and then, once the product is expressed, removing it from the bioreactors and separating cell debris and other impurities to prepare the yield for downstream purification. If the upstream process is not optimized the downstream process will not achieve the desired product safety and efficacy. This chapter describes the upstream systems and the materials used in achieving a highly productive yield of a recombinant biopharmaceutical.

Chapter 6. Downstream process

The target protein synthesized by the cells, either by secretion in the medium (mammalian cells) or intracellularly as inclusion bodies, must be isolated from the host cells and purified to remove or reduce product-related and process-related impurities to produce a final product with sufficient purity. Downstream processing involves several unit operations, including but not limited to centrifugation, filtration, chromatography purification, precipitation, and concentration ultrafiltration/diafiltration (UF/DF). The impurities vary in their nature. The overall downstream process design should be robust and consistent in ensuring that these impurities are removed or maintained within established limits that do not pose a safety concern. In the case of mammalian processes or processes that use raw materials of animal or human origin, adventitious viruses should be addressed in the process design to ensure there is no risk of viral contamination in the final product. Downstream processes may include more than one orthogonal step for impurity removal. Platform processes allow for a starting point for the process design of molecules that belong to the same class, but further optimization specific to the target protein is necessary.

Chapter 7. Process and product lifecycle development

Production starts with a batch record for a commercial product that is primed for economically viable manufacturing. The process evolves from a development laboratory setting step-by-step procedures into a final and mature manufacturing process. Focusing solely on qualification efforts without recognizing the development constraints is an imperfect exercise. A good process design and building quality into the process are of primary importance for ensuring a robust process. The process development is initiated with process design, wherein the target product is identified. The development of the product profile is followed by identifying the process steps, characterization, and validation using a risk assessment, prior knowledge, and process and product understanding to establish a robust system that allows consistent manufacturing of the desired product quality. These activities help establish a well-defined control strategy. Manufacturers must maintain the process in a state of control for the product's lifecycle and manufacture after defining and verifying the process, even as the environment, raw materials, equipment, personnel, and manufacturing process shift.

Chapter 8. Quality and compliance systems

Commercial manufacturing mandates compliance with regulatory requirements that are determined by the quality system in place. While scientists and technicians are familiar with the technology aspect, cGMP and good laboratory practice (GLP) compliance require a different level of understanding that is documentation based and also continued compliance with the systems. This chapter describes how to create a cGMP and GLP compliant system starting with the responsibility of the management, resource allocation, design and qualification of manufacturing systems, and continuous monitoring of quality attributes. The quality assurance system emphasizes the need for a validation master plan that includes elements specific to biopharmaceuticals, including virus validation, virus inactivation, and analytical methods validation. Details of process validation from a statutory and analytical perspective are presented. The discussion of good laboratory practice covers the current regulatory requirements, creating a plan and documentation to stay compliant, and specific analytical methodologies issues. A valuable database of the regulatory inspection results from over 20 years of FDA audits is included to provide a self-audit analysis to assure full compliance.

Chapter 9. Single-use technology
Single-use technology has long been part of biopharmaceutical manufacturing but remained limited to a few components until a couple of decades ago when the regulator imposed requirements to isolate contamination, at a higher capital cost of establishing traditional manufacturing facilities. Improved yield and quality become evident, resulting in the fast growth of single-use components that now range from starting a cell culture to upstream, downstream, and fill and finish. Since big pharma had already invested in fixed-pipe technology, single-use technology remains favored by smaller or newer companies. This chapter presents a detailed outlook of the present and the future of single-use technology, that is expected to become the primary technology in the next couple of decades. This chapter provides introductions to single-use technology companies' sources, the regulatory compliance details, and suggestions for creating a comprehensive single-use train.

Chapter 10. Advancements and trends in biomanufacturing
Recombinant protein therapeutics is the single most important product group driving today's pharmaceutical development with a broad range of indications—oncology to cardiovascular and other infectious diseases. With several developments and blockbuster drugs on the market, there is an increasing need to enable technologies and other supporting functionalities to keep pace with this demand. Monoclonal antibodies have emerged as top-class in the recombinant protein group. Since the first licensed product, manufacturing processes have evolved over the years, starting with improving low productivity processes, expanding the use of host cell systems, and engineering these systems to their full potential to express both simple and complex proteins. These upstream advancements have shifted the process constraints to the downstream processes, mainly due to a lack of flexibility in handling high productivity processes. The projected future demand for recombinant therapeutics will probably shift the emphasis onto developing processes that can accommodate multiple products in a flexible production facility. However, this is likely to be limited in the current operation state due to the current plant designs and a lack of flexibility with facilities. The next-generation facilities will maximize facility utilization, have a reduced footprint, ease scalability, and reduce the downtime between batches or even products in a multiproduct facility, but most importantly, reduce processing time and cost, and improve efficiencies. Single-use technologies, process intensification strategies for upstream and downstream operations, and modular facilities enable technologies to be widely integrated into bioprocessing.

Additionally, a glossary of terms is included for quick reference and teaching of terminologies and concepts.

Acknowledgments

We owe our gratitude to our scientific and professional colleagues, particularly those we only know from the field's seminal literature but have never met. We may have subconsciously quoted their work, assuming it was all in the public domain; I hope they would forgive us for doing so, as it would be impossible to recognize them. An extensive bibliography is not required.

Finally, we would like to admit our mistakes, and we could not find a better statement than that which appeared in the first edition of *Encyclopedia Britannica* (1786):

> WITH regard to errors, in general, whether falling under the denomination of mental, typographical, or accidental, we are conscious of being able to point out a greater number than any critic whatever. Men who are acquainted with the innumerable difficulties attending the execution of a work of such an extensive nature will make proper allowances. To these, we appeal and shall rest satisfied with the judgment they pronounce.

We will appreciate receiving your comments to improve this treatise in the future, and most kindly, if you find any mistakes.

Disclaimer: The authors do not accept responsibility for any technical or legal suggestions or advice provided in this book; all views expressed in this book are those of the authors in their personal capacity and not as the Patent Agent of the US Patent and Trademark Office, as an officer of any company or in any academic positions held, or in any capacity as advisors to regulatory agencies.

Sarfaraz K Niazi, PhD (niazi@niazi.com), Deerfield, IL, USA
Sunitha Lokesh, MS (Sunitha.lokesh@pharmsci.com), Rolling Meadows, IL, USA
December 2020

Author biographies

Sarfaraz K Niazi

Sarfaraz K Niazi, PhD, is an Adjunct Professor at the University of Illinois. He has authored 60+ major books, 100+ research papers, and 100+ patents, mainly in bioprocessing. He has hands-on experience establishing biopharmaceutical projects, from concept to market, and serves as an advisor to several regulatory agencies, including the FDA. As CEO of Pharmaceutical Scientist LLC, he provides consultancy in biosimilars, bioequivalence, regulatory planning, and the adoption of new technologies. He can be contacted at niazi@niazi.com; author's website: www.niazi.com; consulting website www.pharmsci.com.

Sunitha Lokesh

Sunitha Lokesh is the Director of Scientific Affairs at Pharmaceutical Scientist, LLC. She is a biotechnology engineer with an MS degree from the Illinois Institute of Technology. She has 12+ years of experience creating recombinant cell lines, developing upstream and downstream processes, coordinating testing strategies, and coordinating the filing of several regulatory applications with the FDA. She provides consulting services in developing master plans for manufacturing biopharmaceuticals, regulatory guidance, and implementation of innovative technologies in the manufacturing of biopharmaceuticals. She can be contacted at sunitha.lokesh@pharmsci.com; www.pharmsci.com.

Chapter 1

Introduction to biopharmaceuticals

This chapter clearly defines biopharmaceuticals derived as biological medicines, including biosimilars, commercial production, and global regulatory compliance requirements. A detailed description of the nature and properties of biopharmaceutical properties and how these macromolecules have shifted the arena of drug therapy are provided. A classification of the drug types that constitute biopharmaceuticals is defined, and examples are provided. Extensive discussion of protein structure and properties of hormones, peptides, enzymes, protein scaffolds, antibodies, and others, along with structural modifications that can modify biological molecules' disposition characteristics, is described in detail. The critical concern in using biological drugs is their immunogenicity, which is covered with more information to understand the structural elements that can produce an immune response. Factors that affect therapeutic proteins' pharmacokinetics are essential to understand in developing a safe and effective product. Biopharmaceutical manufacturing involves recombinant technology; detailed analysis of the expression systems, their relative advantages and disadvantages are described. Newer gene and cell therapy and gene editing technologies are described in a summary form and covered in chapter 3. This chapter's main objective is to educate scientists and technicians engaged in manufacturing biopharmaceuticals about the essential science elements required to develop and manufacture these products and secure their regulatory approvals.

1.1 Overview

In writing this book we needed to establish a clear definition of biopharmaceuticals, given the wide disparity in biopharmaceuticals' scope. The term biopharmaceuticals is composed of two words, *bio* means produced by biological means and *pharmaceutical* means 'relating to medicinal drugs, or their preparation, use, or sale' as an adjective and 'a compound manufactured for use as a medicinal drug' as a noun, where medicinal means 'relating to or involving medicines or drugs'.

doi:10.1088/978-0-7503-3175-3ch1

According to the official definition, a drug is recognized by an official pharmacopeia or formulary. It is also a substance intended for use in the diagnosis, cure, mitigation, treatment, or prevention of disease, or anything, except for food, that is used to change the function or structure of body.

In some countries, such as the UK, biological macromolecules and nucleic drugs are commonly called 'biological products'.

High-priced medicines, mostly biologics, are classified as specialty drugs by the European Medicines Agency (EMA). According to the EMA, advanced medicinal therapy products (ATMPs) are medicines for human use dependent on gene therapy, somatic cell therapy, or tissue engineering. These are all examples of medicines based on genes, cells, or tissue-engineered products or combinations thereof. The EMA uses the term 'advanced therapies' to refer to ATMPs, but it is less precise outside of those contexts.

Biological drugs are 'licensed' by the FDA (by the Center for Biologics Evaluation and Research (CBER)) instead of 'approved', dating back to when the FDA would test and allow the release of a product while the manufacturer was 'licensed' to manufacture it. For example, CBER licensed therapeutic proteins, and even though CDER now deals with these products the terminology remains the same; CDER now licenses and approves drugs. For this regulatory classification, a protein is defined as a compound with 40 or more amino acids, and a biopharmaceutical peptide with less than 40 amino acids is classified as a non-biologic drug even if it is produced using recombinant engineering.

A biopharmaceutical is referred to in the scientific literature as a biologic(al) medicinal product, or biologic, in addition to the regulatory definitions. Furthermore, many prescription drug products are derived from biological sources, whether made, extracted, or semi-synthesized. Incompletely synthesized pharmaceuticals include vaccines, blood, blood products, and allogenic. Cell therapy uses somatic cells, gene therapies, tissues, recombinant therapeutic proteins, and living medicines.

Biopharmaceuticals are made up of sugars, proteins, nucleic acids, and complex combinations of these compounds, or living cells and tissues. They are isolated from living sources such as humans, animals, plants, fungi, and microbes (or their precursors or components). If a drug is labeled biopharmaceutical, the primary difference is that bacteria, yeast, and mammalian cells express the drugs in living organisms, they are not synthesized chemically. The latter types of pharmaceuticals, on the other hand, are generated through a sequence of chemical reactions. Therefore, excluded from the definition of biopharmaceuticals are blood products, vaccines, and sera.

After a review of the various definitions of biopharmaceuticals, we have decided to formulate a brief and precise definition of 'biopharmaceutical drug product' (BDP):

> A BDP is a therapeutic, preventive or diagnostic entity manufactured in, extracted from, or semi-synthesized from an engineered living entity.

In this context *therapeutic* means to heal disease, while *preventive* means entities administered to prevent a disease's occurrence, but excludes preventive entities such as live or attenuated vaccines, but not the recombinant, DNA, or mRNA vaccines.

In our definition *diagnostic* means an entity used in the diagnosis of disease using components produced from engineered living entities.

An *engineered entity* means an entity whose DNA structure has been modified using recombination technology. A significant concern while using engineered entities comes from the possibility of creating new species when using transgenic animals and plants. ATryn, the first such drug made from the milk of a genetically modified goat, created controversy before its approval by the EMA in June 2006. Cross-contamination of biopharmaceutical crops with non-engineered crops or crops engineered for non-medical purposes is also a concern.

The safety issue of engineered entities does not arise when using cells or organisms since they are discarded at the end of the manufacturing cycle and do not survive outside the bioreactor environment.

A *living entity* means bacteria, mammalian cells, fungi, plants, animals, or any other source capable of replicating itself.

The above description of biopharmaceuticals, therefore, excludes:

- Plant extracts.
- Human or animal tissues or tissue extracts.
- Blood, blood components, plasma, and milk.
- Naturally produced enzymes, antibiotics, and other chemicals in microbes.
- Living entities administered to immunize, such as live or attenuated vector vaccines.

Examples of biopharmaceutical products include:

- Substances that are (almost) identical to the signaling proteins produced by the body. Examples include the protein erythropoietin, which stimulates blood production, the (single) 'growth hormone', and the biosynthesis of human insulin and its analogs.
- Antibodies produced by a single cell are called monoclonal antibodies. These antibodies are similar to those used by the human immune system to combat bacteria and viruses. However, because they are 'suitable', they can be used to neutralize or obstruct any specific substance or cell type (using hybridoma technology or other methods).
- Typically, the immunoglobulin framework is connected to a naturally occurring receptor in the receptor structures (fusion proteins). The receptor provides an exact structural characteristic in this case, whereas the immunoglobulin structure provides stability and additional pharmacological benefits.

Significant kinds of biopharmaceuticals (with a representative example) include:

- Factors of blood (factor VIII and factor IX).
- Thrombolytic agents (tissue plasminogen activators).
- Hormones (insulin, glucagon, growth hormone, gonadotrophins).
- Factors of blood production (erythropoietin, colony-stimulating factors).
- Interferons (α-, β-, γ-interferons).
- Products based on interleukin (interleukin-2).
- Vaccines (hepatitis B surface antigen).

- Single-clone antibodies (bevacizumab, ranibizumab).
- Factors of tumor necrosis (adalimumab)
- Enzymes for therapy (L-asparaginase).
- Tissue cultures to produce vaccines (flu vaccine).
- Viral gene therapy is the manufacture of a virus to artificially incorporate a useful genetic material (DNA and mRNA vaccines).

The above list involves a variety of bioreactor configurations, including photobioreactors, microbial cells (for example, recombinant *E. coli* or yeast cultures), mammalian cell lines, plant cell cultures, and moss plants. Commercial production faces two significant obstacles: manufacturing costs (low-volume, high-purity products are desirable) and contamination by microbes (bacteria, viruses, and mycoplasma). Whole plants are one of the alternative production platforms being investigated (plant-made pharmaceuticals). Volume 2 delves deeper into this subject.

1.1.1 Biosimilars

A biosimilar product has no clinically significant differences to a reference biopharmaceutical product for which a full regulatory filing has been approved. As patents for blockbuster biologics expire, interest in biosimilar manufacturing, also known as follow-on biologics, has grown. As a result, we expect that many of this book's readers will be involved in developing and manufacturing biosimilars.

Generic chemical drugs are made up of simple and fully characterized chemically related active ingredients; the active ingredients are much more complex and diverse in biologicals. Owing to their heterogeneity and high process sensitivity, originators and follow-on biosimilars will see variations in specific variants over time. Both originator and biosimilar biopharmaceuticals must, however, have equivalent safety and clinical performance throughout their lifecycle. Modern analytical instruments monitor process variations and define a particular design space for each biologic (e.g. liquid chromatography, immunoassays, mass spectrometry, etc). Biosimilars, unlike small-molecule generics, require a different regulatory framework requiring more intensive testing than needed for small-molecule generics, yet less than for a new biological product.

The EMA launched the Adapted Biosimilars Pathway in 2003. This pathway is based on a thorough demonstration of the 'comparability' of the 'biosimilar' product to an existing approved product. In addition, the Patient Safety and Affordable Care Act of 2010 in the United States developed an expedited approval process for biosimilar approval by comparing it with an FDA-licensed reference biological product.

Chapter 4 provides details on the development and regulatory approval of biosimilars.

1.1.2 Intellectual property

When a new biopharmaceutical is created, a patent application is typically filed to obtain exclusive manufacturing rights to recoup the cost of biopharmaceutical development. The criteria for obtaining a patent in the United States and Europe differ somewhat, with the European standards being deemed more difficult to satisfy. By 1995 there were 15 600 patent applications and by 2001 there were 34 527. The United States creates the most intellectual property (IP) in the biopharmaceuticals industry, accounting for approximately half of all awarded patents worldwide; however, there is still space for development and innovation. Chapter 8 provides details of the intellectual property management of biopharmaceuticals.

The high cost of pharmaceuticals, particularly biologics, is a significant source of concern. They have rekindled interest in learning more about the federal government's position on expensive new therapeutics development. Many FDA-approved biotherapeutics were discovered by scientists at public-sector research institutions, including Remicade, Enbrel, Humira, and Avastin. Additionally, approximately 25% of new molecular entities were developed as a result of publicly funded research. The following platform technologies were developed with public funds: recombinant DNA technology (Cohen–Boyer patents); methods for producing recombinant bacterial DNA (Riggs–Itakura patents); antibody production and chimerization (Cabilly patents); glycosylated recombinant proteins in mammalian cells (Axel patents); and gene silencing with small interfering RNAs (Mello–Fire patents).

Many new medicines would not have been produced without these platform innovations, resulting in a drastically different economic outlook for the pharmaceuticals industry.

1.1.3 Regulations

1.1.3.1 European Medicines Agency (https://www.ema.europa.eu/en/human-regulatory/research-development/scientific-guidelines/biological-guidelines)

A biological medicinal product is an active substance derived from or extracted from a biological (living) system in the European Union. It requires, in addition to physicochemical testing, biological testing for full characterization. A biological medicinal product's characterization combines testing the active substance and the final medicinal product and the production process and its control. For example:

- Biotechnology or other methods may be used to create the production process. In addition, products can be made using more traditional approaches, such as blood or plasma-derived products and various vaccines.
- The active material is derived from a microbial, animal, human, or plant source and consists of whole microorganisms, mammalian cells, nucleic acids, proteinaceous, or polysaccharide components.
- Therapeutic and immunological pharmaceutical products, gene transfer materials, or cell therapy materials are examples of products based on the modes of action.

The EMA regulations are mostly in concurrence with the FDA regulations, except for technical details discussed in a separate chapter.

1.1.3.2 The Food and Drug Administration (USA) (https://www.fda.gov/files/drugs/published/Biological-Product-Definitions.pdf)

Biological products were first regulated by the National Institutes of Health (NIH) and its predecessors. This regulatory authority was transferred to the FDA in 1972.

The FDA generally regulates biologics under its authorities under the Public Health Service Act (PHSA) but regulates some biologics as drugs under powers in the Federal Food, Drug, and Cosmetic Act (FFDCA). Additionally, biologics are regulated by the Center for Biologics Evaluation and Research (CBER) and the Center for Drug Evaluation and Research (CDER). CBER regulates traditional biologics, such as vaccines, blood and blood products, allergenic extracts, and specific devices and test kits. CBER also oversees gene therapy, cellular therapy, human tissue for transplantation, and xenotransplantation tissue used to transplant nonhuman cells, tissues, or organs into humans.

Most therapeutic biologics are regulated by the CDER as are prescription brand-name and generic drugs and over-the-counter drugs. In 2003 CBER handed over responsibility for therapeutic biologics to CDER which now regulates:

- Monoclonal antibodies, proteins that bind to a particular substance or cell in the body. A drug or toxin can be carried by a monoclonal antibody. Infliximab is an antibody used to treat Crohn's disease, ulcerative colitis, rheumatoid arthritis, and psoriasis. It is an example of a monoclonal antibody medication.
- Cytokines, proteins that regulate the immune system (stimulate or slow it down) and are used to combat cancer, viruses, and other diseases. Interleukins, interferons, and colony-stimulating factors such as filgrastim are examples.
- Growth factors, such as the human growth hormone somatropin, are substances created by the body that control cell division and cell survival.
- Enzymes are proteins that make the body's chemical reactions move faster. Many cell functions, such as cell signaling, development, and division, are aided by enzymes. For example, enzyme inhibitors can be used to treat cancer by blocking the enzymes that cancer cells need to expand.
- Immunomodulators, substances that help the body fight cancer, infection, and other diseases by activating or suppressing the immune system.

The PHSA regulates most biological products approved for marketing by the FDA through a biologics license application (BLA). In comparison, the FDA approves chemical products for marketing through a new drug application (NDA) or abbreviated new drug application (ANDA). To receive licensure, the sponsor (usually the manufacturer) must show in the BLA that the biological product and the facility where it is made, handled, packaged, or stored meet requirements to ensure the product is clean, pure, and potent. Any subsequent change to the permitted manufacturing process, such as a change in the raw material supplier or the replacement of a piece of equipment, necessitates a demonstration of the comparison of the quality parameters before and after the change to ensure that the product's safety and effectiveness are preserved. For example, between 1998 and

October 2014, the brand-name biologic Remicade (infliximab) underwent 37 manufacturing changes, each requiring proof of comparability, through chemical, physical, and biological assays.

Certain biological products were previously regulated as medicines, with NDAs licensed under the FFDCA rather than the NIH's biologics under the PHSA. For example, in 1941 Congress gave the FDA authority over insulin marketing, a natural biological product source. Insulin comprises a short chain of 51 amino acids discovered in the 1940s in the same way that many biologics were—by extracting it from animals—hence the word 'natural source'. Despite its similarities to other biologics, the FDA regulated insulin as a drug rather than a biologic until March 2020. Natural source biological products such as glucagon, human growth hormone, hormones to treat infertility, hormones to control menopause and osteoporosis, and certain therapeutic enzymes have been approved as medicines under the FFDCA rather than biologics under the PHSA (hyaluronidase and urokinase). Under the FFDCA some of these products were regulated as medicines (e.g. insulin and human growth hormone created using recombinant DNA technology). In contrast, others were regulated as biologics under the PHSA (e.g. cytokines, proteins involved in the immune response, and blood factors). Biologics licensed under the FFDCA will be considered licenses under the PHSA as of 23 March 2020.

The biotechnology industry started developing its first biologics for human therapeutic agents in the late 1970s and early 1980s (e.g. recombinant proteins and monoclonal antibodies).

1.2 Biopharmaceutical molecules

1.2.1 Overview

In comparison to conventional chemical drugs, biologics are relatively large and complex molecules. They can contain proteins (and their constituent amino acids), carbohydrates (such as sugars), nucleic acids (such as DNA), and a variety of combinations of these substances. The term biologics may also be used to describe cells or tissues used in transplantation.

A generic pharmaceutical product is chemically identical to the brand-name pharmaceutical product from which it is derived. A chemical drug's molecular structure is much simpler and easier to define than that of a biologic. Aspirin, for example, is composed of nine carbon atoms, eight hydrogen atoms, and four oxygen atoms, as shown in table 1.1. Remicade is a large biologic medication with over 6000 carbon atoms, nearly 10 000 hydrogen atoms, and approximately 2000 oxygen atoms.

The cost of specialty drugs, including biologics, is high. Some biologic drugs, such as Soliris (eculizumab) and Vimizim (elosulfase alfa), cost more than $250,000 per patient per year in the United States. Zolgensma, a new drug approved by the FDA, costs more than $2.1 million per dose. It is made by AveXis, a drug maker owned by Novartis. The global spending on biologics exceeded $250 billion in 2019, almost 50% from the United States. By 2026 the biologics market will exceed $700 billion.

Table 1.1. Relative sizes of chemical and biologic drugs.

Drug (nonproprietary name)	Molecular formula
Chemical drugs	
Aspirin	$C_9H_8O_4$
Tylenol (acetaminophen)	$C_8H_9NO_2$
Sovaldi (sofosbuvir)	$C_{22}H_{29}FN_3O_9P$
Small biologic drugs	
Lantus (insulin glargine)	$C_{267}H_{404}N_{72}O_{78}S_6$
Epogen (epoetin alfa)	$C_{809}H_{1301}N_{229}O_{240}S_5$
Neupogen, Zarxio (filgrastim)	$C_{845}H_{1339}N_{223}O_{243}S_9$
Growth hormone (somatropin)	$C_{990}H_{1528}N_{262}O_{300}S_7$
Large biologic drugs	
Enbrel, Erelzi (etanercept)	$C_{2224}H_{3472}N_{618}O_{701}S_{36}$
Remicade, Inflectra (infliximab)	$C_{6428}H_{9912}N_{1694}O_{1987}S_{46}$

Notes: The nonproprietary name of a drug product is used in drug labeling, drug regulation, and the scientific literature to identify a pharmaceutical substance or active pharmaceutical ingredient. C, carbon; H, hydrogen; O, oxygen; N, nitrogen; F, fluorine; P, phosphorus; S, sulfur.

Sources: https://www.everycrsreport.com/reports/R44620.html), Drugs@FDA, https://www.accessdata.fda.gov/scripts/cder/daf/, and Drugs.com.

1.2.2 Molecular mechanism

To stay alive, human body cells depend on a vast array of proteins (approximately two thousand) to perform nearly every functional and structural task. The coding in our DNA molecules allows these proteins to be expressed within each of our body cells. Genes are DNA segments that hold a message that contributes to the development of proteins. They are nucleotide sequences found in all living organisms' genomes (A, T, G, and C). The sequence of each of these genes is unique to a protein (figure 1.1).

For many reasons, the cells of the body often become deficient in producing the required proteins, including genetic mutations that reduce or stop the production of proteins, leading to many life-threatening conditions. The types of proteins that are affected include hormones, enzymes, antibodies, and many more. There are two ways to fix this cellular abnormality. First, administer the deficient proteins, and second, retrain the cells to produce the required proteins. The first category comprises complex biopharmaceuticals with large molecular weight and variable structures. It is not possible to synthesize these proteins, not only because they lack technology but also because of their variable structure—essentially, these molecules are a group of molecules—and it is not possible to replicate their variability. As a result, we have two choices: extracting them from healthy subjects or producing them in living entities whose genetic code has been altered to produce these molecules.

The training of cells is the subject of gene therapy and cell therapy technology. In addition to the naturally occurring proteins we can also design proteins as antibodies

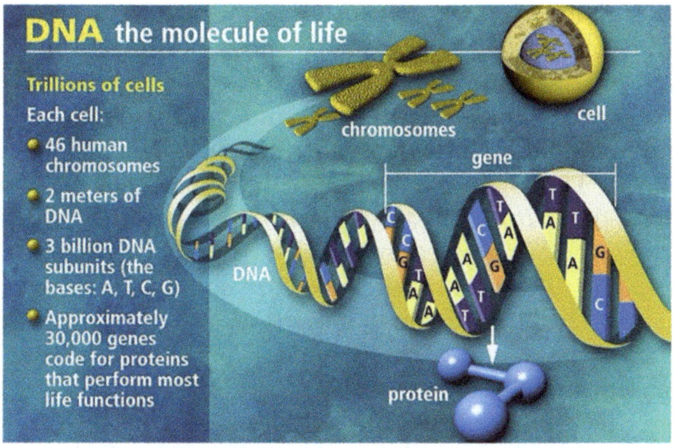

Figure 1.1. Structure of DNA and protein production in cells. Courtesy of US Department of Energy Genomic Science https://public.ornl.gov/site/gallery/detail.cfm?id=397.

to alleviate diseases where a protein-receptor binding is involved in the body. This area of monoclonal antibodies is one of the essential categories of biopharmaceutical products and is discussed in detail in another chapter.

1.2.3 Therapeutic classification

Biopharmaceutical products can be classified based on their pharmacological activity:

- Replacing a deficient or abnormal protein.
- Providing a novel role.
- Interfering with another molecule.
- Delivering a cytotoxic drug, effector protein, or a radionuclide.

Based on the types of molecules, biopharmacautical products can be classified into enzymes, hormones, interleukins, antibodies, anticoagulants, Fc proteins, growth factors, interferons, etc.

Classifications based on the molecular mechanism of action are, for example, mAbs which bind non-covalently to their target, enzymes which impact covalent bonds, and serum albumin which displays behavior without complex interactions.

Bispecific mAbs and multi-specific fusion proteins, mAbs conjugated with small-molecule drugs, and proteins with improved pharmacokinetics are among the forthcoming engineered proteins.

1.2.4 Hormone peptide drugs

The definition of hormones also includes the autocrine and intracrine signaling chemicals produced in cells, and paracrine signaling with nearby cells. Glands contain hormones, which are then sent to the circulatory system to identify target organs to regulate their physiology. Hormones can be classified chemically as eicosanoids, steroids, amino acid derivatives, peptides, and proteins. The endocrine signaling system is composed of glands that secrete hormones.

The mRNA that comes from DNA inside the cell nucleus provides the synthesis of hormones, starting with prohormones and precursors in the endoplasmic reticulum. The N-terminal signal sequence and its glycosylation are removed before it enters cytoplasm. The secretory vesicles in the membrane then receive prohormones from where they are secreted by exocytosis to attend to stimulation, such as an increase in the cAMP concentration of Ca^{2+} in the cytoplasm.

Amino acid residues that were superfluous are contained in the prohormones to guide the hormone receptor's folding into its active configuration but have no function once the hormone folds. Endopeptidases in the prohormone's cleaving just before it is released into the bloodstream generate the molecule's hormone type. Peptide hormones then proceed through the blood to reach their target body cells, where they interact with receptors. Examples of hormones include:

- Human insulin applies to diabetes.
- Growth hormone deficiency, along with development issues and AIDS, require the use of growth hormone.
- Infertility, ovulation regulation, menopause osteoporosis, and others require the use of follicle-stimulating hormone and other hormones, however, this application is not yet widely used.

1.2.5 Human hematopoietic factor

Examples of hematopoietic factors include:

- Recombinant human erythropoietin applies to anemia.
- GM-CSF is used in the treatment of cancer and cancer chemotherapy-induced immunity alteration that may lead to infections.
- Hematopoietic factors are also used in children with dysplasia, malignant hematological disease, or complications in diabetes.

1.2.6 Human cytokines

Cytokines are small proteins (~5–20 kDa) known for their cell signaling ability. Cytokines affect the behavior of body cells as they are involved in the autocrine systems. Examples of cytokines include interferons, chemokines, lymphokines, interleukins, and tumor necrosis factors, but generally not hormones or growth factors.

A wide range of cells produce cytokines, which comprise B lymphocytes, macrophages, T lymphocytes, endothelial cells, mast cells, stroma cells, and fibroblasts. Multiple cells can produce each cytokine acting by receptor interaction and modulate balancing between humor and cell-based responses. Cytokines further regulate growth, responsiveness, and maturation. Cytokines are also known to inhibit or enhance other cytokines. A major difference between the hormones and the cytokines is that while both are cell signaling, hormones are only made by specific cells and are present in much lower concentrations. Examples of cytokines include:

- Alpha interferon for chronic viral hepatitis and certain cancers.
- β interferon for multiple sclerosis (MS).

- Interleukins 1, 2, and 11, for renal cell carcinoma, chemotherapy-induced thrombocytopenia, and chronic granulomatous diseases.

1.2.7 Human plasma protein factor

Examples of plasma protein factors include:

- Hemophilia and hemostasis are treated with recombinant human coagulation factor VII, which is only available from NovoSeven (Novo Nordisk).
- Hemophilia B is treated with recombinant human coagulation factor IX, only available as Renefix (Genetics).
- Tissue plasminogen activator (tPA), one of Activase's first drugs, is used to treat acute myocardial infarction.
- Only the drug Xigris has C-reactive protein, which is used to diagnose extreme sepsis (Eli Lilly).
- The first recombinant drug from transgenic animals, recombinant human antithrombin (ATryn), was approved in 2006.

1.2.8 Human bone formation protein

Examples of bone formation proteins include:

- Recombinant human bone morphogenetic proteins (rhBMP-x; $x = 1–15$ plus 8a 8b) for acute tibial fractures and bone healing.
- Spinal healing, Plexin-B2.

1.2.9 Recombinant enzymes

Examples of recombinant enzymes include:

- Replacement treatment for congenital enzyme deficiency.
- Fusion protein: a limited number of recombinant drugs with an inhibitory function.
- Exogenous recombinant proteins: exogenous proteins have been validated to develop monoclonal antibody drugs to treat human disease. Only recombinant hirudin, which is used to treat thrombotic disease, has been licensed for sale.

1.2.10 Antibodies

Hormones are classified chemically as eicosanoids, steroids, amino acid derivatives, peptides, and proteins. The endocrine signaling system consists of glands that produce hormones. Synthetic antibodies are produced by introducing human genes in mice to produce a targeted antibody that similarly neutralizes the body's proteins; several other methods also produce antibodies, as described in chapter 2. Monoclonal antibodies come from an identical monoclonal cell and have only one type of antigen, while polyclonal antibodies come from several cells producing multiple antigens.

Immunoglobulin (Ig) is a large (150 kDa) Y-shaped protein that recognizes and neutralizes pathogens. A paratope, alternatively called an antigen-binding site, is a region on the surface of an antibody that recognizes and binds to an antigen. It is a short region (between five and ten amino acids in length) of the antibody's Fab region, a component of the fragment antigen-binding domain. It contains fragments of both the heavy and light chains of the antibody.

This enormous diversity of antibody paratopes allows immune systems to identify a broad range of antigens. Chapter 2 is devoted to providing a better understanding of antibodies, the most critical biopharmaceutical category.

The antibody recognizes an antigen via its variable region. The tips of the 'Y' contain a paratope (an antigen-binding site, which recognizes and binds to an antigen) specific to an epitope (the part of an antigen molecule to which an antibody attaches itself), just like a critical fit, as if this were the only lock. This interaction results in the antibody tagging a pathogen or infected cell, and neutralizing the target by locking a part of a pathogen required for body invasion. Antibodies communicate with the immune system through their Fc region which is a glycosylation site (figure 1.2).

Recombinant antibodies are monoclonal antibodies produced *in vitro* using synthetic genes. Recombinant antibody technology entails extracting antibody genes from source cells, amplifying and cloning them into a suitable vector, injecting the vector into a host, and producing a functional antibody. If the appropriate

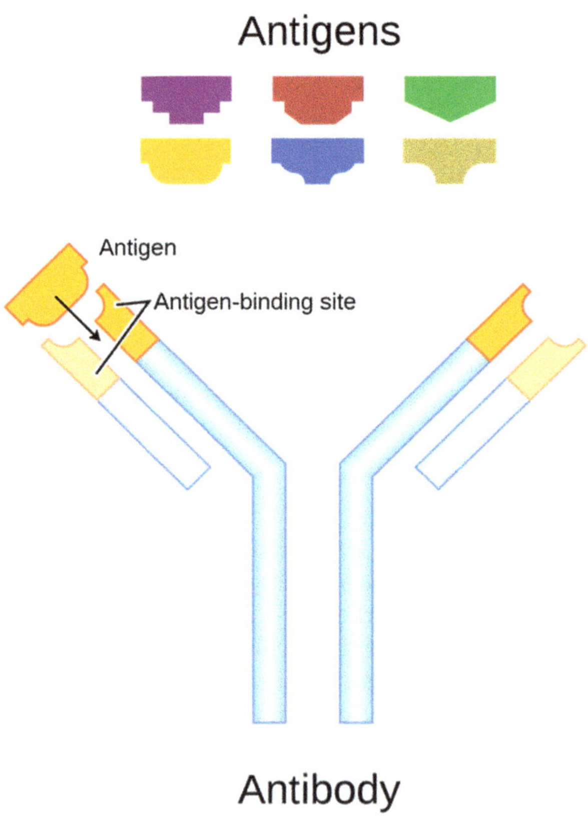

Figure 1.2. The structure of a monoclonal antibody and recombinant antibodies. (Source: Fvasconcellos, originally a work of the United States Government).

oligonucleotide primers or hybridization probes are available, recombinant antibodies can be cloned from any antibody-producing animal species. Modifying antibody genes enables *in vitro* production of new antibodies and antibody fragments such as Fab fragments and scFv. This is accomplished by creating new H and L chain combinations at the combining site stage. To do this, unique CDRs can be mutated. In addition, display libraries, usually expressed in a phage or yeast, may pick desirable characteristics from antibody sequence changes.

1.2.10.1 Synthetic antibodies

Synthetic antibodies are proteins that have been engineered to bind with high affinity and specificity to their target proteins. Synthetic antibodies are affinity reagents generated entirely *in vitro*, without the use of animals. Synthetic antibodies include recombinant antibodies, nucleic acid aptamers, and non-immunoglobulin protein scaffolds. Synthetic antibodies' antigen-recognition sites can be engineered to recognize any desired target due to their *in vitro* manufacturing method. It may go beyond the conventional immune repertoire provided by natural antibodies. Synthetic antibodies have several advantages over antibodies derived from animals, including lower development costs, reproducibility of the reagent, increased affinity, specificity, and stability under a wide variety of experimental conditions.

Synthetic antibodies derived from non-immunoglobulins are typically structurally distinct from antibodies. For example, aptamers are synthesized from nucleic acids, whereas non-immunoglobulin protein scaffolds/peptide aptamers are synthesized from non-immunoglobulin protein scaffolds/peptide aptamers. By inserting hypervariable loops, the antigen-binding site is formed. By constraining the hypervariable binding loop at both ends within the protein scaffold, the synthetic antibody's binding affinity and specificity are increased to levels comparable to or exceeding those of a natural antibody. As a result, these molecules have several advantages over conventional antibody structures, including a smaller size that enables greater tissue penetration, a shorter generation time (weeks rather than months for natural and recombinant antibodies), and lower costs.

Affimer proteins are small, robust affinity reagents with a molecular weight of 12–14 kDa. The cystatin family of cysteine protease inhibitors serves as the scaffold for the affimer protein. Two variable peptide loops and a variable N-terminal sequence within the protein scaffold provide a high-affinity binding surface for the specific target protein. For various molecular recognition applications, affirmer binders for ubiquitin chains, immunoglobulins, and C-reactive protein have been developed.

1.2.10.2 Structural protein scaffolds

In the cytoskeleton and extracellular matrix, molecules such as type IV collagen provide a mechanical scaffold. There is an increased need to make the proteins more robust, enabling long-acting formulations. There are fundamental stability problems with mAbs, more particularly in the hinge region. Engineered protein scaffolds are protein families with non-IgG architecture that have been developed with novel binding. A scaffold is a single chain polypeptide framework that contains a highly structured core associated with various elements of high conformational tolerance, permitting insertions, deletions, and other substitutions. Scaffolds have reduced molecular weights compared to the molecular weights of mAbs, and while they have similar features as antibodies, scaffolds remain unrelated to mAbs. Protein scaffolds possess enhanced solubility and thermal equilibrium, and better tissue penetration. The scaffolds include a single polypeptide chain structure and provide high bacterial expression. Scaffolds can be IgG- (e.g. scFv, single domain names) or non-IgG-like molecules (antibodies, anticline, DARPins, and dual-affinity retargeting molecules). These groups provide smaller molecules with epitope binding and specificity properties of mAbs. As an example, non-IgG-like molecules have higher stability, cysteine-free strings, and elastic pharmacokinetic properties.

1.3 Protein structure and properties

Understanding biopharmaceuticals begins with understanding proteins and their three- and four-dimensional structures, which enable complex interactions with thousands of atoms in the receptor sites that initiate immune system responses. The multidimensional nature of these molecules distinguishes them from small-molecule chemical structures, which are invariably fixed in their spatial arrangement due to fixed covalent bonds; in the case of proteins, an abundance of hydrogen bonds forms the higher order of structure, and protein activity is determined by more than just which functional group is available for reassembly.

Proteins and antibodies have a primary structure composed of a fixed amino acid sequence. The peptide chain twists into an alpha-helix, a type of secondary structure. This helical segment is incorporated into a tertiary structure via polypeptide chain folding. Thus, multiple polypeptide chains make up the quaternary structure. Another feature of this structure is the four-dimensional structure, which includes protein groups' interaction with formulation components that can change the protein structure. Figure 1.3 shows the four types of possible protein structures.

Protein synthesis necessitates many pieces of cellular machinery, the most important of which are ribosomes. Proteins are synthesized from the N-terminus to the C-terminus in a sequential manner at a rate of 50–300 amino acids per minute; the folding begins once the chain has acquired 50–60 amino acids—co-translational protein folding that constraints and limits the pathways a protein can take into a higher-order structure (HOS), and this may explain why Levinthal calculations come short. Theoretically, the process by which the protein is folded into the lowest activation energy state is paradoxical, as it takes between 0.1 and 1000 s for the protein to fold into the correct form given the millions of possible configurations—this is referred to as the Levinthal paradox—as it would otherwise take thousands of years of trial and error to reach the lowest energy state. There appears to be a brain-like structure and a central control mechanism—something we do not understand today, but may understand in the future.

1.3.1 Primary structure

During the protein synthesis process, polypeptide chains are formed on the ribosome (called the translational phase). A peptide (or amide) bond is a chemical bond that

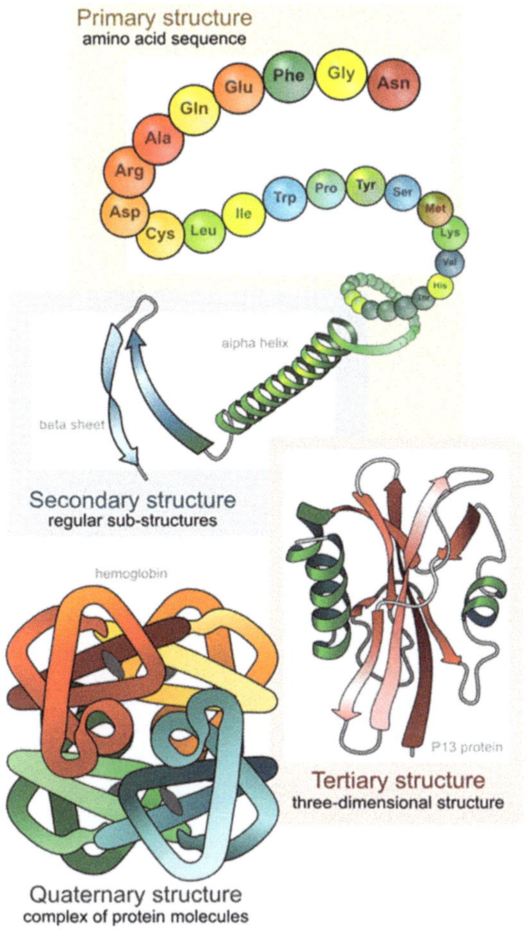

Figure 1.3. Four elements of possible protein structures. (Source: Text adapted from OpenStax, *Concepts of Biology*, 18 May 2016 http://cnx.org/contents/b3c1e1d2-839c-42b0-a314-e119a8aafbdd@9.10; licensed under CC-BY 4.0 by OpenStax.)

forms between amino acid groups. The side chain of each amino acid dictates how it is inserted into the protein structure and thus defines its essential properties. This classification is based on the way the functional groups on amino acid components interact with the water surrounding the protein molecule. Water is a bipolar molecule and amino acids are classified as hydrophilic or hydrophobic, except for glycine, which is found on the surface of proteins, primarily inside chains, where it exhibits a high degree of versatility. In contrast, proline provides rigidity to the protein structure by introducing torsion angles into the polypeptide chain section. These two residues are abundant because they are required for the three-dimensional structure to be formed.

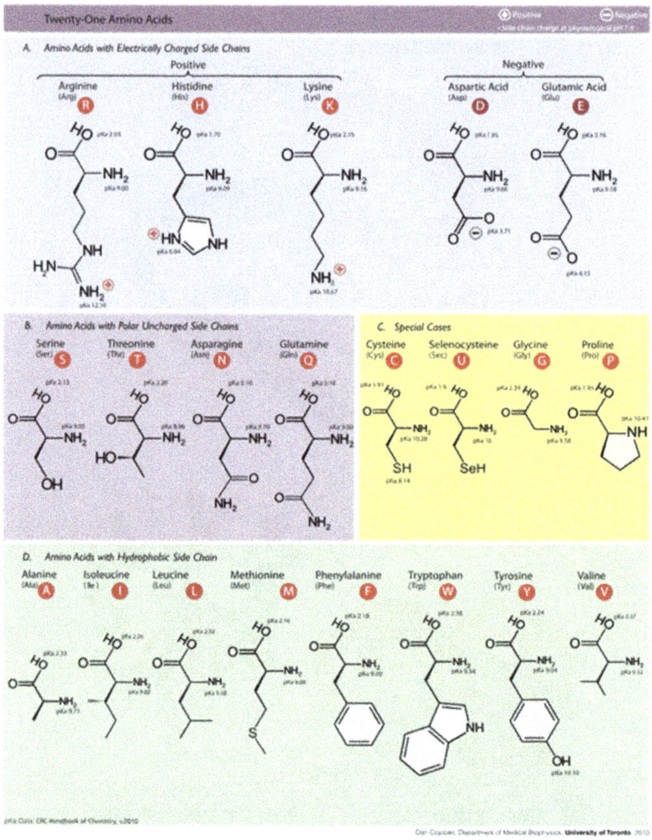

Figure 1.4. The essential amino acids. (Source: Dan Cojocari, Creative Commons.)

The first level of protein structure is the amino acid sequence, comprising only the 20 naturally occurring essential amino acids (figure 1.4).

The chemical link between amino acids is called a peptide bond. It is formed between the carbonyl oxygen and the α-carbons on each side of the peptide bond. The amide nitrogen and hydrogen are forced to bind by the partial-double-bond character between the amide nitrogen and the carbonyl (figure 1.4).

The peptide bond's planar structure restricts the angular range of bond rotation around the C–N and C–C bonds, expressed as ϕ (phi) and ψ (psi), respectively. These restrictions are summarized in a two-dimensional graphical plot called the Ramachandran plot. A Ramachandran plot (alternatively called a Ramachandran diagram) is a technique for visualizing the backbone dihedral angles of proteins about their amino acid residues. Because the peptide is planar due to the partial-double-bond character, the angle at the peptide bond is typically 180°. Because dihedral angle values are circular and 0° equals 360°, the Ramachandran plot's edges 'wrap' right to left and bottom to top (figure 1.5). This graph illustrates how certain structural features of proteins, such as the alpha-helix, are restricted to a particular range of angles (figure 1.6).

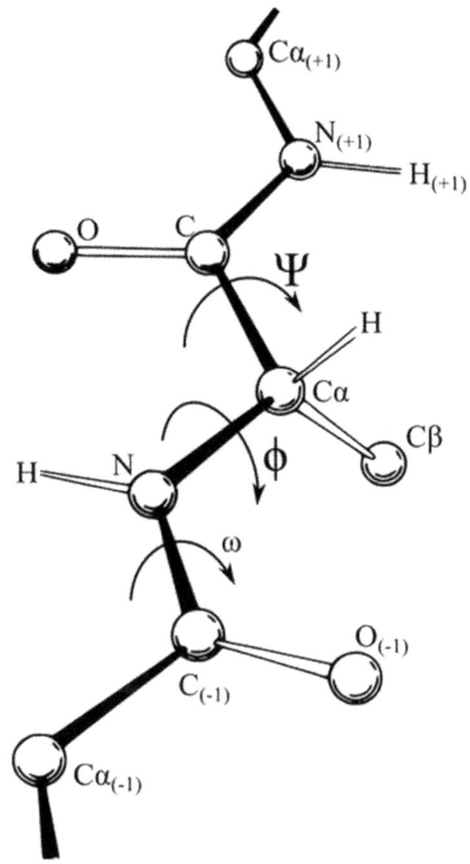

Figure 1.5. Peptide bond; the double-bond character is about 40% due to resonance. (Source: Dcrjsr, vectorized Adam Rędzikowski, Creative Commons.)

To form a hydrogen bond, two electronegative atoms (in the case of an α-helix, the amide N and the carbonyl O) must interact with the same hydrogen. Although the hydrogen is covalently bound to one of the atoms (the hydrogen-bond donor), it interacts electrostatically with the other atoms (the hydrogen-bond acceptor, O).

Most amino acids are found in polar or hydrophobic states (table 1.2).

The primary structure comprising an amino acid sequence creates the three-dimensional structure, with the folding occurring due to the distribution of polar and non-polar side chains. The folding is triggered by incorporating hydrophobic side chains into the molecule's interior, thereby reducing contact with the aqueous environment and lowering the free energy of the carbon molecule. As a result, proteins are composed of a hydrophobic core surrounded by hydrophilic residues. Hydrogen bonds polarize peptide bonds in a hydrophobic environment, resulting in polypeptide regions forming normal three-dimensional patterns called secondary structures. There are two types of secondary structures: helices and sheets. The three-dimensional structure is formed by dividing the primary structure, a string of amino

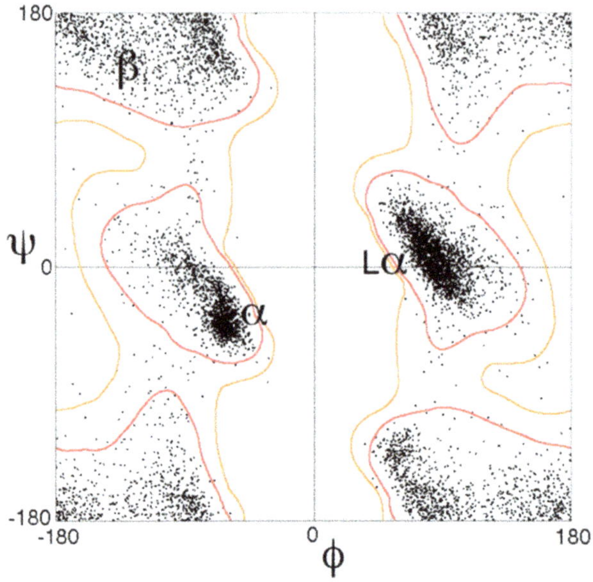

Figure 1.6. Ramachandran plot for glycine. (Source: Dcrjsr, Creative Commons.)

Table 1.2. The physical properties of the essential amino acids.

Property	Amino acids
Charged amino acids	Arginine, lysine, aspartic acid, glutamic acid
Polar (may participate in hydrogen bonds	Glutamine, asparagine, histidine, serine, threonine, tyrosine, cysteine, methionine, tryptophan
Hydrophobic (normally buried inside the protein core)	Alanine, isoleucine, leucine, phenylalanine, valine, proline, glycine

acids, into polar and non-polar side chains. Hydrophobic side chains are buried within the molecule, reducing interaction with the aqueous environment, and lowering the free energy of the carbon molecule. As a result, hydrophilic residues surround proteins with a hydrophobic core. Polarity is a property of peptide bonds. Hydrophobic environments neutralize hydrogen bonds, allowing polypeptide regions to form regular three-dimensional patterns known as secondary structures; helices and sheets are two secondary structures.

Insulin, a hormone, exemplifies these structural elements (figure 1.7).

1.3.2 Secondary structure

The second level of organization in proteins is determined by the unique amino acid sequence within a polypeptide chain; this is the second level of organization that contributes to structural motifs and folds, which are the third level of organization.

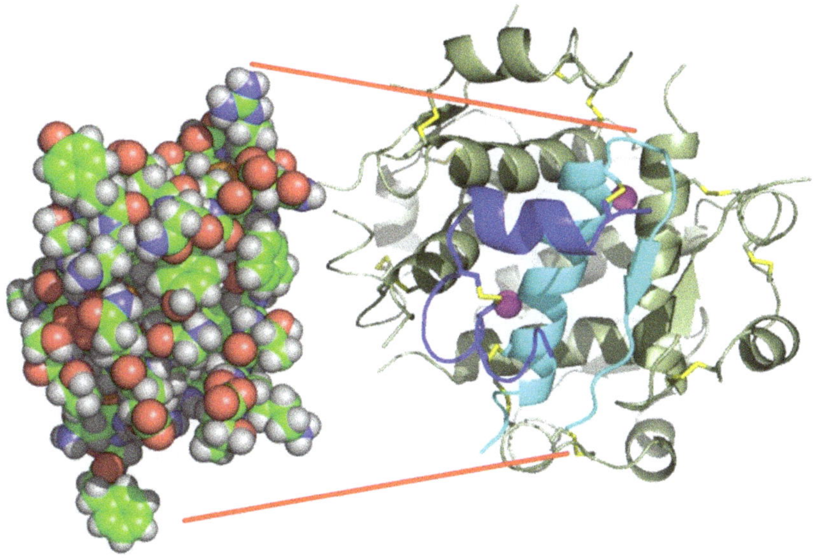

Figure 1.7. The amino acid sequence of insulin. (Source: Isaac Yonemoto, Creatives Commons.)

The protein folds into a motif (repeating element) within a three-dimensional structure determined by the α-helix, β-sheet, β turns, and other noncovalent interactions.

Each amino acid contains a carboxylic and an amine group, which are linked together to form a chain via a dehydration reaction that links one amino acid to another via its carboxyl group, resulting in a polypeptide chain with an unbound carboxyl group at the C-terminus and an amine group at the N-terminus.

A structural motif is a secondary structure, whereas domains are compact, local, and semi-independent units composed of numerous motifs. Because motifs are found in proteins and enzymes with various functions, they cannot be used to predict biological functions. A domain is a three-dimensional protein structure's folding unit; it is a conserved segment of a particular protein sequence and (tertiary) structure that can evolve, function, and exist independently of the rest of the protein string. Each domain forms a stable and foldable three-dimensional structure. Domains are the fundamental units of proteins. A single domain can be found in a wide variety of proteins. The use of domains in molecular production enables the recombination of proteins. Domains can be between approximately 25 and 500 amino acids in length. As zinc fingers (a small protein structural motif defined by the coordination of one or more zinc ions (Zn^{2+}) to stabilize the fold) and disulfide bridges, domains are stabilized by these structures. The domain of the EF-hand (a helix–loop–helix structural domain or motif found in a large family of calcium-binding proteins) is exemplified by calmodulin.

Due to their distinct nature, the domains can be 'swapped' between proteins via a genetic modification to create chimeric proteins. Because domains can be cloned, expressed, and purified independently of the rest of the protein, they can also exhibit activity when linked to a known activity. Certain proteins contain only a single domain or fold, but the folds are unconnected.

The polypeptide chain's tertiary structure forms an individual hydrophobic core built from the secondary structure connected to loop regions. Core residues are more conserved than residues in the loops. The tertiary structure is described in secondary structure as:

- The α domains are built exclusively from α-helices with small folds, often simple bundles with helices running up and down.
- The β domains with an antiparallel β sheet core commonly comprise two sheets that are packed against each other, resulting in patterns that appear as an arrangement of strands.
- The $\alpha + \beta$ domains comprise a combination of α and β motifs. Since it is difficult to classify proteins into this class due to the other three classes' overlaps, they are not used in the CATH domain database (https://www.cathdb.info).
- The α/β domains are made from a combination of β–α–β motifs that predominantly form a parallel β-sheet surrounded by amphipathic α-helices.

Domains have size constraints and convert E-selectin (also known as CD62 antigen-like family member E (CD62E), endothelial-leukocyte adhesion molecule 1 (ELAM-1), or leukocyte-endothelial cell adhesion molecule 2 (LECAM2)) into 692 residues in lipoxygenase-1; however, disulfide bonds are typically used to stabilize domain names with fewer than 40 residues. Significant domains with a length greater than 300 residues are likely to contain multiple hydrophobic cores. Figure 1.8 illustrates the average number of disulfide bonds formed during the domain name creation process.

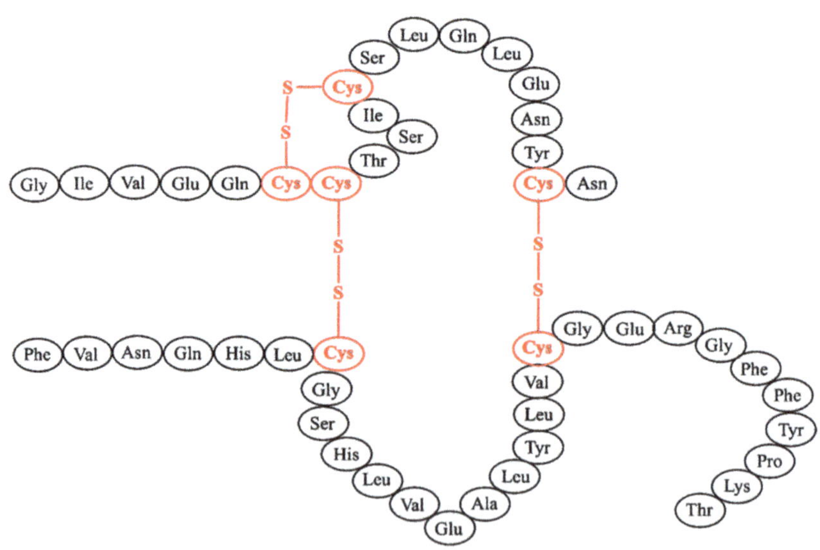

Figure 1.8. Disulfide bonds in the insulin protein domain.

When two domains are covalently bonded, there is a functional and structural advantage because the structures are more stable than when they are not covalently bonded.

1.3.2.1 Alpha-helix

The most common type of secondary structure in proteins is the α-helix (figure 1.9).

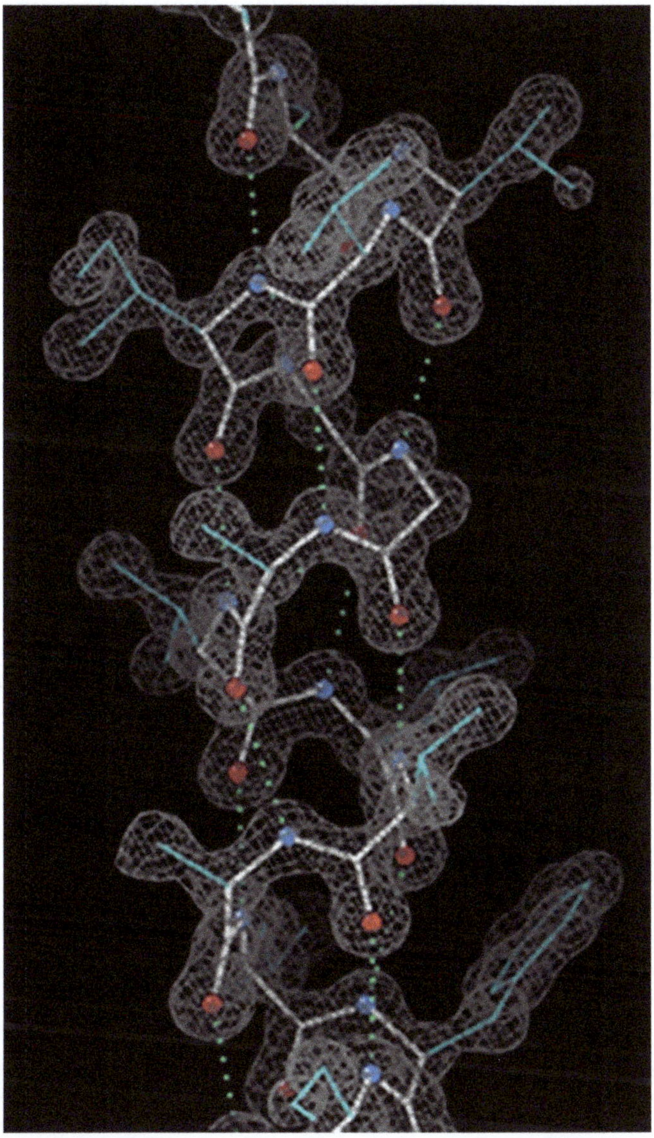

Figure 1.9. An α-helix in ultrahigh-resolution electron density contours, with oxygen atoms in red, nitrogen atoms in blue, and hydrogen bonds as green dotted lines (PDB file 2NRL, 17-32). The N-terminus is at the top. (Source: Dcrjsr, Creative Commons.)

All functional groups in proteins can form H bonds, regardless of whether the residues are arranged in a secondary structure or not. These are H-bonded to one another or water molecules. Water's bipolarity enables it to accept two hydrogen bonds and thus stabilize the structure of proteins by forming hydrogen bonds between the side chain and the main chain and side chain groups that connect distinct protein groups collectively. Additionally, water is shown to be involved in binding ligands to proteins, mediating the interactions of charged groups. The energy of a hydrogen bond varies between 2 and 10 kcal mol^{-1}, depending on the angle formed and the distance between the donor and acceptor.

1.3.2.2 β-sheets

The secondary structure of proteins, the β-sheets, is also stabilized by hydrogen bonds. Figure 1.10 is an example of a beta-sheet with stabilizing hydrogen bonds as dotted lines. As in α-helices, these hydrogen bonds are not always formed between adjacent residues. Instead, a β-sheet is formed by joining different amino acid sequence segments, known as β-strands. As a result, the β-sheet comprises a network of hydrogen bonds that tie multiple β-strands together.

The same β-sheet is shown in figure 1.10 in a 3D structure in a so-called 'ribbon' representation (figure 1.11).

β-sheets and helices, which can be replicated and alternated throughout the amino acid chain, characterize the organization.

1.3.3 Tertiary structure

Polypeptide chains frequently fold spontaneously into stable tertiary protein structures. The forces responsible for folding are unknown. Chaperones are proteins that assist other proteins in folding correctly using the proteolytic apparatus found in cells.

The arrangement of large proteins from structural domains defines a protein fold edge. Each domain can fold independently, which accelerates the folding process

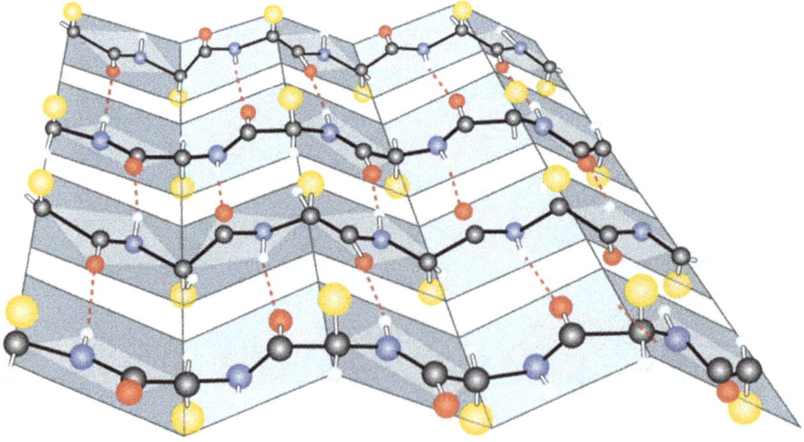

Figure 1.10. β-sheets of protein. (Source: Preston Manor School + JFL, Creative Commons).

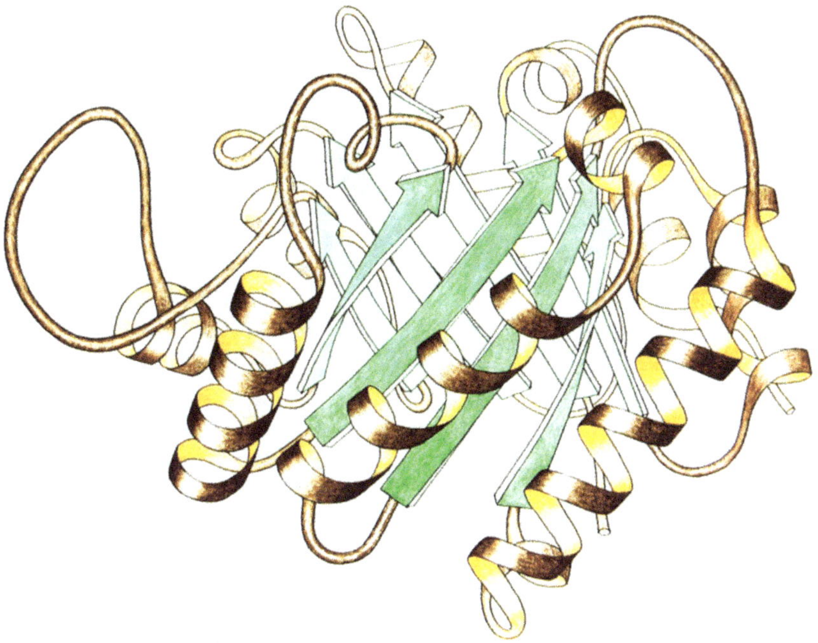

Figure 1.11. Ribbon schematic of a triose P isomerase monomer. The arrows show the direction of the β-sheet, which is from the N- to the C-terminus. The parallel sheet is indicated by the arrows pointing in the same direction and the antiparallel sheet by the arrows pointing in opposite directions. (Source: Jane Richardson, Creatives Commons.)

and minimizes the potential for a large number of residue connections. Additionally, given the random distribution of hydrophobic residues in proteins, domain formation optimally buries hydrophobic residues while leaving hydrophilic residues on the protein's surface such as human interferon gamma (figure 1.12).

1.3.4 Quaternary structure

Quaternary construction is the fourth degree. Many polypeptide chains (subunits) are arranged in a quaternary structure, which can be identical (homo-oligomer) or different (alternative) (hetero-oligomer). The subunits interact with one another and with the same complex target protein, forming an active site and interacting with several target proteins (figure 1.13).

1.3.5 Post-translational modification (PTM)

The primary to quaternary structure of a protein and the additional properties acquired during the cellular phase of protein synthesis are referred to as 'post-translational modifications' because they occur after the gene (nucleic acid sequence) is translated into the protein sequence (the amino acid chain). These modifications, collectively referred to as the 'maturation process', occur prior to cell protein secretion. These modifications include the grafting of one or more chemical groups,

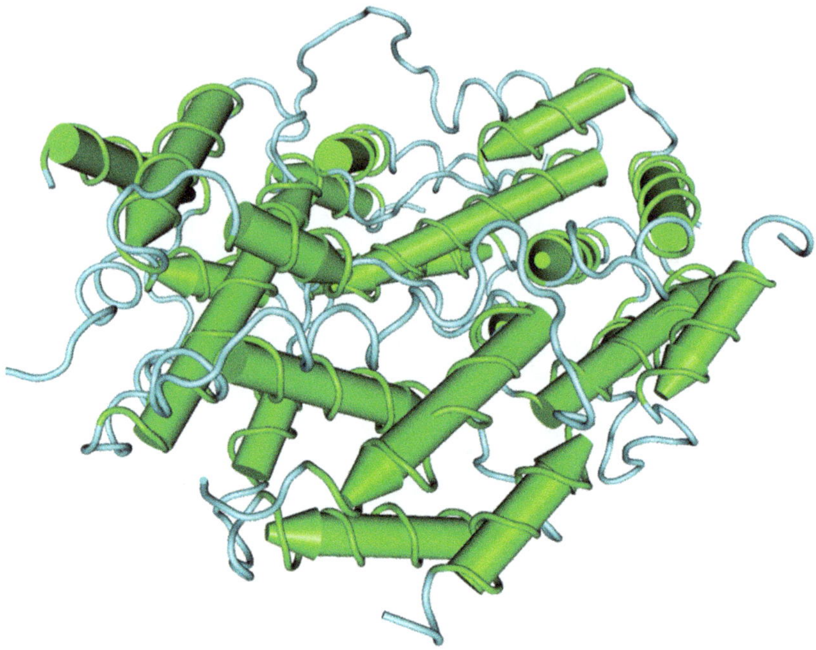

Figure 1.12. The three-dimensional structure of human interferon-gamma. (Source: Nevit Dilmen, Creative Commons.)

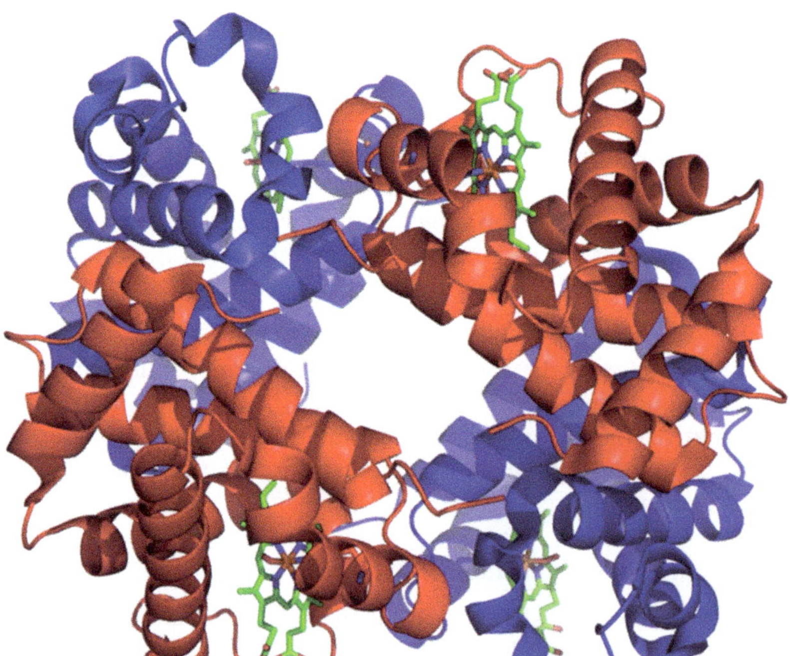

Figure 1.13. Diagram of a dimer of *Escherichia coli* galactose-1-phosphate uridylyltransferase (GALT) in complex with UDP-galactose (stick models), a quaternary structure. (Source: https://commons.wikimedia.org/wiki/File:Galactose-1-phosphate_uridylyltransferase_1GUP.png.)

such as phosphate or sulfate groups, or sugars (in the case of glycosylation), onto specified amino acids to alter the global charge and physicochemical or biological properties of these 'mature' proteins in their active forms.

These post-translational modifications, which occur at specific sites on the protein, are not regulated by the gene that encodes the protein sequence; rather, each cellular form presents a unique combination of interior milieu conditions, such as the presence of enzymes and the thermodynamic conditions of the reaction; therefore, many of these complex chemistries are so diverse. This does not apply to prokaryotic organisms (bacteria) or very primitive eukaryotes such as yeasts. Because these are native to 'mammalian' cells such as CHO cells, upstream processes for these cells are complex and time-consuming to develop, in particular if the goal is to match the post-translation modification profile of biosimilars.

Table 1.3 lists common PTMs by specific amino acids. Modifications occur on the side chain unless indicated otherwise.

Table 1.3. Common PTMS by specific amino acid residues.

Amino Acid	Abbreviation	Modification
Alanine	Ala	N-acetylation (N-terminus)
Arginine	Arg	Deimination to citrulline, methylation
Asparagine	Asn	Deamidation to Asp or iso(Asp), N-linked glycosylation
Aspartic acid	Asp	Isomerization to isoaspartic acid
Cysteine	Cys	The disulfide-bond formation, oxidation to sulfenic, sulfinic, or sulfonic acid, palmitoylation, N-acetylation (N-terminus), S-nitrosylation
Glutamine	Gln	Cyclization to pyroglutamic acid (N-terminus), deamidation to glutamic acid, or isopeptide bond formation to a lysine by a transglutaminase
Glutamic acid	Glu	Cyclization to pyroglutamic acid (N-terminus), gamma-carboxylation
Glycine	Gly	N-myristoylation (N-terminus), N-acetylation (N-terminus)
Histidine	His	Phosphorylation
Isoleucine	Ile	
Leucine	Leu	
Lysine	Lys	Acetylation, Ubiquitination, SUMOylation, methylation, hydroxylation
Methionine	Met	N-acetylation (N-terminus), N-linked ubiquitination, oxidation to sulfoxide or sulfone
Phenylalanine	Phe	
Proline	Pro	Hydroxylation
Serine	Ser	Phosphorylation, O-linked glycosylation, N-acetylation (N-terminus)
Threonine	The	Phosphorylation, O-linked glycosylation, N-acetylation (N-terminus)
Tryptophan	Trp	Mono-or di-oxidation, the formation of kynurenine
Tyrosine	Tyr	Sulfation, phosphorylation
Valine	Val	N-acetylation (N-terminus)

Source: Wikipedia.

Table 1.4. Frequency of the most common modifications in protein structure derived from reported protein structures.

Frequency	Modification
58383	Phosphorylation
6751	Acetylation
5526	N-linked glycosylation
2844	Amidation
1619	Hydroxylation
1523	Methylation
1133	O-linked glycosylation
878	Ubiquitylation
826	Pyrrolidone carboxylic acid
504	Sulfation

Source: Wikipedia.

The ten most common modifications are listed in table 1.4.

The effect of PTMs on protein activity is well-documented, as shown in table 1.5.

1.3.5.1 Glycosylation

The IUPAC dictionary defines glycan and polysaccharides as 'compounds composed of a large number of monosaccharides that are glycosidically related'. However, the carbohydrate component of a glycol-conjugated molecule, such as a glycoprotein, glycolipid, or proteoglycan, may also be referred to as a glycan, even if the carbohydrate is merely an oligosaccharide. The most common type of glycan is monosaccharide O-glycosidic linkages. For example, cellulose is a 1,4-linked D-glucose glycan (or, more precisely, a glucan), whereas chitin is a 1,4-linked N-acetyl-D-glucosamine glycan.

Glycans can be linear or branched and contain monosaccharide residues in homo- or heteropolymers. Chemical modifications are extremely complex because the glycan structures are attached to the protein backbone. Glycosylation of proteins take place in the endoplasmic reticulum and Golgi apparatuses. Glycosylation is a process that involves sugar groups such as mannose, fructose, or galactose branching on the protein on specific amino acids (for example, Asn in the Asn–X–Thr sequence) and sugar groups such as mannose, fructose, or galactose branching in a well-defined order. Even when specific mandatory sequences are present in each structure, glycosylation chemical reactions result in forming 'sugar chains' that are complex and diverse when all possible attaching combinations (number of antennae on a glycosylation site and nature of sugars that comprise this antenna) are considered.

Glycosylation, a co-translational and post-translational modification, is one of the most common post-translational modifications. In the membrane and secreted proteins, glycans play several structural and functional functions. Most proteins made in the rough endoplasmic reticulum are glycosylated. Figure 1.14 gives an example of how glycosylation occurs and its impact on protein properties and the reproducibility of recombinant proteins.

Table 1.5. Effect of PTMs on protein function and physiological processes.

Function	Phosphorylation	Glycosylation	S-nitrosylation	Methylation	N-acetylation	Palmitoylation	N-myristoylation	Prenylation	Sumoylation	Ubiquitination
Apostosis			X		X		X			
Protein stability		X	X		X				X	X
Protein–protein	X	X					X	X		
Protein–membrane						X	X	X		
Protein trafficking	X	X				X				
Thermodynamic, kinetic		X								
Activity		X	X							
Extracellular export							X		X	
Cell signaling			X			X	X		X	
Transcription				X	X					
DNA repair					X				X	
Cell cycle division	X									X
Immune response	X									X
Chromosome maintenance				X						
Chromosome assembly									X	

Source: Martina Audagnotto and Matteo Dal Peraro, Creatives Commons https://en.wikipedia.org/wiki/Post-translational_modification#/media/File:Image_for_Wiki_1.jpg.

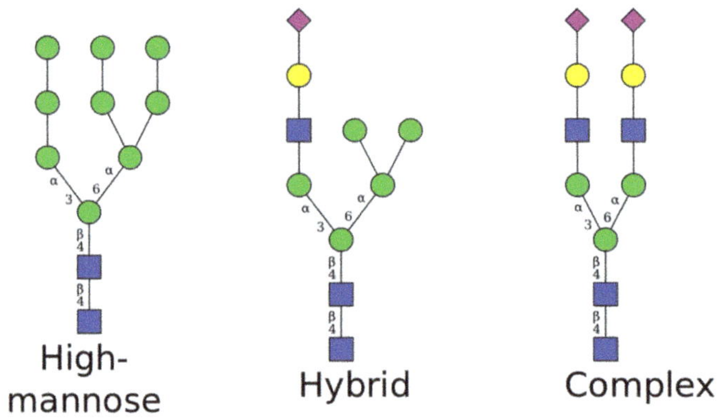

Figure 1.14. Schematic drawings of types of N-glycans present on some protein sequences. (Source: Dna 621, Creative Commons.)

There are four types of glycosylation links:

- The most common type of glycosidic bond is N-linked glycosylation, which is required to fold eukaryotic proteins and cell–cell and cell–matrix attachment. In eukaryotes and the lumen of the endoplasmic reticulum of archaea, the N-linked glycosylation process can be observed, but only very rarely in bacteria.
- O-linked glycosylation is a type of glycosylation found in the Golgi apparatus of eukaryotes, archaea, and bacteria. Xylose, fucose, mannose, and GlcNAc phosphoserine glycans are additional glycans.
- In C-mannosylation a mannose sugar is added to the first tryptophan residue in the W–X–X–W series (W indicates tryptophan, where X is any amino acid). It is an odd reaction since the sugar is bound to carbon rather than nitrogen.
- Glypiation, the formation of GPI anchors, is a particular form of glycosylation. A protein is connected to a lipid anchor through a glycan chain in this form of glycosylation.

1.3.5.2 Sialic acid

Finally, sialic acid in the form of neuraminic N-acetyl acid (NANA) caps the end of the sugar chain in human cells, whereas in many mammals a portion of the sialic acid is in the form of neuraminic N-glycolyl acid (NGNA), as the gene encoding the enzyme responsible for the conversion of NANA to NGNA is silenced and inactive. This species specificity is critical for selecting systems for carbohydrate expression of the desired recombinant protein. It ensures that the sialylation is as close as possible to the human type. The mature protein, which has been partially 'glycolyzed' and 'sialylated', acquires acidic properties and a different isoelectric point (pI). As a result of post-translational modifications, the protein appears as a mixture, a molecular population with the same basic protein structure (primary sequence imposed by the gene sequence) but with various types of sugar chains attached, providing each protein molecule with a unique pI and allowing for their separation in analytical testing based on their charge (isoelectric focusing).

1.3.6 Association and aggregation

The higher-order structures of proteins are stabilized through many weak and strong bonds, including weak noncovalent bonds, formed ionic, dipoles (hydrogen bonds), non-polar (hydrophobic), and van der Waals interactions. These bonds involve the interaction of amino acid side chains and the polypeptide chain. Since the transition from a polypeptide chain to a higher-order structure requires a significant loss of structuring, it must be compensated by enthalpy released from the forming of the bond (energy is released when a bond is formed); as a result, the protein structure can remain in a dynamic state of structuring that may affect its activity as well as its stability. In most instances, the changes are transitory, and the protein returns to its native structure. However, the possibility of dynamic changes to protein structure makes it possible for a molecule to have a different activity if its physicochemical properties are altered. Additionally, if there is aggregation, this may lead to loss of activity and probably an increase in the immunogenicity of the protein.

Protein aggregation is caused by two factors: colloidal and conformational instability. The attractions on the surface of proteins can make colloidal dispersions that can be dynamic and significantly affect the safety and efficacy of proteins under stress conditions. The conformational changes are brought about by the buried functional groups' hydrophobic interactions. Thus there is a likelihood of both types of aggregates and, in some instances, one leading to another.

There is also a likelihood of aggregation due to molecular crowding when the drug is exposed to the high concentration and high plasma protein concentration. Recently, biopharmaceuticals have been formulated in higher concentration formulations to reduce the injection volume to change the route of administration from intravenous administration to subcutaneous administration, such as MabThera (rituximab).

1.4 Pharmacokinetic manipulations

Therapeutic proteins have a short life in the body that is prolonged by FcN-mediated recycling—changing the half-life from hours to days. Monoclonal antibodies have restricted penetration into tissue because of their large size. They are transported from blood circulation to peripheral tissues by passive transfer through fenestrae pores in capillary walls or by transcellular pathways of endothelial cells. The latter pathway can be an active transport in surrounding tissues like a fluid by fluid-phase pinocytosis, receptors through receptor-mediated endocytosis (FcγR-mediated), and into immune cells by phagocytosis. Passive diffusion depends on the size, hydrophobicity, and charge of the molecule. In the subcutaneous route, macromolecules are restricted to the interstitial space and reach blood through capillaries or lymphatic vessels. Transport through capillaries is passive and limited to molecular weight no more than 16 kDa leaving most large molecule protein drugs dependent on the lymphatic system through the thoracic duct for entry and distribution. Their distribution is also limited to plasma rather than body tissues for the same reasons.

The elimination of antibodies can be affected by their immunogenicity and anti-drug antibodies that can competitively bind to the active portions of the therapeutic proteins such as receptor binding sites, to make the antibody less effective and alter its pharmacokinetic profile. Conversely, aggregation of antibodies enhances their clearance, in particular when used in high doses that promote phagocytosis in the reticuloendothelial system. Endocytosis is a receptor-mediate process where receptor saturation influences bioavailability targeting surface receptors—an antigen sink.

1.4.1 Protein modification to increase the duration of action

Therapeutic proteins are often modified chemically or by recombinant engineering to prolong their half-life, using albumin, Fc fusion, and pegylation as major reactions. Proteins can be labeled as 'unnatural' construction—the fusion proteins or the conjugate (e.g. pegylated) proteins and a huge assembly of virus particles or nanoparticle delivery systems.

1.4.1.1 Albumin–protein fusions

Albumin has three domains, each having two subdomains connected by a flexible loop leading to seven binding sites for fatty acids. It has a half-life of 19–22 days that is prolonged because of its FcRn-mediated recycling like the antibodies. Albiglutide (Tanzeum®), a recombinant fusion protein fused to albumin, has been approved as the only such product.

1.4.1.2 Fc-fusion proteins

The Fc fusion is proteins or peptides fused with the Fc found in antibodies altering the pharmacokinetics properties. Fc-fusion proteins serve as antagonists or agonists in receptor-ligand interactions, blocking receptor binding or stimulating receptor functions. A wide range of fusion products have been approved include etanercept (Enbrel®, Amgen/Pfizer), Alefacept (Amevive®), abatacept (Orencia®), rilonacept (Arcalyst®), romiplostim (Nplate®), belatacept (Nulojix®), aflibercept (Eylea®), and ziv-aflibercept (Zaltrap®). Biobetter fusion proteins include denileukin diftitox (Ontak®), corifollitropin-α (Elonva®), eftrenonacog-α (Alprolix®), albiglutide (Tanzeum®), efraloctocog-α (Eloctate®), and dulaglutide (Trulicity®).

The fusion of an Fc (fragment crystallizable) part of an antibody (typically an IgGI antibody) with that of another pharmaceutically relevant protein through recombinant genetic technology results in fusion proteins. The Fc portion of an antibody increases circulation time just as does the pegylation; examples include Fc fusion to the blood-clotting factor VIII and factor IX. The fusion of two relatively large proteins, each being over 5.0 kDa, raises the question of whether this would affect either protein. While potential variance is possible, the existing science reveals no significant impact.

1.4.1.3 Long-acting connectivity

When a biologically active protein is combined with a long-acting protein, the resulting product has a longer half-life based on several techniques that combine:

- Elastin-like peptide,
- XTEN—recombinant peptide with protein/polymer properties,

- Immunoglobulin Fc fragment,
- Recombinant serum albumin,
- Transferring, and
- Carboxy-terminal peptide

1.4.1.4 Protein PEGylation

Pegylation involved polyethylene binding covalently to proteins forming a conjugate that is considered a new drug. The new drug will have a longer half-life as the higher molecular weight reduces glomerular excretion while reducing immunogenicity the protein molecule is covered under the PEG molecule. The first PEGylated product that was approved is pegademase (Adagen®); other approved products are mainly protein conjugates except for peginesatide (Omontys®) that is a peptide, and pegaptanib (Macugen®) that is an RNA aptamer.

Pegylation is one of the more popular for prolonging the half-life for drugs used to treat hepatitis, multiple sclerosis, cancer, hemophilia, chronic kidney disease, and many more. The first pegylated product approved by the FDA was pegylated ademase bovine in 1990. Many more drugs have been approved, as listed in table 1.6. It is worth mentioning that all the FDA-approved drugs contain ethylene glycol or mPEG. The PEGylated molecules have a longer half-life and are often less immunogenic because of the protection of protein structure domains. Table 1.6 lists the currently approved PEGylated products.

1.4.1.5 Drug–antibody conjugates

Another class of proteins involves combining a drug with an antibody. With the approvals of antibody–drug conjugates, such as Kadcyla®, Adcetris®, and Besponsa®, an increasing number of such products are entering clinical trials, the targeted delivery of cytotoxic agents to cancer cells is a robust method for the treatment of both hematological malignancies and solid tumors.

1.5 Immunogenicity

1.5.1 The immune system

Every living species is provided with a mechanism for defending itself against foreign and potentially harmful agents. This is a component of our internal pharmacy, which has helped us survive millions of years of evolution. An innate system and an adaptive system incorporate humoral and cellular activity and make up the immune system.

The humoral system or antibody-mediated beta cellular immunity involves macromolecules in extracellular fluids—antibodies, complement proteins, and antimicrobial peptides (‘humor’ refers to body fluids). Humoral immunity involves antibodies, the activation of cytokines, isotype switching, germinal center formation, cell generation, and affinity maturation. Additionally, it includes the effector function of antibodies, including toxic and pathogen neutralization, opsonin promotion of phagocytosis, and complement activation.

Table 1.6. Currently approved PEGylated biopharmaceutical products.

Brand	INN	Indication	Functionalized PEG	PEG size (kDa)	Number of PEGs	Year
Adagen	Pegademase bovine	Combined immunodeficiency disease associated with adenosine deaminase	mPEG succinimidyl succinate	5	11–17	1990
Oncaspar	Pegaspargase	Precursor cell lymphoblastic leukemia	mPEG succinimidyl succinate	5	69–82	1994
Pegasys	Peginterferon α-2a	Hepatitis B, C	mPEG succinimidyl carbonate	40	1	2001
Neulasta	Pegfilgrastim	Cancer neutropenia	mPEG aldehyde	20	1	2002
Somavert	Pegvisomant	Acromegaly	mPEG succinimidyl succinate	5	4–6	2003
Macugen	Pegaptanib	Age-related macular degeneration	mPEG amino	40	1	2004
Mircera	Methoxy polyethylene glycol epoetin	Anemia associated with chronic kidney disease	mPEG succinimidyl succinate	30	1	2007
Cimzia	Certolizumab pegol	Rheumatoid arthritis	mPEG maleimide	40	1	2008
Kristexxa	Pegloticase	Chronic gout	mPEG p-nitrophenyl carbonate	10	36	2010
Palynziq	Pegvalise	Phenylketonuria	mPEG succinimidyl succinate	20	9	2010
Movantik	Naloxegol	Opioid-induced constipation	mPEG7 (hexamethylene glycol)	0.339	1	2014
PegIntron	Peginterferon α-2b	Hepatitis C	mPEG succinimidyl carbonate	12	1	2014
Plegridy	Peginterferon β-1a	Multiple sclerosis	mPEG aldehyde	20	1	2014
Adynovate	Rurioctocog α pegol	Hemophilia A	–	20	2	2016
Rebinyl	Nonacog β pegol	Hemophilia B	–	40	1	2017
Asparlas	Calaspargase pegol	Acute lymphoblastic leukemia	mPEG succinimidyl carbonate	5	31–39	2018
Revcovi	Elapegademase	Adenosine deaminase severe combined immunodeficiency	mPEG succinimidyl carbamate	5.6	13	2018

Source: https://www.accessdata.fda.gov/scripts/cder/daf/.

1.5.2 Antibodies

Antibodies are Y-shaped proteins, with two large, heavy chains and two small light chains, and are formed in plasma cells to provide the immune system—antibodies identify and neutralize pathogens. Antibodies are glycoproteins with a molecular weight of around 150 kDa, with sugar chains added to some amino acid residues. Immunoglobulins are another name for antibodies. There are several antibody isotypes, IgA (dimeric), IgD, IgE, IgG, and IgM (tetrameric or pentameric). The Ig prefix is an abbreviation of immunoglobulin.

Chapter 2 describes the structure, function, development, and manufacturing of antibodies in detail.

1.5.3 Antigens

Antigens are molecules that stimulate the production of antibodies and thus initiate an adaptive immune response. Antigens bind to lymphocytes and their receptors, antibodies, and T-cell receptors, among other immune response components. Antigens are structural molecules or fragments of molecules that bind directly to antibodies and are recognized by adaptive immune system antigen receptors (B-cell receptors or T-cell receptors). Without the aid of an immunologic adjuvant, antigens do not induce an immune response. Antigens are proteins or polysaccharides, with lipids from pathogen cell walls, coats, capsules, flagella, toxins, or fimbriae being less common. Only when nucleic acids and lipids are mixed with proteins and polysaccharides do they become antigenic. Antigens may be egg white, pollen, proteins from transplanted tissues, or components of the surface of transfused blood cells, in addition to pathogens. Vaccines, which are used to elicit an immune response, are an excellent example of antigens. Antibodies are generated or engineered to interact with antigens based on the antibody's complementary deciding region (figure 1.2).

The T-cell receptor (TCR) recognition must be broken down into small pieces within the cell and introduced to a T-cell receptor by the main histocompatibility complex (MHC). For example, a hapten is a small molecule bound to a large carrier molecule, such as a protein, to make it antigenic.

The immune system is usually not reactive to antigens produced in the body, known as self or endogenous antigens, but rather to antigens outside the body, known as exogenous antigens. The distinction between exogenous and endogenous proteins, however, has many exceptions. The body's autoimmune response in treating type I diabetes caused by an autoimmune reaction against pancreatic cells is an excellent illustration of this.

Numerous immune cells are activated when cells present the immune system with their antigenic structures via histocompatibility. Additionally, antigens and immunogens have a high molecular weight, a high molecular complexity, and the ability to degrade into fragments that can bind to MHC proteins (or MHC antigens) on the surface of the antigen-presenting cell (APC), where the entire complex then binds to T-cells. Carbohydrate antigens are not processed or displayed because they can directly bind to B-cells and activate them to produce antibodies. The type of antibodies produced is route-dependent; for example, antigens that meet mucous membranes typically produce one type of antibody, whereas intramuscular and intravenous immunizations frequently produce another.

1.5.4 Biopharmaceutical immunogenicity

Because the human immune system recognizes it as non-self, a therapeutic protein, whether recombinant or not, maybe immunogenic. Antigen-presenting cells will recognize and degrade a protein injected into patients. Peptides presented on the surface of antigen-presenting cells in the grooves of major histocompatibility complex molecules will bind to T-cells produced in the thymus. T-cells stimulate the proliferation of B-cells by recognizing these peptides as foreign. Additionally, B- and T-cells are included in the adaptive immune system. Due to the immunoglobulins on their cell surfaces, activated B-cells recruit the complement system and macrophages from the innate immune system to destroy and eliminate the antigen after binding to the protein's three-dimensional structure. A classical immune response to foreign protein results in the development of high-affinity antibodies of various isotypes and memory cells, which contribute to an improved response following repeated antigen exposure (the principle behind vaccination).

When antibodies are neutralizing in nature, this immune response can have serious consequences ranging from mild tolerance to therapeutic inefficiency. Antibodies to therapeutic proteins such as erythropoietin (EPO), hematopoietic growth factors (GM-CSF), and thrombopoietic and megakaryocyte differentiation factors (TPO/MGDF) may have serious consequences, including blocking not only the exogenous protein's function but also the endogenous protein's, with the attendant complications. The development of neutralizing antibodies can be attributed to various mechanisms, including direct binding to a biological activity site or indirect binding to a site that is unrelated to the biological activity site but inhibits its activity due to a structural conformational change. Non-neutralizing antibodies bind to the therapeutic protein site but do not affect the biological activity site. Assuming that they do not directly neutralize the biological target, in this case increasing the clearance of the bonding complex to the same extent as biological activity neutralization may alter the drug's bioavailability.

Antibodies directed against biotechnology-derived proteins such as insulin, factor VIII or IX, or interferons do not have the same severe consequences. Therapies can be continued in the presence of antibodies by adjusting therapeutic protein doses. Generally, if the immunogenicity is unrelated to a change in the drug's pharmacokinetic/pharmacodynamic profile (PK/PD), no dose adjustment should be made.

Since the invention of monoclonal antibodies, the consequences of antibodies generated against them have been documented, particularly when derived from animal or bacterial proteins. The observed reactions may be systemic, such as those observed during injection of these materials, and local or acute hypersensitivity reactions (generally not due to the antibodies). Due to the enhanced purification of proteins enabled by recombinant DNA technology and the humanization of the protein skeletons of monoclonal antibodies, anaphylactic or allergic immune reactions are rare.

If a biopharmaceutical product is immunogenic, repeated administration to patients over time increases the risk of antibodies being raised that can cause anaphylaxis, alter the protein's pharmacokinetic properties, or prevent the medication from binding to its target receptor, rendering it ineffective. Anaphylaxis is a life-threatening allergic reaction mediated by immunoglobulin E (IgE) antibodies to a substance. Immune responses with high titers of neutralizing IgG antibodies significantly impair therapeutic efficacy. Another potentially fatal clinical outcome of antibody production is cross-reactivity with the patient's endogenous protein.

Only when immunogenicity is clinically important—when it affects the therapeutic protein's protection or efficacy—is it a concern. When antibodies change how a drug responds in the body, when antibodies make a protein less therapeutically successful, when antibodies change natural proteins in the body, or when antibodies trigger a serious allergic reaction are all examples of clinically significant immunogenicity. Clinically important immunogenicity is uncommon, but it must be controlled for all therapies.

One of the most important issues and considerations in the manufacture of drugs is their safety. These questions have emerged because of the intrinsic structure of proteins, which can cause pronounced effects such as immunogenicity responses. On the other hand, biological drugs do not always cause an immune response, and many have a lower immunogenicity risk than even food proteins. Nonetheless, in all regulatory filings, the developer must show that manufacturing does not result in any structural modifications that may be immunogenic; one such element is protein aggregation, which is a common side effect of these drugs' manufacturing.

Over the last few decades, regulatory agencies' guidelines have shifted significantly, tightening the safety criteria. Regulatory agencies provide detailed descriptions of adverse drug reactions and side effects. According to the MHRA, adverse drug reactions (ADRs) or adverse drug effects (ADEs) are a harmful and unintentional response to a drug that occurs at usual therapeutic doses used in humans for the prevention, diagnosis, or treatment of disease or the alteration of a physiological activity (the UK Agency). The terms 'impact' and 'reaction' are interchangeable. ADRs are available in a variety of formats:

- Type A: An exaggerated pharmacological response, such as a pharmacodynamic effect (e.g. beta-blocker-induced bronchospasm) or a toxic response (e.g. deafness from aminoglycoside overdose).
- Type B: Drug-induced diseases (e.g. antibiotic-associated colitis), allergic reactions (e.g. penicillin anaphylaxis), and idiosyncratic reactions (e.g. aplastic anemia with chloramphenicol).
- Type C: Disorders that persist over time, such as osteoporosis treated with oral steroids.
- Type D: Anticonvulsants and lisinopril, for example, have delayed (lag time) teratogenic effects.
- Type E: Cessation of use (withdrawal), such as benzodiazepine withdrawal syndrome.
- Type F: Efficacy failure (no response), such as antimicrobial resistance.

In general, side effects are predicted responses at normal doses due to the molecule's pharmacology. Such side effects can be well-known and even anticipated, necessitating little to no improvement inpatient care. All adverse effects attributable to drug administration, regardless of etiology, are classified as drug reactions. They are divided into two categories: immunologic and non-immunologic etiology. Unpredictable effects account for 20%–25% of all adverse drug reactions, both immune and non-immune mediated, while predictable non-immune events account for 75%–80%.

1.5.5 Immunogenicity testing

Detecting, quantifying, and characterizing antibodies generated against the product is the most practical and widely used method for determining unintended immunogenicity. It is anticipated that improved models for assessing immunogenicity will become available in the future, including DNA microarrays. Generally, no single assay can provide all the necessary information about the immunogenicity profile of a biological product. As a result, a panel of carefully chosen and suitable antibody detection and quantification assays is used. These antibody responses are then correlated with the PK/PD and clinical effects when the antibodies are evaluated to determine if they make a 'significant difference' in terms of safety and efficacy. The testing plans are based on a risk assessment that goes through multiple stages, starting with a screening assay to determine the binding of a biopharmaceutical in serum samples of animals and humans. At this point we are detecting anti-drug antibodies. This evaluation is then subjected to confirmation to ensure there are no false positives. Once established, anti-drug antibodies are characterized.

1.6 Recombinant expression

All living organisms use gene coding to manufacture proteins within their cells. As a result, it is possible to modify the genetic code of living organisms in order to produce a desired protein. Because the genetic code is universal, it can be read by all cellular systems in the animal, plant, and bacterial kingdoms. This technology is referred to as recombinant protein expression. The introduction of recombinant therapeutic proteins derived from the human sequence into heterologous host systems (bacteria, yeast, plant, mammalian cell, or transgenic animals) causes the host system to 'produce' a protein derived from the given sequence. This is the basis for gene universality.

Biopharmaceutical products are synthesized through the expression of genes in engineered living organisms. The genes are composed of DNA segments that carry a message that ultimately results in the production of proteins. They are nucleotide sequences found in the genomes of all living organisms (A, T, G, and C). Each of these genes has a sequence that is unique to a particular protein. The first engine to produce biopharmaceuticals was the fungus used to manufacture penicillin by Fleming in 1928. We are still using filamentous fungi to produce drugs and other chemicals. Many more biological manufacturing engines have been introduced, bacteria, yeast, animal cells, plant cells, and viruses, to manufacture biopharmaceuticals, ranging from therapeutic proteins, antibodies, anticancer cytotoxic drugs and vaccines, anti-infectious disease antibiotics and vaccines, to hormonal disorder therapy and many other indications. Details on the use of recombinant technology are described in chapter 2 of volume 2, *Unit Processes.*

1.6.1 Understanding DNA and RNA

To remain alive, living organisms must follow a code called DNA, or deoxyribonucleic acid. The RNA, or ribonucleic acid, assists the DNA code in carrying out its instructions. Nucleic acids are large macromolecules composed of four nucleobases, two purine bases, and one pyrimidine base. According to one hypothesis, RNA existed before DNA, and DNA is a mutation of RNA. DNA may also be found in the nucleus (nuclear DNA) and mitochondria (mitochondrial DNA) of a cell. The phosphate group, a five-carbon sugar (the stable 2-deoxyribose), and four nitrogen-containing nucleobases (adenine, thymine, cytosine, and guanine) combine to form two nucleotide chains (figure 1.15).

Transcription results in the formation of RNA, a single-stranded, linear molecule. It works in conjunction with DNA, assisting it in completing the tasks assigned to it by DNA. The phosphate group, a five-carbon sugar (the less stable ribose), and four

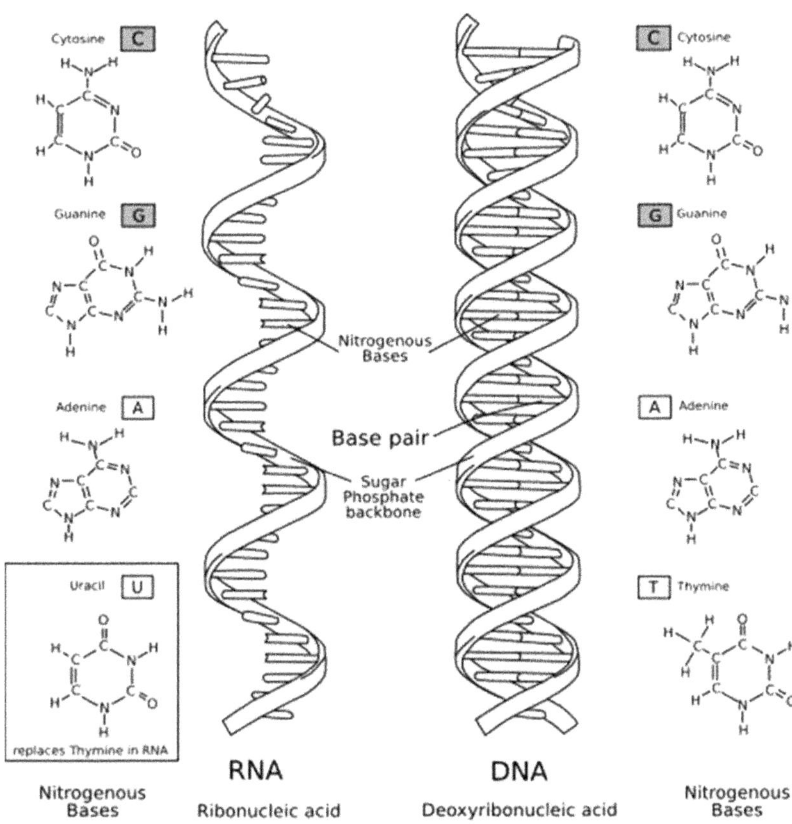

Figure 1.15. Structural differences between DNA and RNA. (Source: Antilived, Fabiolib, Turnstep, Westcairo, Creative Commons.)

nitrogen-containing nucleobases are found in RNA. These are adenine, uracil (not thymine), guanine, and cytosine (figure 1.16).

The nucleobases are bound to their sugar-phosphate backbone and adenine is linked to thymine, while cytosine is linked to guanine. Thus, each nucleobase on a DNA strand binds to the nucleobase on the strand opposite it. The two strands of DNA twist and wind around each other due to this linking, creating various shapes such as the familiar double helix (DNA's 'relaxed' form), circles, and supercoils.

Adenine and uracil (not thymine) form an RNA connection, while cytosine remains connected to guanine. RNA folds in on itself as a single-stranded molecule to bind up its nucleobases, but not all become partnered. The most popular of these three-dimensional shapes is the hairpin loop, which helps decide whether the RNA molecule is messenger RNA (mRNA), transfer RNA (tRNA), or ribosomal RNA (rRNA). Table 1.7 provides a comparison of the two nucleic acids.

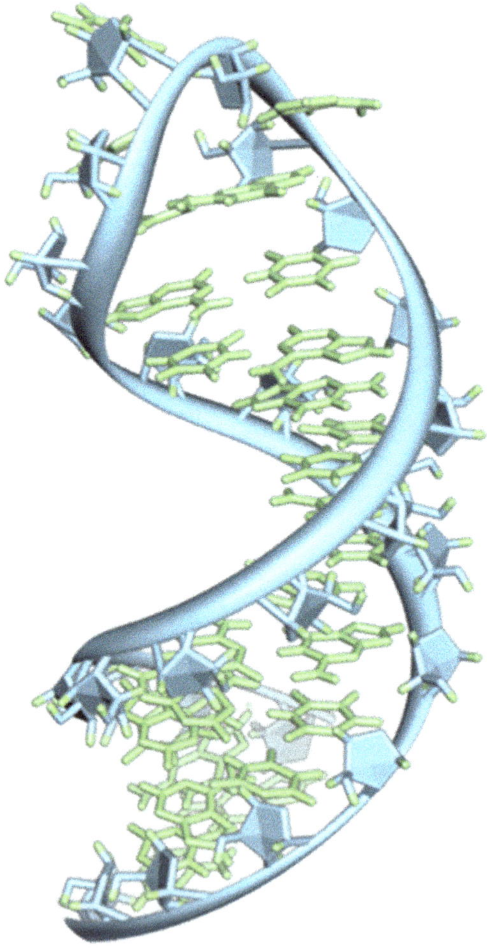

Figure 1.16. RNA curls up into a hairpin loop. (Source: Vossman, Creative Commons CCBY 3.0.)

Table 1.7. Comparison of DNA and RNA.

	DNA	RNA
Full name	Deoxyribonucleic acid.	Ribonucleic acid.
Definition	A nucleic acid is a molecule that contains the genetic instructions that all modern living organisms use to evolve and work. The proteins that DNA's nucleotides generate with the aid of RNA express or manifest DNA's genes.	The information contained in DNA dictates which traits should be created, activated, or deactivated, while the various forms of RNA carry out the task.
Function	The set of rules that a living organism must follow to survive, storing and transmitting genetic information. Evolution is defined by the gradual, consistent changes in DNA observed over time. Mutations can be detrimental to an organism, neutral, or beneficial. Around 19 000 genes comprise the human genome which are contained in small fragments of long DNA strands. The distinctions between distinct living organisms and related living organisms result from the precise instructions contained in genes, which are determined by the order of the nucleobases in DNA.	DNA controls the formation, activation, and deactivation of traits, whereas RNA performs the work. It aids in the execution of DNA's blueprint instructions. It is responsible for transporting genetic material from the nucleus to the ribosome, which is required for protein synthesis. mRNA is responsible for transporting genetic information from the nucleus of a cell to the cytoplasm and ribosome. tRNA is found in the cytoplasm of cells and acts as a scaffold for mRNA. For example, a ribosome's tRNA transports amino acids to mRNA, which are the building blocks of proteins. rRNA is found in the cytoplasm of a cell. The ribosome translates the information provided by the mRNA and tRNA. This information is used to determine if a polypeptide or protein should be synthesized or manufactured.
Structure	A double-strand separates the strands. The phosphate group is followed by a five-carbon sugar (the stable 2-deoxyribose) and four nitrogen-containing nucleobases: adenine, thymine, cytosine, and guanine.	The phosphate group, a five-carbon sugar (the less stable ribose), and four nitrogen-containing nucleobases are found in single-stranded DNA: adenine, uracil (not thymine), guanine, and cytosine.

(*Continued*)

Table 1.7. (*Continued*)

	DNA	RNA
Base pairing	Adenine connects to thymine (A–T), while cytosine connects to guanine (C–G) (C–G).	Adenine combines with uracil (A–U), while cytosine combines with guanine (C–G).
Location	DNA is found in a cell's nucleus and mitochondria.	In a cell mRNA is a type of RNA that transports genetic information from the nucleus to the cytoplasm and ribosome. A cell's cytoplasm contains tRNA, which acts as a scaffold for mRNA. A ribosome's tRNA transports amino acids to the mRNA, which are the building blocks of proteins.
Stability	Because of the C–H bonds, the deoxyribose sugar in DNA is less reactive. In alkaline conditions, it is stable. Enzymes have a harder time 'attacking' DNA since the grooves are smaller.	Because of the C–OH (hydroxyl) bonds, ribose sugar is more reactive. In alkaline conditions it is not stable. Since RNA has wider grooves, enzymes can 'attack' it more easily strike it.
Propagation	DNA is self-replicating.	When RNA is necessary it is synthesized from DNA.
Unique features	The DNA helix has a β-form helix geometry. Since DNA is tightly packed in the nucleus, it is covered. Therefore ultraviolet rays have the potential to harm DNA.	In RNA an α-form helix geometry is found. In addition RNA strands are synthesized, degraded, and reused continuously. As a result, RNA is less likely to be affected by ultraviolet rays.

1.7 Gene and cell therapy

Gene therapy is modifying a patient's genetic code to cure a disease. The genetic material transferred then regulates how cells manufacture proteins. Cell therapy treats a patient with live cells that may come from a donor (allogeneic cells) or the patient (autologous cells). These transferred cells can give rise to any cell type when pluripotent, and multipotent cells can give rise to other cell types with limitations when pluripotent.

Figures 1.17 and 1.18 describe the gene and cell therapy design; the patient's product is biopharmaceutical.

1.7.1 Gene editing

Gene editing changes an organism's DNA by adding, removing, or altering a genome's particular location. As shown in figure 1.19 several technologies are available. Clustered regularly interspaced short palindromic repeats (CRISPRs),

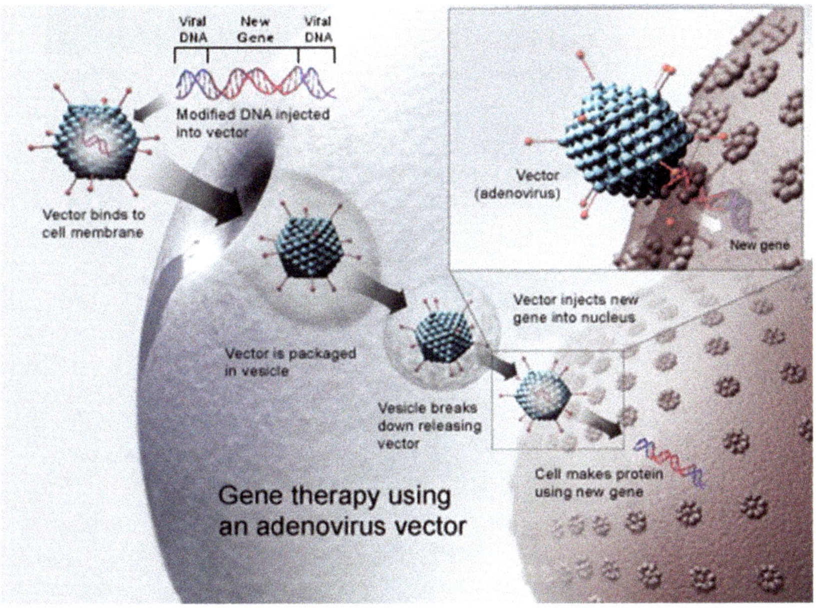

Figure 1.17. Strategies for delivering therapeutic transgenes into patients. (Source: US Federal Government (National Institutes of Health) (Public Domain).)

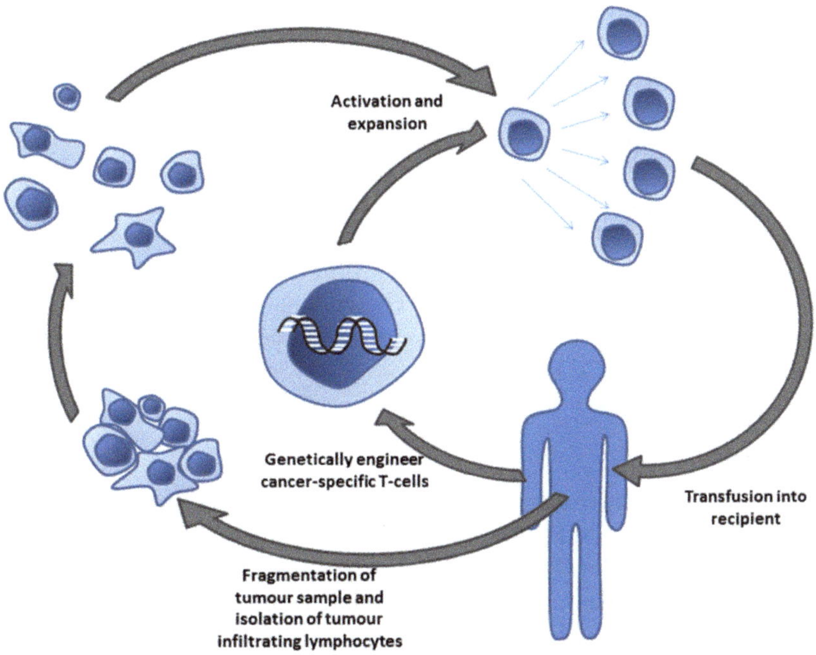

Figure 1.18. Adoptive T-cell therapy. Unique to cancer, T-cells can be obtained by fragmenting and isolating tumor-infiltrating lymphocytes or genetically engineering peripheral blood cells. Before transfusion into the recipient (the tumor bearer), the cells are activated and grown.) (Source: Simon Caulton, Creative Commons.)

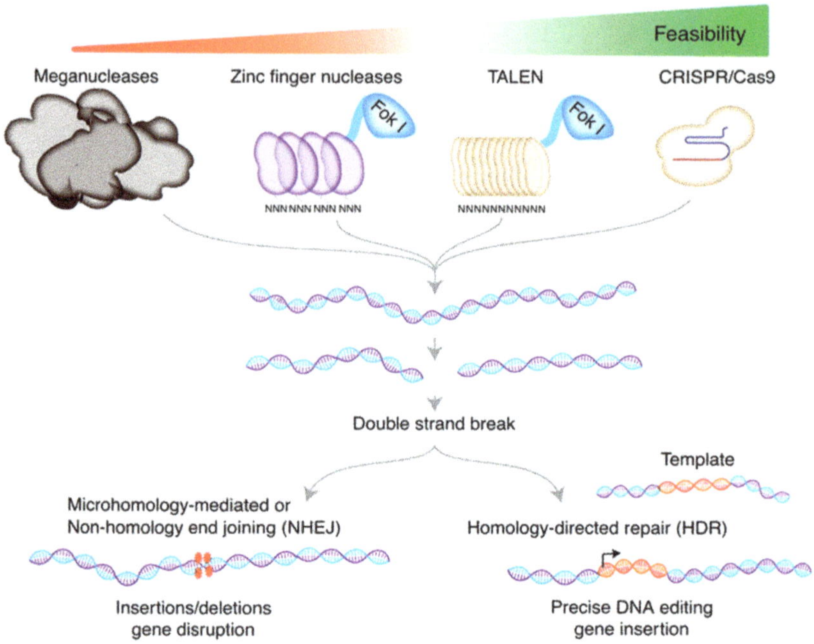

Figure 1.19. Schematics of the use of meganuclease, zinc finger nuclease, TALEN, and CRISPR nucleases for genome editing. Reprinted by permission from Macmillan Publishers Ltd: Adli M 2018 The CRISPR tool kit for genome editing and beyond *Nat. Commun.* **9** 1911, copyright (2018).

which bacteria use to develop immunity against viruses, are the most effective technology available today. They are formed by the integration of short viral genome sequences into the bacterial genome. CRISPR-associated proteins (Cas) cleave the viral DNA sequences that match them, allowing the eukaryotic genome to be cut at any desired location via plasmid injection of Cas genes. The FDA has not yet approved any gene-editing products, however, the agency recently approved a CRISPR-based diagnostic kit for COVID-19 monitoring. In 2020 the Nobel Prize in Chemistry was shared by Emmanuelle Charpentier (Max Planck Unit for the Science of Pathogens) and Jennifer A Doudna (University of California, Berkeley, USA) for the discovery of CRISPR technology.

1.8 Conclusion

Biopharmaceuticals comprise a broad category of products, primarily recombinant therapeutic proteins and this chapter offers teaching about the changes in the scientific approaches to develop these products. Given the high cost of developing biopharmaceuticals, intellectual property considerations are important and require understanding by scientists at all levels of development and manufacturing designed and executed based on regulatory regulations across the globe. Understanding biopharmaceuticals starts with studying the molecular structure and how the molecules interact with receptors in various biopharmaceuticals, hormones, hematopoietic factors, cytokines, plasma protein factors, bone formation factors, recombinant enzymes, and antibodies. Detailed understanding of the primary, secondary, tertiary, and quaternary structure of proteins, including post-translational modifications, is needed to design a robust manufacturing process. Structural designs for pharmacokinetic modulation and increased action duration, such as pegylation, the design of fusion proteins, and unnatural construction, are essential to new drug development. Most critical for biopharmaceuticals is their immunogenicity and how to control it in designing the product and the manufacturing conditions and practices. Recombinant expression requires an understanding of DNA and RNA. The new science of gene and cell therapies is also part of the field of biopharmaceuticals.

IOP Publishing

Biopharmaceutical Manufacturing, Volume 1
Regulatory processes
Sarfaraz Niazi and Sunitha Lokesh

Chapter 2

Antibody biopharmaceuticals

Antibodies are one of the most significant commercial categories of biopharmaceutical products; this chapter describes in detail their structure and mode of action, a listing of 83 FDA approved monoclonal antibodies, their types and their targets, including bispecific, multi-specific, Fab fragments, single chain, humanized, chimeric, and fully human monoclonal antibodies; related topics include affinity maturation, antigenized antibodies, IgG1 fusion proteins, drug or toxin conjugates. Development technologies of mouse hybridoma, transgenic animals, phage display, and single B-cells are detailed and compared for clinical testing and cost advantages. The commercial production of monoclonal antibodies is summarized, and an extensive list of online databases to find antibody properties is provided.

2.1 Overview

In 1900 Paul Ehrlich suggested the concept of 'magic bullets', suggesting that if a compound could be created that selectively selects disease-causing species a toxin for that organism could destroy the organism. The Nobel Prize in Physics went to Paul Ehrich and Élie Metchnikoff in 1908. The B-cell in the multiple cancer myeloma was discovered to generate a single type of antibody (a paraprotein—a protein present in the body only in the precancer stage) in the 1970s, and this knowledge was used to investigate the structure of antibodies. In 1973 Jerrold Schwaber identified antibody production using human–mouse hybrid cells. In 1975 Georges Köhler and César Milstein succeeded in fusing myeloma cell lines with B-cells to create hybridomas that generated antibodies unique to known antigens, immortalizing them (because of fusion with myeloma cells). They shared the Nobel Prize with Niels Kaj Jerne in 1984. Furthermore, Greg Winter and his team pioneered strategies to humanize mAbs, removing unwanted side effects. In 2018 Jam Allison and Tasuku Honjo were awarded the Nobel Prize for discovering cancer therapy through negative immune regulation inhibition mAbs that avoid inhibitory linkages.

Antibodies, also known as immunoglobulins, are glycoprotein molecules in our blood and tissue fluids that assist in the battle against infection. Antibody

doi:10.1088/978-0-7503-3175-3ch2

molecules are available in a range of shapes and sizes. However, the basic structure is a 'Y' shape, with the two tips designed to identify and bind foreign agents (such as bacteria, viruses, and other pathogens), foreign substances, or other harmful cells. The majority of the polyclonal antibodies detect a variety of epitopes (an antigenic portion) and thus identify antigens in various orientations, which is helpful in specific assays where a single epitope would weaken the identification of an analyte. Continuous culture of B-cell hybridomas provides a reliable and potentially inexhaustible source of high-specificity antibodies. There are also other ways to make antibodies, such as 'display' systems, chemical synthesis, and B-cell extraction.

Antibodies can bind to almost any non-self-surface antigen (self-antigens are found on body cells). Antibodies are important for biomedical research, diagnostics, and therapy because they are the key to immunity. Antibody structure–function relationships provide a platform for protein engineering, which is used to create various biologics for various therapeutic indications. The antigen affinity, effector function, and biophysical properties of antibodies are all discussed in this chapter.

Understanding antibodies requires a thorough understanding of the immune system that produces them. Antibodies are formed against different sections or regions of an immunogen called antigenic determinants, or epitopes, typically consisting of six to eight amino acids, when an immunogen triggers a humoral immune response. Antibodies recognize and interact with three-dimensional shapes (discontinuous epitopes) in folded proteins, as well as linear amino acid stretches of continuous epitopes. Each antibody molecule recognizes only one epitope, and a single B-cell clone forms each antibody. Monoclonal antibodies are antibodies that have a single affinity and are produced from a single B-cell clone.

2.1.1 Naming

The naming of antibodies practiced until 2017 followed the below-described system with a random prefix, followed by a target class (substem A), and then followed by the description of the species (substem B) and finally, the stem, 'mab':

- -ba- bacterial.
- -ami- serum amyloid protein (SAP)/amyloidosis (pre-substem).
- -ci- cardiovascular.
- -fung- fungal.
- -gros- skeletal mass related growth factors and receptors (pre-substem).
- -ki- interleukin.
- -li- immunomodulating.
- -ne- neural.
- -os- bone.
- -toxa- toxin.
- -ta- tumor.
- -vet- veterinary use (pre-stem).
- -vi- viral.

After 2017 substem B has included the same random prefix and only a target class description, followed by the stem 'mab'. The target class is defined as follows, but it is still under revision:

- -a- rat.
- -axo- rat–mouse (pre-substem).
- -e- hamster.
- -i- primate.
- -o- mouse.
- -u- human.
- -vet- veterinary use (pre-substem).
- -xi- chimeric.
- -xizu- chimeric-humanized.
- -zu- humanized.

The FDA now requires a random four-letter suffix for all-new biological drugs, including biosimilars.

2.1.2 Commercial antibodies

The approved monoclonal antibody therapies include a wide variety of diseases, including cancer, multiple sclerosis, organ rejection, anticoagulant, infections, asthma, macular degeneration, arthritis, psoriasis, Cohen's disease, bone loss, lupus, Castleman disease, hypercholesteremia, Sezary syndrome, angioedema, HIV, Castleman disease, dermatitis, hemophilia, hemoglobinuria, Muckle–Wells syndrome, Merkle cell disease, migraine, hypophosphatemia, hemophagocytic lymph histiocytosis, sickle cell anemia, thyroid eye disease, and osteoporosis.

The first monoclonal antibody approved in 1986 by the FDA was Orthoclone OKT3. By mid-2020, there were 83 approved therapeutic mAbs and derivatives such as antibody fragments (Fab), fusion proteins (Fc-fusion proteins), single-chain variable fragments (scFv), antibody–drug conjugates (ADCs), and bispecific antibodies, most of which are used in treating cancer (table 2.1).

There are over 600 therapeutic mAbs under development, making this field of therapeutics the most active and it is anticipated to produce many lifesaving therapies.

2.2 The immune system

The human immune system has developed into two distinct forms of protection against infection and disease that include the innate and adaptive immune systems, both of which are made up of various cell types originating in the hematopoietic stem cells in the bone marrow. Lymphocytic and myeloid progenitors are formed by hematopoietic stem cells (figure 2.1).

To mount and coordinate an effective immune response, lymphocytes, inflammatory cells, and hematopoietic cells must communicate with one another. This is done by cytokines. Cytokines are a large class of tiny proteins or glycoproteins with a wide range of functions, usually less than 30 kDa in size. They were first

Table 2.1. US FDA-approved mAbs as of June 2020.

No	mAb	Brand	Expression	Target	Format	Technology	Indication	Date approved
1	Muromonab-CD3	Orthoclone OKT3	Hybridoma	CD3	Murine IgG2a	Hybridoma	Kidney transplant rejection	1986
2	Abciximab	Reopro	Sp2/0	GPIIb/IIIa	Chimeric IgG1 Fab	Hybridoma	Prevention of blood clots in angioplasty	1994
3	Rituximab	MabThera, Rituxan	CHO	CD20	Chimeric IgG1	Hybridoma	Non-Hodgkin lymphoma	1997
4	Trastuzumab	Herceptin	CHO	HER2	Humanized IgG1	Hybridoma	Breast cancer	1998
5	Infliximab	Remicade	Sp2/0	TNFα	Chimeric IgG1	Hybridoma	Crohn's disease	1998
6	Palivizumab	Synagis	NSO	RSV	Humanized IgG1	Hybridoma	Prevention of respiratory syncytial virus infection	1998
7	Alemtuzumab	Campath, Lemtrada	CHO	CD52	Humanized IgG1	Hybridoma	Chronic myeloid leukemia	2001
8	Adalimumab	Humira	CHO	TNFα	Human IgG1	Phage display	Rheumatoid arthritis	2002
9	Ibritumomab tiuxetan (ADC)	Zevalin	CHO	CD20	Murine IgG1	Hybridoma	Non-Hodgkin lymphoma	2002
10	Omalizumab	Xolair	CHO	IgE	Humanized IgG1	Hybridoma	Asthma	2003
11	Bevacizumab	Avastin	CHO	VEGF-A	Humanized IgG1	Hybridoma	Colorectal cancer	2004
12	Cetuximab	Erbitux	Sp2/0	EGFR	Chimeric IgG1	Hybridoma	Colorectal cancer	2004
13	Natalizumab	Tysabri	NSO	ITGA4	Humanized IgG4	Hybridoma	Multiple sclerosis	2004
14	Ranibizumab	Lucentis	*E. coli*	VEGF-A	Humanized IgG1 Fab	Hybridoma	Macular degeneration	2006
15	Panitumumab	Vectibix	CHO	EGFR	Human IgG2	Transgenic mice	Colorectal cancer	2006

16	Eculizumab	Soliris	NSO	C5	Humanized IgG2/4	Hybridoma	Paroxysmal nocturnal hemoglobinuria	2007
17	Certolizumab pegol (ADC)	Cimzia	*E. coli*	TNFα	Humanized Fab, pegylated	Hybridoma	Crohn's disease	2008
18	Ofatumumab	Arzerra	NSO	CD20	Human IgG1	Transgenic mice	Chronic lymphocytic leukemia	2009
19	Canakinumab	Ilaris	SP2/0	IL-1β	Human IgG1	Transgenic mice	Muckle–Wells syndrome	2009
2o	Golimumab	Simponi	Sp2/0	TNFα	Human IgG1	Transgenic mice	Rheumatoid and psoriatic arthritis, ankylosing spondylitis	2009
21	Ustekinumab	Stelara	Sp2/0	IL-12/23	Human IgG1	Transgenic mice	Psoriasis	2009
22	Tocilizumab	RoActemra, Actemra	CHO	IL6R	Humanized IgG1	Hybridoma	Rheumatoid arthritis	2010
23	Denosumab	Xgeva, Prolia	CHO	RANKL	Human IgG2	Transgenic mice	Bone loss	2010
24	Brentuximab vedotin (ADC)	Adcetris	CHO	CD30	Chimeric IgG1; ADC	Hybridoma	Hodgkin lymphoma, systemic anaplastic large cell lymphoma	2011
25	Belimumab	Benlysta	NSO	BLyS	Human IgG1	Phage display mice	Systemic lupus erythematosus	2011
26	Ipilimumab	Yervoy	CHO	CTLA-4	Human IgG1	Transgenic	Metastatic melanoma	2011
27	Raxibacumab	Abthrax	NS0	*B. anthracis* protective antigen	Human IgG1	Transgenic mice	Anthrax infection	2012
28	Trastuzumab emtansine	Kadcyla	CHO	HER2	Humanized IgG1; ADC	Hybridoma	Breast cancer	2012
29	Pertuzumab	Perjeta	CHO	HER2	Humanized IgG1	Hybridoma	Breast cancer	2012

(Continued)

Table 2.1. (*Continued*)

No	mAb	Brand	Expression	Target	Format	Technology	Indication	Date approved
30	Obinutuzumab	Gazyva, Gazyvaro	CHO	CD20	Humanized IgG1 glycoengineered	Hybridoma	Chronic lymphocytic leukemia	2013
31	Blinatumomab	Blincyto	CHO	CD19, CD3	Murine bispecific tandem scFv	Hybridoma	Acute lymphoblastic leukemia	2014
32	Ramucirumab	Cyramza	NSO	VEGFR2	Human IgG1	Phage display	Gastric cancer	2014
33	Vedolizumab	Entyvio	CHO	$\alpha 4\beta 7$ integrin	Humanized IgG1	Hybridoma	Ulcerative colitis, Crohn's disease	2014
34	Pembrolizumab	Keytruda	CHO	PD-1	Humanized IgG4	Hybridoma	Melanoma	2014
35	Nivolumab	Opdivo	CHO	PD-1	Human IgG4	Transgenic mice	Melanoma, non-small cell lung cancer	2014
36	Siltuximab	Sylvant	CHO	IL-6	Chimeric IgG1	Hybridoma	Castleman disease	2014
37	Secukinumab	Cosentyx	CHO	IL-17α	Human IgG1	Transgenic mice	Psoriasis	2015
38	Daratumumab	Darzalex	CHO	CD38	Human IgG1	Transgenic mice	Multiple myeloma	2015
39	Elotuzumab	Empliciti	NSO	SLAMF7	Humanized IgG1	Hybridoma	Multiple myeloma	2015
40	Mepolizumab	Nucala	CHO	IL-5	Humanized IgG1	Hybridoma	Severe eosinophilic asthma	2015
41	Necitumumab	Portrazza	NS0	EGFR	Human IgG1	Phage display	Non-small cell lung cancer	2015
42	Alirocumab	Praluent	CHO	PCSK9	Human IgG1	Transgenic mice	High cholesterol	2015
43	Idarucizumab	Praxbind	CHO	Dabigatran	Humanized Fab	Hybridoma	Reversal of dabigatran-induced anticoagulation	2015
44	Evolocumab	Repatha	CHO	PCSK9	Human IgG2	Transgenic mice	High cholesterol	2015
45	Dinutuximab	Unituxin	Sp2/0	GD2	Chimeric IgG1	Hybridoma	Neuroblastoma	2015
46	Obiltoxaximab	Anthim	GS-NS0	*B. anthrasis* PA	Chimeric IgG1	Hybridoma	Inhalational anthrax	2016

47	Reslizumab	Cinqaero, Cinqair	NS0	IL-5	Humanized IgG4	Hybridoma	Asthma	2016
48	Olaratumab	Lartruvo	NS0	PDGFRα	Human IgG1	Transgenic mice	Soft tissue sarcoma	2016
49	Ixekizumab	Taltz	CHO	IL-17α	Humanized IgG4	Hybridoma	Psoriasis	2016
50	Atezolizumab	Tecentriq	CHO	PD-L1	Humanized IgG1	Hybridoma	Bladder cancer	2016
51	Bezlotoxumab	Zinplava	CHO	*Clostridium difficile* enterotoxin B	Human IgG1	Transgenic mice	Prevention of *Clostridium difficile* infection recurrence	2016
52	Avelumab	Bavencio	CHO	PD-L1	Human IgG1	Phage display	Merkel cell carcinoma	2017
53	Inotuzumab ozogamicin	Besponsa	CHO	CD22	Humanized IgG4	Hybridoma	Acute lymphoblastic leukemia	2017
54	Dupilumab	Dupixent	CHO	IL-4Rα	Human IgG4	Transgenic mice	Atopic dermatitis	2017
55	Benralizumab	Fasenra	CHO	IL-5Rα	Humanized IgG1	Hybridoma	Asthma	2017
56	Emicizumab	Hemlibra	CHO	Factor IXa, X	Humanized IgG4, bispecific	Hybridoma	Hemophilia A	2017
57	Durvalumab	Imfinzi	CHO	PD-L1	Human IgG1	Transgenic mice	Bladder cancer	2017
58	Sarilumab	Kevzara	CHO	IL6R	Human IgG1	Transgenic mice	Rheumatoid arthritis	2017
59	Gemtuzumab ozogamicin	Mylotarg	NS0	CD33	Humanized IgG4; ADC	Hybridoma	Acute myeloid leukemia	2017
60	Ocrelizumab	Ocrevus	CHO	CD20	Humanized IgG1	Hybridoma	Multiple sclerosis	2017
61	Brodalumab	Siliq, Lumicef	CHO	IL-17R	Human IgG2	Transgenic mice	Plaque psoriasis	2017
62	Guselkumab	Tremfya	CHO	IL-23 p19	Human IgG1	Phage display	Plaque psoriasis	2017

(*Continued*)

Table 2.1. *(Continued)*

No	mAb	Brand	Expression	Target	Format	Technology	Indication	Date approved
63	Erenumab	Aimovig	CHO	CGRPR	Human IgG2	Transgenic mice	Migraine prevention	2018
64	Fremanezumab	Ajovy	CHO	CGRP	Humanized IgG2	Hybridoma	Migraine prevention	2018
65	Burosumab	Crysvita	CHO	FGF23	Human IgG1	Transgenic mice	X-linked hypophosphatemia	2018
66	Galcanezumab	Emgality	CHO	CGRP	Humanized IgG4	Hybridoma	Migraine prevention	2018
67	Emapalumab	Gamifant	CHO	IFNγ	Human IgG1	Phage display	Primary hemophagocytic lymphohistiocytosis	2018
68	Tildrakizumab	Ilumya	CHO	IL-23 p19	Humanized IgG1	Hybridoma	Plaque psoriasis	2018
69	Cemiplimab	Libtayo	CHO	PD-1	Human mAb	Transgenic mice	Cutaneous squamous cell carcinoma	2018
70	Moxetumomab pasudodox	Lumoxiti	*E. coli*	CD22	Murine IgG1 dsFv	Phage display	Hairy cell leukemia	2018
71	Mogamulizumab	Poteligeo	CHO	CCR4	Humanized IgG1	Hybridoma	Mycosis fungoides or Sézary syndrome	2018
72	Lanadelumab	Takhzyro	CHO	Plasma kallikrein	Human IgG1	Phage display	Hereditary angioedema attacks	2018
73	Ibalizumab	Trogarzo	NS0	CD4	Humanized IgG4	Hybridoma	HIV infection	2018
74	Ravulizumab	Ultomiris	CHO	C5	Humanized IgG2/4	Hybridoma	Paroxysmal nocturnal hemoglobinuria	2018
75	Crizanlizumab	Adakveo	CHO	P-selectin	Humanized IgG2	Hybridoma	Sickle cell disease	2019
76	Brolucizumab degeneration	Beovu 2019	*E. coli*	VEGF-A	Humanized scFv	Rabbit	hybridoma	Macular
77	Caplacizumab	Cablivi	*E. coli*	von Willebrand factor	Humanized nanobody	Hybridoma	Acquired thrombotic thrombocytopenic purpura	2019

78	Romosozumab	Evenity	CHO	Sclerostin	Humanized IgG2	Hybridoma	Osteoporosis in postmenopausal women at increased risk of fracture	2019
79	Polatuzumab vedotin	Polivy	CHO	CD79β	Humanized IgG1 ADC	Hybridoma	Diffuse large B-cell lymphoma	2019
80	Risankizumab	Skyrizi	CHO	IL-23 p19	Humanized IgG1	Hybridoma	Plaque psoriasis	2019
81	Isatuximab	Sarclisa	CHO	CD38	Humanized IgG1	Hybridoma	Multiple myeloma	2020
82	Teprotumumab-trbw	Teoezza	CHO	IFG-1R	Humanized IgG1k	Hybridorma	Thyroid eye disease	2020
83	Sacituzumab govitecan-hziy	Trodelvy	Sp2/0	CL2A	Humanized IgG1k	Hybridoma	Breast cancer	2020

Notes: Unituxin (dinutuximab) is withdrawn from use in the European Union; Raptiva® (efalizumab) was approved in 2003 by the FDA, and in 2004 by the EMA, and withdrawn in 2009; Mylotarg® (gemtuzumab ozogamicin) was approved in 2000 by the FDA, and withdrawn from the market in 2020; Zenapax® (daclizumab) was approved in 1997 by the FDA, and in 1999 by the EMA, it was withdrawn from the market for commercial reasons. Before mid-2017, a mAb that had been humanized was designated by the inclusion of the stem 'zu' in its name (e.g. trastuzumab), and chimeric mAbs were designated as chimeric by the addition of 'xi' (e.g. rituximab). However, antibodies named after mid-2017 will not contain the 'zu' and 'xi' stems in their generic names.

Source: https://www.fda.gov/vaccines-blood-biologics/development-approval-process-cber/biological-approvals-year.

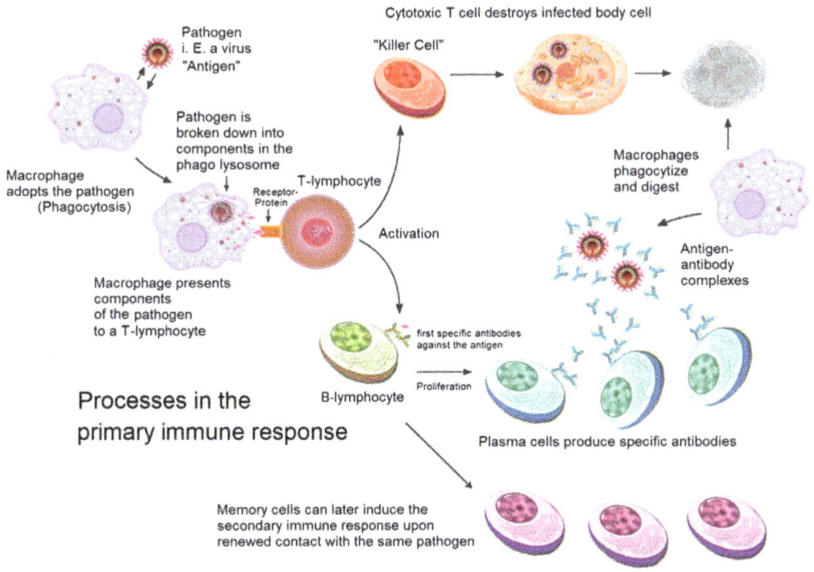

Figure 2.1. Primary immune response. (Source: Wikipedia: Sciencia58 and Domdomegg, Fæ, Petr94, Manu5; Creative Commons Attribution https://commons.wikimedia.org/wiki/File:Primary_immune_response_1.png.)

found to have immunomodulatory properties, but they have since been discovered to perform various functions in developmental processes such as cell differentiation and directed migration. The cells and macrophages are two of the most important cytokines makers, influencing both innate and adaptive immune responses. They can, however, be transiently generated and secreted by nearly all nucleated cells.

The high-affinity binding of a cytokine's receptor expressed on the surface of a target cell mediates the cytokine's downstream effects. This action could be autocrine (on the same cell), paracrine (on an adjacent cell), or endocrine (on a cell outside the body) (on a distant cell; not the usual way for cytokine responses). Receptor interaction activates intracellular signaling cascades, resulting in altered gene expression in the target cell and a biological impact (figure 2.1). The target cell undergoes differentiation, proliferation, and activation after cytokine stimulation, among other things. Some types of cytokines are as follows:

- SCF and TPO are self-renewal cytokines, and Flt3Ligand, SCF, TPO, IL-3, and IL-6 are expansion cytokines generated by hematopoietic stem cells.
- Neutrophils, eosinophils, basophils (called for their staining properties), mast cells, and monocytes are all produced by myeloid progenitors, which can subsequently be differentiated into dendritic cells (DCs) and macrophages.
- Because of their multilobed nuclei, neutrophils, such as eosinophils and basophils, are granulocytes (cells with granules) that belong to the polymorphonuclear (PMN) class of leukocytes.

2.2.1 Innate immune system

The innate immune system, the body's initial line of defense, offers a quick and complete immunological response. The adaptive immune system, on the other hand, detects and eliminates dangerous microorganisms from the body. The adaptive immune system is slower than the innate immune system in keeping you healthy.

Myeloid progenitors generate neutrophils, eosinophils, basophils (named for their staining properties), mast cells, and monocytes, which can be differentiates into dendritic cells (DCs) and macrophages. Mast cells are tissue-resident granulocytes that release histamine, heparin, and other chemicals that help protect against parasites, heal wounds, and promote angiogenesis. Activated macrophages divide and contribute in wound healing as well as bacterial and viral defenses.

Perforins and granzymes are small granules found in the cytoplasm of cytotoxic NK cells that destroy their target cells. They are formed from the same lymphoid progenitor as B- and T-cells.

2.2.2 Adaptive immune system

The adaptive immune system has the ability to remember things for a long time. This mechanism allows for a faster and more effective immune response in the event of a possible pathogen interaction.

When activated, mature B-cells differentiate into memory cells and plasma cells, which secrete pathogen-specific antibodies and play a key role in the protective immune response (figure 2.2). (While B-cells are generally referred to with the label B because they come from bone, the term was first applied to the bursa of Fabricius, the site of hematopoiesis in birds that is necessary for the development of B-cells.) One of three types of antigen-presenting cells in the body are B-cells (APCs). MHC class I proteins can be found on the surfaces of all nucleated cells in the body. MHC class II proteins, as well as other co-stimulatory molecules, are commonly expressed on the surfaces of APCs, such as macrophages, B-cells, and dendritic cells. MHC class II cells activate T-cells by expressing peptide fragments of processed antigens.

Cellular crosstalk is critical in adaptive immune responses, such as naive B-cells, which require CD4+ T-cells to establish an efficient response to antigens. Activated cells, such as neutrophils, release chemokines and cytokines, which cause innate immune system interaction. This operation influences DC recruitment and activation.

The immune system's two arms work together to provide the most efficient defense for the body through cellular interaction and chemical signals such as cytokines and other released chemicals.

2.3 Monoclonal antibodies

The majority of antibodies formed as part of a normal immune response are polyclonal, meaning they are made up of many different B-lymphocytes. As a result, their antigen specificities are significantly different (e.g. by binding different epitopes

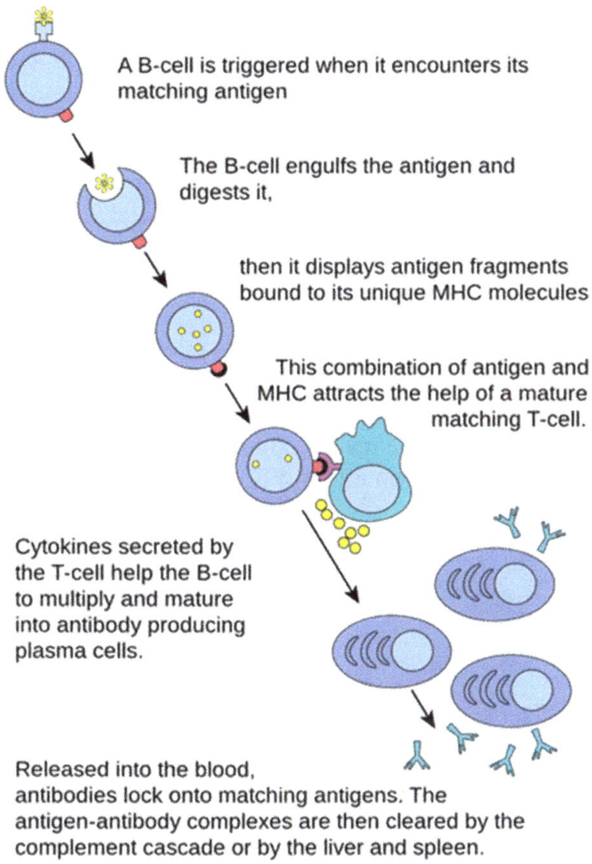

Figure 2.2. B-cell activation and antibody production. (Source: Wikipedia: Fred the Oyster - The Immune System https://commons.wikimedia.org/wiki/File:B_cell_activation.svg.)

or binding the same epitope with different affinities). On the other side, a single B-cell clone may produce a significant amount of antibodies.

Monoclonal antibodies are found in immune cells that are clones of a single parent cell (mAbs). Because they attach to the same epitope they have a monovalent affinity. Antibodies are important tools in biochemistry, molecular biology, and medicine because they can produce monoclonal antibodies that can bind directly to any material and detect or purify it. Polyclonal antibodies, on the other hand, are generated naturally by B-cells and can bind to multiple epitopes. Bispecific monoclonal antibodies are monoclonal antibodies that target two multi-specific antibodies that can target a variety of epitopes.

Monoclonal antibodies are used in many molecular immunology studies. When used in conjunction with epitope mapping and molecular modeling, monoclonal antibodies allow antigenic profiling and visualization of macromolecular surfaces. Monoclonal antibodies are also being used in a number of clinical laboratory diagnostic studies as well. Because of their high specificity, these reagents are widely used in the detection and recognition of serum analytes, cell markers, and pathogenic agents. In addition, an infinite supply of reagent can be obtained by maintaining a continuous culture of hybridoma cells containing these antibodies. In comparison to the comparatively small availability of polyclonal antibody reagents, a continuous supply function allows for the standardization of both the reagent and the assay technique. Polyclonal and monoclonal antibodies have advantages and disadvantages in terms of development, expense, and general applications.

2.3.1 Mode of action

The antibody's variable region the Fab (antigen-binding fragment) is the smallest unit of an antibody with antigen-binding capability, resulting in conformational changes in the contact surface areas of both the antibody and the antigen in a lock and key fit model that minimizes changes in surface conformation of the unbound and bound states (figure 2.3). Thus in the unbound and bound states the antibody and antigen's

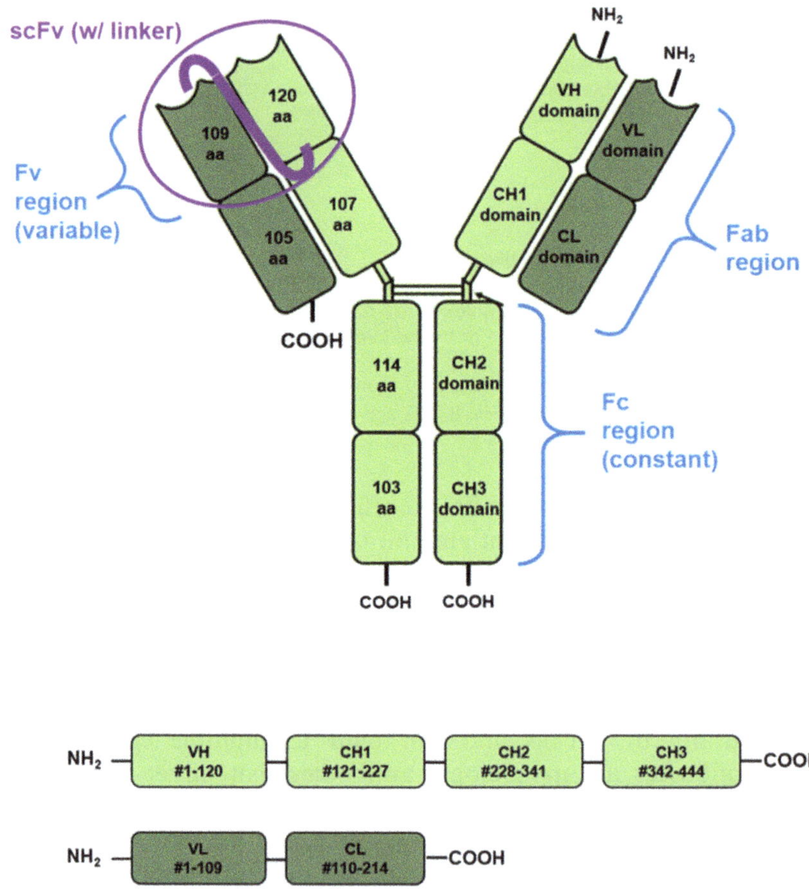

Figure 2.3. Schematic of IgG antibody structure. The residue numbering is approximate. (Source: https://upload.wikimedia.org/wikipedia/commons/2/27/Anatomy_of_an_IgG.png; AJVincelli, Public domain, via Wikimedia Commons.)

backbone conformations are the same. Conversely, in the induced mode, the antibody and antigen's conformational changes may be quite extensive. Both the side chain and backbone atoms in the contact region may experience conformational changes after binding, particularly in the complementarity-determining regions (CDRs). The antigen samples a population of specific conformational states in the conformational selection model before binding. Antibody binding may depend on the antigen's pre-activation states, which may be affected by the antigen's microenvironment. Sorting out the target interaction kinetics also guides how their pharmacology is optimized.

If the molecular underpinnings of disease are understood and the key molecules implicated in pathogenesis are discovered, antibodies can be a viable treatment option, as evidenced by several antibodies currently in clinical use:

- Migraine prevention with anti-CGRP receptor antibodies (erenumab, galcanezumab, or fremanezumab).
- PCSK9 anti-protein convertase subtilisin/kexin type 9 (evolocumab or alirocumab) antibodies, hypercholesterolemia.
- Anti-fibroblast growth factor 23 is a protein that inhibits the growth of fibroblasts (FGF23), burosumab antibody: X-linked hypophosphatemia (XLH).
- Anti-IL6R antibodies (sarilumab and tocilizumab) are used to treat rheumatoid arthritis.
- Hemophilia A is treated with an anti-IXa/Xa factor antibody (emicizumab).
- Anti-Willebrand factor antibodies (caplacizumab): thrombocytopenic purpura.

The rapid development of antibody engineering points to many new potential therapies utilizing multiple mechanisms for treating a disease. As an example, anticancer monoclonal antibodies work on several mechanisms: rituximab acts by opsonization (making cancer cells more recognizable by the body immune system; ADCC, CDC), cetuximab by blocking epidermal growth signals, bevacizumab by stopping blood vessel formation, and ibritumomab is used to deliver radiation to cancer cells.

The affinity of mAbs for the target antigen is determined by the variable region and CDR. To assess affinity, the interaction constant for binding between the antibody and a single monovalent antigen *in vitro* is used. This affinity is amplified (e.g. 10^{18}, a nearly irreversible binding reaction, rather than 109 L/mol) when the antibody is bivalent (e.g. full-length). Antibody affinities are frequently in the 10^5 to 10^{11} L/mol range (picomolar to nanomolar affinity). Another important feature of mAbs is their ability to attract other immune cells and molecules (such as complement), all of which can kill target cells thanks to the Fc component of the antibody.

The desired effect of an mAb directed toward a cell surface antigen includes blocking a cell surface receptor function or killing the target cell. In certain cases, the target antigen is a cell surface receptor, and mAb binding may disrupt the receptor's normal or physiological role, preventing cell proliferation or survival. Examples include mAbs targeted against the epidermal growth factor (EGFR) or the receptor tyrosine kinase erbB-2 (also called HER2).

In other cases, a tumor cell or a B-cell clone that produces an autoantibody may be the target (e.g. an antiplatelet antibody in immune thrombocytopenia (ITP)). As part of the cell-killing process, complement proteins, phagocytes, or natural killer (NK) cells may be recruited, promoting immune-mediated destruction of the cell(s) expressing the target antigen on their surface.

Interactions with the Fc part of the mAb are commonly used to recruit immune mediators. Fc receptors can affect antibody-dependent cellular cytotoxicity (ADCC) or antibody-mediated phagocytosis by monocytes/macrophages by recruiting effector cells, which can modulate the cell-killing effects of mAbs. Fc receptors may also induce cell death via complement-dependent cytotoxicity (CDC), which occurs when a mAb binds to a target cell and triggers the complement cascade. Some antibodies have characteristics of both ADCC and CDC, and some mAbs are engineered to change their Fc binding to increase cell death. On CDC and ADCC, complement activation can have both agonistic and antagonistic effects, and it is unclear which mechanisms are responsible for killing cancer cells. Using an antibody as a vehicle to deliver a toxin or cytotoxic drug directly to the target cell using a mAb-drug or mAb-toxin conjugate may also increase target cell killing.

Fc receptors are present on lymphocytes, neutrophils, monocytes, dendritic cells, and epithelial cells. Fc receptors are engineered to bind specific receptors on cells' subpopulations or have specific glycoprotein modifications. The Fc portion properties can vary depending on antibody isotype (e.g. immunoglobulin G (IgG), IgA, or IgM). The majority of therapeutic mAbs are IgG1, which has a well-known half-life and effector functions, including complement fixation. The Fc component may bind to complement C1q and activate the classical complement pathway, or it may bind to antigen-presenting cell receptors, causing phagocytosis. The Fc portion can activate or suppress B-cells, depending on the antigen and timing of interactions.

For the efficacy of an mAb directed against a soluble molecule such as a plasma protein or a drug, antigen binding and sequestration of the protein away from its usual binding partners may be sufficient.

mAbs can target a variety of plasma proteins, including:

- Adalimumab, afelimomab, certolizumab pegol, golimumab, infliximab, and other TNF inhibitors.
- Bevacizumab—a vascular endothelial growth factor.

Examples of drugs include:

- Dabigatran (anticoagulant)—Idarucizumab.
- Digoxin (antiarrhythmic agent)—Digoxin immune Fab.

These drugs are effectively neutralized when bound to the mAb because they are unable to interfere with their usual targets. Macrophages eventually remove them from the body through Fc-mediated uptake and lysosomal degradation.

The method by which a therapeutic mAb protects against infectious diseases is like natural humoral immunity, although it does not fully describe the nature of microbe elimination. Potential uses include the treatment or detection of infections. Most mAbs target proteins on a virus's surface, thus neutralizing the virus from getting into cells. Palivizumab is an antibody against the fusion (F) glycoprotein from the respiratory syncytial virus (RSV); it prevents viral entry into host cells. The antiviral preventive mAbs act on *Hemophilus* influenza's conserved hemagglutinin A platform. Such treatment may be of help in situations where vaccination produces inadequate humoral immunity. Most mAbs against bacteria can function both prophylactically and therapeutically (e.g. attacking the *Bacillus anthracis* protective antigen domain or one of the *Clostridioides difficile* toxins).

COVID-19 brought many new development technologies to the surface; one of them uses an antibody cocktail which was administered to President Trump within hours of his testing positive for COVID-19. A combination of two strong neutralizing antibodies (REGN10987+REGN10933) that target non-overlapping epitopes on the SARS-CoV-2 spike protein appears to hold promise for a novel class of antibody biopharmaceuticals.

2.4 Types of antibodies

'Types of antibodies include:'

- Migraine prevention with anti-CGRP receptor antibodies (erenumab, galcanezumab, or fremanezumab).
- Type 9 subtilisin/kexin anti-protein convertase (PCSK9) hypercholesterolemia, antibodies (evolocumab or alirocumab).
- Anti-fibroblast growth factor 23 is a protein that inhibits the growth of fibroblasts (FGF23) X-linked hypophosphatemia antibody (burosumab) (XLH).
- Rheumatoid arthritis is treated with anti-IL6R antibodies (sarilumab and tocilizumab).
- Hemophilia anti-IXa/Xa factor antibody (emicizumab): A.
- Purpura thrombocytopenic, anti-Willebrand factor antibodies (caplacizumab).

Since non-linear epitopes bind to traditional antibodies, they are monospecific and usually recognize only one antigen. Some antibodies have multi-specificity, which happens when a very similar epitope is detected on different antigens. Antibodies that identify orthologous proteins in different organisms or antibodies that interact with members of different conserved protein families are examples of species cross-reactive antibodies. Antibody efficacy as a therapeutic agent in different animal models has also been hindered by species specificity. Antibody cross-reactivity is modified using the same combinatorial techniques that are used to enhance antibody affinity. Specificity engineering also requires in silico design techniques and the availability of experimental structural expertise (figure 2.4).

The antibodies used for therapeutic purposes are divided into two categories. The naked antibody is the first type of antibody that can work by multiple mechanisms, such as mediated pathways (e.g. ADCC/CDC), direct cancer cell targeting to cause apoptosis, targeting the tumor microenvironment, and targeting immune control points. Antibodies attack cancer cells in controlled pathways by enlisting the help of natural killer cells or other immune cells.

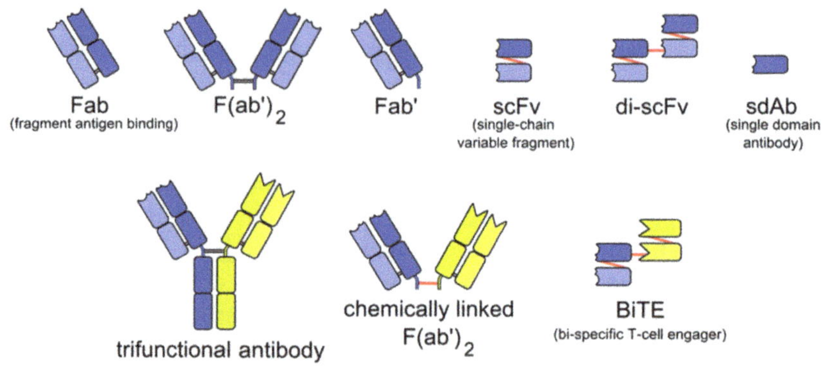

Figure 2.4. Types of monoclonal antibodies with other structures than naturally occurring antibodies. Top row: monospecific antibodies (fragment antigen-binding, F(ab′)2 fragment, Fab′ fragment, single-chain variable fragment, di-scFv, single-domain antibody). Bottom row: bispecific antibodies (trifunctional antibody, chemically linked F(ab′)2, bispecific T-cell engager). Heavy chains have a darker color, light chains a lighter color. Parts of antibodies with different targets are colored differently. Constant regions are shown as regular round-edged boxes, variable regions as boxes with an irregularly shaped end. Artificial links between fragments are colored red. (Source: Wikipedia: Anypedotos https://commons.wikimedia.org/wiki/file:engineered_monoclonal_antibodies.svg?uselang=jaBy.)

Antibody Fc point mutations or glycosylation alteration are employed to improve ADCC or CDC therapeutic effects, such as cancer cell-killing capabilities. The conventional preferred mechanism for therapeutic antibodies is channel-mediated apoptosis of cancer cells. Antibodies limit carcinogenesis by containing components implicated in cancer cell growth. For example, the bevacizumab antibody targets VEGF to reduce the blood arteries around the tumor, effectively cutting down the nutritional supply essential for cancer cell growth.

Immune levels are important objectives for cancer treatment. Novel biopharmaceuticals are those that have the synergistic effects of antibodies and chemotherapeutic medicines, radiation, or other biological agents. The development of novel biomarkers will improve the efficacy and specificity of antibody-based treatment for human diseases. Immune cytokine antibody–drug conjugates, antibody–radionuclide conjugates, bispecific antibodies, immunoliposomes, and T-cell chimeric antigen receptors (CAR-T) are all common methods for increasing antibody efficacy. A cytokine is combined with an antibody to generate an immunocytokine, which improves delivery specificity. Antibody–drug conjugates consist of an antibody that targets a cancer-specific antigen and a small-molecule medicine; the antibody increases delivery to the tumor site, enhancing the efficacy of the small molecule while lowering non-specific toxicity to non-target tissues and side effects. The antibody can also be used with a radionuclide to provide more precise tumor treatment.

2.4.1 Bispecific (BsAbs)

If targeting single components fails, a superior treatment method comprises more than one target related to a single action mechanism. The use of monospecific antibody forms has limitations in that certain individuals will no longer respond to such treatment after a certain amount of time. Resistance may occur during sick tissue growth due to interaction between signaling pathways. As a result, BsAbs are being used to improve the therapeutic profile in cases when numerous illnesses are present.

There are three approved BsAbs: Catumaxomab (Trion Pharma), which can carry T-cells or T-lymphocytes closer to cells expressing EpCAM through CD3 binding; blinatumomab (Micromet/Amgen), which also has a CD3-binding arm to B lymphomas with CD19; and Helimbra or emicizumab-kxwh (Roche–Chugai), which mimics the cofactor VIII .

BsAbs can attach to distinct antigens or epitopes by combining the specificities of two antibodies, which is particularly advantageous when spatial-temporal connections cannot be utilized by combining or overlapping antibodies. Theoretical advantages of using BsAbs include assigning different immune cells to tumor cells to increase tumor killing. BsAbs also allow simultaneous blocking of two different mediators or pathways that play special or overlapping roles in pathogenesis and potentially increase binding specificity by dealing with two different antigens on the cell surface instead of one, and BsAbs can potentially irradiate tumor cells.

Bispecific antibodies are designed to improve the therapeutic efficacy of monospecific antibodies. The actions of antibody-engaged effector cells boost the therapeutic efficiency of bispecific antibodies. The binding site of the antibody (scFv or Fab) is cleaved from the constant region and then conjugated to different locations in immunoliposomes.

Finally, CAR-T involves injecting the gene into T-cells in order to generate a chimeric T-cell receptor antibody that assaults a cancer antigen. The immune system recognizes and destroys cancer cells. To improve the therapeutic effects of bispecific antibodies, antibodies that target two receptors are being developed. The actions of antibody-engaged effector cells can boost the therapeutic efficiency of bispecific antibodies. The binding site of the antibody (scFv or Fab) is cleaved from the constant region and then conjugated to different areas in immunoliposomes. For example, liposomal medicines are nano-drug delivery methods that allow for more specific targeting. It opens up exciting new possibilities for the creation of innovative protein treatments.

2.4.2 Fab fragments and single-chain antibodies

Instead of using full-length antibodies, the use of antibody fragments can enhance pharmacokinetics and tumor mass penetration (because fragments are smaller). In contrast to full-length antibodies, which have two antigen valences (binding sites), fragments usually only have one.

Antigen-binding fragments (Fab) consist of a variable domain and the first constant region of each heavy and light chain. Single-domain antibody linker peptide (sdAb)–sdAb is an antibody fragment consisting of a variable portion of the light chain or hFab fragments lack the antibody Fc component (the remainder of the heavy chain) and hence cannot interact with Fc receptors or activate the complement. As a result, when taken alone, they are usually ineffective for indications that rely on cell death. The following are some examples of clinical applications:

- Caplacizumab is a single-domain antibody (sdAb) made up of a bivalent variable-domain-only mAb fragment with a high affinity for the von Willebrand factor X receptors (VWF). Its binding prevents VWF from interacting with platelets, which is essential in microvascular thromboses like those seen in thrombotic thrombocytopenic purpura patients (TTP).
- Instead of full-length antibodies, antibody fragments may enhance pharmacokinetics and tumor mass penetration (because fragments are smaller). Instead of the two valences found in full-length antibodies, fragments normally have only one valence for the antigen (binding site).
- Antigen-binding fragments (Fab) are made up of a variable domain and the first constant region of each heavy and light chain and are known as Fab fragments. The single-domain antibody linker peptide (sdAb)–sdAb is an antibody fragment consisting of a light chain variable region or hFab fragments lack the antibody Fc component (the remainder of the heavy chain) and are therefore unable to interact with Fc receptors or activate complement.

2.4.3 Humanized and chimeric mAbs

Antibodies for human targets are produced in several non-human animals, including rodents, pigs, and rabbits. Humanization is necessary for all of these non-human antibodies. The easiest method is to create a chimera is by combining the variable domains of non-human antibodies with constant human domains to generate 70% human material molecules. In some instances, chimeric antibodies have demonstrated decreased immunogenicity, but they have also elicited certain anti-therapeutic responses in humans. CDRs have transferred from a non-human (usually murine) 'parental' antibody to the scaffold of a human antibody in a CDR-grafting technique to reduce immunogenicity further. Alternative humanization approaches, in addition to CDR grafting, including resurfacing, super-humanization, or optimization of the human string content, all of which necessitate amino acid sequence analysis to determine the possible impact of amino acid substitutions as well as structure and function antibodies.

mAbs derived from non-human animals (e.g. mouse, rat) are 'humanized' to varying degrees by engineering amino acid substitutions to make them more human-like. This is achieved by the use of recombinant DNA technologies. In theory, the closer an mAb is to human-derived sequences that are exchanged by many people, the less likely it is to evoke an immune response. Infusion reactions and decreased effectiveness are two possible immunogenicity side effects, but these are difficult to predict.

However, the immunogenicity of not all amino acid residues or classes of residues is the same. Furthermore, defining what constitutes a chimeric antibody versus a humanized antibody (e.g. how many amino acid residues must be modified for an antibody to count as humanized) has become extremely difficult, and meanings have developed. Non-human-derived small but important parts of the complementarity-determining region (CDR) are present in humanized mAbs, but human-derived large constant regions of the immunoglobulin heavy and light chains are missing.

Chimeric antibodies are those in which the Fc component of the immunoglobulin molecule is human (but not the CDR). Chimeric mAbs and humanized antibodies, in general, contain between 65% and 90% human sequence. As a result, there are many methods for producing fully humanized antibodies.

The fully human mAbs technique is based on phage display, in which a library is composed of various exogenous genes inserted into filamentous bacteriophage. The library proteins are then introduced as fusions with a phage coat protein on the phage surface, allowing for selecting specific binders and characterization of affinities.

Daclizumab, which stimulates the IL-2 receptor and reduces transplant rejection, was the first CDR-grafted humanized mAb licensed by the FDA in 1997 and has been shown to reduce antigen identification failure. Daclizumab is manufactured through CDR grafting and a human structure that is most similar to the murine system. In some circumstances, amino acids are required for the murine system to promote antibody binding activity. These residues frequently work in tandem with CDRs to provide a paratopic antibody or to interact directly with antigens. X-ray crystallography, cryo-electron microscopy, and computer-aided protein homology modeling of the antibody–antigen complex structure are used to identify specific matrix residues. Using 'human back to mouse' mutations, the amino acid locations in the matrix can subsequently be restored in CDR-grafted humanized antibodies, resulting in better affinity and end-product stability.

Several approaches exist for calculating the humanness of the component area of mAbs. The 'degree of humanization' of antibody sequences is calculated using the 'H-score', which calculates the mean sequence identity relative to the human vector field database sequences subset. A germinal index characterizes the after-assist germline humanization of a macaque antibody. The germline structure series is defined by the G-score, which is obtained from the H-score. The T20 score analyzer (a tool that calculates the humanization of monoclonal antibody variable region sequences; an analyzer that can consistently differentiate human antibody variable region sequences from those of other species, such as murine sequences) is set up in a large database of peptides to distinguish human antibody variable region sequences from those of other species. It is used to distinguish between fully human antibodies and humanized antibodies.

Humanized antibodies aid in the development of clinical tolerance to mAb therapy. Chimeric or humanized mAbs make up 50% of all mAbs used in human medicine today. Trastuzumab (Herceptin), one of the most well-known humanized antibodies, was approved in 1998. Trastuzumab is the only medicine on the WHO List of Important Medicines, and it is used to treat patients with human epidermal growth factor receptor 2 (HER2)-positive metastatic breast cancer and adenocarcinoma gastroesophageal junction. Murine antibodies are no longer created; the last one was generated in 2003, the same year that adalimumab, the first entirely human antibody, was launched.

2.4.4 Affinity maturation

Antibodies discovered by humanized, phage, or transgenic methods are often improved further, including the substitution of residues in the binding area. Antibody gene diversification is the first step for *in vitro* affinity maturation and it can be achieved by a variety of methods, including random mutations, targeted mutations, and chain shuffling. In the variable regions of *E. coli* antibody genes, mutations are introduced randomly by error-pronounced PCR in mutator E. In chain shuffling methods, one of two chains, VH and VL, is fixed and recombined to create a next-generation library with different partner chains. Mutations are also introduced into unique regions of the antibody gene. This method of targeting mutations aids in the diversification of CDR resides, which improves antibody affinity. As a result, since mutations accumulate more quickly in the CDR than residues in the system, this method is more applicable to *in vivo* somatic mutations during B-cell evolution.

Human and non-human antibodies often lack the binding properties needed for therapeutic purposes. Increasing binding affinity is a crucial step in lead candidate development because it affects the medication dosage and therapeutic efficacy. In the engineering of antibodies directed against different antigens, there are numerous methods, tools, and strategies divided into two classes based on the antibody variant generation process. The first is the logical nature of the variations, followed by their representation within the option method. The other alternative is to create a library of variants that differ multiple positions, then display them in a choice system with the required selection process. Due to many variants in a library covering the entire combinatorial space, the latter approach is most widely used for affinity maturation. When only a few positions and amino acids are being checked, the former method may be adequate because it is simple and inexpensive. Whatever approach is used, structure-based computational design can aid by evaluating candidates *in silico* to reduce the library's size or the number of mutants that must be expressed.

2.4.5 Antigenized antibodies

Antigenization is a research method in which a mAb is designed to convey an antigen (e.g. a vaccine). It is performed by substituting a portion of the antibody polypeptide with a microbial antigen fragment. A sequence is introduced to various parts of the antibody molecule. Antigenized mAbs have a longer serum half-life than isolated antigen fragments and may be more tolerable than other microbe fragments, potentially making them effective as vaccinations.

The successful presentation of microbial peptides in antibody molecules has been demonstrated in a variety of animal systems (e.g. for influenza viruses in mice). This potentially game-changing technique, however, has only been tested on animals. For example, recombinant DNA techniques were employed to graft a bovine herpesvirus B-cell epitope onto a cow immunoglobulin protein. Cows were inoculated with this antigenized antibody, which resulted in the development of antibodies against the virus.

2.4.6 IgG1 fusion proteins

IgG1 fusion proteins (also known as Fc-fusion proteins) are biopharmaceuticals that take advantage of some of the immunoglobulin Fc region's properties, such as increased half-life. Antigen-binding complementarity is not determined in IgG1 fusion proteins (CDR). As a result, they lack a biologic target in the same way that mAbs do, despite the fact that the protein to which Fc is fused often has a particular biologic role that is manipulated. Some of these fusion proteins are identified by the suffix '-cept', others contain 'Fc' in their names.

Examples of clinical-use IgG1 fusion proteins are as follows:

- Etanercept is a fusion of two soluble TNF-alpha receptors with the Fc part of an IgG molecule. It has two TNF receptors, making it bivalent (i.e. one etanercept molecule binds two TNF molecules). TNF-alpha inhibition is used to treat a number of immunologic and rheumatologic disorders.
- A type of factor VIII supplement is recombinant human factor VIII fused to the Fc portion of IgG (rFVIII-Fc). A patient's antigen-specific antibody is isolated. Since extracting a tumor and regional lymph nodes is widely used in routine cancer care, this procedure is particularly applicable to cancer therapeutics.

2.4.7 Drug or toxin conjugation

An antibody–drug conjugate (ADC) is a novel therapeutic format that combines a humanized or human monoclonal antibody with highly cytotoxic small molecules (payloads) through chemical linkers. It has the potential to revolutionize cancer chemotherapy. As compared to conventional chemotherapy, a new molecular framework based on antibodies enables selective delivery of a potent cytotoxic load to cancer cells, resulting in increased efficacy, decreased systemic toxicity, and preferred pharmacokinetics (PK), pharmacodynamics (PD), and biodistribution. Following the FDA approval of Adcetris® and Kadcyla®, a class of drugs known as ADCs has expanded rapidly, with over 60 ADCs currently in clinical trials.

Here are some examples:

- Moxetumomab pasudotox is a humanized mouse monoclonal antibody (mAb) that targets CD22 and is conjugated to a *Pseudomonas* exotoxin A toxic fragment.
- Polatuzumab vedotin is a CD79b-targeting humanized monoclonal antibody (the B-cell antigen receptor complex-associated protein beta chain). It is linked to the monomethyl auristatin E (MMAE), a dolastatin analog, by a protease-cleavable linker that improves plasma stability.
- Brentuximab vedotin is a CD30-targeting monoclonal antibody that is linked to MMAE through a cleavable linker.

Binding to internalizing receptors is advantageous for many ADCs because it allows the conjugate to enter the cell and the active moiety to exert its effects. The first licensed ADC, gemtuzumab ozogamicin, demonstrated ablation of cells with acute myeloid leukemia (AML). Gemtuzumab (anti-CD33) is coupled to

N-acetyl—calicheamicin dimethyl hydrazide via non-specific lysine conjugation and a 4-butanoic acid spacer to generate the ADC (4-acetylphenoxy). The drug–antibody ratio (DAR) is typically between two and three. Toxins have been conjugated or fused to antibodies, in addition to natural products, to develop tumor-targeting immunotoxins. Antibody–radionuclide conjugates (ARCs) containing radionuclides are a novel type of ARC.

2.5 Development of antibodies

Antibodies are created to find solutions for:

- Neutralizing infecting organisms.
- Binding to receptors to elicit a pharmacologic response.
- Binding to endogenous chemicals to alleviate a disease.
- Creating analytical methods of separating and identifying a large molecule.

2.5.1 Endogenous method: single B-cell

To start, a patient's antigen-specific antibody is isolated. Since extracting a tumor and regional lymph nodes is widely used in routine cancer care, this procedure is particularly applicable to cancer therapeutics. These tissues are used to harvest tumor-infiltrating lymphocytes. Existing antibodies are also isolated from peripheral blood, bone marrow, or other lymphoid tissues such as the spleen or tonsils. Various investigational mAbs against viruses such as the human immunodeficiency virus (HIV) and hepatitis C virus (HCV) are examples of this method.

This method for producing monoclonal antibodies from single human B-the cells uses reverse transcription polymerase chain reaction (RT-PCR) and expression vector cloning to examine the immunoglobulin gene's repertoire and reactivity at the single-cell level.

Up to now, this method has not developed any therapeutic mAbs approved by the US FDA.

2.5.2 Exogenous methods

Different influenza viruses cause epidemics each year, and influenza vaccines are the most effective approach to avoid seasonal influenza. Single B-cell isolation has become routine practice for producing powerful and broadly neutralizing anti-influenza antibodies.

The primary techniques involved in developing antibodies are immunizing an animal, using a phage display, or using a single B-cell from humans to identify the antibody (figure 2.5). For commercial production of an identified antibody, two techniques are used: the classic mouse hybridoma technology and recombinant manufacturing, regardless of how they were identified.

Of the approved antibodies, the distribution of technologies to produce the antibody to establish its structure and activity, the most common method is hybridoma (figure 2.6)

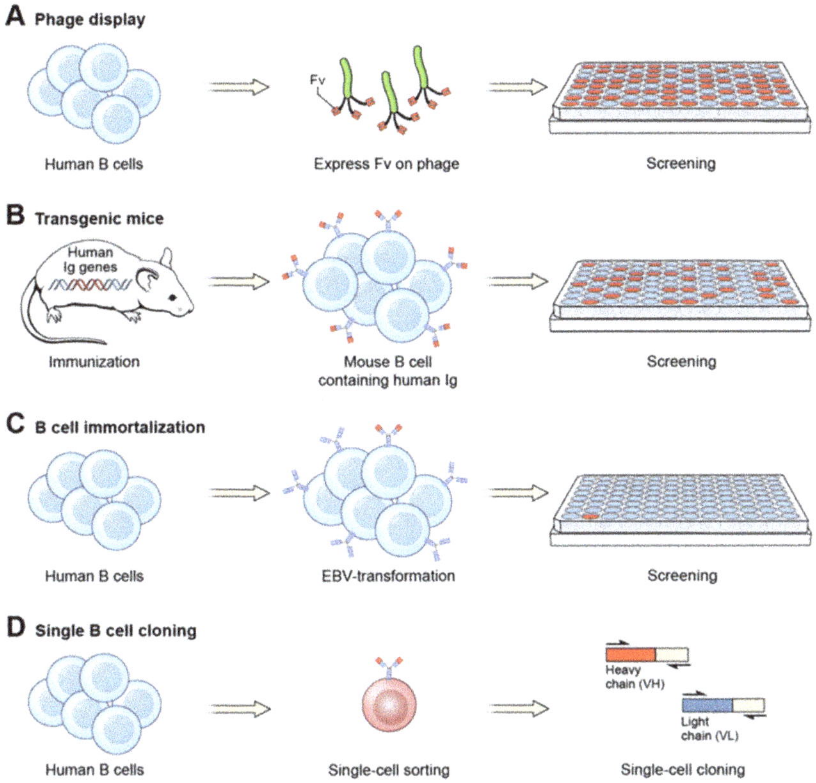

Figure 2.5. Approaches to develop therapeutic antibodies. (A) A phage. The human antibody collection shown in the phage is used to identify antigens of interest. ELISA screens immune-positive phage clones after 3–5 rounds of biopanning. Then DNA sequences are processed to build and human IgGs are released. (B) The mouse is transgenic, compared to the strategy of mouse hybridoma or single B-cell approaches. (C) B-cell immortalization. (D) Methodology with a single B-cell. PBMCs are prepared for the separation by flow cytometry of suitable B-cells from contaminated or vaccinated donors. Following the RT-PCR, each B-cell's VH and VL details tell about the generation of human mAbs. (Source: Wikipedia: Bioshore https://commons.wikimedia.org/wiki/File:Isolation_of_human_monoclonal_antibodies._.tif.)

2.5.3 Mouse hybridoma

Köhler and Milstein created the first monoclonal antibodies in mice in 1975 using a hybridoma method. Hybridomas are created by immunizing an animal against a specific epitope on an antigen and collecting B-lymphocytes from the animal's spleen (figure 2.7). The B-lymphocytes are then united with an immortal myeloma cell line that lacks the gene hypoxanthine-guanine-phosphoribosyltransferase (HGPRT) and has no additional immunoglobulin-producing cells (by chemical or virus-induced methods). The hybridoma cells are then grown *in vitro* in a hybridoma-specific media (hypoxanthine-aminopterin-thymidine-containing medium) (primary fusion). B-lymphocytes and myeloma cells live because they inherit the myeloma cells' immortality and primary B-lymphocytes' selective resistance (the myeloma cells cannot synthesize myeloma cells because they lack HGPRT).

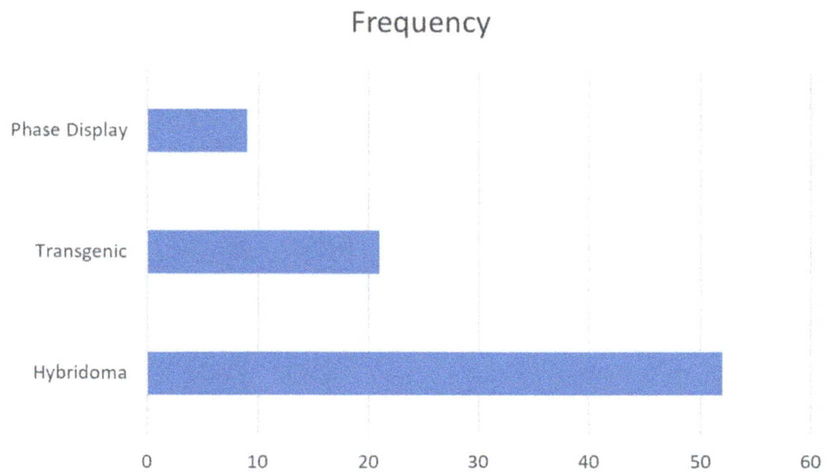

Figure 2.6. Methods of producing antibodies for identification and testing for structure elucidation for approved antibodies.

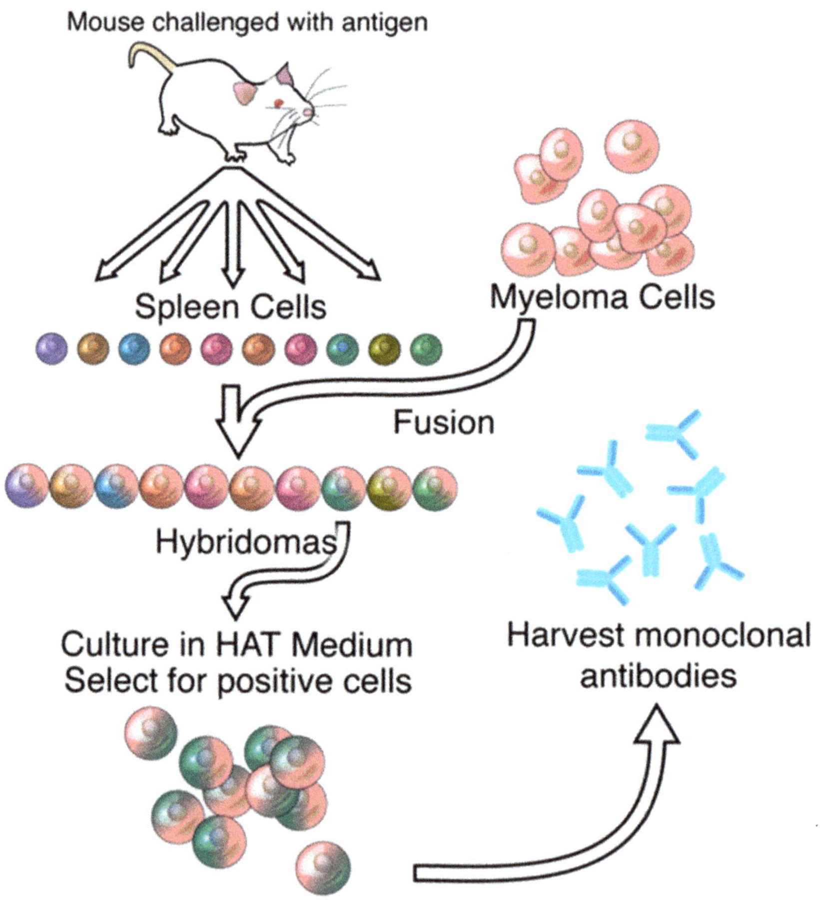

Figure 2.7. Hybridoma technology. The traditional mouse hybridoma technique begins with immunization of mice with appropriate antigens to cause an immune response. Harvested splenocytes are fused to create hybridoma cells that persistently secrete antibodies with myeloma cells. Selected leads are used after the screening to produce chimeric or humanized antibodies. (Source: Wikipedia: Adenosine https://upload.wikimedia.org/wikipedia/commons/9/9a/Monoclonals.png.)

A majority of approved antibodies were first created in a hybridoma, but only antibody muromonab-CD3 (Orthoclone OKT3) continues to be manufactured in a hybridoma; all other antibodies are produced in a recombinant cell line, primarily CHO cells. Some people who are exposed to mouse antibodies develop an immune response to the mouse antibody sequence, which is why hybridoma technology is no longer used for commercial manufacturing. As a result, methods for engineering modifications to the immunoglobulin molecule, such as humanizing the antibody or creating a chimeric antibody, have been created, and these are used in the majority of mAbs that were initially selected in animals. Human immunoglobulin loci have been engineered into mice in place of endogenous mouse sequences, resulting in the generation of human antibodies in mice.

The initial hybridoma culture contains antibodies formed by a variety of primary B-lymph cyclones, each of which secretes its antibody into the culture medium (i.e. the antibodies are still polyclone). Dilution is used to divide each clone into different culture wells. It follows a multi-well screening of the cell culture medium for the particular antibody activity needed. The right B-lymphocytes were then recloned and retested for growth after being grown from the positive wells. The positive hybridomas and monoclonal antibodies that have formed may then be preserved in liquid nitrogen.

Laboratory animals are initially exposed to the antigen against which an antibody is to be generated (mammals, such as mice). This is normally done over a period of weeks with a series of antigen injections. In some situations, these injections are coupled by *in vivo* electroporation, which dramatically improves the immune response. After isolating splenocytes from the patient's spleen, the B-cells are mixed with immortalized myeloma cells. Electrofusion is a technique for fusing B-lymphocytes with myeloma cells. Electrofusion uses an electromagnetic field to align and fuse B-cells and myeloma cells. Polyethylene glycol, on the other hand, may be employed to help B-cells and myelomas fuse in the majority of cases. Because myeloma cells do not secrete antibodies and lack the gene for hypoxanthine-guanine phosphoribosyltransferase (HGPRT), they are resistant to the HAT (hypoxanthine-aminopterin-thymidine) media.

The combined cells are cultured in HAT media for 10 to 14 days. Aminopterin inhibits nucleotide synthesis by inhibiting the route. As a result, unfused myeloma cells perish. They are unable to synthesize nucleotides via a new or salvaged pathway because they lack HGPRT. Unfused myeloma cells must be removed because they have the potential to overrun other cells, particularly weakly formed hybridomas. Unfused B-cells die due to their short lifespan. Only the B-cell–myeloma hybrids survive because the HGPRT gene is derived from B-cells. These cells are immortal and generate antibodies (a characteristic of B-cells; a myeloma-cell property). In multi-well plates, the incubated liquid is subsequently diluted to the point where each well contains only one cell. Because all the antibodies in a well are produced by the same B-cell, they are all monoclonal antibodies directed to the same epitope.

The hybridomas that produce specific antibodies are identified and selected in the next step, a rapid primary screening procedure. The first test was performed using the ELISA process. The incubation of the supernatant hybridoma culture, secondary enzyme-labeled conjugate, and chromogenic substrate yield a colored product, indicating a positive hybridoma. Immunocytochemical, western blot, and immunoprecipitation-mass spectrometry screening can also be used. Unlike western blot assays, immunoprecipitation-mass spectrometry allows for the screening and ranking of clones that bind to native (non-denatured) protein forms.

Other media containing interleukin-6 are necessary to produce several identical daughter clones. A hybridoma colony will continue to grow and produce antibodies once developed in a culture medium containing antibiotics and fetal bovine serum. The hybridomas are grown in multi-well plates and then transferred to larger tissue culture flasks after selection. It maintains hybridomas' integrity and provides suitable cells for cryopreservation as well as supernatant for further study. The supernatant culture will produce 1 to 60 g ml^{-1} of a monoclonal antibody, which should be stored at –20 °C or lower until required. Using culture supernatant or pure immunoglobulin preparedness, a potential monoclonal antibody-forming hybridoma is further investigated for reactivity, specificity, and cross-reactivity.

One of the major disadvantages of this technology was that the sequence and post-translational modifications of a monoclonal antibody developed using this method were of rodent origin, making the molecule immunogenic in human patients and resulting in the development of human anti-mouse antibodies (HAMA). As a result of these breakthroughs in genetic engineering, chimeric and humanized antibodies were created using phage display technology or transgenic mice. The mouse variable light and heavy chain sequences were grafted onto a human IgG scaffold or the same complementarity-determining regions (CDR) to make chimeric antibodies and humanized antibodies.

High-affinity human antibodies are developed by further selecting hybridoma clones derived from inoculated transgenic mice, depending on the immunization schedule. Human neutralizing antibodies produced from human B-cells have also shown promise in the treatment of infectious diseases using a theoretically similar strategy.

2.5.4 Transgenic mice

To boost antibody drug production, several transgenic animals, including fully humanized mice and second-generation human chimeric mice, have been employed. Human antibodies with high affinity are created depending on the immunization technique and the clones of the animals. Transgenic animals are being fine-tuned in order to produce fully human mAbs. The FDA authorized panitumumab, an anti-epidermal growth factor receptor antibody produced in a transgenic mouse, in 2006. The number of totally human antibodies produced by transgenic mice has grown significantly, with 19 medicines already on the market (table 2.1).

Transgenic mice provide a stable substrate for the creation of antibody–drug conjugates. Humanization is unnecessary. Transgenic animals have several advantages over existing human antibody creation approaches, including increased variety, maturation of *in vivo* affinity, and clonal selection to enhance antibody yield. The large size of human Ig loci, on the other hand, presents a challenge in the development of the transgenic mouse. Furthermore, the repertoire development of transgenic mice is comparable to that of humans, with a wide range of rearrangements and high expression of human segments V, D, and J. To address these significant obstacles, various methods are used to produce animals that transmit human antibody repertoires. A list of transgenic mice used in commercial products is given below:

- Tetravalent bispecific tandem lg (TBTI) (synonymous with DVD-lgTM), mice that have been genetically modified.
- Transgenic mice: XenoMouse® from Abgenix, HuMAb-Mouse® from Medarex, UltiMAb® from Medarex, and Veloclmmune®.

2.5.5 Surface display libraries

Exogenous proteins or peptides are expressed on the host (from the simplest virus to mammalian cells) by fusing exogenous genes with membrane protein genes or modifying proteins or peptides to be anchored by host surface elements, according to surface display technology. A significant advantage of this technology is that it establishes a clear connection between genotype and phenotype. It facilitates the study of proteins' and peptides' affinity for their targets and protein functions as enzymes by saving the protein purification process. High throughput screening for proteins and peptides bound to a particular ligand and blocking their activities is a beneficial drug discovery technique. Figure 2.8 shows the four types of libraries that can be constructed to produce antibodies. A phage library can be produced using yeast, mRNA, and ribosomes. However, only the phage display has been used to develop currently approved antibodies. Each of these innovations has its own set of benefits and drawbacks, which are detailed below.

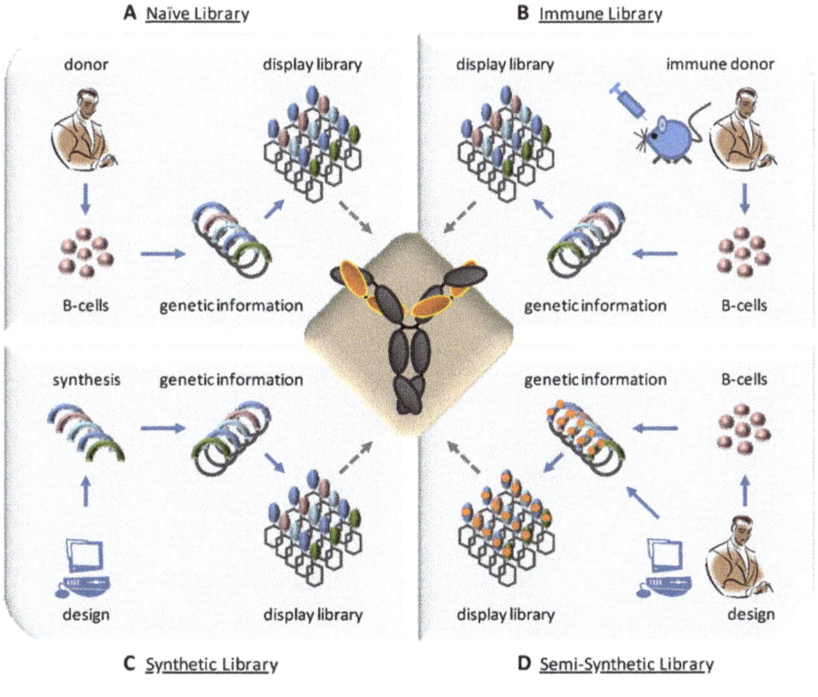

Figure 2.8. Types of libraries for making antibodies. (Reprinted with permission of MDPI from Ponsel D *et al* 2011 *Molecules* **16** 3675–700; https://doi.org/10.3390/molecules16053675; Creative Commons license.)

Yeast display, bacterial display, ribosome display, and mRNA display are all competing methods for *in vitro* protein evolution.

2.5.5.1 Phage display

Antibody phage displays ushered in a new era in antibody drug development. Phage display has been used commonly in protein or antibody characterization and, most notably, antibody discovery since the invention of phage display technology in 1985 by G Smith and the publication of the first series of antibody libraries on phages in 1990. Immunization of host animals and subsequent purification steps are necessary for conventionally produced antibodies. Furthermore, antibodies used in clinical applications must be of human origin. Since it is not feasible to immunize humans with specific antigens, downstream sequencing and humanization of animal origin antibodies are needed, which is normally expensive and results in inconsistent results (figure 2.9).

A phage, also known as a bacteriophage, is a virus that has DNA or RNA encapsulated by coat proteins. They will use bacteria as a host to insert their genes into the bacteria's genome and produce coat proteins and genes that are used to reform phage particles in the cytoplasm and secrete them into the periplasm. Phagemid is a plasmid that can create phages after being transformed into bacteria. Phages with exogenous proteins/peptides shown on the capsid are created when exogenous genes are inserted into a phagemid and transformed into bacteria. A phagemid is a plasmid with a fl replication origin from an fl phage. When combined with filamentous phage M13, it can be used as a form of the cloning vector. A phagemid can be replicated as a plasmid and packaged in viral particles as single-stranded DNA. *E. coli* filamentous bacteriophages (f1, fd, M13), T4, T7, and phage are the most commonly used phages. A phage is a virus that can infect and replicate bacteria and archaea after being injected into the host (bacteria). Phages

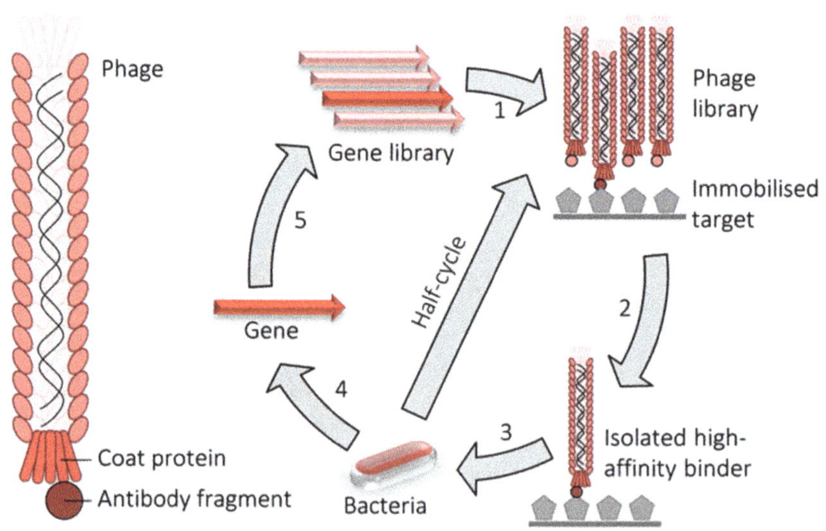

Figure 2.9. Phage display library. (Source: Wikipedia: Thomas Shafee https://commons.wikimedia.org/wiki/File:Phage_display.png.)

are proteins that encase a DNA or RNA genome, and they may have simple or complex structures.

Nonimmunized human B-cells, antigen immunized human or animal B-cells, and *in vitro* gene synthesis by randomizing various V(D)J genes of humans are among the libraries' sources. A recombination is a form of somatic recombination that occurs only in developing lymphocytes during the T- and B-cell maturation process. It gives rise to the vast array of antibodies or immunoglobulins and T-cell receptors present in B- and T-cells, respectively. The gene pool is then subcloned into phage display vectors and transformed into sufficient bacteria to harvest the phage libraries. Following that, the library is subjected to multiple rounds of affinity panning to capture the relevant antigen-binding antibodies seen on the phage. The monoclonal antibodies are then detected using an ELISA test, sequenced, and cloned into mammalian cells such as CHO for mass production.

These libraries contain single-event variables such as variable heavy and light chains, as well as antibody fragments (scFv- wherein the VH and VL are connected using a polypeptide linker). The antibody fragments bind to the bacteriophage's surface protein (e.g. PIII or PVIII) and are displayed(e.g. M13). A procedure known as panning is used to select the antibodies (based on the antigen's affinity). To enrich antigen-specific antibody fragment phages, several rounds of panning can be used. The chosen monoclonal Ab is cloned into an appropriate vector to generate antibodies and then transfected into a suitable host cell.

Isolating B-lymphocytes from human blood, isolating the mRNA, and converting it to cDNA using PCR to amplify both VH and VL segments are steps in the phage display process. These segments are then cloned into a vector alongside a bacteriophage's PIII protein (usually as scFv) before being inoculated with a supplementary helper phage and infecting *E. coli* to create a library containing approximately 10^{10} cells. The M13 and fd filamentous phages are the most widely used bacteriophages in phage display, but T4, T7, and phage are also used. The bacteriophage coat of *E. coli* will then secrete the bacteriophage, which includes VH and VL segments. The antigen used to re-inoculate *E. coli* can then be used to pick different VH and VL segments. The plasmid-containing cells are then separated and sequenced. After the library has been constructed, it is used to produce new antibodies. It is not in need of rehabilitation. Since the entire process is performed *in vitro*, no immunizations are needed, and antibodies are produced much faster than with traditional hybridoma technology, which is particularly useful when toxic antigens cannot be used to vaccinate.

Gene repertoires are obtained from naive or immunized organisms for phage display libraries, or libraries are manufactured employing randomized CDR sequences within predefined frames. To create phage view naive antibody libraries, IgM repertoires with altered V genes are employed. Because the gene sequences are taken from human donor B-cells, the naive libraries are reasonably similar to the germline of human antibodies, with a minimal risk of immunogenicity. The fundamental advantage of an immunized library over a naive library is that *in vivo* antibody genes have undergone natural affinity maturation in the immunized library, allowing for the development of high-affinity antibodies against the target. To generate fresh libraries for each new target, however, the antigen of interest must stimulate immunogenic responses successfully. Using single and naive libraries, as well as synthetic libraries, high-affinity (sub-nanomolar size) antibodies against a wide range of targets have been produced.

The following is a general phage display screening methodology for discovering polypeptides with high affinity for the specified target protein or DNA sequence:

1. In microtiter plate wells, immobilized proteins or DNA sequences are immobilized.
2. Several genetic sequences are produced as fusions with the bacteriophage coat protein in a bacteriophage library. They can be seen on the surface of the viral particle. The protein on display is the phage's genetic sequence.
3. After the phage has had time to bond, the phage display library is attached to the bowl, and the dish is rinsed.
4. Except for the phage-displaying proteins that bind to the target molecules, the rest is washed away.
5. The associated phage is eluted and utilized to infect suitable bacterial hosts in order to manufacture additional phages. The new phage creates an enriched mixture with considerably fewer non-binding (i.e. irrelevant) phages than the original mixture.
6. Steps 3–5 can be performed one or more times to improve the binding protein content of the phage library.
7. The DNA inside the interacting phage is sequenced after further bacterial-based amplification to classify the interacting proteins or protein fragments.
8. To generate a protein or antibody in large amounts, the gene encoding the binding to a target protein is sequenced and inserted into a recombinant expression system such as CHO.

Antibody libraries containing millions of distinct antibodies on the phage are used in the pharmaceutical industry to isolate highly specific therapeutic antibody leads for synthesis into antibody medicines, primarily as anticancer or anti-inflammatory therapies. Adalimumab was the first fully human anti-tumor necrosis factor (TNF) antibody licensed by the FDA in 2002 for the treatment of rheumatoid arthritis. To date, the FDA has approved nine human antibody medicines developed via phage display (table 2.1)

The phage display analysis is primarily reliant on databases and mimotope analytical procedures. Databases, tools, and web servers are used to exclude target-unrelated peptides, characterize small-molecule–protein interactions, and map protein–protein interactions. Researchers map conformational epitopes using a three-dimensional structure of a protein and peptides from a phage display experiment. Many of the quick and effective computational methods are available on the Internet.

A list of selected libraries includes:

- Bispecific ADAPTIRTM platform, monospecific ADAPTIRTM platform phage display antibody library.
- Dyax human antibody phage display library and Cambridge Antibody Technology (CAT) human antibody phage display library.
- Human scFv phage display library HuCAL®. The phage library from MorphoSys is an antibody phage display library. Peptide phage display library versus antibody phage display library.

- Human G1-kappa antibody phage display library, a combinatorial phage display library using Fab POTELLIGENT® AzymetricTM technology (afucosylation). The presentation of phages guides selection.

Table 2.2 lists the commercial platforms available for phase display development.

Obtaining patent protection for products and inventions while preventing infringement of patents granted to others is critical in developing medicines. As a result, phage display antibody discovery sites' intellectual property rights are shifting, affecting drug development significantly. All phage display patents have expired, including Breitling/(EP0440147), Dübel's McCafferty/(EP0774511), Winter's (EP0589877), Dyax's phage antibody libraries, and Cambridge Antibody Technologies' phage antibody libraries.

2.5.5.2 Yeast display

Yeast display is commonly used in the fermentation industry, aside from its use in antibody discovery. The fermentation method uses yeast cells that contain a full enzyme or the enzyme activity site. Yeast display aided fermentation has many advantages over conventional methods that use distilled enzymes, including lower enzyme costs, easier fermentation control, and a simpler downstream purification stage.

In certain cases, yeast display is superior to phage display, but it has some disadvantages.

Yeast display can be used for fluorescence-activated cell sorting (FACS) in the screening/panning stage, allowing for precise antibodies with desired binding properties. However, changing the washing buffer compositions can only approximately stratify antibodies' affinity to antigens in phage show.

Table 2.2. Commercial display phage platforms.

ABDEG™
Palivizumab BiTE® (Bispecific T-cell engager) technology matured affinity
T-cell engineering with chimeric antigen receptors (CAR) CrossMAb technology
DARPin® dual variable-domain immunoglobulin (DVD-IgTM)
DART®, an abbreviation of dual-affinity re-targeting
EBV immortalization GPEx®
GlycoExpress™, GlycoMAb® Glutamine Synthetase GS Gene Expression System™, i-body Glymax technology (defucosylation) human hybridoma
Humaneering® technology, humanization by CDR grafting BALB/c LAPSCOVERY™, MORPHODOMA®, n-CoDeR®, Nanobody®, no glycosylation site CH2 N84.4>A PENTRA® stable scFv PETization™, SIMPLE antibody™, Triomab® (Trifunctional Ab)
True Human™ antibody discovery platform
The VelociGene® XOMA Metabolism (XMet) platform XmAb®
Antibody Engineering Technology YB2/0 cell line: low fucosylation
YB2/0 cell line: low fucosylation
ZMapp™

Compared to antibodies expressed by *E. coli* in phage show, yeast displayed antibodies may be folded correctly and modified, such as glycosylation, to be similar to their native structure in mammals.

Since yeast has a lower transformation efficiency than a phage, the diversity of antibodies expressed by yeast may be lower (10^7–10^9 for yeast, up to 10^{11} for phage).

The yeast display system will select low-affinity antibodies based on antibody avidity since there are far more copies of antibodies on the yeast surface (10^4–10^5) than there are on a phage.

Exogenous proteins or peptides are expressed on the cell surface by being bound or anchored to the yeast cell wall composition, as seen in the yeast display. Exogenous genes are fused with cell wall protein genes to do this, with Aga2p being the most widely used for antibody display. Aga2p is anchored to the cell surface through two disulfide bonds to Aga1p, a member of the yeast agglutinins protein family responsible for mating events. Aga2p is relatively far from the cell wall which is one benefit of using it as the fusion protein. The fused antibodies are more flexible in a vacuum, preventing steric hindrance from reducing activity. Another benefit is that Aga2p is expressed after cell growth under the guidance of the GAL1 promoter, shielding yeast cells from potentially harmful antibodies and ensuring that all antibodies in the library are displayed. Antigen-coated magnetic beads are used to scan yeast with antibodies or other proteins displayed, followed by multiple rounds of FACS screening to identify antibody/proteins with desirable properties.

Yeast display has some benefits over phage display, but it also has some disadvantages. It is suitable for fluorescence-activated cell sorting (FACS) in the screening/panning stage, enabling precise sorting of antibodies with desired binding properties. However, by changing the washing buffer compositions, antibodies' preference for antigens in phage display can only be approximately stratified.

In contrast to antibodies expressed by *E. coli* in phage display, yeast displayed antibodies can be correctly folded and adjusted, such as glycosylation, to be similar to their native structure in mammals, thanks to their eukaryotic expression mechanism.

2.5.5.3 mRNA display

The display is employed in *in vitro* protein and peptide evolution using mRNA to build molecules that can attach to the intended target. Puromycin-linked translated peptides or proteins are produced, which are linked to their mRNA progenitor. The compound attaches to an immobilized target in a selection stage (affinity chromatography). The well-binding mRNA–protein fusions are then reverse transcribed to cDNA and sequenced using a polymerase chain reaction. As a result, a nucleotide sequence encoding a peptide with high affinity for the target molecule is created.

Puromycin is a tyrosyl-tRNA analog having a structure that is similar to that of an adenosine molecule. The other half appears to be a tyrosine molecule. Puromycin has a non-hydrolyzable amide bond, as opposed to tyrosyl-cleavable tRNA's ester link. Puromycin disrupts translation, leading translation products to be issued prematurely. Puromycin can be found at the 3′ end of mRNA templates utilized in mRNA display technology. As translation proceeds, the ribosome circles the mRNA template. When the prototype reaches the 3′ end, the fused puromycin will join the ribosome's A site and be integrated into the nascent peptide. The mRNA–polypeptide fusion is then released by the ribosome. The addition of fused puromycin to the mRNA template is not the sole modification necessary to form an mRNA–polypeptide fusion. Puromycin must be recruited in conjunction with oligonucleotides and other spacers in order to give stability and the appropriate duration for puromycin to reach the A site.

Despite the fact that both ribosome and mRNA display are *in vitro* selection procedures, mRNA display has a modest advantage over ribosome display. Puromycin-linked covalent mRNA–polypeptide complexes are used in mRNA display, whereas stalled, noncovalent ribosome–mRNA–polypeptide complexes are used in ribosome display. Ribosome–mRNA–polypeptide complexes must be retained in a complex for ribosome display because they are noncovalent. This may make it more difficult to minimize background binding during the selection stage. Polypeptides are attached to a massive rRNA–protein complex termed a ribosome, which has a molecular weight of nearly 2 000 000 Da in a ribosome display device. It is likely that the selection target and the ribosome will interact in an unpredictable manner, resulting in the loss of potential binders during the selection stage.

In contrast, the puromycin DNA spacer linker utilized in mRNA display technology is a quarter of the size of a ribosome. This linker will have a lesser chance of communicating with an immobilized selection target. As a result, mRNA display technology is more likely to produce less skewed results.

2.5.6 Recombinant expression

Once the antibody has been identified, its commercial manufacturing platform is primarily recombinant expression (figure 2.10)

Recombinant DNA technology comprises:

1. Cloning heavy chains and light chains and generating the full-length cDNA from the respective mRNA (e.g. cDNA libraries, synthetic gene synthesis).
2. Ligating the gene of interest into the expression vector (the expression vector is designed to have elements that will enable expression in the host cell).
3. The plasmid is then produced in competent *E. coli* cells (via transformation).
4. The recombinant plasmid DNA is extracted and purified in preparation for transfection into a suitable host cell for production (CHO, HEK, etc).
5. In the presence of a selectable marker, positive clones are selected and further screened for productivity, protein quality, etc, to establish the cell line for producing the protein of interest.

The production of monoclonal antibodies from single human B-cells is based on the reverse transcription polymerase chain reaction (RT-PCR) and expression vector cloning analysis of the immunoglobulin gene repertoire and reactivity at the single-cell level. mAbs are large multimeric proteins (with a typical molecular weight of 150 kDa) that require a number of post-translational modifications to function properly,

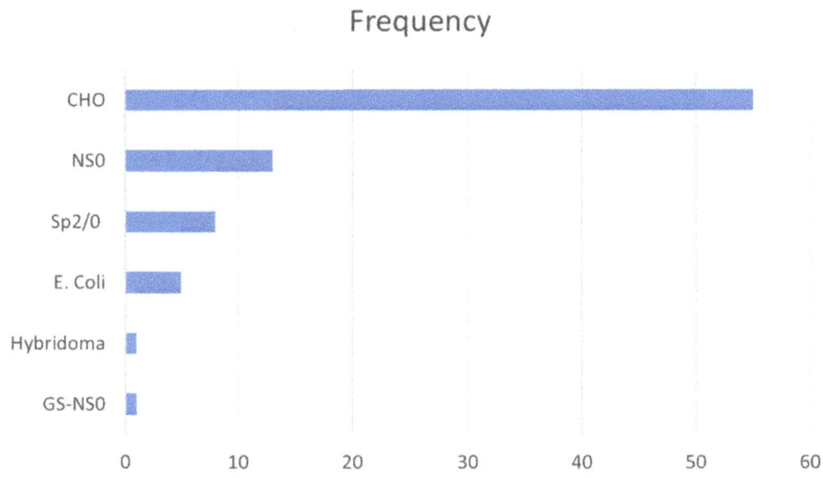

Figure 2.10. Expression system used for commercial manufacturing of approved antibodies.

including glycosylation and disulfide bond formation. As a result, a eukaryotic production scheme with typical eukaryotic post-translational modifications is employed.

CHO cells are primarily used in modern mAb production (figure 2.7). A possible alternative EKT cell line is being evaluated, including yeast cells that grow as rapidly as mammalian cells.

2.6 Conclusion

Antibodies constitute the largest category of biopharmaceuticals made possible by the advent of recombinant technology. Being a newer type of drug category, monoclonal antibodies have detailed guidance from regulatory agencies requiring a clear understanding of how the human immune system operates. The types of antibodies are varied and offer broad treatment of diseases from cancer to rheumatoid arthritis to asthma. The development technology of monoclonal antibodies is also emerging using novel technologies that require a careful review of databases that report the properties of these complex molecules, as provided in this chapter.

2.7 Databases relevant to antibodies

- GenBank® is a large database of publicly accessible DNA sequences for over 205 000 species. More than 60 000 are within the embryophyte, collected by submissions from individual laboratories and batch submissions from large-scale sequencing projects.
- The NIH genetic sequence database is an annotated list of all publicly available DNA sequences (Benson D A *et al* 2013 *Nucleic Acids Research* **41** D36–42).

- GenBank is a member of the International Nucleotide Sequence Database Collaboration, which also includes Japan's DNA DataBank (DDBJ), the European Nucleotide Archive (ENA), and the National Center for Biotechnology Information's GenBank. These three entities exchange data on a daily basis. GenBank is released every two months and can be accessed via the FTP server. The release notes for the current version of GenBank give thorough information about the release as well as notice of future modifications to GenBank. There are various methods for searching and retrieving data from GenBank:
  - With Entrez Nucleotide, you may search GenBank for sequence identifiers and annotations (https://www.ncbi.nlm.nih.gov/nucleotide/).
  - Using the Basic Local Alignment Search Tool (BLAST), find and align GenBank sequences to a query sequence (http://www.ncbi.nlm.nih.gov/blast/).
  - Using NCBI e-utilities, you may search, link, and download sequences programmatically (https://www.ncbi.nlm.nih.gov/books/NBK25501/).
  - The ASN.1 and flat file formats are available through the NCBI's anonymous FTP server at ftp:/ftp.ncbi.nlm.nih.gov/ncbi-asn1 and ftp:/ftp.ncbi.nlm.nih.gov/genbank.
- As of May 2020, the Protein Data Bank (PDB) (www.rcsb.org) had about 164 840 antibody fragment structures (Fabs, Fvs, scFvs, and Fcs) and a limited number of full antibody structures. The complexes of these proteins, as well as other macromolecules, peptides, and haptens molecules, are included in the structural data.
- The Drug Bank database (https://www.drugbank.ca/drugs/) is a one-of-a-kind resource in bioinformatics and cheminformatics that integrates specific drug data with complete drug target information. The most recent release of DrugBank (version 5.1.6, published 22 April 2020) comprises 13 577 drug entries, including 2634 approved small-molecule pharmaceuticals, 1377 licensed biologics (proteins, peptides, vaccines, and allergens), 131 nutraceuticals, and over 6375 experimental (discovery phase) drugs. Furthermore, these drug entries are connected to 5229 non-redundant protein sequences (i.e. drug target/enzyme/transporter/carrier). Each entry contains about 200 data fields, half of which contain drug/chemical information and the other half contain drug target or protein information.
- IMGT/mAb-DB is a component of IMGT®, the international immunogenetics information system®, an integrated information system specializing in immunoglobulins (IG), T-cell receptors (TR), the major histocompatibility complex (MHC) of humans and other vertebrates, the immunoglobulin superfamily (IgSF), the MHC superfamily (MhcSF), and related proteins of the immune system (RPI) of vertebrates and invertebrates (http://www.imgt.org/mAb-DB/).
- The European Collection of Cell Cultures is a cell culture collection service for the research community, with approximately 40 000 cell lines and 450 antibodies (http://www.hpacultures.org.uk/collections/ecacc.jsp).

- The Hybridoma Databank (HDB) contains information about hybridomas and their immunoreactive products. There is information on the development of a hybridoma as well as the reactivity and non-reactivity of its secretory product. There is also information about the availability of a specific hybridoma and its mAb product. The HDB's information is taken from books, catalogs, and survey forms (http://www.atcc.org/).
- The Monoclonal Antibody Index is a biotechnology database that provides annual updates on over 9000 monoclonal antibodies used to diagnose and treat cancer, transplants, infection, heart disease, and other diseases (http://www.gallartinternet.com/mai/index.htm).

IOP Publishing

Biopharmaceutical Manufacturing, Volume 1
Regulatory processes
Sarfaraz Niazi and Sunitha Lokesh

Chapter 3

Gene and cell therapy

Gene and cell therapy constitute the most recent advances in the line of new biopharmaceutical products. An overview of the diseases and risks in selecting treatment modalities, development, and ethical issues, and regulatory challenges are introduced in this chapter. A description of approved products is provided to reflect on the future. Among the notable technologies, gene therapy and cell therapy, the DNA and mRNA vaccines and CAR-T techniques, are discussed in detail and their methodologies and relative advantages discussed. Upstream and downstream technologies for gene and cell therapy products and other allogenic products are described, including understanding regulatory controls that include characterization of the cell population, release testing, and radioisotope tagging. Issues related to vectors and vector preparation are also presented. Preclinical evaluation methods and challenges in the commercialization of gene and cell therapy products are also pointed out.

3.1 Overview

Living tissues are made up of cells, trillions of them, and the thousands of genes in the nucleus of the cells provide the instructions for producing proteins and enzymes that shape muscles, bones, and blood and support body functions, including digestion, energy production, and development. Cell therapy is the process of replacing defective or nonfunctional cells with fresh ones. Although it is not a biopharmaceutical product, blood transfusion is the best example of cell therapy. The administration of autologous (cells obtained from the same person), allogeneic (tissue cells from another person), or xenogeneic (foreign) live cells that have been modified or treated *ex vivo* is known as somatic cell therapy. *Ex vivo* propagation, extension, selection, or chemical treatment of cells, as well as other biological alterations, are all part of the production process for somatic cell therapy products. These cellular products are also utilized to diagnose and prevent a wide range of illnesses.

Gene therapy involves replacing a disease-causing gene with one that does not and adding or turning off genes that help the body combat or cure diseases. The vehicle for

doi:10.1088/978-0-7503-3175-3ch3

inserting new genes into cells is called a 'vector' and is genetically modified to carry the gene (figure 3.1). Viruses, for example, can deliver genetic material into cells spontaneously and can be used as vectors. A virus's capacity to induce infectious disease is blocked, enabling the passing of therapeutic genes into human cells without any side effects. A combination of cell and gene therapy occurs when a cell is genetically engineered.

The gene and cell therapy products approved as of April 2021 are listed in table 3.1. Readers can update the information on FDA-approved products by visiting https://www.fda.gov/vaccines-blood-biologics/cellular-gene-therapy-products/approved-cellular-and-gene-therapy-products.

3.2 Gene therapy

Gene therapy is a form of genetic therapy that involves gene manipulation both inside and outside the body. A gene may be defective or absent in whole or part from birth, or it can change or mutate during adulthood. Any of these variations may cause problems with protein synthesis, resulting in diseases (section 3.1).

Gene therapy involving genetic alteration outside of the body necessitates the injection of a vector containing the gene directly into the body part with the defective cells. Blood, bone marrow, or other tissue can be taken from a patient for gene therapy to alter cells outside of the body. The different types of cells generated are isolated and separated in the lab. The desired cells are administered using a vector that carries the desired gene.

3.2.1 Methodologies

The following are some of the main approaches used in gene therapy to prevent or cure diseases:

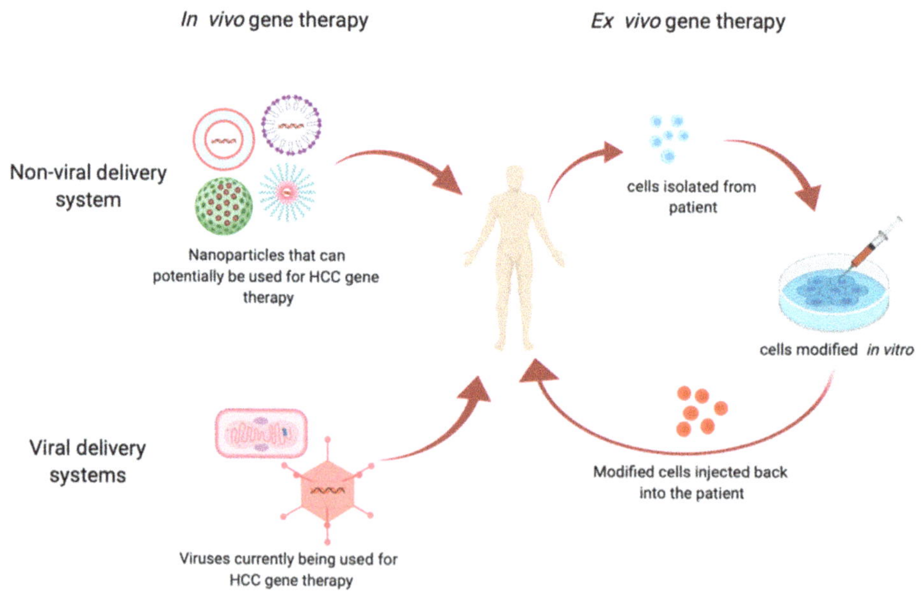

Figure 3.1. Mechanism of gene therapy. (Reprinted with permission of MDPI from Reghupaty S C and Sarkar D 2019 *Cancers* **11** 1265; https://doi.org/10.3390/cancers11091265, Creative Commons License.)

Table 3.1. Approved gene therapy and cell therapy products.

Product	Treatment/indication
Alipogene tiparvovec (Glybera)	Lipoprotein lipase deficiency (LPLD), a rare hereditary condition that can lead to severe pancreatitis, is reversible.
Allogeneic cultured keratinocytes and fibroblasts in bovine collagen (Gintuit)	A topical (non-submerged) application of an allogeneic cellularized scaffold product to treat mucogingival conditions in adults in a surgically formed vascular wound bed.
Autologous CD34+ enriched cell fraction (Strimvelis)	A topical (non-submerged) application of an allogeneic cellularized scaffold product to treat mucogingival conditions in adults in a surgically formed vascular wound bed.
Autologous cultured chondrocytes on a porcine collagen membrane (Maci)	A cellularized autologous scaffold product which patches single or multiple symptomatic full-thickness cartilage defects with or without bone involvement.
Axicabtagene ciloleucel (KTE-C19, Axi-cel, Yescarta)	Treating B-cell lymphoma that has not responded to conventional treatment. T-cells from the patient are genetically engineered and returned to the patient to populate the bone marrow.
Azficel-T (Laviv)	Adults with mild to extreme nasolabial fold wrinkles may have their appearance improved.
Betibeglogene autotemcel (Zynteglo)	Treating beta-thalassemia, a rare and potentially debilitating blood disorder.
Eteplirsen (Exondys 51)	A type of antisense therapy, it treat types of Duchenne muscular dystrophy (DMD) caused by a specific mutation.
Gendicine	A recombinant adenovirus engineered to express wildtype-p53 (rad-p53) designed to treat patients with mutated p53 genes.
Idecabtagene vicleucel (Ibecma)	Treatment of adult patients with relapsed or refractory multiple myeloma after four or more prior lines of therapy including an immunomodulatory agent, a proteasome inhibitor, and an anti-CD38 monoclonal antibody.
Macugen (pegaptanib sodium injection)	An anti-angiogenic treatment of neovascular (wet) age-related macular degeneration (AMD).
Mipomersen (Kynamro)	For treating homozygous familial hypercholesterolemia, administered by subcutaneous injection.
Neovasculgen Cambiogen plasmid	The gene encoding for a vascular endothelial growth factor is delivered during the treatment of peripheral artery disease, including essential limb ischemia (VEGF). It was developed by the Human

(*Continued*)

Table 3.1. (*Continued*)

Product	Treatment/indication
	Stem Cells Institute in Russia and was approved in Russia in 2011.
Nusinersen (Spinraza)	Treats spinal muscular atrophy (SMA), a rare disease.
Onasemnogene abeparvovec (Zolgensma)	Treats SMA, a rare disease.
Patisiran (Onpattro)	A double-stranded small interfering ribonucleic acid (siRNA) formulated for targeted delivery to hepatocytes, the primary source of transthyretin (TTR) protein production.
Provenge (Sipuleucel-T)	For treating asymptomatic or minimally symptomatic metastatic castrate-resistant (hormone-refractory) prostate cancer.
Single-stranded oligonucleotide (Defitelio)	Single-stranded oligonucleotides isolated from the mucosa of pigs' intestines. It is used to treat veno-occlusive disease in the liver after a bone marrow transplant.
Talimogene laherparepvec (Imlygic)	Indicated in the local treatment of unresectable cutaneous, subcutaneous, and nodal lesions in patients with melanoma recurrent after the initial surgery.
Tisagenlecleucel (Kymriah)	B-cell acute lymphoblastic leukemia (ALL) is treated using the body's T-cells to battle cancer (adoptive cell transfer). It is a herpes virus that has been genetically modified (an oncolytic herpes virus). Two genes are deleted—one that shuts down a cell's defenses and another that helps the virus escape the immune system—and a human GM-CSF gene was inserted in their place.
Voretigene neparvovec (Luxturna)	For treating Leber's congenital amaurosis—the first FDA-approved *in vivo* gene therapy. Biallelic RPE65-mediated hereditary retinal disease, also known as Leber's congenital amaurosis, is an inherited condition that causes progressive blindness.

- In order to increase protein production, a new copy of a gene is inserted into target cells. To deliver the genome into the cells, a mutated virus, such as an adeno-associated virus (AAV), is frequently used. Gene-insertion therapies are used to treat a variety of disorders, including acute combination immunodeficiency adenosine deaminase (ADA-SCID), hemophilia, Leber's congenital amaurosis, congenital blindness, lysosomal storage diseases, X-linked chronic granulomatous disease, and others.
- Using a gene-altering technique (e.g. CRISPR/cas9, TALEN, or ZFN), gene repair is accomplished by deleting repetitive or abnormal elements of a gene or replacing an impaired or defective DNA region. Gene correction causes the production of a protein that, in most cases, works to cure the disease. A wide range of diseases can be diagnosed using gene repair. HIV is deleted

from the genomes of contaminated laboratory mice using gene-editing techniques, and the expanded region responsible for Huntington's disease is removed from the human gene using gene-editing techniques.

- Gene silencing prevents the development of a specific protein by targeting messenger RNA (mRNA). Both human and animal cells contain single-stranded mRNA, whereas viruses contain double-stranded RNA. Human and animal cells recognize double-stranded RNA as viral and destroy it to prevent viral spread. In genome silencing, small RNA sequences bind to double-stranded special sequences in the target mRNA. It enables the elimination of mRNA using the same cellular machinery that destroys viral RNA. Gene silencing can reduce the amount of alpha tumor necrosis factor (TNF) in rheumatoid arthritis patients' joints, and gene silencing affects only those targets and not other tissues.
- Reprogramming alters the properties of a cell by introducing one or more genes into it. This procedure is particularly effective in tissues where mutations result in a variety of cell types. TYpe I diabetes develops when many of the pancreas' insulin-producing islet cells are damaged. At the same time, the pancreatic cells that contain digestive enzymes are unaffected. Reprogramming such cells to produce insulin will benefit patients with type I diabetes.
- Cell removal methods are used to kill malignant (cancer) tumor cells, whereas stable (non-cancer) tumor cells are targeted for overgrowth. Tumor cells can be eliminated by injecting them with 'suicide genes' and then releasing a prodrug that causes the cells to die. Viruses could be genetically modified to preferentially attach to tumor cells. Oncotropic viruses can carry therapeutic genes that increase tumor cell toxicity, causing the immune system to fight the tumor or preventing the growth of blood vessels that feed the tumor.
- Germline gene therapy is a laboratory technique that modifies human reproductive cells (eggs).

3.2.2 Risks of gene therapy

While there are many benefits to using gene therapy, modifying genetic code can produce many risks, some very serious:

- Gene therapy or cell therapy may not be as effective as anticipated, frequently prolonging or exacerbating symptoms or complicating the disease with adverse therapy effects.
- The expression of genetic material or the survival of stem cells may be insufficient. It may be too short-lived to completely heal or improve the disease, or it may be too short-lived to replace proteins that cause a strong immune response to the protein.
- As in autoimmune diseases, the immune response becomes uncontrolled, resulting in attacks on normal proteins or cells.
- Cancer or viral/fungal/bacterial infections are involved, resulting in an insufficient immune response that leads to therapy resistance.
- There is no way to 'turn off' gene expression in clinical trials with the current generation of vectors if it appears to be harmful.
- The integration of genetic material into DNA from retroviral or lentiviral vectors could occur alongside a gene involved in regulating cell growth. Through insertion mutagenesis, this insertion may eventually cause a tumor.

- There is little data on how it affects the embryo during pregnancy.
- Due to a lack of tight control over stem cell division, transplanted stem cells gain a growth advantage and progress to a type of cancer or teratomas.

3.2.3 Gene-editing technologies

Genome engineering came into existence in the 1970s, when it was shown that yeast and bacteria could take up exogenous DNA and be randomly integrated into the genome. A variety of techniques are currently used for gene-editing.

- Cre-Lox recombination is a method that uses a site-specific recombinase to perform deletions, insertions, translocations, and cell DNA reversals at specific locations. This method entails modifying DNA to be targeted to a particular cell or triggered by an external stimulus. It is used in both eukaryotic and prokaryotic processes.
- LoxP recognition sequences (locus of X(cross)-over locus in P1) are 34 base-pair long recognition sequences composed of two 13 bp long palindromic repeats separated by an 8 bp asymmetric core spacer sequence.
- FLP-FRT recombination, like Cre-Lox recombination, is a type of site-driven recombination. The *Saccharomyces cerevisiae* baker's yeast recombinase flippase (Flp) recombines sequences between the short flippase recognition target (FRT) sites.
- A type of genetic recombination is homologous recombination. Nucleotide sequences are shared by two related or identical double-stranded or single-stranded nucleic acid molecules (usually DNA in cellular organisms, but may also be RNA in viruses).
- Zinc finger nucleases (ZFNs) are restriction enzymes formed by fusing a zinc finger's DNA-binding domain to a DNA cleavage.
- Transcription activator-like effector nucleases (TALENs) are restriction enzymes that have been programmed to cut specific DNA sequences. They are created by fusing the DNA-binding domain of a TAL effector with a DNA cleavage domain (a nuclease that breaks DNA strands).
- The Sleeping Beauty transposon is a synthetic DNA transposon that is designed to insert precisely defined DNA sequences into vertebrate animal chromosomes in order to introduce new traits and discover new genes and functions.
- Recombineering (recombination-mediated genetic engineering) is a genetic and molecular biology technique based on homologous recombination systems, as opposed to the older/newer method of mixing DNA sequences in a defined order using restrictive enzymes and leagues.
- The CRISPR-Cas system is a prokaryotic immune system that protects foreign genetic elements present in plasmids and phages, which provide immunity. The spacer gene in RNA helps the Cas proteins to identify and break foreign pathogenic DNA (CRISPR-associated). International RNA is broken by specific Cas proteins that are directed by RNA. CRISPR is used in around half of all sequenced bacterial genomes and almost all sequenced archaeal genomes. Cas9 (or CRISPR-associated protein 9′) is an enzyme that identifies and cleaves specific DNA strands complementing the CRISPR gene

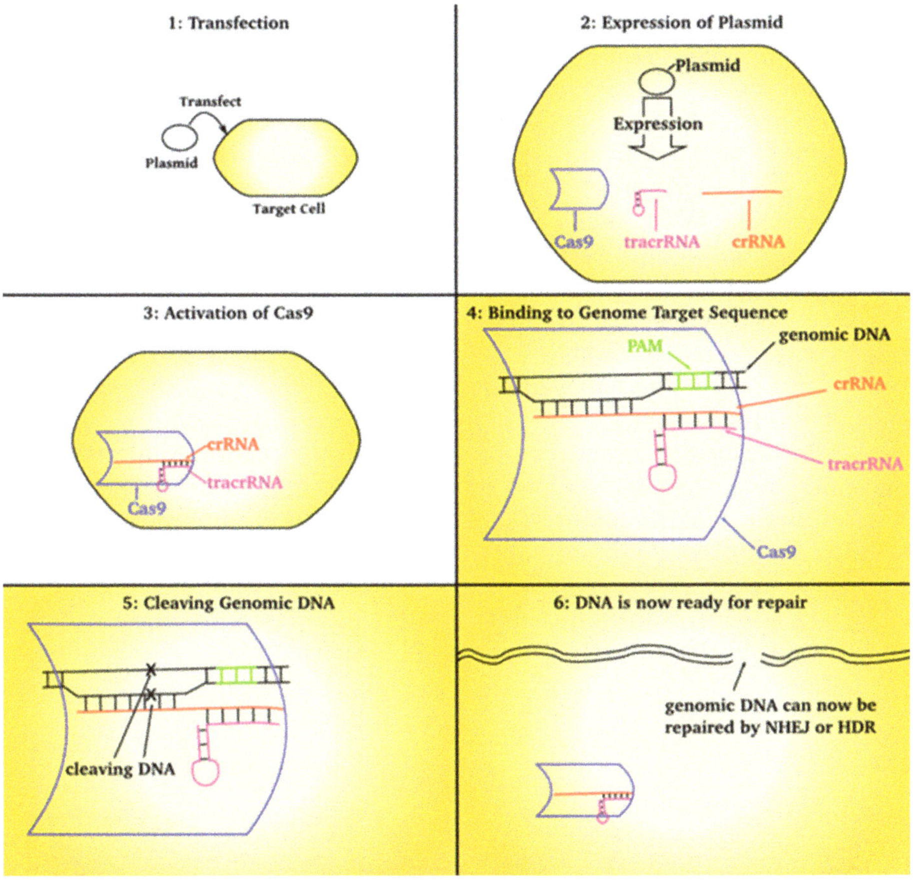

Figure 3.2. The CRISPR system. (Source: Wikipedia: Nielsrca, based on figures in *Nature Protocols*, Creative Commons.)

using CRISPR sequences. Cas9 enzymes are the foundation of a technology known as CRISPR-Cas9, which can be used to manipulate genes in humans using CRISPR sequences. This editing method has a broad variety of applications, from basic biological science to biomedical product creation and disease treatment (section 3.2). Its ability to alter various DNA sequences makes it a useful tool for repairing disease-causing mutations (figure 3.2).

3.2.4 Vector manufacturing

Viruses that have been updated are used as vectors for gene delivery and oncolytic viruses in gene therapy. Viral vectors shield the new gene from blood-borne enzymes that would otherwise destroy it and deliver it to the correct cells. Viral vectors effectively force cells to accept the virus's new gene into their nucleus. The transduced cells then use the most recent gene to produce a new protein that performs its normal function. Viral vectors have been genetically modified to remove the majority of their essential genes, preventing uncontrolled virus replication and allowing for gene injection.

Viral vectors for gene therapy are chosen based on (i) the size of the DNA or gene that can be packaged, (ii) the quality of absorption by the desired therapy cells, (iii) the duration of gene expression, (iv) the effect on the immune response, (v) the ease of incorporation into the DNA or the ability to remain as a stable DNA, and (vi) the ease of incorporation into the DNA or the ability to remain as a stable DNA.

Oncolytic viruses are programmed to replicate in cancer cells rather than in normal human cells. When oncolytic viruses multiply in cancer cells they cause more oncolytic viruses to infect the cancer cells around them.

In the past, retroviral and adenoviral vectors have received more attention. Nonetheless, vectors derived from adeno-associated virus (AAV) and lentiviral vectors have advanced in terms of defense and target tissue expression profiles. Herpes simplex virus and pox/vaccinia vectors are also promising as oncolytic vaccines due to their efficacy.

The AAV is the most widely used *in vivo* genetic alteration vector, with many serotypes that can target various tissues in the body (naturally occurring or recombinant). The AAV is not considered pathogenic and is well tolerated, resulting in a decreased inflammatory response. As a result, AAV is thought to be primarily episomal, as opposed to other viral vectors such as lentivirus, which integrate into the host cell's genome.

Table 3.2 summarizes the physical properties of the various viruses that are used for gene therapy

Table 3.2. Overview of common viruses used for generating gene therapy viral vectors.

Parameter	Adenovirus	AAV	Retrovirus	Lentivirus
Coat	Non-enveloped	Non-enveloped	Enveloped	Enveloped
Genome	dsDNA	ssDNA	ssRNA	ssRNA
Insert capacity (Kb)	7.5	~4.5	8	8
Inflammatory, infection	Broad host range	Broad excluding hematopoietic cells	Transduces only dividing cells	Broad host range
Hot genome interaction	Non-integrating	Integrating and non-integrating	Integrating (mostly random)	Integrating
Transgene expression	Transient or long-lasting depending on immunogenicity	Potentially long-lasting but slow on the onset	Long-term	Long-term
Size	~90 nm	~20 nm	~90–120 nm	~90–120 nm
Stability	High	High	Low	Low
Buoyant Density	1.34 CsCl	1.41 CsCl	1.16 CsCl (sucrose)	1.16 CsCl (sucrose)
Example	Ad5	AAV2, AAV3, AAV6	MMSV, MSCV	HIV-1, HIV-2

3.2.4.1 Viral vector manufacturing

There are several methods for creating viral vectors. The system's decision is influenced by tissue tropism, the ability to incorporate or non-integrated change, *in vivo* or *ex vivo* phase, previous immune response, and whether the target cell replicates. In viral vector applications, high-efficiency transduction and stable transgenic expression levels are desirable outcomes.

Viruses, including oncolytic viruses, are modified and used as gene delivery vectors in gene therapy. The newly formed gene is protected from blood-borne enzymes that could destroy it and prevent it from reaching the appropriate cells by viral vectors. Viral vectors force cells to accept the virus's new gene into their nuclei. Transduced cells then use the new gene to produce a new protein that functions in the same way as the original genes. Because most of the critical genes in viral vectors have been removed, uncontrolled virus replication and gene modification are prevented.

Viral vectors for gene therapy are chosen based on (i) the size of the packaged DNA or gene, (ii) the quality of absorption by the desired therapy cells, (iii) the duration of gene expression, (iv) the effect on the immune response, (v) the ease of manufacturing, and (vi) the ease of integration into the DNA or the ability to remain as a stable DNA.

Oncolytic viruses have been genetically modified so that they only replicate in cancer cells and not in normal human cells. Oncolytic viruses replicate in cancer cells, causing them to erupt and infect nearby cancer cells with more oncolytic viruses.

In the past, retroviral and adenoviral vectors received more attention. Vectors derived from adeno-associated virus (AAV) and lentiviral vectors, on the other hand, have improved target tissue expression profiles and evolution safety. Oncolytic vaccines based on the herpes simplex virus and pox/vaccinia vectors show promise as well.

The AAV is the most widely used *in vivo* genetic alteration vector, with many serotypes that can target various tissues in the body (naturally occurring or recombinant). The AAV is not pathogenic and is well tolerated, resulting in a reduced inflammatory response. As a result, unlike other viral vectors such as lentivirus, which are thought to integrate into the host cell's genome, AAV is thought to remain primarily episomal.

The goal of viral vector design is to effectively bundle the therapeutic gene or nucleotides into infectious viral particles in order to prevent the production of wild-type particles or empty particles. As a result, many vector systems are created for safety, convenience, or necessity by transiently co-transfecting multiple plasmids to avoid the toxicity of a vector component in producer cells (figure 3.3).

In most cases, viral vectors are composed of several expressible units to reduce the possibility of recombination and the formation of a wild-type particle. The expressible units are typically plasmids (a genetic structure in a cell that can replicate independently of the chromosomes, usually a short circular DNA strand in the cytoplasm of a bacterium or protozoan) injected into producer cells via transfection or a virus-transducing 'helper'. If a recombination event occurs, the expression units are usually mutated to deactivate the function of the wild-type virus. Newer HIV lentiviral vectors, for example, lack the genes for the virulence factors tat, vpr, vpu, and nef, and have gag and pol on different plasmids from rev and env1, as well as other mutations, such as deletions in the 3 LTR.

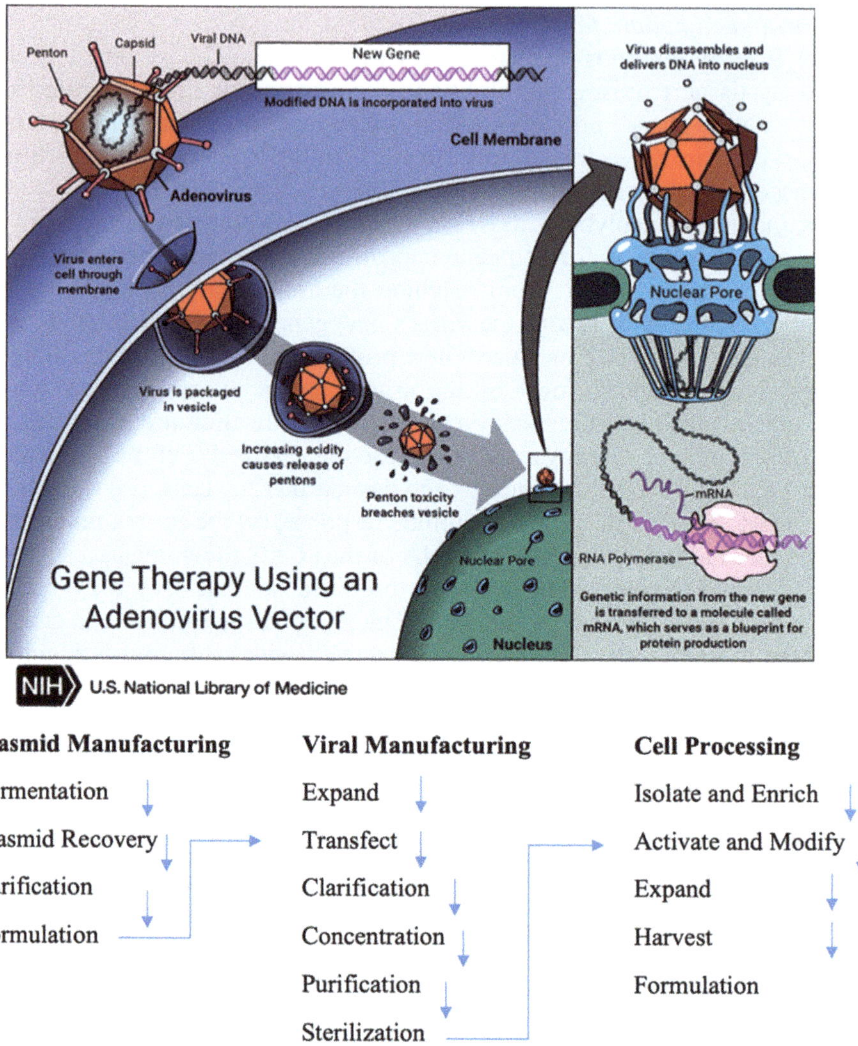

Figure 3.3. Schematic of AAV vector development through transfection. (Source: U.S. National Library of Medicine https://medlineplus.gov/genetics/understanding/therapy/procedures/.)

Figure 3.3 shows how an AAV vector is generated by transfection.

Normally, the AAV vector and therapeutic gene are transfected as expression cassettes from various plasmids. Inside the cell expression cassettes generate viral proteins and genetic DNA containing the therapeutic gene(s). Transgene-containing viral particles assemble in the cytoplasm of the AAV system. Particles are released via cellular lysis before being filtered and characterized for public consumption. Creating viral vectors necessitates a number of fabrication steps or platforms. Plasmids encoding the helper virus and the therapeutic gene, cell lines used to create the vector, and other products are required to create the therapeutic viral vector. Plasmids can be replaced by helper transducing viruses in some cases. To improve

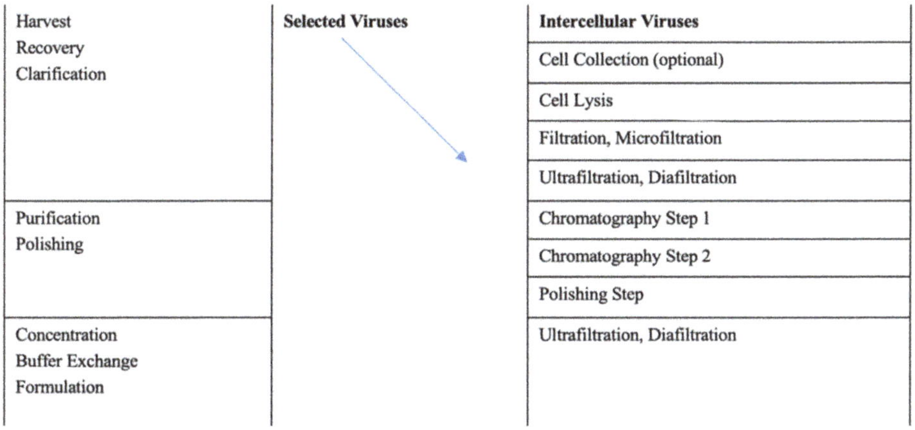

Figure 3.4. Example of three manufacturing steps for the generation of modified cells for *ex vivo* gene therapy via viral vectors produced by transfection.

the production process, reliable producer cell lines that reduce or eliminate transfection and transduction measures can be developed.

The next step in the vector manufacturing process is the creation of an infectious viral vector (figure 3.4). Cells are transfected with plasmids and then collected to create the viral vector. After the cells are lysed, the AAV viral particles accumulate in the media and cytoplasm, resulting in a higher overall yield. It is then distilled, filtered, titrated, characterized, and processed for use in *ex vivo* or *in vivo* applications. In *ex vivo* transduction, the viral vector collects and updates host cells (figure 3.4). The cells are harvested, characterized, and formulated after modification and pre-transplantation. Transduced cells can be extended in cell culture before being re-infused into patients.

Despite the fact that the number of hollow fiber and fixed-bed bioreactor systems is increasing, the variety is currently limited. Microcarriers in rocking bioreactors or stirring tank bioreactors are another option for adherent cells because they have a large surface area for cell attachment. Rocking bioreactors with a microcarrier or single-cell suspension culture are an option for producing smaller vector volumes for research or clinical trials because their current capacity at low cell density is 500 L. Rocking bioreactors that can be reused can also be used to supply cell seed to larger bioreactors. Virus production in the vaccine industry has been successfully scaled up to 2000+ L using microcarriers in stirred tank bioreactors. A single 2000 L bioreactor with a microcarrier density of 10 cm^2 mL^{-1} has a surface area of 2000 m^2.

The main disadvantages of microcarriers are development efforts to optimize cell attachment, growth, and viability. The ability to transfect/transduce cells growing on a microcarrier must be recognized early in the development process. A stirred tank bioreactor is the most common bioreactor used to commercialize mAbs and recombinant proteins. As a result, the framework has matured, and both industry and regulatory authorities are familiar with its application. These are the most efficient methods for mass production.

Suspension cell development works best in bioreactors. Single-cell suspension culture has a significant advantage over adherent cell culture. They can be easily scaled up in the lab from a spinner to a bioreactor for bioreactor production without requiring cell detachment. Animal-free media are beneficial to the output because they reduce the risk of adventitious agents and simplify purification. Adapting custom cells to suspension culture, on the other hand, can be time-consuming and extend process development times.

The efficiency of downstream purification is influenced by the upstream manufacturing process. Shear forces, for example, may increase cell debris on adhered cells in microcarrier systems or other fluid-motion systems. Another theory is that introducing fewer impurities into the bed of a fixed-bed bioreactor could benefit it.

3.2.4.2 Downstream manufacturing

The goal of downstream processing is to isolate the viral vector from the various impurities produced during upstream processing and to get the virus into the proper state to formulate for patient administration.

In conventional purification steps, ultracentrifugation, flocculation, and other methods of varying complexity are used. Recent advances in highly scalable chromatography and tangential flow filtration have enabled the same degree of impurity removal as more traditional methods.

Purification presents several challenges due to the differences in physical properties between different vectors, necessitating a variety of novel techniques and solutions. AAV, for example, has been collected in the medium and is being extracted from cells in part via a lytic process. AAV viruses are typically harvested by cell lysis to increase yield. The amount of contaminants released by host cell cell lysis, such as DNA and protein, will increase the purification burden. The transgene will be present in up to 95% of total particles in AAV vectors, with a significant portion of 'soft' particles being transgene-free. Despite the ongoing debate over whether empty particles have a potential benefit in AAV therapies, they are still regarded as a significant contaminant. Downstream RV/LV vector processing presents its own set of challenges.

The upstream manufacturing method chosen adds to the complexity of vector purification.

Despite differences in physical properties, current downstream protocols typically include a confirmation stage (which may include cell lysis), purification (ion-exchange/affinity centrifugation or chromatography), and polishing phase (additional step/size exclusion). Filtration/dialysis/centrifugation steps are frequently used in addition to the aforementioned to focus, turn filters, or finalize the formulation.

To improve viral yield, the clarification stage involves removing large debris and macromolecular complexes from the initial crude suspension as well as cell lysis (AdV, AAV). Nucleases (for example, benzonase) are commonly used during or after cell lysis to degrade nucleic acids and disrupt macromolecular complexes. Because lysis is highly dependent on cell concentration, incubation time, incubation temperature, cell type, and virus generation, each cell lysis method must be optimized. Many filters with pore sizes as small as 0.2 μm are commonly used during clarification to remove cell debris and impurities, which is an important part of a bioburden control strategy.

Traditionally, ultracentrifugation techniques have been used to purify small amounts of virus for research or limited clinical trials. Typically, viral particles are pelleted and condensed through a matrix, then resuspended and filtered through the buoyant mass using gradient ultracentrifugation. To further purify the particles, multiple rounds of gradient ultracentrifugation are required. CsCl was once widely used in AAV and AdV ultracentrifugation. Iodixanol, on the other hand, is now preferred over CsCl because it is less toxic, requires less dialysis, eliminates oxidation, and reduces run times from 24 hours to a few hours. Because enveloped viruses, such as RV/LV, are unstable, sucrose is commonly used to mitigate this risk.

Chromatography can be used in several downstream steps to purify vaccines and vectors for gene therapy, including aggregation, concentration, purification, and polishing. Chromatography is a well-known method for purifying vectors based on physical properties such as net charge, hydrophobicity, ligand affinity, size, and others. Compared to centrifugation, chromatography is more versatile and cost-effective. Columns can be used repeatedly and in a variety of ways in conjunction or combination. Furthermore, chromatography is an effective tool for removing potentially harmful adventitious forces, which is critical in late-stage growth.

Ion-exchange chromatography (IEC) is a simple and inexpensive method for detecting a wide range of viruses. Positively and negatively charged viruses can be bound using anion and cation exchanges. IEC would theoretically require the isolation of empty particles from all AAV serotypes; however, the resins and conditions for each serotype would need to be optimized. One disadvantage of open particle separation is that it takes a long time to create, often more than two months, due to the shallow gradients required for separation. Any factor that influences vector charge, such as serotype, vector height, or transgene insert length, necessitates redevelopment of the gradient conditions.

Hydrophobic reaction chromatography (HIC) has long been used in the vector and recombinant protein industries for viral capture/clearance. HIC has been used in the purification of both AdV and AAV2. It has not, however, been tested with different AAV serotypes.

The interaction of a viral particle with a ligand is used in affinity chromatography (AC). Heparin affinity chromatography, which can handle larger volumes, was used to purify LV and other viruses. Mild salt solutions may help to keep the vectors intact. There are some drawbacks, such as non-specific impurity binding, which necessitates additional purification steps. One issue that arises when using AAV is that full particles cannot be distinguished from empty particles, necessitating the use of additional methods such as ion-exchange chromatography or ultracentrifugation. There is a problem with ligand leakage into the vector preparation in affinity chromatography. To detect ligand leakage, additional analytical steps, as well as potential purification steps, may be required.

To reduce the possibility of microbial contamination in the final product, bioburden management is critical for products that must comply with GMP regulations. While its placement in downstream protocols may vary, the vector's final preparation is filtered as a controlling factor by 0.2 μm filtration in most protocols. Certain vector products are not sterilized due to their small scale. Sterile filtration can be avoided if the process is certified as fully aseptic; however, this requires validation and operations to be performed in a cleanroom.

3.2.5 DNA vaccines

One type of gene therapy is gene therapy in the form of a DNA vaccine. This involves the injection of a genetically engineered plasmid containing the DNA sequence encoding the antigen(s) to which an immune response is desired, causing cells to produce the antigen directly and elicite a defensive immune response. DNA vaccines have advantages over traditional vaccines in that they can elicit a broader range of immune responses. Although there are several DNA vaccines for veterinary use, including ones for West Nile Virus in horses and canine melanoma, no DNA vaccines for human use have been approved. In contrast, the COVID-19 pandemic has prompted the development of several novel vaccines, including DNA vaccines.

DNA vaccination creates an effective immune memory by displaying antigen–antibody complexes on follicular dendritic cells (FDC), which are active B-cell stimulators. T-cells may be stimulated by similar germinal center dendritic cells. Because antibody development 'overlaps' long-term antigen expression, antigen–antibody immunocomplexes form and are displayed by FDC, FDC can generate immune memory.

The DNA is injected into the body and taken up by cells, which then use their natural metabolic processes to synthesize proteins based on the plasmid's genetic code. These proteins are classified as foreign because they contain regions of amino acid sequences that are characteristic of bacteria or viruses. When host cells process and reflect on their surface, the immune system is notified, triggering immune responses.

Alternatively, to facilitate cell entry, the DNA could be encased in protein; if this capsid protein is used in the DNA, the resulting vaccine would have the potency of a live vaccine without the risk of reversion (figure 3.5).

A DNA vaccine must contain both an antigen and an adjuvant in order to elicit an adaptive immune response. DNA is applied systemically or topically, depending on the plasmid, for example, via intramuscular injection. Transfected keratinocytes or myocytes express the transgene, and the peptide/protein derived from it is released by exosomes or apoptotic bodies. This material is endocytozed by immature dendritic cells (iDC), which then present antigen to CD4+ T-cells in draining lymph nodes via the major histocompatibility complex system (MHCII). Direct transfection of APC, including iDC, results in endogenous transgene expression and simultaneous presentation via MHCI and MHCII, resulting in parallel CD8+ and CD4+ T-cell responses. A humoral immune response is triggered when the B-cell receptor recognizes the protein antigen and receives assistance from pre-activated antigen-specific CD4+ T-cells.

3.2.5.1 Advantages of DNA vaccines

- There is a low chance of infection.
- MHC class I and MHC class II molecules both present the antigen.
- Type 1 or type 2 T-cell responses may be polarized.

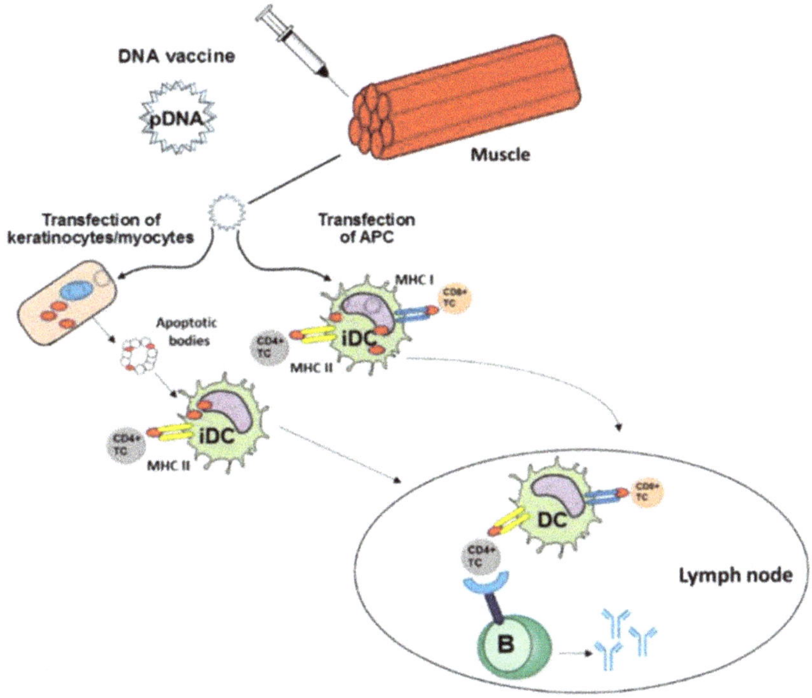

Figure 3.5. DNA vaccines induce adaptive immune responses. (Source: Hobernik D and Bros M 2018 *Int. J. Mol. Sci.* **19** 3605; https://doi.org/10.3390/ijms19113605, Creative Commons License.)

- The immune response is concentrated on the antigen in question.
- Creation and production are easy.
- Storage and transport stability are good.
- Efficiency in terms of cost.
- They eliminate the need for peptide synthesis, recombinant protein expression, and purification, as well as the use of potentially toxic adjuvants.
- Long-term immunogen persistence.
- *In vivo* expression ensures that the protein has a more eukaryotic cell-like structure and undergoes post-translational modifications.

3.2.5.2 Undesirable effects of DNA vaccines

- Only protein immunogens are permitted (not useful for non-protein-based antigens such as bacterial polysaccharides).
- There is a chance that genes that regulate cell growth will be harmed.
- There is a possibility of inducing anti-DNA antibody formation.
- Tolerance to the produced antigen (protein) is a possibility.
- Bacterial and parasitic atypical protein production is a possibility.
- Plasmid vectors.
- Vector design.

3.2.5.3 Delivery routes

DNA vaccines are given systemically via intravenous injection to enter secondary lymphatic organs, orally via (attenuated) bacteria as a vehicle to induce DNA uptake by intestinal antigen-presenting cells (APCs), and pulmonaryly via nebulized DNA to induce uptake by lung cells. In transdermal delivery, needle-free and needle-based administration of particle-adsorbed DNA vaccines via helium pressure (gene gun, particle mediated epidermal delivery (PMED)) and needle-based administration via microneedles and tattoo devices have both been clinically tested. Immediate electroporation improves intradermally administered DNA vaccine transfection of cutaneous APCs and non-APCs. Subcutaneous injection is the most common method for transfecting fibroblasts and keratinocytes, which express transgenes and release antigen for APC uptake. Similarly, intramuscular injection of DNA vaccines results in myocyte transfection, which expresses/releases antigen for APC uptake, and electroporation at the injection site improves myocyte transfection rates (see table 3.3).

Table 3.3. Summary of plasmid DNA delivery methods.

Method of delivery	Formulation of DNA	Target tissue	Amount of DNA
Parenteral Injection (hypodermic needle)	An aqueous solution in saline	IM (skeletal); ID; IV (subcutaneous and intraperitoneal with variable success)	Large amounts (approximately 100–200 μg)
Gene gun	DNA-coated gold beads	ED (abdominal skin); vaginal mucosa; surgically exposed muscle and other organs	Small amounts (as little as 16 ng)
Pneumatic (jet) injection	Aqueous solution	ED	Very high (as much as 300 μg)
Topical application	Aqueous solution	Ocular; intravaginal	Small amounts (up to 100 μg)
Cytofectin-mediated	Liposomes (cationic); microspheres; recombinant adenovirus vectors; attenuated Shigella vector; aerosolized cationic lipid formulations	• DNA-coated gold beads in aqueous solution in saline • Solution (aqueous) • Solution (aqueous) • Aerosolized cationic lipid formulations; cationic liposomes; microspheres; recombinant adenovirus vectors; attenuated Shigella vector	Variable

Source: https://handwiki.org/wiki/Biology:DNA_vaccination.

Table 3.4. Advantages and disadvantages of commonly used DNA vaccine delivery methods.

Method of delivery	Advantages	Disadvantages
IM or ID injection	There is no specialized delivery system in place. Permanently or semi-permanently expressed pDNA spreads quickly across the body.	Inefficient site for uptake due to morphology of muscle tissue. Relatively large amounts of DNA used. The Th1 response may not be the response required
Gene gun	Direct DNA bombardment of cells. DNA is contained in small quantities.	DNA is sprayed directly into the cells. DNA in trace quantities. It is possible that the Th2 answer is not what is needed. As a carrier, inert particles are needed.
Jet injection	No particles are required. DNA may be administered to cells a few millimeters to a few centimeters under the skin's surface.	Significant shearing of DNA after high-pressure expulsion. Ten-fold lower expression and lower immune response. Requires large amounts of DNA (up to 300 μg).
Liposome-mediated delivery	It is possible to elicit high levels of immune response. Increases the transfection of pDNA administered intravenously. Liposomes administered intravenously—DNA complexes can transfect any tissue. Liposome–DNA complexes administered intranasally can cause expression in the distal mucosa and the nasal mucosa and the generation of IgA antibodies.	Toxicity. Ineffectiveness in serum. Risk of disease or immune reactions.

Source: https://sites.google.com/site/dnareplicationsystem/dna-vaccination/delivery-methods.

Table 3.4 lists the relative advantages of delivery methods

Antigen type, antigen location (intracellular versus secreted), antigen number, frequency, immunization dose, site, and antigen delivery method all influence the antibody responses elicited by DNA vaccinations.

Humoral responses to a single DNA injection can last significantly longer than those to a single recombinant protein injection. Comparisons of antibody responses elicited by natural (viral) infection, immunization with recombinant protein, and immunization with plasmid DNA are summarized in table 3.5.

Table 3.5. Comparison of T-dependent antibody responses raised by DNA immunizations, protein inoculations, and viral infections.

	Method of immunization		
	DNA vaccine	Recombinant protein	Natural infection
Amount of inducing antigen	ng	μg	? (ng–μg)
Duration of antigen presentation	Several weeks	<1 week	Several weeks
Kinetics of antibody response	Slow rise	Rapid rise	Rapid rise
Number of inoculations to obtain high avidity IgG and migration of ASC to bone marrow	One	Two	One
Ab isotype (murine models)	C'-dependent or C'-independent	C'-dependent	C'-independent

Source: https://handwiki.org/wiki/Biology:DNA_vaccination.

Furthermore, DNA vaccination produces lower titers of specific antibodies than recombinant protein vaccination. In contrast, DNA immunization-induced antibodies have a higher affinity for native epitopes than recombinant protein-induced antibodies. In other words, DNA immunization produces a better response. Antibodies can form after just one DNA vaccination, but recombinant protein vaccines usually require a booster shot. DNA immunization can affect the T-helper profile of the immune response, and thus the antibody isotype, in ways that neither natural infection nor recombinant protein immunization can.

3.2.6 mRNA vaccines

An mRNA vaccine, also known as messenger RNA vaccine, is a relatively new type of vaccine that uses mRNA-containing vectors, such as lipid nanoparticles, to provide acquired immunity. The RNA sequence codes for antigens and identical proteins are similar to those of the pathogen. After the vaccine is injected into the body the host cells convert this sequence into encoded antigens, stimulating the body's adaptive immune system, causing antibodies to be produced against the pathogen.

Another type of mRNA vaccine is used where the mRNA encodes for a fully human IgG antibody. Antibodies that are identical or close to those found in a patient with a prior history of potent immunity in this type are coded for by the mRNA.

The first mRNA vaccines were approved by the FDA and EMA late 2020 to prevent COVID-19 infections. Both vaccines provided a higher level of efficacy and safety compared to traditional vaccines. It is now anticipated that more mRNA vaccines will replace existing vaccines such as the influenza vaccine that may not be repeated every year.

RNA vaccines have several advantages over DNA vaccines in terms of production, administration, and safety, and they have shown promise in human clinical trials. Although much of the early research on mRNA vaccines focused on cancer prevention, the focus has recently shifted to pathogen prevention, which includes influenza, Ebola, Zika, *Streptococcus* spp., *T. gondii*, and coronavirus.

The *in vitro* transcription of a cDNA template, typically plasmid DNA (pDNA), with a bacteriophage RNA polymerase results in functional synthetic mRNA. Unpolished pDNA contains traces of bacterial genomic DNA as well as three types of pDNA (supercoiled, relaxed circle, or linear) in varying proportions. Recent developments also point to the possibility of creating the gene pool using PCR.

All components, including plasmid DNA, phage polymerases, capping enzymes, and NTPs, are readily available as GMP-grade traceable components. As mRNA therapeutics progress toward commercialization, more cost-effective GMP source materials may become available.

The first step in producing GMP mRNA is the creation of a DNA template, followed by enzymatic IVT. It adheres to the same multistep protocol as study-scale synthesis, but with additional safeguards and potency checks. The process may be modified to accommodate modified nucleosides, capping techniques, or template removal, depending on the mRNA build and chemistry. *E. coli* template plasmid DNA is linearized with a restriction enzyme to begin the manufacturing process, and runoff transcripts with a poly(A) tract at the 3′ end are synthesized. NTPs are then converted to mRNA by a bacteriophage's DNA-dependent RNA polymerase (T7, SP6, or T3). The template DNA is then degraded using DNase. Finally, the mRNA is capped, either chemically or enzymatically, to allow for efficient translation *in vivo*. Under ideal conditions, multi-gram scale mRNA synthesis can yield more than two g/l of full-length mRNA.

Purification of mRNA after synthesis removes reaction components such as enzymes, free nucleotides, residual DNA, and truncated RNA fragments, among others. Although LiCl precipitation is commonly used for laboratory-scale preparation, clinical purification is done with derivatized microbeads in batch or column formats, which are more convenient to use on a large scale. Because dsRNA is a strong inducer of interferon-dependent translation inhibition, it is critical to remove it and other contaminants from certain mRNA platforms to ensure the final product's potency.

RNA can be degraded by both enzymatic and chemical degradation pathways. To reduce the effects of reactive oxygen species and divalent metal ions on mRNA stability, formulation buffers are screened for contaminating RNases and can include buffer components such as antioxidants and chelators. Until being taken up by cells, mRNA is quickly degraded by extracellular ribonucleases. As a result, complexing agents that protect RNA from degradation may significantly improve the efficacy of mRNA vaccines. Complexation can also improve cellular uptake and transmission to the cytoplasmic translation machinery.

As a result, mRNA often forms complexes with lipids or polymers. It is also possible to boost the stability of mRNA products by packaging them in nanoparticles or co-formulating them with RNase inhibitors. However, attempts have been made to keep such mRNA–lipid complexes unfrozen for longer periods of time.

The Food and Drug Administration (FDA) and the European Medicines Agency (EMA) have yet to issue a definitive recommendation for mRNA vaccine products. The increasing number of clinical trials conducted under the supervision of the EMA and FDA, on the other hand, suggests that regulators have recognized different organizations' approaches to demonstrating that drugs are safe and suitable for human testing. As mRNA belongs to the diverse vaccine community of genetic immunogens, many of the guiding principles that have been established for DNA vaccines and gene therapy vectors can be applied to mRNA with some modifications to reflect mRNA's unique features. As mRNA products become more common in the vaccine industry, detailed guidelines for manufacturing and testing new mRNA vaccines will most likely be developed.

3.2.6.1 Safety

Other vaccine platforms, such as live viruses, viral vectors, inactivated viruses, and subunit protein vaccines, are not as safe as mRNA architecture. It contains no toxic chemicals or cell cultures that could be contaminated with unwanted viruses. Furthermore, because mRNA is produced quickly, contaminating microorganisms have few chances to join. The theoretical risks of infection or vector integration into host cell DNA are not a concern for mRNA in the vaccinated population. For the reasons stated above, mRNA vaccines are thought to be a reasonably safe vaccine format.

Preclinical and clinical studies are being conducted to investigate the effects of non-native nucleotides and delivery system components on local and systemic inflammation, biodistribution and persistence of expressed immunogen, activation of autoreactive antibodies, and potential toxic effects of non-native nucleotides and delivery system components. Certain mRNA-based vaccine platforms have been linked to inflammation and possibly autoimmunity, and type I interferon responses have been linked to inflammation and possibly autoimmunity. Another potential risk associated with the presence of extracellular RNA during mRNA vaccination is the presence of extracellular RNA. Extracellular naked RNA has been shown to increase the permeability of densely packed endothelial cells, contributing to edema. Extracellular RNA is also frequently responsible for blood coagulation and abnormal thrombus formation. As multiple mRNA modalities and delivery mechanisms are used in humans for the first time and tested in larger patient populations, protection will need to be controlled.

3.2.6.2 Delivery

The methods of RNA administration are classified based on whether the RNA is passed to cells within the organism (*in vivo*) or outside the organism (*ex vivo*). The use of RNA as a vaccine tool in the form of self-amplifying mRNA was discovered in the 1990s, and the mode of mRNA uptake has been recognized for over a decade. It was also discovered that different injection routes, such as skin, blood, or muscles, resulted in different levels of mRNA absorption, highlighting the importance of the administration route in delivery. As a result of their fundamentally different sizes, the mechanisms and evaluation of self-amplifying mRNA will differ.

Encapsulating mRNA in lipid nanoparticles has been appealing for a variety of reasons. The lipid primarily acts as a barrier against degradation, allowing for more consistent translational results. The ability to customize the outer lipid layer also enables ligand interactions to target specific cell types.

Non-viral delivery methods have been developed to achieve similar immunological responses as RNA viruses. The most common RNA viruses used as vectors are retroviruses, lentiviruses, alphaviruses, and rhabdoviruses, each with its own structure and function.

3.3 Cell therapy

Cell therapy removes cells in a patient's body to cure diseases and produce specialized cells continuously. Scientists started experimenting with injecting animal tissue to prevent and treat disease in the nineteenth century. Even though such efforts yielded no positive benefit, studies in the mid-twentieth century revealed that human cells could be used to help prevent donor organs from being rejected by the human body, eventually leading to successful bone marrow transplantation, which is now a common practice in treating patients with damaged bone marrow as a result of disease, infection, radiation, or chemotherapy. In recent decades, scientists have become more interested in stem cells and cell transplantation as a potential new therapeutic strategy for various diseases, primarily degenerative and immunogenic pathologies.

Donor transfusion of red blood cells, white blood cells, and platelets is the most common type of cell therapy and blood transfusion. Hematopoietic stem cell transplantation, which produces bone marrow and has been used for over 40 years, is another popular cell therapy.

Somatic cell therapy is the administration of *ex vivo* manipulated or processed autologous (cells from the same person), allogeneic (tissue cells from another person), or xenogeneic (foreign) living cells. *Ex vivo* propagation, expansion, selection, or pharmacologic treatment of cells, as well as other biological changes, are used in the production of somatic cell therapy products. These cellular products could also be used to prevent or diagnose diseases.

In cell treatment, pluripotent cells are employed, which can turn into any type of cell in the body, whereas multipotent cells can only convert into one type of cell. Somatic cell therapy involves the injection of activated lymphoid cells, such as lymphokine-activated killer cells and tumor-infiltrating lymphocytes, as well as the implantation of programmed-to-die cell types, such as hepatocytes, myoblasts, or pancreatic islet cells.

Infusion, injection at various places, or surgical implanting are used to disperse cells in aggregated form or in combination with solid supports or encapsulated components for therapeutic purposes. The matrices, fibers, beads, or other materials employed, as well as the cells, are excipients, external active ingredients, or medicinal tools. The most prevalent type of cell treatment is blood transfusion.

Gene therapy and cell therapy are frequently used in tandem, with stem cells being removed from a patient, genetically modified in tissue culture to produce a new gene, expanded to sufficient numbers, and then returned to the patient to relieve the underlying cause of genetic and acquired diseases by replacing the defective protein(s) or cells that trigger the disease's symptoms and to re-establish the disease's symptoms.

Prosthetics, medications, organ transplantation, and protein replacement, among other therapies, face many of the same ethical issues as genetic and cell therapy. The issues include novel DNA sequences contaminating the human genome and, when using somatic (adult) cells for gene or cell therapy, there is the possibility of unintentional genome intervention. To rule out these possibilities, vectors are tested in lab animals to make sure they do not get into the germline. In clinical trials, human sperm is examined to ensure that the gene has not been incorporated into the genome. Second, there is the question of disease eradication by germline manipulation. Even though modern gene-editing techniques have simplified the procedure, there is still much controversy on whether or not editing is ethical.

The use of embryonic stem cells or human fetal tissue as a stem cell source is still controversial.

3.3.1 Types of cell therapies

There is no uniform classification scheme for cell therapies. However, a classification based on cell types and cell potency can be made:

- *Embryonic stem cells* (ESCs) are cells that are formed during the development of an embryo. Pluripotent embryonic stem cells are one such example. The embryos used to isolate stem cells are typically unused embryos created for IVF-assisted reproduction (IVF). Because ESCs are pluripotent, they can self-renew and form any cell in the body. Because of their pluripotency, ESCs have the advantage of versatility in designing therapeutic methods, which raises ethical concerns. Furthermore, stem cell lines derived from embryos are not genetically matched to the patient, increasing the likelihood of the transplanted cell being rejected by the patient's immune system.
- *Induced pluripotent stem cells* (iPSCs) are a type of pluripotent stem cell. From a divided (somatic) state, a pluripotent adult cell, such as a skin cell, is reprogrammed. These cells benefit from pluripotency without raising the ethical concerns associated with embryonic stem cells. Similarly, iPSCs can be removed from a patient to avoid immune rejection. iPSCs are created by injecting a cocktail of genes into an adult cell, typically via a viral vector. While the process's efficiency has greatly improved, the low rate of reprogramming remains an issue. One issue is that iPSCs are derived from adult cells, making them 'older' than embryonic stem cells, as evidenced by a higher rate of programmed cell death, lower rates of DNA damage repair, and more point mutations.
- *Nuclear transfer embryonic stem cells* (ntESCs). To generate pluripotent cells, the nucleus of an adult cell from a patient is transferred to an oocyte (egg cell) derived from a donor. Using the nucleus transfer technique an egg cell is

transformed into a pluripotent stem cell. The derived cells, including iPSCs, are likely to be recognized by the body because they match the patient's nuclear genome. Because the resulting ntESCs contain both the patient's nuclear DNA and the donor mitochondria, this technique is particularly well suited to diseases with weakened or dysfunctional mitochondria. The production of ntESCs is time-consuming and necessitates the use of a donor oocyte.

- *Parthenogenetic embryonic stem cells* (pESs) are chemically treated pluripotent cells that promote the formation of embryos without the use of sperm (parthenogenesis) and the extraction of ESCs from developing embryos using unfertilized oocytes.
- *Hematopoietic stem cells* (HSCs) are multipotent blood stem cells capable of transforming into any type of blood cell. Adult bone marrow, peripheral blood, and umbilical cord blood all contain HSCs.
- *Multipotent stem cells* (MSCs) are multipotent tissue cells found in the umbilical cord, bone marrow, and fat tissue. MSCs promote adipose marrow tissue formation by forming bone, cartilage, muscle, and adipocytes (fat cells).
- *Nerve stem cells* (NSCs) are stem cells derived from the nervous system. Adult neural stem cells can be found in small numbers in certain areas of the mammalian brain. By recharging neurons and providing encouragement they can be rejuvinated.

3.3.2 CAR-T treatment (cellular immunotherapy)

CAR-T cells are T-cells that have been genetically modified to produce an artificial T-cell receptor for use in immunotherapy (chimeric antigen receptor T-cells). Chimeric liposomes (fluid sacs surrounded by a fatty membrane) can be absorbed by cells more easily than naked DNA/RNA. Liposomes of various types are used to bind preferentially to specific tissues. Small RNA sequences have been delivered to specific tissues by a type of membrane vesicle that cells form and release endogenously (extracellular vesicles or 'exosomes').

CARs (chimeric immunoreceptors, chimeric T-cell receptors, or artificial T-cell receptors) are protein receptors that allow T-cells to interact with a different protein pathway. The receptors are chimeric because they combine antigen-binding and T-cell binding functions into a single receptor.

The goal of CAR-T immunotherapy is to control T-cells so that they can target and kill cancer cells more effectively. CAR-T cells are created using T-cells from the patient's own blood (autologous) or healthy donor T-cells (allogeneic). These T-cells are genetically engineered after being extracted from a human to express a specific CAR that instructs them to target an antigen on the tumor surface. CAR-T cells are programmed to recognize an antigen on a tumor that is not present in healthy cells. CAR-T cells act as a 'living vaccine' against cancer cells after being injected into a patient. When CAR-T cells come into contact with antigen-containing cells, they activate, proliferate, and become cytotoxic. Cells are destroyed by CAR-T cells via a variety of mechanisms, including increased cell death, cytotoxicity (toxicity to other living cells), and widespread induced cell proliferation.

The first step in the synthesis of CAR-T cells is the isolation of T-cells from human blood. The leukocytes are then separated using a blood cell separator in a process known as leukocyte apheresis. Peripheral blood mononuclear cells (PBMCs) are then isolated and stored. The products of leukocyte apheresis are then transported to a cell processing facility. In the cell processing core, specific T-cells are induced to proliferate and multiply in large numbers. The cytokine interleukin 2 (IL-2) and anti-CD3 antibodies are commonly used to promote T-cell expansion.

The expanded T-cells are then processed before being transduced with a retroviral vector containing a gene encoding the programmed CAR, which is typically an integrative gamma retrovirus (RV) or a lentiviral (LV) vector. These vectors are healthy because the U3 region has been partially deleted. Instead of using retroviral vectors, the CRISPR/Cas9 gene-editing tool will insert the CAR gene into specific locations within the genome. Until the modified CAR-T cells are activated, the patient is given lymphodepletion chemotherapy. As the number of circulating leukocytes in the patient decreases, so does the number of cytokines released. It lessens resource competition, allowing the CAR-T cells to develop. T-cells are genetically modified to produce chimeric antigen receptors that target antigens found in tumor cells, after which they are injected into the patient, where they invade and destroy cancer cells.

The CD19 antigen is found in many B-cell cancers, and the FDA has approved the first two CAR-T therapies that target it. Tisagenlecleucel (Kymriah) is a drug that has been approved to treat relapsed/refractory B-cell precursor acute lymphoblastic leukemia (ALL). Axicabtagene ciloleucel (Yescarta/Kite Pharma) on the other hand, is approved to treat relapsed/refractory diffuse B-cell lymphoma (DLBCL).

Two newer technologies are transposon and transposase based methods that use DNA plasmids in conjunction with a process called electroporation to pass genes into T-cells. Cytokines (interleukin 15) are genetically tethered to the T-cell membrane during the CAR T-cell development process to limit genetic alteration to infusion to two days or less (mbIL-15). Multiple antigen receptors may also be used. In the future, cells from a donor graft could be used to create an off-the-shelf, allogeneic T-cell therapy made from donor cells that does not require gene-editing.

CAR-T cells are unquestionably a game changer in cancer treatment. The introduction of CAR-T cells into the body, on the other hand, has serious side effects, including cytokine release syndrome and neurological toxicity. Because CAR-T cell therapy is a newer treatment, there is little data on its long-term effects. In female patients treated with CAR-T cells, the main issues are patient long-term survival and pregnancy complications.

3.3.3 Allogenic cell therapy

Allogeneic and autologous therapy differ in terms of the source of the therapy cells. Allogeneic treatments are created in large batches from various donor tissues (such as bone marrow), whereas autologous medicines are created from a single patient lot. Certain autologous operations are performed on the patient's cells in the clinic or hospital. These treatments are not biological products and are not produced in accordance with good manufacturing practices (cGMP). While both treatments promote cell growth using similar technologies, the size of the treatments differs. Allogeneic therapies are 'off-the-shelf' treatments that are used to treat many patients (sometimes thousands), and there is more time available before delivery for medication quality control. Autologous therapies are considered 'normal' products for each patient, and the patient sample identification chain is critical to ensuring that the patient receives the correct product. Manufacturing allogeneic cells is similar to manufacturing protein drugs and other large-scale cell-derived components. Autologous cells, on the other hand, require scaling up to produce a large number of individual products at the same time.

For cell therapy scale-up manufacturing solutions, batches of up to 2000 L in single-use disposable formats with corresponding downstream process hardware would most likely be required.

The newer allogeneic cell therapies are the modified mesenchymal stem cell (MSC) or fibroblasts, which are being tested for both their capacity for lineage differentiation and their effects on immunomodulation and paracrine signaling. A limited number of approved MSC-based products include Prochymal and stem cells for graft versus host disease (GvHD) diagnosis and Cartistem for osteoarthritis. Therapies derived from a pluripotent stem cell (PSC) are currently under development for retinal pigment epithelium (RPE) for macular degeneration treatment, neural lineage cells for spinal cord injury, and pancreatic beta cells for insulin-dependent diabetes treatment. Until now, PSC-derived cell drugs have been developed on a small scale using manual tissue culture techniques. Preclinical production of cardiomyocytes, neurons, and hepatocytes derived from PSC is also gaining traction, with the potential to supply wide markets.

Cell dosing requirements for these PSC therapies are unknown; however, dosing estimates suggest up to 10^9 cells per patient. With market sizes estimated to be in the tens of thousands of doses for certain treatments, industrial production would likely necessitate batch sizes of 200 to 2000 L.

The manufacturing of cell therapy is distinct for several reasons. Criteria for properly maintaining or controlling cell recognition and viability, the need to recover functional cells rather than soluble components at the end of fermentation, and the inability to render the final drug product sterile are only a few examples.

The most important difference from conventional bioprocessing is the downstream processing of drugs for cell therapy. Although upstream cell therapy product cultivation can be done on equipment designed to generate bacterial or CHO cells with only minor hardware modifications, many of the cell harvests, washes, and formulation requirements are very different. For cell therapeutics, the cells themselves are the most important drug product, and they must be recovered with consistent high viability, efficiency, and potency. In contrast, the culture media product may be retained by protein and virus processing equipment, resulting in cell death, disruption, or pre-recovery. The biological conditions of the culture, on the other hand, can cause additional issues, such as the need to disaggregate normally adherent cells before harvesting. While enzymes are frequently used to dissociate aggregates in culture vessels, most harvesting solutions require that cells be reduced to a single cell suspension before being harvested. This raises issues with enzyme incorporation, mixing, sufficient post-incubation destruction, and enzyme removal or inactivation in a short period.

3.4 Regulatory considerations

As of June 2020, there were approximately 3300 interventional clinical trials involving gene therapy. Because every gene in the human genome can be targeted, the potential for new therapies is enormous. However, no federal law establishes guidelines or limits for human genetic engineering. Concurrent legislation governs the subject at the Department of Health and Human Services, the FDA's Recombinant DNA Advisory Committee, and the National Institutes of Health (NIH). The NIH serves as the primary agent for federally funded gene therapy research. These guidelines are recommended for privately funded projects. The NIH also funds research that advances or improves genetic engineering techniques, as well as evaluates current research ethics and outcomes. A set of recommendations on gene manipulation has been issued by an NIH advisory committee. The guidelines cover laboratory and human test subject safety, as well as various forms of genetic experimentation. A clinical trial of gene therapy, unlike any other type of clinical research, must be approved by the NIH's Recombinant DNA Advisory Committee before it can begin.

The FDA publishes a comprehensive set of guidelines for gene and cell therapy: https://www.fda.gov/vaccines-blood-biologics/biologics-guidances/cellular-gene-therapy-guidances.

The WHO also provides guidance on gene therapy products: https://www.who.int/ethics/topics/human-genome-editing/WHO-Commissioned-Governance-1-paper-March-19.pdf.

Biological products are often complicated, hard-to-describe combinations. Both the manufacturing process and the completed product require quality control. Poor production process management can lead to the introduction of noxious agents or other pollutants, as well as unintended changes in the biological product's characteristics or stability that may not be evident during final product testing. As a result, the procedures and reagents utilized in the manufacturing process must be defined. Quality control should also be applied to cell banks and important intermediates in the production process. Sensitive materials, such as vector-containing supernatants, should be examined for lot-to-lot repeatability in the end product and process. Current general regulations (21 CFR 210, 211, 312, and 600) may apply and serve as a guide.

For phase I exploratory trials of somatic cell and gene therapy products, the evidence should demonstrate sufficient protection and rationale. Early exploratory studies in serious or life-threatening disorders could have fewer data submissions than later stages of product development. Although some proof of reason should be supplied, the data analysis necessary to support the start of phase I trials is largely concerned with safety.

At a later stage of product development, data from additional product testing should be accessible. A quantitative potency assay that represents bioactivity *in vivo* should be devised, and product stability should be examined, to assure product integrity. Licensure requires confirmation of clinical effectiveness in addition to protection.

If a product's formulation is modified throughout development, quantitative biological potency testing and, if necessary, preclinical safety evaluation should be utilized to compare the alternative formulations. Product comparability should be established, or the sponsor should examine whether previous trials should be repeated, if the product used in later phase trials differs significantly from that used in earlier trials and the results of the earlier trials are essential to the final product evaluation.

To make the filing procedure easier, manufacturing information for vectors that will be utilized in numerous INDs can be submitted in a single file. A master file is neither 'approved' nor 'disapproved', just like the product it describes. The master file, on the other hand, includes more information and data to the IND. When the same substance is used on different patient populations, it may cause distinct concerns or signal different risk levels. Nonetheless, the master file can aid in the identification of frequent concerns and the efficient resolution of such issues. Permission to cross-reference a master file can be granted to multiple IND sponsors, minimizing unnecessary submissions while retaining the desired level of confidentiality.

3.4.1 Development and characterization of cell populations for administration

3.4.1.1 Collection of cells

The following details should be included:

- Cell type: Autologous, allogeneic, or xenogeneic cells should be used. The origin of the tissue, as well as any other identifying information, should be provided.
- When choosing a donor, keep the following considerations in mind: Any pertinent features of the donor(s), such as age and gender, should be stated. Allogeneic donors should follow the criteria for blood donors (21 CFR 640.3), the testing and acceptance procedures should be identified, and any variations should be justified, according to the 'Points to Remember in the Collection, Processing, and Testing of Ex-Vivo-Activated Mononuclear Leukocytes for Administration to Humans'. Additional Public Health Service recommendations on organ and tissue donation should be implemented as necessary. HIV-1 and HIV-2, hepatitis B and C viruses, HTLV-1, and other infectious agents should also be considered exclusion criteria. The serological, diagnostic, and clinical history of the donor should all be recorded. In some circumstances, donor follow-up may be necessary, and the methods for gathering donor information and retaining records should be described thoroughly.

'A Proposed Approach to the Control of Cellular and Tissue-based Products,' released on 28 February 1997 contains more instructions on adventitious agent testing and labeling for autologous cells, FR 9721 ((https://www.govinfo.gov/app/details/FR-1997-03-04/97-5240)). Assume that non-human animals are included in your calculations. An overview of the herd's or colony's origins, associated genetic traits, husbandry, and health status should be presented in that scenario (for more details, see 'PHS Guidelines on Infectious Disease Issues in Xenotransplantation', August 1996, 61 FR 49920 and January 1997, 61 FR 49920).

3.4.1.2 Tissue typing

Polymorphisms such as blood type should be typed as needed when using allogeneic donors. The importance of matching donor–recipient histocompatibility antigens (HLA groups I and II, and maybe minor antigens in some circumstances) as well as typing processes and acceptance criteria should be discussed.

Assume or suggest that cell mixes from multiple donors are adequate. In that instance, any cell interactions that could result in immunological responses or other alterations that diminish cell efficiency should be given specific consideration.

Cell mixes with several donors are notoriously difficult to define. As outlined in the 28 February 1997 (62 FR 9721) proposed approach to the regulation of cellular and tissue-based products, multiple-donor cell mixture products will not meet the requirements for regulation as human cellular or tissue-based products under section 361 of the Public Health Service Act (the PHS Act). The Federal Food, Drug, and Cosmetic Act, as well as section 351 of the Public Health Service Act, will refer to such items.

3.4.1.3 Procedures

The procedures for extracting cells should be submitted, along with the facility's location and any equipment or materials used.

3.4.1.3.1 Procedures for cell culture

- *Quality assurance procedures*: Cell culture activities in general should be closely monitored in terms of material quality, production controls, and equipment validation and monitoring.
- Validation of serum additives and growth factors, as well as verification of freedom from adventitious agents, should be included in all media and components for culture coverage. Culture media items should be documented, including their provenance and lot numbers. Sensitizing medium elements should be avoided, such as specific animal sera, certain proteins, and blood group compounds. To assure the reproducibility of cell culture features, identification, purity, and potency measures for growth agents should be created. Penicillin and other beta-lactam antibiotics should be avoided during development due to the potential of serious hypersensitivity reactions in individuals.
- In cell culture, antagonistic substances including cells are treated, multiplied, and submitted to laboratory operations in circumstances that limit contamination with adventitious agents, according to the documentation. Cells should be examined for contamination on a frequent basis throughout long-term culture. Bacteria, yeast, mold, mycoplasma, and accidentally transmitted viruses can all be found in cells.
- *Monitoring cell identity and heterogeneity*: Cell culture identity and heterogeneity control should be ensured through manufacturing and testing

methods. To avoid contamination of one cell culture with another, cell culture procedures and facilities should be organized carefully. Extensive drift in the characteristics of a cell population or overgrowth by a distinct cell type previously present in low numbers can occur during cell culture. Cell identity should be quantitatively examined to detect such changes, such as by monitoring cell surface antigens or biochemical markers. Contamination or substitution by other cells in use in the facility should be detected during the identification procedure. Boundaries for acceptable culture composition must be established. In some circumstances, quantitative functional potency assays can be utilized to phenotype populations. The desired function should be managed while altering cells, and tests should be performed on a regular basis to ensure that the desired phenotype is maintained. In some circumstances, identity testing should involve donor–recipient matching as well as immunological phenotyping.

- If the targeted therapeutic effect is centered on a specific molecular species generated by cells, sufficient structural and biological evidence should be supplied to indicate the existence of an effective and biologically active type.
- Identification of important aspects of the cultured cell population is critical for culture lifespan (phenotypic markers such as cell surface antigens, functional properties, and bioassay activity, as appropriate). The cultural stability of these features has changed over time. This profile should be used to establish the cultural period's limits.

3.4.1.3.2 Cell banking system

For certain somatic cell treatment products generated repeatedly from the same cells, cell banking systems are allowed. Bacterial cells make plasmids, while mammalian cells make recombinant viral vectors when packaging or producer cells make gene therapy vectors. A well-structured cell banking system should handle these cell stocks (often a two-tiered system).

The following is a description of a cell bank device that was used:

- A summary of the origins and evolution of cells is required.
- The cell freezing and recovery methods are among the techniques that must be explained. It is necessary to specify the components that will be used (such as DMSO or glycerol). It is necessary to provide the number of vials preserved in a single lot as well as the storage conditions.
- *Characterization*: Acceptable genotypic and phenotypic markers should be used to establish the identity of the cells, with the percentage of the cell population possessing such markers acting as a measure of purity. Restriction mapping or assay of the protein's bioactivity expressed by an inserted gene should be employed to evaluate vector retention and identity in transduced or vector-producing cells.
- Contaminating biological agents should be checked for, including fungus, viruses other than vectors, mycoplasma, bacteria other than an expected bacterial host strain, and replication-competent viruses such as vectors.
- In the case of MCBs harboring bacteria bearing plasmids of interest, bacteriophage testing is not required. Nonetheless, bacteriophage should be examined because it has the potential to reduce yield and stability.

- Data on how long and under what conditions cells can be frozen and still be usable when thawed should be collected as part of the product development process.
- *Tests on thawed cells*: After thawing and expansion, viability, cell identification, and function tests should be repeated. Until the cells are frozen, equate the yield of viable cells and quantitative functional equivalents to those values. Using aliquots of the frozen cells, sterility should be validated.
- Working cell banks should be identified via selective phenotypic or genotypic testing if they are utilised. Restriction mapping or an assessment of secreted protein activity, as in MCBs, should be used to validate vector retention and identification. They should be free of bacterial and viral infections as well.
- Only one extended culture of end-of-production cells for producer cells should be performed to assess if the growing conditions induce additional contaminants or if vector integrity is compromised. Sponsors should propose a research plan that is as precise and responsive as possible.
- For cell therapies created particularly for each patient, such as autologous cells to treat specific patients, cell banking procedures may not be possible. On the other hand, checking the final cellular product for key properties should be considered.

3.4.1.3.3 Materials used in the production process

Antibodies, cytokines, serum, protein A, toxins, antibiotics, other chemicals, or solid supports like beads used during *in vitro* manipulation procedures can all have an impact on the final therapeutic product's protection, purity, and potency. These components must be defined, and a certification program with specific requirements for each component must be developed to determine whether it is suitable for use in the manufacturing process. When using reagent-grade material, the certification program should include, as needed, testing for the component's protection, purity, and potency.

Limits should be set for the concentrations of all manufacturing components that may remain in the finished product. The methods used to remove them, as well as quantitative testing (including a description of methods and sensitivity) to demonstrate the effectiveness of their removal, should be provided. When cells are administered via binding or uptake, some additional components may be present in measurable amounts. In such cases, the toxicity of these components should be evaluated in animals or other appropriate systems.

3.4.2 Characterization and release testing of cellular gene therapy products

These specifications apply to cellular products, including *ex vivo* transduced cells used in gene therapy.

Quality control testing should be performed on the final biological product to be administered, as well as the manufacturing process and materials used. The specifications to be applied to the final product and other production elements, as well as the range of acceptable values for each, should be specified.

A biological product lot is an amount of material thoroughly combined in a single vessel. This approach can be utilized for somatic cell and gene therapy for the sake of planning lot testing processes. This means that any cell population, vector preparation, or other product for such therapies prepared as a unique final mixture should be subjected to appropriate lot release testing. Individual-recipient preparations differ from large-batch preparations, and the appropriate lot release criteria for each protocol's practical limits should be chosen. The repeatability of methods is measured by lot-to-lot variation.

3.4.2.1 Cellular identification

Quantitative testing employing phenotypic and chemical assays should be utilized to confirm cell identity and measure heterogeneity (21 CFR 610.14).

3.4.2.1.1 Potency

The relevant function of the cells and the relevant biosynthetic products biosynthesized by the cells should be characterized and quantified as a measure of potency if they are known (21 CFR 610.10).

3.4.2.1.2 Viability

The viability of the cells should be determined, and a safe lower limit should be set.

3.4.2.1.3 Adversary agents testing

Bacteria, fungi, mycoplasma, and viruses should not be found in the cells, according to the tests. In circumstances where the final cell therapy product is too short-lived to undergo adequate sensitivity testing before delivery to patients, the FDA will soon authorize validation of a mycoplasma-free manufacturing process.

Validation of endotoxin testing by LAL or other appropriate assays, as well as purity (21 CFR 610.13), should be established. Endotoxin testing procedures should be examined on a case-by-case basis for suitability and appropriateness. To demonstrate that cell preparation does not interfere with endotoxin detection, the test should be validated.

3.4.2.2 General safety inspection

The completed product must pass a general safety inspection (21 CFR 610.11). When appropriate, modified procedures in conformity with 21 CFR 610.9 may be devised.

3.4.2.2.1 Banks of frozen cells

Lot release testing on thawed cells is required when cell populations frozen for later distribution are thawed, grown, and then supplied to patients.

3.4.3 Additional possible applications: radioisotopes or toxins added to cell preparations

Cells modified by radiolabeling or pre-loading with bioactive compounds such as toxins could be used for therapeutic or diagnostic purposes. As a result, the cell implant can be utilized to distribute its own products, functions, and other items. As a result of cell implantation and the localisation of the radionuclide or toxin, as well as the metabolic features of the cells, new safety concerns may arise. These should be foreseen and addressed as soon as possible. Previously, the use of radiolabeled or toxin-conjugated antibodies was fraught with similar issues.

3.4.4 Vector production, characterization, and release testing for gene therapy

Obtaining the information required below may be problematic in some systems. Alternative methodologies and data may be submitted to CBER officials for assessment. Depending on the nature of the proposed clinical trial, such as the route and frequency of administration, as well as the intended patient group, different sorts of information are required to assure appropriate safety.

3.4.5 Construction and characterization of vectors

Source materials for vectors should be characterized and documented thoroughly. On viral vectors or plasmids derived from cloned and described constructions, confirmatory identity tests should be performed. Descriptions of any vectors, helper viruses, and producer cell lines utilized to prepare the final construct should be included in the vector derivation information. Within the construct, known regulatory components such as promoters and enhancers should be identified.

Early in the product development process, vector characterisation consisting of sequence data from relevant regions of vectors and an indication mapping complemented by protein characterization is acceptable. For subsequent stages of product development and licensure, more specific sequencing information should be provided. Sequencing the genetic insert plus flanking regions, as well as any major alterations to the vector backbone or places known to be prone to alteration during molecular manipulations, may suffice when sequencing the complete vector is not possible due to the size of the construct. Vector sequences that modify vector–host interactions should be specified if they are known, and the stability of the host cell/vector system should be considered.

3.4.6 Vector manufacturing system

The host cell, final gene construct, or vector intermediate used to manufacture the vector make up the vector production system (for example, retroviral producer cell). The selection and characterization of the recombinant host cell clone, including vector copy number and the physical condition of the final vector, the construct inside the host cell (i.e. integrated, or extra-chromosomal), should be described in detail. It is necessary to offer a detailed description of the techniques for propagating and increasing the recombinant host cell clone, establishing seed stock, and certifying seed stock.

3.4.7 Viral master banks

A master viral bank (MVB) should be generated and defined when a virus, with or without a therapeutic gene, is utilized as a seed in the development of a therapeutic vector. Adenovirus, adeno-associated virus, herpes virus, poxviruses, and other lytic and non-lytic viruses would all be included as vectors. The sponsor should specify the source materials (plasmids, vectors, oligomers, etc) and molecular processes employed to construct the source or seed vector. The genetic integrity and stability (i.e. identity) of the seed vector should be validated, as well as the vector seed's bioactivity. The gene's expression should be assessed in the absence of bioactivity data.

Avirulent agents such as viruses, bacteria, fungi, and mycoplasma should be absent from master seed stocks. MVBs must be proved to be devoid of replication-competent viruses that may occur as a result of contamination or recombination during MVB creation in the case of replication-defective or replication-selective vectors. The vector and the viability of tests in the presence of the vector virus will determine the testing for other unsuitable viruses.

3.4.8 Lot-to-lot distribution vector testing and specifications

The following are some general testing guidelines. Not all of the tests provided will apply to every vector class. Sponsors should choose acceptable testing methodologies and ask CBER if they have any queries about a test's applicability. Assume that a drug material (defined as a bulk product that is not necessarily in its final formulation) and a drug product (defined as a finished product) are the same thing. Only one set of tests is required in this situation.

Any of the standard assays described below can evaluate quantitative specificity and sensitivity with sufficient precision. Before submitting data confirming assay performance to the IND, assay procedures should be confirmed by testing known amounts of reference lots, spiked samples, and other relevant measures.

Testing for drug material (bulk product, not necessarily final formulation):

- The absence of impurity (21 CFR 610.13).
- Test for total DNA or RNA content if the vector composition, such as A260/A280, needs it.
- Using agarose gel electrophoresis, for example, check for size and structure homogeneity, supercoiled versus linear.
- Gel electrophoresis, for example, can be used to check for contamination with RNA or host DNA, as well as a test with the bacterial host-specific probe.
- Proteins should be tested for if they are present as a contaminant, such as in a silver-stained gel.
- In cases where a non-infectious virus, such as empty capsids, would be a contaminant, test for it.
- Toxic materials in the manufacturing process are tested.
- Individuality (21 CFR 610.14).

The drug substance should be tested for vector identity using methods such as restriction enzyme mapping with multiple enzymes or PCR. If a facility produces multiple constructs, identity testing should be performed to ensure that the constructs can be distinguished and cross-contaminated.

3.4.8.1 Agents of chance

Sponsors are urged to collect data validating testing techniques other than those stated as adventitious agent testing methods become more sensitive and specific over time, allowing for future policy modifications. Some information can be obtained in cases where a vector product interferes with appropriate assays, such as a lytic viral vector that kills indicator cells in an assay for an adventitious virus, parallel mock cultures using the same media and other reagents to allow contaminant outgrowth, or assays in the presence of neutralizing antibody.

3.4.8.2 Aerobic and anaerobic bacteria and fungi sterility test (21 CFR 610.12).

Mycoplasma testing should be carried out as described in attachment #2 of the FDA memorandum 'Points to Consider in the Characterization of Cell Lines Used to Produce Biologicals (1993),' which outlines the protocols for detecting mycoplasma contamination.

In some circumstances, testing for adventitious viruses is required because vector source materials or cell lines may be contaminated with adventitious viruses. The adventitious virus may also be introduced during the production process in some situations. It is recommended that you test for a wide spectrum of possibly infecting viruses.

Potency assays should be validated during the product development process. To determine expression of the inserted gene, transfection of appropriate cells and demonstration of an active gene product using an appropriate assay with high sensitivity and specificity can be used.

Whenever possible, a potency assay should assess the biological activity of the produced gene product rather than just its presence. An enzyme activity assay detecting substrate conversion to product, for example, might be preferable to an immunological assay detecting epitopes on the enzyme if the suggested therapy is based on enzymatic activity. A qualitative potency test should be performed if a quantitative potency assay is not available.

The drug's product testing (product in its final formulation) is tested for the qualities described below and should be checked using quantitative, validated assays on the vector product in its final container form. On a drug product (final product), endotoxin and general safety tests are not required, but they are required on a drug substance (bulk product).

Limulus amebocyte lysate or another approved assay should be used to validate endotoxin testing. Because therapeutic DNA plasmid products are among the specified biotechnology products, according to 21 CFR 610.11, general safety is not necessary, even if liposomes are incorporated.

3.4.9 Issues in gene therapy related to specific vector classes

3.4.9.1 Additional factors to consider when using plasmid vector products

Many of the above-mentioned products and quality-control requirements apply to plasmid DNA products. The entire plasmid should be sequenced in general.

Plasmids should be tested for RNA, protein, and bacterial host DNA toxins, as well as quantities of linear and supercoiled DNA, throughout manufacturing and in the presence of toxic substances. Toxic compounds such as ethidium bromide should be avoided during the manufacturing process.

Because of the potential of severe hypersensitivity responses in patients, penicillin and other beta-lactam antibiotics should be avoided during development. To prevent antibiotic resistance traits from propagating to environmental microorganisms, selection markers that influence antibiotic resistance in widespread clinical usage should be avoided if antibiotic selection is used throughout development. Where appropriate, the residual antibiotic in the finished product, as well as the possibility for allergy, should be taken into account.

3.5 Conclusion

Biotechnology has evolved quickly over the past few decades, allowing the emergence of techniques that revolutionized DNA manipulation to treat and diagnose disease. Cell therapy and gene therapy have transformed the future of medicine. Biopharmaceuticals overlap the cell and gene therapy fields where recombinant manipulation is involved. Given the complexity and risks of gene manipulation, the regulatory agencies have provided extensive guidance and requirements for developing and manufacturing them. This chapter provides details of cell and gene therapies pertinent to development, manufacturing, and regulatory compliance.

Sarfaraz Niazi and Sunitha Lokesh

Chapter 4

Formulation of biopharmaceuticals

The delivery of biopharmaceutical products is a complex exercise due to their chemical structural instability, rapid physical degradation, and complex chemical interactions with excipients. In addition to being inherently immunogenic, their degradation can enhance immunogenicity and raise other safety issues. This chapter introduces the structural elements of biopharmaceuticals that lead to chemical interactions and degradation and suggests methods to stabilize the formulated products. Commercial formulations are also presented. While the current methodology for the delivery of biopharmaceuticals remains parenteral, many other delivery forms are emerging, including oral, nasal, transdermal, pulmonary, and ocular products in polymer-based, hydrogel, lipid-based emulsion, liposome, and nanoparticle systems, which are also discussed with their future perspective.

4.1 Overview

The development requirements for biopharmaceuticals are different from the small chemical molecule drugs because of their larger molecular size and the variability of molecular structure and how these properties impact the body's immune systems.

First, because most protein drugs are administered via parenteral routes, most of the science of protein drug formulation is concerned with the art of injectable formulations. The choice of delivery route is restricted because of (i) instability of the protein structures in many sites of the administration environment, such as the acidity in the gastro-intestinal tract, (ii) the large molecular size and high hydrophilicity preventing absorption across some biological membranes, leading to faster elimination, and (iii) high dose-response sensitivity that does not allow significant variation in the bioavailability.

Second, proteins in general have many structural features, such as (i) the functional groups methionine, cysteine, histidine, tryptophan, and tyrosine, all of which are susceptible to oxidation and require several common approaches to stabilize, and (ii) conformational changes and aggregation, which are properties unique to large molecules and necessitate the inclusion of product-specific formulation components.

doi:10.1088/978-0-7503-3175-3ch4

Third, all proteins are temperature, light, and agitation sensitive during storage, shipping, and handling. In some cases, the label warns against shaking the product. With so many variables that can affect the quality and efficacy, the formulation challenges are heightened to address many aspects not generally considered in small-molecule drugs.

Biopharmaceuticals' formulations vary largely depending on the delivery route, which is mostly parenteral, but new advances are extending the routes to many other noninvasive routes. The choice of administration route is made based on the practicality and probability factors, requiring a detailed understanding of the interaction of biopharmaceuticals with the environment of the route of administration. Over the past two decades, many computation tools have become available that allow the fast creation of decision matrices to optimize the formulations.

Because of the anatomical and physiological characteristics of the site of administration, each route of administration faces distinct biological barriers. To overcome these barriers, many formulation approaches have been developed, including advances in information technology, biotechnology, and nanotechnology combined with sophisticated medical devices, electric or magnetic forces, or sonic waves to maximize the noninvasive drug delivery system effectiveness.

A formulation is intended to deliver a biopharmaceutical active to the site of administration, from where it crosses biological barriers, entering the bloodstream, and finally, the site of action that may or may not be known. Given the high likelihood of the degradation of products during the shelf-life, a large volume of proprietary technology has been developed, including thousands of patents, to enable a dosage form to deliver the drug to the site of action.

An appendix to this chapter details the physicochemical properties proteins and peptides, an important source of information in designing formulations.

Since biopharmaceuticals' development is costly, scientists inevitably become engaged with evaluating the intellectual property relating to biopharmaceutical manufacturing and delivery systems. Chapter 8 provides details about intellectual property and its management and forms a required reading for the scientists engaged in the formulation of biopharmaceuticals.

Biopharmaceuticals' stability is managed by either chemically modifying the molecule or selecting proper excipients. These combinations will assure stability without affecting the safety or efficacy of the product. Structure modifications include creating protein scaffolds and PEGylating the molecules and other technologies elaborated in detail in chapter 1.

4.2 Structural instability

4.2.1 Basis

A good understanding of protein structure and the common structural risks in formulating these products is needed to appreciate the complexities of formulating biopharmaceutical products. Proteins are made up of a chain of amino acids with reactant groups that allow them to link together to form multidimensional structures; their chemical and physical properties are based on this reactivity.

Table 4.1. Impact of formulation and environmental factors on the degradation of proteins.

Factor	Impact
Buffer species	Deamidation
Light	Photo decomposition
Metal ions	Hydrolysis, oxidation
Other excipients	Maillard reaction
Oxygen	Oxidation
pH	Hydrolysis, deamidation
Temperature	Most routes

Protein instability is caused by both chemical and physical reactions. Chemical instability is the formation or dissolution of covalent bonds within a polypeptide or protein structure, resulting in oxidation, deamidation, reduction, and hydrolysis. Physical or conformational instabilities, dissociation, denaturation, accumulation, and precipitation are all terms used to describe how things unfold. When a chemical event, such as oxidation, causes a physical reaction, such as aggregation, the protein degradation pathways become synergistic. Physical changes in small-molecule drugs, except for the PK profile, may not pose a significant clinical risk in general. However, in the case of biopharmaceuticals, these are critical for determining the safety of these drugs.

Table 4.1 lists several formulation variables that can trigger uncertainty. Based on knowledge of the chemistry of chemical and biopharmaceutical molecules, these factors are well confirmed. Biopharmaceuticals, on the other hand, have a unique effect on protection and efficacy.

Although it was difficult in the past to understand the complex degradation mechanisms that cause both chemical and physical instability and to evaluate approaches to stabilize biopharmaceutical formulations, analytical chemistry has advanced significantly. Appendix A of this chapter (Physicochemical properties of proteins and peptides approved by FDA) is a compilation of the critical properties of products approved by the FDA as of March 2020. This extensive database will allow formulators to take a focused scientific approach to develop formulations by comparing them with similar molecules in terms of their properties to select stable formulations.

Biopharmaceuticals are usually less resilient to minor variations in solution chemistry than small-molecule drugs. They are only stable in terms of composition and conformation within a small pH and osmolarity range. To remain in solution throughout the product's shelf-life, several molecules need supportive formulation components. Unlike small-molecule drugs, which are extremely stable in most lyophilized formulations, even lyophilized protein products are subject to substantial degradation.

Table 4.2 summarizes common stability issues encountered during protein formulation development, as well as potential solutions. Given the unique complexity of each molecule, this list should only be used as a guideline.

Table 4.2. Stability issues in protein biopharmaceuticals.

Problems	Potential causes	Possible solutions
Cleavages	Protease impurity, other unknown mechanisms	pH, product purity, inhibitors
Covalent aggregation	Disulfide scrambling, other unknown mechanisms	pH, inhibit non-covalent aggregation
Cyclic imide	pH around 5	pH optimization
Deamidation	pH < 5.0 or pH > 6.0	pH optimization
Non-covalent aggregation	Solubility, structural changes, heat, shear, surface, denaturants, impurities	pH, ionic additives, amino acids, surfactants, protein concentration, raw material purity
Oxidation	Active oxygen species, free radicals, metals, light, impurity	Excipient purity, a free-radical scavenger, active oxygen scavengers, methionine
Surface denaturation, adsorption	Low protein concentration, specific affinity, protein hydrophobicity	Surfactants, protein concentration, pH

4.2.2 Physical degradation

Protein degradation is caused by hydrophobic surfaces, boiling, lyophilization, reconstitution, interaction with organic solvents, shaking, and other physical and chemical factors. Physical stress can cause denaturation, adsorption, accumulation, or deposition on the container walls.

4.2.2.1 Structural changes

Biological macromolecules, such as antibodies, have a three-dimensional tertiary structure that defines the folded structure, resulting in a complex mix of intramolecular and intermolecular interactions between functional amino acid groups and external environments. Non-covalent interactions such as electrostatic interactions, van der Waals backbone interactions, side-chain residue interactions, hydrogen bonding, and hydrophobic interactions play an important role in maintaining the folded native structure. Because the folded structure is in dynamic equilibrium, any sources that disrupt the interaction balance will cause the structure to change, resulting in the large molecules being unstable. In an aqueous solution, for example, more soluble amino acid residues are exposed and interact with solvent molecules, whereas non-polar residues are encapsulated, resulting in a hydrophobic center.

Protein folding is determined by the amino acid sequence, which results in the biologically active form of the protein. A protein, on the other hand, unfolds either in an intermediate state or directly to a denatured state in its native structure. When the intermediate or unfolding species is formed, the variants are more likely to assemble more stable complexes such as aggregation due to their higher free energy. Aggregates are composed of multiple monomers in various forms that are held together by covalent or non-covalent bonds. Dilution, for example, can dissociate native monomer cluster aggregates due to the form of monomer association. Precipitation or irreversible aggregates can result from the nucleation of different monomers. Aggregation is caused by a variety of stressors, the most common of which are temperature, mechanical, and freeze/thaw stress in manufacturing. Aggregation can be caused by a pH change, and high temperatures can cause conformational destabilization or partial/complete unfolding. IgG4 forms more soluble aggregates than IgG1 at lower pH and higher temperatures, for example, due to lower conformational stability caused by lower unfolding temperature and changes in tertiary structures.

Because protein tertiary structures are vulnerable to environmental physical stress, structural changes in mAbs can occur at any point during the manufacturing process, from protein expression to processing and storage. Structural transformation is attributed to non-physiological conditions during processes that can drive the adaptation of structural variants in the finished product. Stresses such as buffer selection, fabrication techniques, and container selection, for example, may exacerbate this problem.

4.2.2.2 Aggregation

Aggregated proteins are a significant source of concern for biopharmaceutical products because they can result in decreased bioactivity and increased immunogenicity. Complexes of macromolecular proteins can activate the immune system, causing it to recognize a protein as 'non-self' and mount an antigenic response.

The buffers used in the formulation can have a significant impact on the product's stability. The use of the acetate buffer, for example, causes precipitation in IgG3 formulations that is not observed when arginine and histidine are used. As a result, adjusting the concentration of buffer salts may help prevent phase changes. The pH of lyophilizate is also important for preventing the aggregation of recombinant vaccine antigens. Additional stabilizers, such as trehalose, are required to keep the antigen intact.

Non-native proteins must also be avoided in any finished mAb-based biopharmaceutical products. If protein solutions are drawn from vials, aggregation can result in non-uniform dosing. Aggregating IFN–, for example, promotes the development of neutralizing anti-drug Abs (NAb), which prevent the IFN receptor from binding and thus reduces clinical efficacy. Induced NAbs can interfere with the normal function of endogenous proteins, particularly hormones and cytokines. NAb blocking endogenous erythropoietin mediated against Eprex® resulted in severe anemia in patients receiving Eprex®, recombinant human erythropoietin, and pure-red cell aplasia (PRCA). The NAb was caused by aggregation, which was commonly aided by prefilled syringes of formulating ingredients such as polysorbate 80 and silicone oil. As a result, any advancement in the formulation or manufacturing process of biopharmaceuticals necessitates additional research in accordance with the ICH Q5E guidelines (https://database.ich.org/sites/default/files/Q5E%20Guideline.pdf). As previously stated, the creator of Eprex® did not conduct the necessary studies and believed that minor changes in the formulation would not affect the protection or efficacy of erythropoietin.

Aggregation is a common problem in protein production and storage. When a protein is exposed to liquid–air, liquid–solid, or even liquid–liquid interfaces, its ability to aggregate is enhanced. Mechanical agitation stresses cause protein aggregation (shaking, stirring, pipetting, or pumping through tubes). It will also develop as a result of freezing and thawing. Temperature, protein concentration, pH, and ionic force are all solvent conditions that can influence the strength and volume of aggregates observed.

Some of the mechanisms that cause protein aggregates to form are domain swapping (ds), strand association (sa), edge–edge-association (ee), and beta-strand stacking. Protein multimers such as dimers, trimers, and tetramers, as well as large polymers, are commonly referred to as aggregators. It is also possible to form non-covalent or covalent aggregates (disulfide-linked). These are completely soluble in a clear solution, partially soluble in a turbid solution, and mostly insoluble as sediment at the bottom of the jar. Non-specific protein-to-protein interactions are triggered by interactions between solvent-exposed hydrophobic groups that form aggregates. The covalent aggregation process cannot be reversed. It is possible for the weakly related non-covalent aggregate to be reversible. It usually proceeds in the same direction as dimers to become multimers; the strongly associated non-covalent aggregates are not reversible by dilution and can precipitate.

Aggregates are classified into three types based on their size range: (i) submicron particles (1 μm size), also known as soluble particles; (ii) sub-visible particles (1–100 m size); and (iii) visible particles (> 100 m size).

Aggregation models come in many different shapes and sizes. Hydrophobic interactions cause denatured or unfolded molecules to aggregate in the 'native to unfolded to aggregate' model. Hydrophobic effects occur when normally submerged hydrophobic regions are released, causing the aggregate to form. Because unfolding increases with temperature and reactions usually follow first-order kinetics, the reaction rate increases in this model.

In the 'native to intermediate to unfolded' model, the intermediate yields the aggregate. The misfolded intermediates are thermodynamically stable and are part of the native state ensemble. As a result, protein aggregation is not a 'abnormal' state, and it can occur even in conditions that favor the native state.

The aggregation of proteins is divided into two stages. The nuclei expand to a critical mass after the nucleation process. Although turbidity measurements are used to track the extent of aggregation, they are not always a reliable predictor of aggregation. When native and folded proteins interact, irreversible non-native structures with high levels of non-native intermolecular-sheet structures can form. The aggregate's onset, rate, and final morphology are influenced by solution conditions such as pH, salt species, salt concentration, co-solutes, excipients, and surfactants. The relative intrinsic thermodynamic stability of the native state determines the exact existence of an aggregate.

Biopharmaceuticals can aggregate at nearly any stage of the process, including hold points, shipping, and long-term storage, due to the various physical and chemical manipulations required in upstream production and downstream processing, followed by formulation and filling operations. Protein aggregation can be aided by agitation of protein solutions (such as shaking, stirring, or shearing) at air–liquid interfaces, where protein molecules can join and unfold, exposing their hydrophobic regions to the load-based interaction.

Agitation-induced aggregation has been observed in proteins such as recombinant factor XIII, human growth hormone, hemoglobin, erythropoietin, and insulin. To avoid significant loss of protein activity or the formation of visible particulate matter, it is critical to minimize foaming caused by agitation during manufacturing (and during product use). Protein aggregation is caused by antimicrobial preservatives used in multidose formulations. Benzyl alcohol accelerates filgrastim aggregation by producing more partially unfolded protein conformations. Increasing the amount of antimicrobial preservatives in a formulation causes it to become more hydrophobic, reducing its aqueous solubility. Proteins can be significantly destabilized by phenol and *m*-Cresol. *m*-Cresol can precipitate protein, and phenol promotes the formation of both soluble and insoluble aggregates.

Freezing and thawing can occur several times during growth, reducing the shelf-life of the product and having a significant impact on protein aggregation. When water-ice crystals form at the container's rim, the effect is known as 'salting out' (where heat transfer is greatest). Protein and excipients become increasingly concentrated in the eventually freezing center of a container.

Precipitation and accumulation will occur during freezing due to high salt and protein concentrations, which will not be completely reversed upon thawing. Thyroid-stimulating hormone, for example, retains its potency for up to 90 days when stored at −80 °C, 4 °C, or 24 °C, but loses more than 40% of its potency when frozen at −20 °C due to subunit dissociation.

Repeated freezing and thawing cycles amplify the effect, having a cumulative effect on the generation and production of sub-visible and visible particulates. The crystallization of buffer components during freezing may cause a pH change. When frozen, potassium phosphate buffers have a much smaller pH transition than sodium phosphate buffers. After oxidation, the pH-dependent reaction can result in a faster reaction and the appearance of precipitates within minutes.

The causes of aggregation related to the manufacturing process are listed in table 4.3.

Table 4.3. Process-related causes of protein aggregation.

Process	Factor
Administration	Diluents, component materials and surfaces, leachable
Fermentation/expression	Inclusion bodies
Fill/finish	Surface interaction, shear, contamination (e.g. silicone oil)
Filtration	Surface interaction, shear
Freeze/thaw	Cryo-concentration, pH changes, ice–solution interfaces
Lyophilization	Cryo-concentration, pH changes, ice–solution interfaces, dehydration
Purification	Shear, pH, ionic strength
Shipping	Agitation, temperature cycling

Notice that the liquid is frozen in cryo-concentration, and the water is removed as ice as it thaws.

Because proteins are dispensed in prefilled syringes with silicone oil lubrication, silicone oil lubrication is one of the most difficult aspects of the formulating process. A significant variation in absorbance at 350 nm in proteins packaged in different syringe brands is not uncommon, depending on the amount and type of silicone used. A properly folded mAb will form an intermediate upon hydrophobic contact with silicone in the syringe, resulting in partial unfolding that yields first a soluble aggregate, then visible precipitation, depending on the concentration of silicone to which the mAb is exposed. When sub-visible and visible particles are combined, they are referred to as insoluble particles. Small sub-visible particles are particles with a diameter of 0.1–1 μm.

Because protein glycation occurs when sucrose is hydrolyzed, a sucrose-based formulation may cause aggregation over time. The presence of such ligands, such as ions, may aid in aggregate formation. Interactions with metal surfaces can result in epitaxic denaturation (the formation of an aggregate by growing a crystal layer of one mineral on the crystal base of another mineral with the same crystalline orientation as the substrate). Foreign particles from the atmosphere, manufacturing processes, or container-closure systems can also cause aggregation (e.g. silicone oil). Furthermore, handling protein products at compounding pharmacies results in multiple-fold aggregation above the initial stage.

A thorough examination of the optimum pH and osmotic conditions is required for the development of a formulation that prevents protein aggregation or precipitation. Surfactants, polyols, or sugars can be added to prevent irreversible denaturation aggregation.

Nonionic detergents (surfactants) are used to improve the stability of a product and prevent aggregation. Because the interaction between proteins and surfactants is hydrophobic, these compounds stabilize proteins by lowering the solution surface tension and binding to hydrophobic sites on their surfaces, reducing the potential for protein–protein interactions that can lead to aggregate formation. Tween 20 and Tween 80, nonionic detergents, can prevent the formation of soluble protein aggregates at surfactant concentrations below the critical micelle concentration (CMC); chelating agents can prevent the aggregation of metal-induced proteins.

4.2.3 Chemical degradation

The formation or destruction of covalent bonds within a polypeptide or protein structure is referred to as chemical instability. Oxidation, deamidation, reduction, and hydrolysis are the most common chemical modifications of proteins. Although a protein molecule has thousands of active functional groups, the identity of these groups is limited since they are made up of a fixed number of amino acids. The hot spots for protein degradation are shown in table 4.4.

Figure 4.1 depicts the most common chemical degradation routes for antibodies used to make most biopharmaceuticals.

Diketopiperazine is an amino acid degradation intermediate found in dipeptide esters and amides. A nucleophilic attack of N-terminal nitrogen on the amide carbonyl occurs between the second and third residues. As a result, peptides with

Table 4.4. Protein sequence hot spots for degradation.

Pathway	Site
Beta-elimination	Ser, Thr
Deamidation	Asn, Gln, -Asn-X-, -Asn-Gly-, Asn-His-, -Asn-Ser-, -Asn-Ala-, -Asn-Asp-, -Asn-Thr-
Diketopiperazine	X-X-Gly, X-Pro-
Glycation	Lys
Hydrolysis	Asp, -Asp-Pro-, -Asp-Gln-, -Pro-Asp, -Asp-Tyr-, -X-Ser-, -X-Thr-
Isomerization	Asn, Asp, -Asn-Gly-, -Asp-Ser-, -Asn-Ser-, -Asp-Ser-
Oxidation	Met, Cys, His, Trp, Tyr
Pyroglutamic acid	H2N-Gln-, H2N-Gln-Gly-, H2N-Glu-

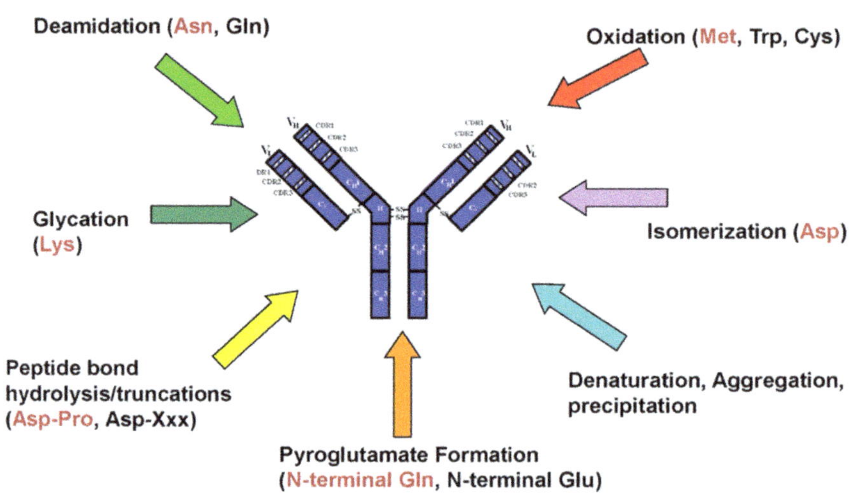

Figure 4.1. Common routes of protein degradation.

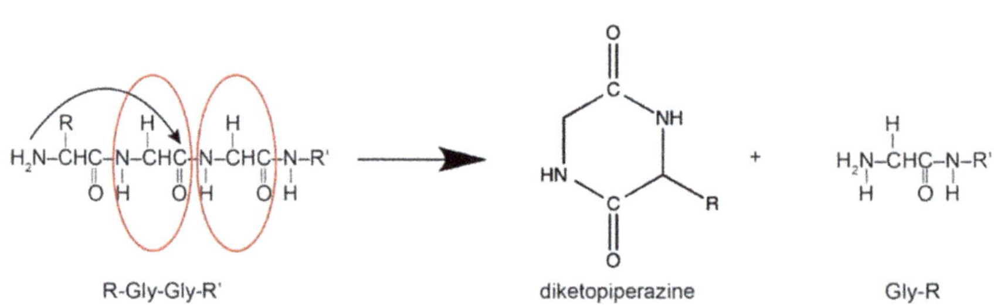

Figure 4.2. Diketopiperazine formation in the degradation of recombinant growth hormone.

Gly as the third or Pro as the second residue from the N-terminus are particularly vulnerable; this reaction occurs best in a neutral or alkaline medium. The diketopiperazine from X-Pro (X = Gly, Ala, Val, Phe, beta-cyclohexylamine, and Arg) is influenced by the cis–trans equilibrium of Pro and the charge distribution

Table 4.5. The effect of chemical modifications of proteins on acidity function and charge heterogeneity (see also appendix A).

Chemical modification	Acidity/charge heterogeneity
Deamidation	More acidic ($z = -1$)
Succinimide formation	More basic or neutral
Glycation	More acidic
Pyroglutamate formation	More acidic ($z = -1$)
Peptide bond hydrolysis	Either acidic or basic

around the peptide bond. The cis type facilitates ring closure. Figure 4.2 shows an example of diketopiperazine formulation. Changing the amino acid preceding the proline residue has a significant impact on the rate of DKP formation at pH 7.0.

Another chemical alteration intermediate is pyrogalacmic acid. Because the terminal Gln will spontaneously cyclize to form PyroE, Gln-Glyn responds much faster than Gln-Glyn. The N-terminal Glu cyclization results in a −17 Da mass change and a −1 or loss of critical residue; the N-terminal Glu cyclization results in no charge change but an 18 Da mass shift.

Although the precise effects of various chemical reactions are difficult to predict, their overall influence on acidity is consistent, as shown in table 4.5.

4.2.3.1 Deamidation

Deamidation, a chemical reaction in which an amino acid's amide functional group is extracted, is the most common chemical degradation in biopharmaceutical products based on mAbs, with isomerization, racemization, and truncation of proteins as possible outcomes.

The rate of deamidation of a protein is greatly influenced by side-chain branching in the $(n + 1)$th residue. The charge of the residue is unimportant except in His, Ser, and Thr. The $(n - 1)$th residue, on the other hand, has only a minor effect on the deamination rate. The half-life of deamidation can range from one day to over a thousand days, depending on the identity of the carboxyl side residue. Protein secondary structure influences the rate of deamidation. When the helical structure of certain proteins changes, the rate of deamidation slows. Deamidation and the resulting structural findings can have a variety of effects on the physicochemical and functional stability of proteins.

Steric hindrance may also influence the rate of deamidation. Bulky asparagine residues can prevent the formation of the intermediate succinimide during the deamidation reaction. By replacing a glycine residue with more voluminous leucine or proline residues, the rate is reduced by 30 to 50 times. Deamidation rates are typically lower in lyophilized formulations due to the limited supply of free water in which the reaction can occur.

The key features of the deamidation reaction include:

- Non-enzymatic reaction that takes place at Asn and Gln.
- Water is needed for the hydrolysis reaction.

- pH affects the pathways.
- Under slightly acidic to essential conditions, the cyclic imide pathway is activated.
- Under acidic conditions, direct hydrolysis occurs.
- In the pH range of 3–4, it is the minimum rate.
- In the pH range of 7–12, the buffer has a catalytic effect.
- For Asn, the rate is 5–10 times higher than for Gln.
- Both normal and beta-peptide bonds are formed.
- The N + 1 residue affects the rate.
- isoAsp-Y to Asp-Y interconvertibility.

Table 4.6 lists some examples of the effect of deamidation, demonstrating the heterogeneity of the effects observed.

4.2.3.1.1 pH

The pH, buffer, and ionic strength all have a significant impact on deamidation. Using formulations with a pH of 3–5 reduces peptide deamidation. Insulin

Table 4.6. Examples of the effects of deamidation of various biopharmaceutical molecules.

Biological molecules	Effect of deamidation
Growth hormone releasing factor analogs	Methanol raises the degree of helicity, thus decreasing the rate of deamidation.
RNAase	The relatively rigid backbone in the loop, which is stabilized by a disulfide between Cys-8 and Cys-12 and by the – turn at residues 66–68, helps prevent the cyclic formation of imide. However, when the enzyme is reduced and denatured before being refolded, aspartic and isoaspartic forms are produced, demonstrating different enzymatic activities. As Asp-67 is replaced with Iso-Asp-67, the isoaspartic form refolds at half the rate of the completely amidated form.
Human growth hormone	Alters proteolytic cleavage decreasing biological activity.
IFN-beta	Increase in biological activity.
Peptide growth hormone releasing factor	As compared to the native peptide, converting to aspartyl and iso-aspartyl types decreases bioactivity by 25 and 500 times, respectively.
Hemoglobin	Hemoglobin's propensity for oxygen improves when an Asn-Gly site is deamidated.
Class II major histocompatibility complex molecules	Asparagine deamidation perturbs antigen presentation.
Human epidermal growth factor	Isomerization of Asp 11 decreases its mitogenic activity by five-fold.
Triose-phosphate isomerase	Subunit dissociation occurs when two Asn-Gly sequences are deamidated.

isoaspartate or aspartate forms of AsnA-21 and AsnB-3 are used depending on the pH of the solution.

Although there are examples of deamidation occurring at lower pH, this occurs primarily through a mechanism independent of succinimide formation. For example, deamidation of Asn in the A insulin chain is favored at pH 5, which is mediated by the appearance of intermediate cyclic anhydride. Although there are examples of deamidation occurring at lower pH, this occurs primarily through a mechanism independent of succinimide formation. Other than pH, factors such as sequence and local structure influence deamidation rate (the steric effect). The presence of amino acids at Asn's carboxyl end has a significant impact on the deamidation rate, with the rate decreasing as side-chain size and branching increased.

As the pH rises, deamidation rates increase, and phosphate and carbonate buffers act as deamidation catalysts. At low pH the amide group on the asparagine side chain of peptides and proteins deprotonates. The asparagine residue's nitrogen atom is attacked nucleophilically by the nitrogen atom of the amide anion peptide carbonyl carbon. In this series of reactions, the peptide chain is cleaved, yielding a fragment of succinimide peptides. Asparaginyl and asparaginyl peptides are formed after the succinamide ring is hydrolyzed.

At pH > 5.0 deamidation occurs via the intermediate unstable cyclic imide formation, which is hydrolyzed spontaneously. In highly acidic conditions (pH 1–2) direct hydrolysis of the amide side chain predominates over cyclic imide formation. The amide's indirect hydrolysis is primarily responsible for peptide bond cleavage.

4.2.3.1.2 Racemization and isomerization

Isomerization is a common chemical degradation pathway that produces the same result as deamidation. Isomerization can occur directly from the Asp residue or indirectly through the intermediate succinimide. Isomerization, like deamidation, occurs more frequently at neutral and basic pH levels, and the rate is influenced by the steric effect.

One of the most common non-enzymatic degradation outcomes is structural isomerization and racemization, particularly isomerization of aspartate to isoaspartate residues caused by neutral pH deamidation. The most common isomerization groups are Asn, Asp, and Gln, all of which contain an intermediate succinimide. Because of this mechanism, Asp degrades much more slowly than Asn. Deamidation produces iso-Asp and Asp in a 3:1 ratio depending on the group's location and mobility. Because both involve a cyclic intramolecular imide intermediate, the mechanisms for aspartate–isoaspartate deamidation and isomerization are similar.

Isomerization can result in racemization as part of the deamidation reaction. Succinimide intermediates formed during asparagine deamidation are easily racemized and deamidated, yielding d-asparagine residues. Except for glycine, other amino acids racemize at alkaline pH. The rates of deamidation of individual amide residues are determined by their primary sequence and three-dimensional (3D) structure, as well as the properties of the solution in which they are immersed, such as pH, temperature, ionic pressure, and buffer ions.

The deamidation rate of glutamine residues, for example, is typically lower than that of asparagine residues.

4.2.3.1.3 Temperature

Storage temperatures influence the rate of deamidation, which varies depending on the buffers used in the formulations. Because amine buffers with high temperature coefficients (such as tris and histidine) change the pH when stored at temperatures other than the preparation temperature, changes in formulation pH can affect the rate of deamidation because deamidation and isomerization are pH-dependent reactions. The water dissociation constant is another indirect temperature effect—the water concentration of hydroxyl ions can vary as a function of temperature, affecting deamidation rates.

4.2.3.1.4 Excipients

Organic solvents reduce deamidation rates because they lower the dielectric constants of a solution. When co-solutes such as glycerol, sucrose, and ethanol are added to a protein solution, the dielectric strength of the solvent is reduced, resulting in significantly lower rates of isomerization and deamidation. Lowering the dielectric strength of the medium from 80 (water) to 35 (PVP/glycerol/water formulations) results in a six-fold reduction in peptide deamidation concentrations, owing to less stabilized ionic intermediates produced during cyclization in the asparagine deamidation pathway. Insulin prepared in neutral solutions containing phenol, for example, eliminates deamidation by stabilizing the tertiary structure (–helix formation) around the deamidizing residue, reducing the possibility of intermediate imide formation.

4.2.3.2 Hydrolysis

Hydrolysis, like deamidation, results in the formation of succinimides, racemization, and isomerization. In the hydrolysis cycle, -x-Asp-y is known as labile, and the reaction in dilute acids is 100 times faster than other bonds. Asp–Pro bonds are 8–20 times more vulnerable to hydrolysis than other Asp-x or x-Asp bonds. The Asp–Gly is vulnerable in highly acidic conditions (pH 0.3 to 3). Other peptide bonds cleave faster at the N-terminus than the –x-Ser, –x=Ther.

Hydrolysis of peptide bonds such as Asp–Gly and Asp–Pro bonds causes protein fragmentation. Asp-Y is 100 times more vulnerable to hydrolysis than any other peptide bond. Hydrolysis is a reaction that occurs after Asn residues have been deamidated, as seen in insulin formulations.

The Asp–Gly and Asp–Pro peptide bonds are the most vulnerable to cleavage in hydrolytic proteins. The hinge area, an antibody's most flexible domain, is also susceptible to hydrolysis. Changing the pH from 9 to 5 can, however, cause the hydrolysis sites of a recombinant monoclonal antibody peptide to change, resulting in increased cleavage outside of that area.

Multimeric proteins can degrade to peptide fragments after dissociating into monomers (or single peptide chain proteins) with two or more subunits. Non-enzymatic fragmentation of peptide bonds between amino acids results in lower molecular weight polypeptides than the intact parent protein.

To minimize hydrolytic fragmentation, formulations must be properly buffered to keep their solution pH within an appropriate range for each protein type. Calcitonin, for example, undergoes hydrolysis at a simple pH, but not at pH 7, even at room temperature. The composition of the buffer also influences hydrolysis. Recombinant human macrophage colony-stimulating factor fragments appear in phosphate-buffered solutions but not in histidine-buffered solutions at the same pH and ionic strength.

It is also critical to limit the role of proteases in protein purification, whether they originate within the manufacturing process (e.g. host-cell proteins) or from outside contamination sources (e.g. adventitious microbes).

4.2.3.3 Disulfide bond

The disulfide bond is an important part of the three-dimensional structure that defines activity and antigenicity in biopharmaceutical products. Several chemical stresses cause these bonds to break, resulting in significant activity loss.

Cross-linking, which may or may not be mediated by disulfide bond formation, can also cause chemical degradation. When disulfide bonds are involved in cross-linking, the reaction takes place by either forming a new disulfide bond or exchanging disulfide bonds. Intramolecular disulfide bonding can change the tertiary structure, whereas intermolecular (or inter-domain) disulfide bonding can change the quaternary structure or cause covalent aggregates to form. At higher pH, the formation of reactive thiolate ions (S) from the Cys residue thiol group (–SH) is preferred, which may increase the likelihood of disulfide bond formation.

4.2.3.4 Glycation

The presence and location of oligosaccharides affect the rate of peptide hydrolysis at low pH levels. Position has no effect on hinge-region cleavage, but it does reduce fragmentation in the CH2 domain (for more information, see figure 4.1). The results of acidic and simple hydrolytic cleavage of peptide bonds are not always the same. Recombinant human macrophage-stimulating factor yields different peptide fragments in acidic and simple pH solutions. Enzymatic protein fragmentation can be caused by the proteolytic activity of residual or contaminating proteases, or, in some cases, autoproteolysis of an enzymatic protein.

Most proteins are glycosylated, and some undergo post-translational modifications such as phosphorylation, which affect their potential degradation pathways and kinetics. Reducing sugars, on the other hand, interact with the most susceptible amino groups, such as the Lys -amino group and the N-terminal -amino group, via the classic Maillard reaction. Glycation affects acidification (loss of positive charge), insolubility, cross-linking, and chromophore formation. A change in color in a solution may indicate an unexpected glycation reaction that is highly sensitive to pH, temperature, amino acid group pKa, adjacent amino acid, sugar reduction concentration, and protein concentration. The dependence of glycation on concentration is a major source of concern. Several mAbs are now formulated in high concentrations to reduce dose-volume, allowing for subcutaneous injection or infusion rather than intravenous injection or infusion.

4.2.3.5 Oxidation

Reactive oxygen intermediates or other oxidants such as hydrogen peroxide (found in formulations as a contaminant of polysorbates, leaching from disposable tubing, and so on), oxygen, metal ions, and other excipients cause protein oxidation. The residues Met, Cys, Trp, His, and Tyr are all affected by oxidation. Oxidation changes the physicochemical properties of a protein (folding and subunit association, for example), resulting in aggregation or fragmentation. The location of oxidized amino acids in a protein in relation to its functional or epitope-like domain can have an impact on its potency and immunogenicity.

Because amino acids in the system interact with oxygen radicals, proteins and peptides are vulnerable to oxidative damage. Methionine (Met), cysteine (Cys), histidine (His), tryptophan (Trp), and tyrosine are the most prone to oxidation (Tyr). Met and Cys contain sulfur atoms, whereas His, Trp, and Tyr contain aromatic atoms.

The key products of oxidation are listed in table 4.7.

Table 4.8 lists some of the key catalysts for the oxidation of proteins.

External factors that cause oxidation include light exposure, interaction with a trace amount of transition metal ion, and the presence of an excipient's degradation

Table 4.7. Key degradation products of oxidation.

Residues	Oxidation products
Met	Met-sulfoxide
Cys	-S-S-disulfide cross-links, sulfenic acid/sulfinic acid/sulfonic acid
His	2-oxoimidasoline, aspartate/asparagine
Trp	N-formylkynurinine, kynurenine
Tyr	Try-Try cross-link, 3–4-dihydroxy phenylalanine

Table 4.8. Oxidation source and process.

The source	Element that contributes
Reagents for chemical analysis	Oxidative burst operation (H_2O_2, Fe^{2+}, Cu^{1+}, Glutathione, HOCl, HOBr, $1O_2$, ONOO)
Irradiation of activated phagocytes in the presence of oxygen UV light, Ozone	Potential for oxidation
Peroxides of lipids	Radicals with a lot of capacity
Mitochondria are the cells that make up the mitochondria (electron transport chain leakage)	HNE, MDA, and acrolein are all terms for the same thing.
Medications and their metabolites	Xanthine oxidase, Myeloperoxidase, and P-450 enzymes are examples of oxidoreductase enzymes Oxidative products, free radicals

product (e.g. polysorbate-degraded hydrogen peroxide). As a result of oxidation, an increase in aggregation can occur. The Met residue is the most oxidized, and sulfoxide and sulfone are the main by-products of this reaction. The oxidation rate is influenced by both the local structure around the oxidation-prone group (e.g. surface exposure and steric hindrance) and the solution pH. While increasing the pH of a solution has been shown to increase the oxidation rate in some cases, this is not a common occurrence.

Oxidation can reduce potency and cause conformational changes that lead to aggregation depending on the oxidation site. Clinical protection cannot usually be extrapolated from knowledge of degradation pathways. Parathyroid hormone, for example, has different biological activity (*in vitro* bioassay) depending on whether it undergoes single or double oxidation of Met-8 or Met-18 (Met-8 with Met-18). Similarly, oxidation of Met-36 and Met-48 in human stem cell factor (SCF) by *Escherichia coli* reduces potency by 40% and 60%, respectively, increasing the constant dissociation rate SCF dimer by 2–3 times, implying improved binding and tertiary structure of the subunit. These examples highlight the importance of thoroughly evaluating any changes in a molecule's behavior due to degradation.

Met, Tyr, Trp, His, and Cys are all oxidation-sensitive residues commonly found in mAbs.

Although it is less common than deamidation and isomerization, oxidation is a major mode of protein degradation for some proteins, such as OKT3 (IgG2), which exhibits oxidation at several Met residues and free Cys when stored at 5 °C.

SThe main effects of buffer species are contaminants, metal chelating ability, and potential interactions with reactive oxygen species. Tris is an iron and copper chelator as well as a hydroxyl radical scavenger; citrate is also a chelator, and phosphoric acid (weak) and tartaric acid are both buffers. In Fenton chemistry, a bicarbonate ion is required, but there are no simple generalizations.

Chelators are as follows. The metal ion concentration in the atmosphere affects metal-ion-catalyzed oxidation. The presence of Fe^{3+}, Ca^{2+}, Cu^{2+}, Mg^{2+}, or Zn^{2+} 0.15 ppm chloride salts has no effect on the oxidation rate of human insulin-like growth factor-1. There is a significant increase in oxidation when the metal concentration is increased to 1 ppm.

Cys, Met, and His are particularly prone to oxidation in metallics. Potent 1:1 chelators (e.g. EDTA, desferrioxamine, nitrilotriacetic acid) can affect oxidation rates based on stoichiometry (chelator/Fe ratio). Oxidation rates increase before all Fe is sequestered at a maximum ratio of 1; at a ratio of 1.1 to 1.2, almost complete protein inhibition is seen. A significant excess (5–20) is required to inhibit the effects of bad chelators (e.g. o-phenanthroline). Chelate–Fe (III) complexes inhibit site-specific oxidation while increasing non-site-specific oxidation. Chelates prevent iron from binding to potential binding sites, and the chelate–Fe (III) complex promotes the formation of ROS species in a way that metal-binding sites do not. Peptide bond hydrolysis is caused by pyroglutamate (more acidic) ($z = -1$), deamidation (more acidic) ($z = -1$), succinimide (more simple or neutral), glycation (more acidic) (acidic or essential).

Cysteine oxidation is more common at alkaline pH, which deprotonates thiol groups. Cysteine is oxidized to form intermolecular or intramolecular disulfide bonds, which are then further oxidized to form sulfenic acid. Transition metals strongly catalyze cysteine oxidation at higher pH (optimum around 6). However, this oxidation reaction is less susceptible to mild oxidants such as hydrogen peroxide than Met or Cys. In the absence of a nearby thiol for disulfide formation, oxidation can produce sulfenic, sulfinic, and sulfonic acids. In a reducing environment, cysteine oxidation results in a nucleophilic attack on disulfide bonds by thiolate ions, resulting in new disulfide bonds and separate thiolate ions. The new thiolate can then react with another disulfide bond to form cysteine.

Protein degradation creates intermolecular disulfide connections, accumulating mispaired disulfide bonds and scrambled disulfide bridges, and changing the conformation and subunit associations of the protein. Molecular by-products of cysteine residues that spontaneously oxidize in metal ions or adjacent thiol groups are sulfinic acid and cysteic acid. The human fibroblast growth factor (FGF-1) undergoes copper-catalyzed oxidation, which can result in homodimers.

Cysteine oxidation affects the spatial orientation of thiol groups in proteins. The distance between those thiol groups determines the rate of oxidation. Large oligomers or non-functional monomers form in the case of basic fibroblast growth factor (bFGF), which contains three easily oxidized cysteines and forms intermolecular or intramolecular disulfide bonds that induce conformational protein modifications. Cysteine oxidation has a significant impact on the spatial orientation of thiol groups in proteins. The rate of oxidation is inversely proportional to the distance between such thiol groups. This can result in the development of large oligomers or non-functional monomers in the long run, as in the case of basic fibroblast growth factor (bFGF), which contains three easily oxidized cysteines and forms intermolecular or intramolecular disulfide bonds that trigger conformational protein modifications. The volume of the protein's side chains is increased by cysteine disulfide bonds, resulting in unfavorable van der Waals interactions.

Peroxides (for example, in Polysorbates) and trace metals, primarily Fe and Cu, have a significant impact on common excipient organisms. High concentrations of sugars and polyols inhibit peroxide contamination caused by polysorbates and polyethylene glycols (PEGs) (possibly through a metal complexation mechanism). Peroxide contamination from polysorbates and PEGs, which are both commonly used as pharmaceutical excipients in the processing and storage of proteins, can lead to oxidation. The degree of oxidation in rhG-CSF was found to be related to the amount of peroxide in Tween 80, with peroxide-induced oxidation appearing to be more intense than that caused by atmospheric oxygen. Plastic or elastomeric materials used in primary packaging container-closure systems, such as prefilled syringes, may also leach peroxide. The degree of oxidation in rhG-CSF was related to the amount of peroxide in Tween 80, with peroxide-induced oxidation appearing to be more extreme than atmospheric oxygen. Peroxide is leached from the plastics and elastomers used in packaging.

4.2.3.5.1 Exogenous variables: pH, temperature, and excipients buffer

Histidine residues are highly susceptible to oxidation when they react with their imidazole rings (metal or photocatalyzed catalyzed oxidation), resulting in the generation of additional hydroxyl species. During light and metal oxidation, oxidized histidine can produce asparagine/aspartate and 2-oxo-histidine (2-O-His), which act as a transient moiety that can cause protein aggregation and precipitation, making 2-O-His isolation as individual degradants difficult. Photo-oxidation via singlet oxygen is also common in the presence of photosensitizers. The inclination and rate of oxidation are influenced by surrounding residues and amino acid conformation, resulting in colored solutions.

Photo-oxidation levels follow this order: non-site-specific processes (e.g. hydrogen peroxide, light) have a harder time oxidizing buried residues. Photo-induced oxidation primarily targets Trp and His groups. Ultraviolet light is another important component that allows proteins to be light sensitive. Photo-oxidation can alter the structure of primary, secondary, and tertiary proteins, causing changes in their long-term stability, bioactivity, and immunogenicity. Even after the light source has been turned off, light exposure can cause a cascade of biochemical events that continue to affect a protein. The amount of energy provided to a protein and the amount of oxygen available from the atmosphere determine these effects. When a compound absorbs enough light at a specific wavelength, it gains enough energy to lift the molecule into an excited state known as photo-oxidation. The excited molecule then transfers its energy to molecular oxygen, causing it to split into singlet oxygen atoms, which are reactive. In the presence of oxygen, tryptophan, histidine, and tyrosine are light-modified. Mono-, di-, tri-, and tetrahydroxy tyrosine are all by-products of tyrosine photo-oxidation. The cross-linking of oxidized tyrosine residues causes the aggregation of such proteins. The photo-oxidation reaction has been pinpointed with pinpoint accuracy. In human growth hormone treated with intense light, for example, oxidation occurs primarily at histidine-21. Photodegradation of the backbone peptide is also a possibility. Alternatively, the energetic protein can directly photosensitize another protein molecule by reacting with it, most commonly through low pH methionine and tryptophan residues. Excipients and lixiviants may work together to promote protein oxidation and degradation. In some cases, these formulating components influence the rate of photo-oxidation. For example, the phosphate buffer accelerates the rate of methionine degradation more than other buffer systems. Excipients such as polyols and sugars that help stabilize protein structure can reduce oxidation rates, while denaturing/unfolding reagents can increase protein oxidation in the solution. Ascorbic acid enhances the oxidation of human ciliary neurotrophic factor, so using a reducing agent like ascorbate can help boost oxidation.

Mono-, di-, tri-, and tetrahydroxy tyrosine are all by-products of tyrosine photo-oxidation. Cross-linking between oxidized tyrosine residues causes aggregation in some proteins. The metal ion concentration in the atmosphere affects metal-ion-catalyzed oxidation. The presence of 0.15 ppm Fe^{3+}, Ca^{2+}, Cu^{2+}, Mg^{2+}, or Zn^{2+} chloride salts has no effect on the rate of oxidation of human insulin-like growth factor-1. There is a significant increase in oxidation when the metal concentration is increased to 1 ppm. Photo-oxidation levels folloe the order its > Trp > Met > Tyr. Trp and His are the key targets of photo-induced oxidation. Non-site-specific pathways (e.g. H_2O_2, light) have a harder time oxidizing buried residues. Ultraviolet light is another important component that allows proteins to be light protected. The structure of primary, secondary, and tertiary amines can be altered by photo-oxidation.

4.3 Formulation composition

The route of administration and the product's physicochemical properties, like those of other drugs, determine the composition of biopharmaceutical formulations. Membrane permeability is low due to the high molecular mass and large size of drug molecules, limiting the delivery routes available. Passive diffusion allows drugs with a molecular mass of less than 500 Da to pass through membranes easily in the gastrointestinal tract and the skin. The inner and outer plexiform layers of the human retina limit the diffusion of macromolecules larger than 76 kDa, and macromolecules larger than 150 kDa cannot enter the inner retina. The membrane permeability of the nasal mucosa is also low for molecules larger than one kDa. Biopharmaceuticals' hydrophilicity eliminates most routes of administration because proteins larger than 3–5 kDa are considered peptidyl molecules, and antibodies larger than 150 kDa are considered antibodies.

Because most protein drugs have a log$P$ value less than zero, drug permeation through biological membranes is difficult and protein drug delivery to intracellular targets is complicated. (See appendix A.) Because of the lipophilic nature of biological membranes and the 3–10 paracellular space, the proteins' large size and hydrophilic nature limit their diffusion and passage through paracellular pathways. As a result, rather than passive diffusion, active transport or endocytosis controls the cellular uptake of hydrophilic proteins. One of the major drawbacks of the endocytic pathway for proteins is endosome entrapment, which leads to lysosomal enzyme degradation.

Another physicochemical drug property that influences absorption is the surface charge of a biopharmaceutical, which is determined by the amino acid sequence of the protein and the pH of its surroundings. Deamination, isomerization, or post-translational alteration are common causes of this complex and heterogeneous physicochemical property, which results in a shift in a protein's net charge and acidic and simple variants. Because of their surface charge, protein drugs can interact with molecules on cell surfaces or tissue components, affecting protein absorption, distribution, and elimination in the body.

4.3.1 Route of administration

Each administration route has its own set of limitations based on anatomical size and location, microclimate, complex physiological conditions, and formulations.

pH conditions in various biological environments affect the ionization, chemical stability, and absorption of protein-based drugs and their delivery mechanisms. At physiological pH, protein drugs, for example, are unstable. The highly acidic gastric environment (pH 1–3) causes protein drug destabilization in the stomach, but higher pH reduces chemical degradation in the ileum and colon. Because hyperosmotic solutions cause transient dehydration of anterior chamber tissues, whereas hypotonic solutions can cause edema, the buffering agent in an ocular delivery system is critical.

Oral protein delivery and bioavailability are hampered by enzyme degradation in the gastrointestinal tract. Protease activity is higher in the small intestine but much lower in the colon, so colon-targeted delivery systems have received a lot of attention as a viable delivery mechanism for protein drugs. Furthermore, colon-targeted drug delivery can result in increased drug absorption and duration of action. In addition to systemic delivery of protein drugs, colon-targeted drug delivery systems are useful for treating local bowel disorders such as colon cancer, ulcerative colitis, Crohn's disease, and amoebiasis. Enzymatic barriers can cause a 'pseudo-first-pass effect' even when non-oral routes of administration avoid the hepatic first-pass effect. Low metabolic enzyme activity, for example, may be a barrier to protein drug delivery via nasal and pulmonary routes.

Mucus and epithelial cell membranes, which act as a major absorption barrier, thwart the administration of non-injectable drugs. Mucus coats all mucosal epithelia, serving as a first line of defense against mechanical damage and the entry of harmful substances at the surfaces of the eye, respiratory tract, and gastrointestinal tract. Large molecules are also physically shielded by the mucus layer. The drug interacts with mucus materials due to its hydrophilic nature and negative charge, slowing drug diffusion and limiting drug absorption in the intestine (figure 4.3).

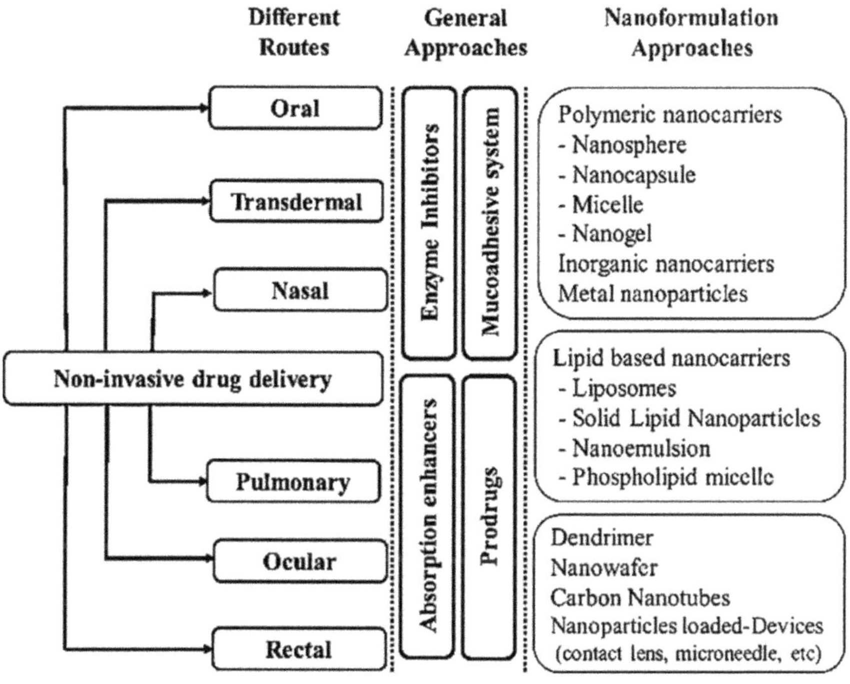

Figure 4.3. Barriers to absorption in various routes of administration.

Mucus contains mucin-type glycoproteins that are secreted, and the thickness of the mucus layer varies greatly throughout the body. For example, airway mucus can be anywhere from 5 to 55 μm thick, but the mucus layer in the nasal tract is very thin. As a result, it has a porous surface compared to other mucosal surfaces. In the eye, the secreted precorneal mucin gel that covers the conjunctiva is 30–40 μm thick. The thickness of the mucus layer in the gastrointestinal tract varies significantly depending on the location and activity of the digestive system. The mucus layer is thickest in the stomach and colon, but it varies in thickness in the ileum and stomach, ranging from 10 to more than 170 μm. As a result, drugs must pass through a thicker mucus layer, despite the colon's lack of proteolytic activity, which makes it a good absorption site for proteins.

4.3.2 Excipients and properties

Pre-formulation is the process of evaluating the biochemical and biophysical characterization of a biopharmaceutical product based on its amino acid sequence, which is influenced by pH, ionic strength, and excipients. Pre-formulation entails using a systematic approach such as the design of experiment (DOE) or the empirical phase diagram (EPD) to establish stability-indicating assays and define a few lead excipients. In the presence of stress, structural (e.g. secondary, tertiary, etc) and functional changes (e.g. movement, potency, binding wherever possible) occur.

Excipients in biopharmaceutical formulations are not licensed as finished products on their own. They range from well-known organic or inorganic molecules to more complex and difficult-to-classify biopharmaceutical structures. The choice of excipient is also influenced by the statutory requirements of different jurisdictions. www.accessdata.fda.gov/scripts/cder/iig/index.cfm is a searchable database (IID) of approved concentrations, dosage forms, and administration routes maintained by the US Food and Drug Administration (FDA). The Japanese Pharmaceutical Excipients Dictionary (JPED) compiles excipients in approved medications, route of administration, and patient treatment in Japan, https://www.yakuji.co.jp/wpyj-002/wp-content/uploads/2020/05/jpe2018orderform.pdf.

Health Canada publishes a list of non-medicinal agents that are permitted. On the other hand, the European Medicines Agency lacks a comparable list or database for European products.

Table 4.9 lists the different types of pharmaceutical excipients that are commonly used in biopharmaceutical products and some examples.

While formulations vary widely in their composition, table 4.10 lists the most common ranges of quantities used.

The dosing criteria determine the amount of active ingredient in a formulation, and the concentration is determined by the product's solubility. Protein solubility is the average amount of protein soluble in the presence of co-solvents when the solution is visibly transparent (i.e. it does not reveal precipitated proteins, crystals, or gels). Ionic strength, salt form, pH, temperature, and certain excipients all affect protein solubility, as evidenced by changes in bulk water surface tension and protein binding to water and ions versus self-association. Protein binding to various

Table 4.9. Common pharmaceutical excipients.

Category	Example
Amino acids	Arginine, aspartic acid, glutamic acid, lysine, proline, glycine, histidine, methionine
Buffer	Phosphate, acetate, histidine, glutamate
Buffers	Acetate, succinate, citrate, histidine, phosphate, tris
Cyclodextrins	Hydroxypropyl beta cyclodextrin
pH	Buffers
Preservatives	Benzyl alcohol, *m*-Cresol, phenol, 2-phenoxyethanol
Solubilizer	Salts, amino acids, surfactants
Stabilizer	Surfactants, sugars, salts, antioxidants
Stabilizers/bulking agents	Lactose, trehalose, dextrose, sucrose, sorbitol, glycerol, albumin, gelatin, mannitol, dextran
Surfactants	Tween 20, Tween 80, Pluronic F68
Tonicity, bulk modifier	Sodium chloride, sorbitol, mannitol, glycine, polyanions, salts

Table 4.10. General ranges of formulation components.

Component	General range
Buffer	5–100 mM
pH	4–8
Salts	0–300 mM
Stabilizers	1%–10%
Surfactants	0.01%–0.1% (w/v)

excipients or salts changes the conformation of the protein or masks certain amino acids involved in self-interaction, which affects solubility. Certain salts, amino acids, and sugars preferentially hydrate (and stabilize as more compact conformations) proteins, contributing to their altered solubility.

A variety of excipients are used to stabilize protein formulations. To improve the product's safety and efficacy, excipients, and buffers for formulation (table 4.8) are used.

The most common ingredients in biopharmaceutical drug formulations, according to an analysis, are:

- Buffering agents such as phosphates, citrates, and acetates ensure that the pH remains as stable as possible.
- Stabilizers such as surfactants and sugars include polysorbates, albumin, mannitol, sucrose, and sorbitol.
- Sugars and electrolytes like sodium chloride are examples of ingredients that change tonicity and conductivity.

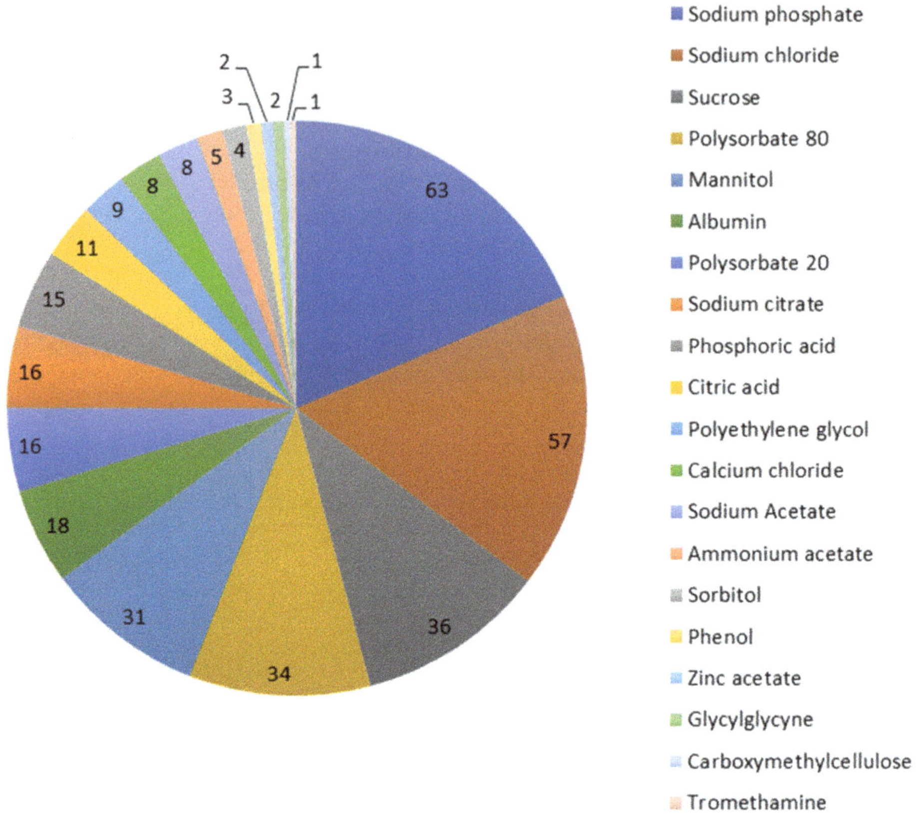

Figure 4.4. Percentage occurrence of common formulation components in biopharmaceutical drugs.

Figure 4.4 shows the percentage of the most common components used in the approximately 200 biopharmaceutical injectable drugs.

4.3.2.1 pH

The pH is the most important of the formulation variables mentioned. Other formulation methods are notoriously difficult to solve physical properties problems, such as precipitation due to solubility and stability. pH optimization is a simple but effective solution for such problems. pH causes chemical reactions such as deamidation, cyclic imide formation, disulfide scrambling, peptide bond cleavage, and oxidation, to name a few. The properties of other functional excipients should be carefully evaluated (e.g. sucrose used to stabilize protein during lyophilization and storage in the dried solid).

Preventing oxidation and precipitation can be accomplished by adjusting the pH and ionic strength of a protein solution, adding sugars, amino acids, and polyols, and employing surfactants. A thorough evaluation of optimal pH and osmotic conditions is a critical element of formulation production to avoid protein aggregation or precipitation—irreversible aggregation caused by denaturation that is prevented with surfactants, polyols, or sugars.

4.3.2.2 Surface tension

A hydrophobic and hydrophilic portion (for example, an alkyl chain) exists in surfactants (e.g. carboxyl and carboxylate groups) such as sodium polyacrylate.

Surfactants (nonionic detergents) are used to improve stability and prevent aggregation. Because proteins and surfactants have a hydrophobic relationship, these compounds stabilize them by lowering the surface tension of their solution and binding to hydrophobic sites on their surfaces, reducing the chance of protein–protein interactions that could lead to the formation of aggregates. The nonionic detergents Tween 20 and Tween 80 will prevent the formation of soluble protein aggregates at surfactant concentrations below the critical concentrations of micelles (CMC). When added to IgG solutions, polysorbate (Tween) 80 stabilizes small aggregates and prevents them from expanding into larger particles.

4.3.2.3 Tonicity

The osmotic pressure of a 'isotonic' solution is the same as that of human blood. Isotonic formulations typically have an osmotic pressure of 250 to 350 mOsmol/kg. Buffering agents are one or more components that are added to an aqueous solution to protect it from pH changes caused by adding acid or alkali or dilution with a solvent. In addition to phosphate buffers, glycinate, carbonate, and citrate buffers are used, with sodium, potassium, or ammonium ions acting as a counterion.

4.3.2.4 Protectants

Novel excipients are used to improve protein/peptide stability because commonly used excipients only achieve a small level of protein stabilization. Only a few examples include resveratrol, a natural phenol, hydroxybutyrate, polyamines, octanoic acid, and quinone-tryptophan derivatives. The viscosity of mAb solutions is reduced ten-fold by hydrophobic salts such as pentane-1,5-diamine salts, and camphor-10-sulfonic acid salts. Extensive safety testing, which may include *in vivo* studies, may be required for uncommon excipients.

When used in conjunction with a protein of interest, lyoprotectants help to avoid or reduce the chemical and physical instability of the protein during lyophilization and subsequent storage.

Preservatives inhibit bacterial activity and can be used as an optional ingredient in formulations. For example, adding a preservative will make developing a multi-use (multiple doses) formulation easier. Octadecyldimethylbenzyl ammonium chloride, hexamethonium chloride, benzalkonium chloride (a mixture of alkyl benzyl dimethylammonium chlorides with long-chain alkyl groups), and benzethonium chloride are all potential preservatives. Preservatives include aromatic alcohols such as phenol, butyl, and benzyl alcohol, alkyl parabens such as methyl or propylparaben, catechol, resorcinol, cyclohexanol, 3-pentanol, and *m*-Cresol, as well as alkyl parabens such as methyl or propylparaben.

One common method of preventing aggregation is to limit protein mobility. This reduces the number of collisions between molecules. Excipients such as surfactants (such as polysorbate 20 and 80), carbohydrates (such as cyclodextrin derivatives), and amino acids (such as arginine and histidine) help prevent aggregation by adsorbing to the air–liquid interface and protecting the protein. Polysorbate 80, on the other hand, can lead to the formation of micelles, which increases the risk of immunogenicity. Cyclodextrin stabilizes commercially available antibody-based drugs in a hydrogel formulation. Pluronic F68, trehalose, glycine, and amino acids such as arginine, glycine, glutamate, and histidine, which are found in many commercial protein therapy products, are all considered safe excipients (GRAS). Avastin® (bevacizumab, 25 mg ml^{-1}) contains ingredients such as trehalose dehydrates, sodium phosphate, and polysorbate 20. Histidine hydrochloride, histidine, trehalose dehydrate, polysorbate 20, methionine, and water for injection are all excipients in subcutaneous Herceptin ® (trastuzumab, 600 mg). Chelating agents are used to prevent metal-induced protein aggregation.

4.3.2.5 Stabilizers

When large polymeric excipients are used, protein stabilization is possible. Neutral polymers can stabilize a protein as crowding agents because the volume effect (steric repulsion) is removed. Neutral polymers such as PVP, Ficoll70, hydroxyethyl (heta) starch, or PEG 4000 are used to stabilize proteins. Newer compositions include functionalized trehalose-containing dextrans and glycopolymers as side-chain units for improved process and storage stability. Glycopolymers made from modified trehalose monomers are used to stabilize the product. Polyanions/polycations, such as heparin and dextran sulfate, pentosan polysulfate, polyphosphoric acid, poly-L-glutamic acid, poly (acrylic acid), and poly (acrylamide) are possible stabilizers due to the delicate development of protein–polyion interactions (methacrylic acid). Activity between polycationic chitosan and negatively charged lactate dehydrogenase (LDH) results in significant stabilization during air-jet nebulization. When agitated, however, negative heparin and keratinocyte growth factor 2 (KGF-2) activity promotes protein aggregation.

Protein destabilization can be caused by strong interactions, as shown by the preferential interaction mechanism. The fibrillation process involves interactions between crystalline and non-native protein species. Under certain conditions and over time, proteins lose their native folded state and form amyloid fibrils, a mechanism linked to a variety of human diseases. Foreign surfaces, such as nanoparticles with different surface properties, can cause the fibrillation process to be disrupted.

The use of recombinant hyaluronidase enzyme in protein formulation to facilitate rapid tissue distribution and, as a result, the administration of a larger-than-normal injection volume is one example. Using polymers or proteins in a protein formulation increases the complexity of the formulation and makes characterization and stability studies of protein formulations more difficult.

4.3.3 Liquid formulations

In designing liquid formulations, electrostatic interactions, van der Waals interactions, hydrogen bonding, and hydrophobicity are all taken into account. High concentrations of saccharides such as sucrose, trehalose, and lactose, as well as polyhydrated alcohols such as sorbitol, mannitol, and polyethylene glycol, can help biologics maintain their native conformation by preventing them from interacting with the protein surface. Saccharose is not used because of the increased surface tension. The process of 'salting in' improves protein solubility by adding small amounts of salt. Salts are used as tonicity inhibitors, but in some cases they may have a negative impact on conformational stability. As a result, counterions and their concentration (as defined by the Hofmeister series) are used to change the stability profile. In pre-formulation screening, this is considered critical. For example, different buffer species at different concentrations (i.e. at different ionic strengths) can achieve the same pH state, affecting the protein's stability. The binding of Zn^{2+} to growth hormone in humans demonstrates how ligand binding can help maintain the native protein state.

Surface-active agents are used to prevent protein adsorption, denaturation, and aggregation at interfaces (both air–water and solid–water) and to affect protein stability by binding to the native and denatured states of the protein in different ways. Surface denaturation can also be caused by agitation and freeze–thaw stress tests, as well as low surfactant concentrations such as Polysorbate 20, Polysorbate 80, Pluronic F68, or others that are good at reducing both soluble and insoluble aggregates. Metal ions, barium from glass vials, vulcanizing agents from stoppers, tungsten oxide from prefilled syringes, and silicone oil penetration are all issues that need to be addressed. EDTA is used to remove metal leachates from stoppers, for example. As a result, counterions and their concentration (as defined by the Hofmeister series) are used to change the stability profile. This is critical during the pre-formulation screening process. Different buffer species at different concentrations (i.e. different ionic strengths) can, for example, produce the same pH state, which can affect protein stability. Binding ligands to proteins, such as Zn^{2+} to human growth hormone, also assists proteins in maintaining their native state.

Antimicrobial preservatives are commonly used in multidose biopharmaceutical products, which account for one-third of all biopharmaceutical products. By causing protein aggregation, preservatives can make biopharmaceuticals more difficult to stabilize. Preservatives such as *m*-Cresol, benzyl alcohol or phenol, phenoxyethanol, and chlorobutanol are commonly used. It is recommended that you screen various preservatives before using them alone or in combination.

4.3.4 Lyophilized formulations

Lyophilization with appropriate excipients improves protein stability against aggregation by reducing protein mobility and restricting conformational flexibility, with the added benefit of minimizing hydrolytic reactions caused by the removal of water. Using appropriate excipients, such as lyoprotectants, can help prevent the formation of aggregates during the lyophilization process and storage of the final product. Effective safety is influenced by the lyoprotectant-to-protein molar ratio. For adequate stability, molar ratios of 300:1 or higher are required, particularly when stored at room temperature. However, such ratios can cause a viscosity increase that is undesirable.

Lyophilization may be required because protein is unstable in an aqueous solution without preservatives. When the required requirements for stability, storage, and shipping align with the target product profile, lyophilization is an important alternative to a liquid formulation, particularly for highly thermolabile products and live virus vaccine products. Freezing is required for lyophilization, followed by vacuum-assisted primary and secondary drying. The drying process presents its own set of challenges. During freezing, denaturation can occur due to the freeze-concentrate condition, frozen surface interfaces, or cold-denaturation. During the freeze-concentrate phase, formulations should consider the effect of local salt and buffer concentrations, as well as increasing trapped oxygen concentrations.

Similarly, for lyophilized products, changes in pH due to buffer crystallization must be factored into the formulation design space. For example, buffering at physiological pH with a low concentration of potassium phosphate buffer (including 10 mM) is preferable to sodium phosphate buffer. This could be due to the pH change as a function of increasing buffer concentration, which causes a significant pH shift during freezing for sodium phosphate. Because of the limited pH changes during freezing, citrate, Tris, and Histidine are good choices for the buffer if the pH range is right. To reduce oxidation, antioxidants (e.g. ascorbic acid) and scavengers (e.g. thiourea) can be used.

External stabilizers known as lyoprotectants may be required in addition to cryoprotectants. For example, liquid pre-formulation screening may reveal that sorbitol is a good stabilizer. However, sorbitol is not preferred in dried formulations due to its low glass transition temperature. Lyoprotectants have high glass transition temperatures to reduce molecular mobility while also preserving the native state in the dry state by acting as a solid water substitute (e.g. sucrose, trehalose). As a result, the use of reduced sugars (e.g. lactose) must be precisely assessed in the risk evaluation analysis. The possibility of phase separation (particularly in a multi-stabilizer system) can, however, limit some excipient combinations (for example, PEG-Dextran). Trehalose is a better alternative to sucrose due to its acid hydrolysis ability at a lower pH.

To prevent 'blowout' of low-concentration products, bulking agents are applied to the lyophilized product (including 1% solid). Amorphous bulking agents (e.g. sucrose, trehalose, lactose, raffinose, dextran, hydroxyethyl starch (HES)) or crystalline bulking agents (e.g. glycine, mannitol) for amorphous excipients (such as HES) or high eutectic temperatures (Teu) for crystalline excipients are examples of excipients (such as glycine and mannitol). Mannitol selection during excipient screening is limited due to the presence of mannitol hydrate and the risk of glass breakage during manufacture (due to high fill volume, incorrect freezing procedure, and high concentration).

Isotonicity aids in pain relief. The isotonicity of a lyophile is difficult to achieve due to the concentration of both the protein and the excipients during the reconstitution period. A 500:1 protein molar ratio can result in hypertonic preparations if the protein concentration is greater than 100 mg ml^{-1}.

Although freeze-drying is one of the most widely used processes for protein drugs, it has several drawbacks, prompting the development of alternative drying methods for proteins such as insulin, trypsin, human growth hormones, and monoclonal antibodies, such as spray-drying, spray-freezing, supercritical fluid drying, and foam drying.

4.4 Delivery routes

Parenteral refers to any route other than the mouth or the alimentary canal (e.g. rectal). The most common administration route for biopharmaceuticals is parenteral injection, which includes intravenous bolus, intravenous infusion, subcutaneous injection, and intramuscular injection. However, due to its invasiveness, discomfort, infection risk, high cost, and low patient compliance, parenteral injection is not always the preferred delivery method.

Noninvasive drug delivery routes such as oral, nasal, pulmonary, ophthalmic, rectal, or transdermal are difficult to develop due to biopharmaceuticals' large molecular size, hydrophilicity, low permeability, and chemical/enzymatic instability. Alternative drug delivery approaches have two major drawbacks. One is that the drug's route of administration is hostile to polypeptides. For example, orally administered proteins are subjected to harsh conditions prior to absorption through the gastrointestinal tract or absorption through the nasal mucosa, which can cause significant metabolism. After administration, adequate drug absorption through the respective barrier layers may be a significant factor in achieving a pharmacological response. Encapsulation in hydrophobic carriers, combination with penetration enhancers, active electrical transport, or chemical modification to increase hydrophobicity are all strategies for improving absorption of hydrophilic (thus poorly absorbed) compounds. Because of advances in nanotechnologies, new nano-formulations with regulated particle size and surface modification are being developed, which improve the target selectivity, systemic half-life, and bioavailability of protein drugs.

Newer approaches also include using a device that injects the drug in the gastrointestinal tract when taken orally. However, there is little likelihood of such products being approved because of the inevitable variability of dosing, a similar situation that caused the withdrawal of inhalation insulin, risk of gastrointestinal bleeding and significant limitation on the dose that can be administered.

4.4.1 Intravenous

Intravenous bolus, intravenous infusion, and subcutaneous delivery are the most common methods for delivering biopharmaceuticals. Most medications with poor or extremely variable bioavailability, such as oncology drugs, are administered intravenously to ensure precise dosing. Drugs that can irritate subcutaneous tissue are better injected intravenously. This path, however, is not recommended for drug self-administration, which is a major cost component of drug therapy. There are not many formulation issues other than the fact that the drug must be in a solution or a very fine emulsion to avoid vein blockage. Some biopharmaceuticals originally designed for intravenous injection or infusion have recently been reformulated for subcutaneous administration, allowing patients to self-administer them.

4.4.2 Subcutaneous

Insulin was the first biopharmaceutical to be accepted as a subcutaneous drug. Intravenous bolus or infusion is the most common route for medications that require a dose calculation, such as oncology drugs, and is delivered by professionals. However, for economic reasons, there has been a recent transition from intravenous to subcutaneous administration. Following the introduction of subcutaneous formulations of trastuzumab and rituximab in Europe in 2013 and 2014, many medications are now reformulated as subcutaneous dosage forms, moving away from intravenous, enabling self-injection of rheumatoid arthritis, multiple sclerosis, or primary immunodeficiency therapies, where mixed dosing (not measured based on body weight) is suggested (table 4.11).

Biopharmaceutical products given subcutaneously have a different pharmacokinetic profile than those given intravenously. The pharmacokinetic profile of biopharmaceutical products injected subcutaneously is typically marked by a slow rate of absorption from the subcutaneous extracellular matrix, with $C_{\max}$ levels lower than those obtained with intravenous dosing.

This absorption pattern into the blood is caused by the decreased permeability of macromolecules through the vascular endothelial. Thus, lymphatics provide an alternative route of absorption into the circulation system. Despite this, lymphatic absorption has been described as a barrier to the complete penetration of molecules injected subcutaneously. Interactions with interstitial glycosaminoglycans and proteins, as well as enzymatic degradation, can all contribute to the incomplete bioavailability of subcutaneously injected molecules.

Unlike small molecules, biopharmaceutical products with molecular weights greater than 20 kDa have limited transport through blood capillaries and enter the circulatory system primarily through lymphatics. Biotherapeutic subcutaneous administration is more immunogenic than intravenous dosing because of the increased lymphatic exposure. If there is an alternative, regulatory agencies currently require immunogenicity testing of the subcutaneous dosage type over intravenous. In those instances where a biological drug is administered via both subcutaneous and intravenous route, the developer is required to demonstrate the safety of subcutaneous administration over the intravenous administration.

Subcutaneous administration of biopharmaceutical products has the disadvantage of incomplete bioavailability of the injected molecule, which ranges from 50% to 80% for mAbs. In terms of the enzymes involved and their translation through species, the underlying pre-systemic catabolism at the subcutaneous administration site or the lymphatic system is still poorly understood. Subcutaneous bioavailability for mAbs is inversely proportional to clearance after intravenous dosing, with mAbs with lower intravenous clearance having higher subcutaneous bioavailability. According to this connection, hematopoietic cells (e.g. macrophages or dendritic cells) can play a role in both subcutaneous first-pass clearance and systemic clearance after intravenous dosing. Due to poor bioavailability, subcutaneous infusions usually require a higher dosage than intravenous infusions, so the cost of products for subcutaneous formulations is higher.

Table 4.11. Examples of biopharmaceuticals delivered in subcutaneous dosage forms.

Molecule	Brand name (originator)	Dosing frequency	Injection volume	Device
Abatacept	Orencia (Bristol-Myers Squibb)	q1w	1 ml	Prefilled syringe, prefilled pen/autoinjector
Adalimumab	Humira (AbbVie)	q2w	0.4–0.8 ml	Prefilled syringe, vial, prefilled pen
Anakinra	Kineret (Swedish Orphan Biovitrum GmbH)[a]	q1d or q2d	0.67 ml	Prefilled syringe
Certolizumab pegol	Cimzia (UCB-Euronext and BEL20)[a]	q2w and q4w	1 ml	Prefilled syringe, vial, prefilled pen
Etanercept	Enbrel (Amgen)	q1w or twice weekly	0.5–1 ml	Prefilled syringe, vial, prefilled pen/ autoinjector, prefilled cartridge for reusable autoinjector
Glatiramer acetate	Copaxone (Teva)	q1d or three times per week	1 ml	Prefilled syringe, pen/autoinjector
Golimumab	Simponi (Janssen)	q1m	0.5–1 ml	Prefilled syringe, prefilled pen/autoinjector
Insulin	Several	PRN	variable	Vials, prefilled pen, syringes
Interferon beta-1a	Rebif (EMD Serono/Pfizer)	Three times per week	0.2–0.5 ml	Prefilled syringe, prefilled pen/ autoinjector, electronic injection system
Interferon beta-1b	Betaseron/Betaferon (Bayer)	q2d	0.25–1 ml	Prefilled syringe, vial, autoinjector
Interferon beta-1b	Extavia (Novartis)	q2d	0.25–1 ml	Prefilled syringe, vial, autoinjector
Peginterferon beta-1a	Plegridy (Biogen)	q2w	0.5 ml	Prefilled syringe, prefilled pen/autoinjector
Rituximab	MabThera/Rituxan Hycela (Roche)	q3w–q3mc	11.7–13.4 ml	Vial and syringe
Sarilumab	Kevzara (Sanofi-Aventis)	q2w	1.14 ml	Prefilled syringe, prefilled pen
Tocilizumab	Actemra (Roche)	q1w and q2w	0.9 ml	Prefilled syringe, prefilled pen
Trastuzumab	Herceptin (Roche)	q3w	5 ml	Vial and syringe

To increase a biotherapeutic's subcutaneous bioavailability, subcutaneous infusions with the dispersion-enhancer hyaluronidase can be administered (or developed as coformulations). The spread of an injected fluid in the subcutaneous tissue is aided by this enzyme. A co-injected molecule can have a higher bioavailability due to the increased dispersion in the interstitial tissue.

Subcutaneous injections are made in a buffered aqueous solution, some of which, including the citrate buffer, are painful. Humira, the best-selling biopharmaceutical drug, was originally launched with a citrate buffer that induces smarting upon injection. After the patent on the gene sequence expired, Abbvie reformulated the medication without the buffer, claiming that the large proteins themselves function as a buffer, a fact that was well known to the developer. The new formulation is more concentrated, resulting in a smaller injection volume and brings less inflammation, based on a new patent for Abbvie, which now protects Abbvie.

Subcutaneous administration serves to provided sustained activity such as in the case of insulin glargine, which precipitates upon subcutaneous injection to allow for prolonged release. Hydrogels with sustained drug release, high tissue biocompatibility, and patient self-administration have been developed because of recent advances in polymer science. Hydrogels produce a deformable drug depot that steadily elutes a high drug concentration to surrounding tissue over time. However, because most hydrogels bind to drugs chemically rather than covalently, drug release is rapid and occurs over a few hours to days, limiting their use for long-term drug delivery.

4.4.3 Oral

While oral administration is the most common method, the high molecular weight of protein drugs makes it nearly impossible due to permeability issues and chemical instability. Bioavailability inconsistency, which is common in oral formulations, is also unacceptably high for biological drugs with a very limited therapeutic range. The use of lipophilic insulin and thyrotropin-releasing hormone derivatives made by fatty acylation with palmitic or lauric acid for oral administration is being investigated. The transformed drug molecules spontaneously form vesicle-like structures (Prosome®, Pharmacosome®), greatly increasing drug bioavailability and circulation time in patients (table 4.12).

Microspheres, liposomes, or nanoparticles encapsulate the polypeptide drug inside a polymeric, phospholipid, or carbohydrate particulate delivery system.

4.4.4 Nasal/pulmonary

The lung allows for rapid and high drug absorption due to its large surface area (approximately 80–140 m^2), very thin alveolar epithelium thickness (0.1–0.5 mm), and large blood supply. Because it avoids the hepatic first-pass effect, is noninvasive, effective at lower doses, and can be used locally or systemically, pulmonary drug delivery is advantageous. Although lung tissue has low enzymatic activity compared to the gastrointestinal tract, the pulmonary epithelium has several immunological properties. However, there are some disadvantages to pulmonary delivery, such as a

Table 4.12. Orally administered biopharmaceuticals.

Product	Drug	Route	Indications
Minirin	Desmopressin	Oral, nasal	Cranial diabetes insipidus or nocturia associated with multiple sclerosis
Sandimmune	Cyclosporine A	Oral	Immunosuppressants
Colomycin	Colistin	Oral	Intestinal infection (caused by sensitive gram-negative organisms)
Cytorest	Cytochrome C	Oral	Leucopenia
Cachexon®	Glutathione	Oral	AIDS-related cachexia
Ceredist OD	Taltirelin	Oral	Spinocerebellar ataxia
Anginovag	Tyrothricin	Oral	Pharyngitis
Vancocin	Vancomycin	Oral	Infection, clostridium difficile-associated diarrhea
Oral-Lyn	Insulin	Buccal	Diabetes mellitus

short duration of action due to the drug's rapid removal. Inhaled drugs are swept from the airways toward the mouth after deposition in the lungs, where they are phagocytosed by alveolar macrophages and removed. As a result, a significant slow-drug release necessitates a method to prevent or suspend the lungs' normal clearance mechanisms before the encapsulated drugs can be administered effectively. In general, proteins with molecular weights ranging from 6000 to 50 000 D have high bioavailability after inhalation. As a result, pulmonary administration has gained a lot of traction as a promising protein drug delivery method.

Only a few of the drugs currently being developed for pulmonary delivery include interleukin-1 receptor (asthma therapy), heparin (blood clotting), human insulin (diabetes), −1 antitrypsin (emphysema and cystic fibrosis), interferons (multiple sclerosis and hepatitis B and C), and calcitonin and other peptides (osteoporosis). Inhalation delivery methods that target specific tissues or organs can be used to deliver gene therapy. Thanks to Inhale's innovative dry powder formulation, packaging, and filling, combined with aerosol system technology, many patients who previously received injections will be able to inhale medication into the deep lung independently and painlessly, where it will be absorbed naturally and effectively into the bloodstream.

Because it plays such an important role in the efficacy of pulmonary drug administration, choosing a delivery system is crucial in the formulation design for pulmonary drug delivery. The most common instruments used to administer therapeutics as aerosols are nebulizers (e.g. jet nebulizers, ultrasonic nebulizers, and vibrating mesh nebulizers), metered-dose inhalers, and dry powder inhalers.

Various nanotechnology-based formulation approaches have been thoroughly investigated for successful protein delivery via the pulmonary route. Nanoparticles, in general, are promising as a protein delivery carrier in the lungs due to their ability to target and release drugs in a controlled manner. Nanoparticles smaller than 200 nm can also avoid detection by alveolar macrophages, resulting in better absorption and drug action. In addition to polymeric nanoparticles, other nanocarriers such as liposomes and solid lipid nanoparticles have been used for pulmonary delivery of protein drugs. The sections that follow take a closer look at these nanocarriers.

Inhalable insulin is a powdered form of insulin that is inhaled and absorbed through the lungs. Inhaled insulins are consumed faster than subcutaneously injected insulins, with a higher peak blood concentration and a faster metabolism. Exubera, the first inhaled insulin medication marketed by Pfizer in 2006, was developed by Inhale Therapeutics (later renamed Nektar Therapeutics), but it was withdrawn in 2007 due to low sales. The FDA approved Mannkind's monomeric inhaled insulin, Afrezza, in 2014. Dypreza inhaled insulin from Highlands Pharmaceuticals was approved for sale in Europe in 2013 and the United States in 2016. The need for precise dosing is critical with inhalable insulin, which is not always possible when the medication is administered through a system.

Although there are some advantages to using the nasal route, such as increased bioavailability and ease of administration, it also results in delivery to the brain. Similarly, macromolecules pass through the lungs very easily, making pulmonary delivery a viable noninvasive protein delivery option. Insulin inhaled is absorbed faster than insulin injected subcutaneously, resulting in a better physiological response to a meal. On the other hand, an inhalation insulin device was recalled due to dose inconsistency. The nasally administered items are listed in table 4.13.

The blood–brain barrier is a major barrier in the treatment of many neuronal degenerative disorders because it regulates the passage of most therapeutics, including proteins, into the central nervous system. In this case, the nasal route may be more effective than the oral or parenteral routes. The incorporation of absorption enhancers to promote permeation through the membrane, mucoadhesive formulations to improve nasal residential time, and prodrug approaches to optimize absorption. Bile salts, surfactants, fluidic acid derivatives, phosphatidylcholines, fatty acids (Tauro dihydro fusidate), cyclodextrins (CDs), cationized polymers, chelators, and cell penetration

Table 4.13. Nasally administered biopharmaceuticals.

Product	Drug	Route	Indications
Antepan	Protirelin	Nasal	Hypothyroidism and acromegaly
Desmospray	Desmopressin	Nasal	Cranial diabetes insipidus or nocturia associated with multiple sclerosis
FluMist® Quadrivalent	Vaccine	Nasal	Influenza
Fortical®	Salmon calcitonin	Nasal	Hypercalcemia or osteoporosis
Kryptocur	LHRH	Nasal	Cryptorchism
Miacalcin	Salmon calcitonin	Nasal	Hypercalcemia or osteoporosis
Minirin	Desmopressin	Oral, nasal	Cranial diabetes insipidus or nocturia associated with multiple sclerosis
Suprecur	Buserelin	Nasal	Prostate cancer, endometriosis
Suprifact	Buserelin	Nasal	Prostate cancer, endometriosis
Synarel	Nafarelin	Nasal	Endometriosis
Syntocinin	Oxytocin	Nasal	To start or strengthen uterine contractions

peptides are all absorption enhancers that help drugs pass through the nasal membrane more easily. Because protein bioavailability is reduced by a short residence time in the nasal cavity, mucoadhesive systems are used to extend nasal retention time. As a result of the use of Carbopol 941 and carboxymethyl cellulose, increased calcitonin and insulin nasal bioavailability was achieved. Mucoadhesive polymers also improve permeation by loosening the tight junctions in the nasal epithelium. As a result, mucoadhesive micro-/nanoparticles, which have a longer residence time and better permeation through the membrane, are useful carriers for protein drug delivery through the nose.

4.4.5 Transdermal

Many cytokines are applied topically but using liposomes to deliver a human epidermal growth factor improves the effect dramatically. Peptide drugs could be delivered via the skin's pilosebaceous pathway using niosomes (liposomes made of nonionic surfactants). When vesicles made of glyceryl dilaurate cholesterol and polyoxyethylene-10-stearyl ether are used, interferon alpha and cyclosporine are better absorbed. Another type of liposome used to deliver insulin through topical application *in vivo* is transfersomes (a phosphatidylcholine/sodium cholate mixture).

Penetration enhancers such as N-alkylazacycloheptanones (Azone) for desglycinamide arginine vasopressin, which temporarily compromise the integrity or physicochemical characteristics of the skin, are used to transmit peptides through the skin. *In vitro*, the nonionic surfactant n-decyl methyl sulfoxide increases Leu-enkephalin penetration through hairless mouse skin, and a urea/ethanol/menthol/camphor/methyl salicylate hydroxypropyl cellulose gel increases absorption of the nonapeptide leuprolide (a luteinizing hormone releasing hormone analog) through hydration and keratolytic effect increases absorption of the nonapeptide leuprolide.

In recent years, many researchers have used iontophoresis, a process that involves electrical stimulation of skin permeability to improve the delivery of short peptides (model tripeptides, vasopressin), growth hormone releasing factor (amino acids 1–44), insulin, and luteinizing hormone releasing hormone. Ultrasonic vibration has been used to deliver insulin *in vivo* with some success.

Microneedles are extremely small needles that can create microchannels in the stratum corneum. Their sizes range from 50 to 900 μm. Microneedles increase patient compliance and provide a flexible medium for hydrophilic and high molecular weight drugs, including protein drugs, to overcome the skin barrier because they are a painless instrument. They are made of silicon, plastics, biodegradable polymers, and carbohydrates. The first generation of microneedles used strong microneedles to perforate the skin membrane and increase drug permeability. While solid microneedles appeared to be effective in delivering insulin, their use is restricted due to poor delivery performance, complicated administration, a lack of precise dosing, and the risk of infection. The drug payload has recently been coated directly onto the microneedle surface, resulting in stronger microneedles. Dip-coating, casting, and deposition are some of the methods used to coat the microneedles. Biodegradable microneedles deliver continuous drug release after the biodegradable polymer matrix is hydrolyzed. Polymers with a high molecular weight and cross-linking density are preferred when making these microneedles.

In sonophoresis, ultrasonic waves are used to improve drug permeability in the skin. The air pockets in the stratum corneum expand and oscillate because of ultrasound waves, disrupting the lipid bilayer and creating cavities that increase drug permeability through the skin. The physicochemical properties of the drug, as well as the net ultrasound exposure period and the pulse 'on' length, determine the degree of drug delivery via sonophoresis. While sonophoresis can help with biopharmaceutical transdermal delivery, the risk of protein instability from ultrasound should be considered. Sonophoresis combined with other enhancement methods such as chemical enhancers, electroporation, and iontophoresis improves drug delivery through the skin more effectively than sonophoresis alone.

The most recent approach in the transdermal delivery of proteins and peptides is electroporation, which uses ultra-short pulses lasting a few milliseconds and a strength of a few hundred volts to cause changes in the skin and enable hydrophilic compounds to move. Both electroporation and iontophoresis use an electric field to propel a drug into the skin. However, electroporation works primarily on the skin to improve drug penetration by changing membrane permeability, whereas iontophoresis works directly on the drug to propel it into the skin.

4.4.6 Ocular

Protein delivery to the eye is obstructed by the blood–retinal barrier and efflux transporters expressed in the posterior segment. The viscosity of formulations has an impact on ocular drug delivery. The cornea spends more time in contact with the eye's surface when the viscosity is high. Nonetheless, it causes reflex weeping and blinking, altering the viscosity of the formulation.

Two products have already been approved for ocular delivery: an anti-vascular endothelial growth factor (anti-VEGF) aptamer and a monoclonal antibody (Lucentis; Ranibizumab). Because traditional topically applied dosage types, such as eye drops, have the major drawback of low bioavailability and thus low therapeutic efficacy, several new strategies have been developed to overcome ocular delivery barriers and increase protein bioavailability via the ocular route of administration. Chemical chaperones and recombinant human hyaluronidase coadministration, for example, aid protein delivery through the ocular path. Because protein aggregation is a major concern in the development of proteins for ocular diseases, a novel strategy involving the use of chemical chaperones (protein aggregation inhibitors) has been developed to prevent protein misfolding and inhibit the self-assembly of aggregation-prone sequences in native protein structures. The use of recombinant hyaluronidases in combination with biopharmaceutical drugs has long been used to improve drug penetration through ocular tissue barriers. Hyaluronidases are enzymes that break down hyaluronic acid, which is an important component of tissue structure. Polymeric micelles, liposomes, nanospheres, nano wafers, and dendrimers are among the nanocarriers being investigated for their ability to deliver proteins in a controlled and targeted manner through the ocular path. Table 4.14 lists the drugs that are administered through the eyes.

Table 4.14. Ocular biopharmaceutical products.

Product	Drug	Route	Indications
Cenegermin	Oxervate	Eye drop	Neurotrophic keratitis
Eylea	Aflibercept	Ocular	Wet age-related macular degeneration (WAMD), diabetic macular edema (DME) or diabetic retinopathy (DR) in DME, macular edema Following retinal vein occlusion (MEtRVO) WAMD, DME or DR in DME, MEtRVO, myopic choroidal neovascularization (mCNV)
Lucentis	Ranibizumab	Ocular	

Nano wafers are small transparent circular or rectangular membranes with arrays of drug-loaded nano reservoirs that release drugs in a more controlled and long-lasting manner than eye drops (a few hours to several days). Among the polymers used are polyvinyl alcohol, polyvinyl pyrrolidone, hydroxypropyl methylcellulose, and carboxymethyl cellulose. Nano wafers are placed on the patient's fingertip and can withstand continuous blinking without being removed. They release the drug slowly, increasing drug resident time and absorption into the ocular tissues, and improving therapeutic efficacy. Furthermore, during drug release, the nano wafer slowly dissolves, leaving the ocular surfaces polymer-free.

To administer ocular drugs, drug-loaded contact lenses can be used. With contact lenses, a longer drug residence period in the eye improves drug permeation into the cornea. Long-term drug release is possible because drug molecules diffuse slowly from the lens matrix. The residence time and drug release rate are increased by encapsulating the drug in nanocarriers and dispersing the drug-loaded nanocarriers in the lens matrix. Drug leaching during storage and delivery, as well as surface roughness-related safety concerns, are all disadvantages of using contact lenses that must be addressed.

4.4.7 Rectal

Protein drugs, which are highly susceptible to physicochemical and enzymatic destabilization, benefit from the use of absorption enhancers, protease inhibitors, prodrugs, and nano-formulations. Absorption enhancers are required for insulin, heparin, calcitonin, recombinant human granulocyte colony-stimulating factor (rhG-CSF), and human chorionic gonadotrophin. Although a variety of absorption enhancers are used in rectal drug delivery, some of them can irritate the mucous membrane and cause damage. Meanwhile, protease inhibitors will improve rectal bioavailability by reducing protein degradation and increasing protein drug enzymatic stability. Protein and peptide absorption can be improved by using prodrugs to protect them from peptidases and other mucosal enzyme degradation. Furthermore, there are nanotechnology-based formulation approaches for improving protein drug delivery via rectal administration.

4.5 Formulation technologies

4.5.1 Hydrogels and *in situ* forming gels

Because of their thermodynamic compatibility with water, hydrogels are three-dimensional polymeric networks made up of crosslinked hydrophilic and biocompatible polymers that swell in aqueous media. Clinical applications for hydrogels include contact lenses, biosensors, tissue engineering components, and drug delivery carriers. Protein drugs can also be delivered more safely and comfortably using hydrogels. Some of the polymers used in protein delivery hydrogels include 2-hydroxyethyl methacrylate, ethylene glycol dimethyl acrylate, N-isopropylacrylamide, acrylic acid, methacrylic acid (MAA), poly (ethylene glycol) (PEG), and poly (vinyl alcohol) (PVA).

For at least three months, a hydrogel associated particulate formulation releases the protein in its active state while maintaining the therapeutic concentration. Hydrogels are polymeric materials that do not dissolve in water and swell dramatically in an aqueous medium under physiological conditions. Cross-linking is a method of forming networks of covalently linked polymer main chains, which can also include strong non-covalent interactions. Polymer chains are crosslinked, which prevents the polymer from fully dissolving. Hydrogels made of hydrophilic polymers, as a result, can absorb water into their structure, causing their network to swell. Because of their high water content, hydrogels are biocompatible and used for tissue regeneration. Despite the fact that hydrogels have potential advantages over other drug delivery methods, their high water content makes designing extended formulations for drug release difficult.

The mechanical properties of hydrogels are also important in pharmaceutical applications because they must maintain their integrity before proteins are released at target sites. To achieve the desired mechanical properties, the degree of hydrogel cross-linking must be adjusted, as a higher degree of cross-linking results in a stronger but more fragile structure. Hydrogels that are both solid and elastic are made using copolymerization.

Hydrogels are frequently designed using physiological stimuli such as pH, ionic strength, and temperature. In response to environmental stimuli, such hydrogels can change their swelling behavior, network structure, permeability, and mechanical strength dramatically. For example, pH-triggered drug release systems are frequently used to protect proteins from the harsh gastric environments, allowing for more efficient oral delivery of protein drugs. Ionic hydrogels with pendant groups that ionize in response to changes in pH in the environment, causing the hydrogel network to swell, are known as pH-responsive hydrogels.

Nanogels are crosslinked polymer nanoparticles with hydrodynamic sizes ranging from 10 to 100 nm that are distributed in an aqueous medium while maintaining their fixed conformation. They are made from both natural and synthetic polymers. Nanogels have the advantage of being able to control and tailor their scale, surface charge, network density, and chemical functional groups to achieve the structural and functional properties that are desired.

Insulin molecules have been covalently bound to light, highly hydrophilic, and multifunctional nanogels for nasal delivery. After intranasal administration, poly

(N-vinyl pyrrolidone)-based nanogels covalently bound to insulin cross the blood–brain barrier and show neuroprotection against amyloid β-induced dysfunction, compared to free insulin.

4.5.2 Nanoparticles

Nanoparticles increase protein physicochemical stability in the gastrointestinal tract by encapsulating proteins in a polymeric matrix with a size range of 10–1000 nm. When used as an oral protein carrier, nanoparticles should be non-toxic and non-immunogenic. Nanoparticles are also important for absorption, distribution, removal, and *in vivo* action in the gastrointestinal tract. For example, nanoparticles smaller than 100 nm are absorbed easily through the intestinal mucosa, but nanoparticles larger than 500 nm have a much lower absorption rate. Unique ligands on the surfaces of nanoparticles may be used to target receptor-mediated transport pathways.

They can be delivered in a variety of ways, including topical, periocular, suprachoroidal, and intravitreal nanoparticles. Intravitreal injection of nanoparticles, on the other hand, causes vitreous clouding due to the light scattering properties of polymeric particles. The loss of bioactivity, low protein stability due to interactions with the nanoparticle matrix, and comprehensive nanoencapsulation methods all complicate the delivery of nano-formulations of proteins. Only a few nanoparticle-mediated ocular delivery systems are being developed at this time.

Natural and synthetic polymers are used to make nanoparticles. The most used materials for nanoparticle preparation include polylactic acid, polylactic-co-glycolic acid, chitosan, gelatin, polymethylmethacrylate, and poly-alkyl-cyanoacrylate. Chitosan is a deacetylated chitin copolymer made up of glucosamine and N-acetyl-glucosamine. Chitosan is a good choice for protein delivery carriers because of its biocompatibility, muco-adhesion, and low toxicity. Chitosan also improves cellular uptake by opening the close junction.

Alginate is a natural anionic polymer that is commonly used as a drug carrier in the pharmaceutical industry. When electrostatically interacting with cationic materials, it can easily form gels due to its anionic surface charge. The high porosity of alginate beads, on the other hand, causes drug leakage. To solve this problem, chitosan or dextran sulfate are commonly used in conjunction with alginate.

In addition to natural polymer-based nanoparticles, various synthetic polymers are used as oral delivery carriers for protein drugs. The polymer polylactic-co-glycolic acid (PLGA) is a representative polymer for oral protein delivery. Lactic acid and glycolic acid form a ring-opening copolymer known as PLGA. Because of its biodegradability and biocompatibility, PLGA is an excellent drug delivery vehicle.

Redox-activated nanocarriers, which are also susceptible to glutathione as a cellular redox regulator, have been proposed as effective drug and gene delivery mechanisms.

The use of nanoparticles to deliver protein drugs through the nose has sparked a lot of interest in recent years. Particularly mucoadhesive nanoparticles spend more time in the nasal cavity. Insulin delivery through the nose has been studied extensively using chitosan nanoparticles.

Because chitosan nanoparticles have a positive charge they spend more time in contact with the nasal mucosal membrane, increasing insulin bioavailability. Intranasal administration of chitosan-N-acetyl-L-cysteine nanoparticles and PEG-g-chitosan nanoparticles improves insulin bioavailability. Intranasal vaccination primarily targets nasal-associated lymphoid tissue, and chitosan, PLGA, and polystyrene polymeric nanoparticles were found to be effective in terms of antigen absorption. By extending the antigen's residence time, trimethyl chitosan nanoparticles improved IgA and IgG efficiency. Similarly, nerve growth factor uptake in the brain is significantly improved by nasal administration of mucoadhesive chitosan nanoparticles.

Polymeric nanoparticles are commonly used as pulmonary drug delivery carriers due to their biocompatibility and ease of surface modification and copolymerization. Chitosan, alginate, and gelatin are the most common natural polymeric carriers. Poloxamer, poly(lactic-co-glycolic) acid, and polyethylene glycol are the most used synthetic nanocarriers for pulmonary drug delivery.

Polymeric nanoparticles, as well as a polymer-based thermo-gelling method, are useful in ocular drug delivery.

Carbon nanotubes are cylinders of carbon with unique physicochemical properties that are easy to manipulate on the surface. They are carbon allotropes that belong to the fullerene family. They are also ideal for targeted or controlled drug delivery, biosensing, and bioimaging due to their superior mechanical properties, high thermal conductivities, and ability to penetrate cell membranes.

Nanoparticles have lipid-fluidizing properties, which affect skin permeability by changing the extracellular lipids in the stratum corneum.

Lipid-based nanocarriers such as liposomes, solid lipid nanoparticles, and nanostructured lipid carriers have been studied for ocular protein delivery. Multiple intravitreal injections may raise the risk of ocular complications such as vitreous hemorrhage, endophthalmitis, retinal detachment, and cataracts; therefore, prolonged drug release may help to lower the risk of these ocular complications.

Solid lipid nanoparticles have several advantages, including physical stability, targetability, controlled release, fast scale-up, and non-toxicity, because they are made up of physiological lipids. They also increase drug absorption through the cornea, which increases the ocular bioavailability of both hydrophilic and lipophilic drugs.

Niosomes, self-assembling nanovesicles composed of nonionic surfactants that behave like liposomes, are also favored for topical ocular drug delivery due to their chemical stability, biodegradability, biocompatibility, lack of immunogenicity, and low toxicity. They can also encapsulate lipophilic and hydrophilic drugs, giving them structural flexibility. Discomes are a type of niosome with wide structures (12–16 mm) as a result of the addition of Solulan C24 (a nonionic surfactant). They are well-suited for ocular administration because they can prevent drainage into the systemic pool due to their large scale. Discomes are also disc-shaped, making it easier for them to fit into the cul-de-sac of the eye. The development of niosomes and discomes is still in its early stages.

Although the benefits of polymeric nanoparticles have been demonstrated for the majority of delivery routes, little progress has been made in developing polymeric nanoparticles for rectal administration. Chitosan and its derivatives, such as PLGA, PLA, and methacrylic acid copolymers, are commonly used to make polymeric nanoparticles for rectal drug delivery. Additionally, the surface of these polymeric nanoparticles is modified to provide additional benefits such as site-specificity or a longer circulation period.

Rectal administration of macromolecules is done with nano-sized liposomes, and there have been a few studies looking into liposomal formulations. For mucosal immunization, updated nanoliposomes containing hepatitis B surface antigen are suggested, consisting of a 1,2-dipalmitoyl-sn-glycerol-3-phosphocholine bilayer engulfing a solid fat center (mainly glyceryl tripalmitate) and using monophosphoryl lipid A as an adjuvant and containing hepatitis B surface antigen. These hybrid liposomes show higher stability and major humoral and cellular immune responses in rats after intracolonic administration. For rectal drug delivery, solid lipid nanoparticles are used as an alternative nanocarrier. Several examples of their rectal application have been published so far, but no evidence that they are superior to traditional formulations has been provided.

Because the stratum corneum is hydrophobic, nanocarriers in lipophilic vehicles should be able to effectively penetrate it. Nano-emulsion is a nanocarrier that is a low viscosity isotropic dispersed system made up of two immiscible liquid phases. Nano-emulsions are created using high-pressure homogenization, phase-inversion temperature, and micro-fluidization. Because nano-emulsion is a thermodynamically unstable product, physical instability such as creaming and flocculation during long-term storage is a major disadvantage. The system achieves a metastable state by optimizing particle size and surfactant composition. While nano-emulsions are less commonly used as a transdermal antigen delivery vehicle due to their thermodynamic instability, transcutaneous immunization of nano-dispersions has shown some promise.

Polymeric micelles, which range in size from 10 to 100 nm, are nanocarriers made up of amphiphilic block copolymers with hydrophilic chains forming a shell and hydrophobic chains forming a heart. They can self-assemble in aqueous media to form an ordered supramolecular structure at concentrations above their essential micellar concentrations. Polymers used to make polymeric micelles for ocular distribution include poloxamer 407, poloxamer 188, methoxy poly (ethylene glycol)-poly(e-caprolactone), poly (butylene oxide)-poly (ethylene oxide)-poly (butylene oxide), polyhydroxyethylaspartamide, and isopropylacrylamide.

Dendrimers are small polymeric carriers that can capture and conjugate high molecular weight molecules. They are tree-like molecules with well-defined, homogeneous, and monodisperse structures that are radially symmetric. Some of the most commonly used dendrimers are polyamidoamines, polyamines, polyamides (polypeptides), poly(aryl ethers), polyesters, and carbohydrates.

4.5.3 Liposome

By encapsulating proteins inside an aqueous center, liposomes are used to increase the membrane permeability of protein drugs. The structural similarity of liposomes to cellular membranes aids in intestinal absorption. Liposomes, on the other hand, have some disadvantages as an oral protein carrier, such as chemical and enzymatic instability in the gastrointestinal tract. Because surface coating is beneficial for oral drug delivery in liposomes to address stability issues, various approaches to altering the surface of liposomes using ligands that interact with specific receptors on the cellular membrane present possibilities. Lectins, a type of glycoprotein found in plants, are a promising ligand for specific binding to mucosal carbohydrate receptors.

Liposomes are bilayer vesicles made of phospholipids with both hydrophilic and hydrophobic compounds encapsulated in the aqueous core or intercalated into the bilayer structure. Liposomes are made up of phospholipid molecules arranged molecularly in water. Hydrophilic phosphate head groups are exposed to the aqueous environment due to their amphiphilic composition. When the hydrocarbon chains link together, they form a lipid film. When water is added and stirred, the lipid layer transforms into covered vesicles. Liposomes can have a single bilayer (unilamellar) or multiple bilayers (multilamellar), with small unilamellar vesicles ranging in size from 20–100 nm and large unilamellar vesicles ranging in size from 100–1000 nm. Liposomes are made using dry lipid hydration, freeze–thawing extrusion, reverse evaporation, and double emulsification. Lipid film hydration, mechanical dispersion to form liposomes, and solvent removal are the main steps. While vigorous shaking is commonly used to disperse liposomes, it results in multilamellar vesicles that are polydispersed. Extrusion through a narrow orifice manipulates liposome size to produce mono-dispersed small unilamellar vesicle liposomes. Several physical stresses (such as heat, organic solvents, and agitation) are present during liposome preparation, affecting protein stability.

In the capsulated form, proteins in liposome polymeric particles have a longer release time. The bilayer destabilization causes liposome breakdown and thus the release of encapsulated agents. Protonation of the phospholipid head group and acid-catalyzed bilayer hydrolysis are two processes that break down biolayers *in vivo*. Changes in the lipid bilayer can change the drug's kinetic profile. PEGylation of the liposomal surface reduces protein interactions with biological fluids, preventing liposomes from aggregating and increasing stability.

Archaeosomes are a lipid-based oral delivery system made up of polar lipids from various archaea. They have unique structural features that enable them to maintain stability at high temperatures, low or high pH, and in the presence of phospholipases and bile salts, potentially resulting in improved gastrointestinal stability. As a result, they have attracted a lot of interest as a protein carrier, including vaccines. Nasal administration of drug-loaded liposomes has proven to be effective in distributing biopharmaceuticals and peptides intranasally because it allows direct nose-to-brain drug delivery via nanoparticles.

Liposomes are efficient carriers for delivering therapeutics into the skin. Because their components are identical to skin lipids, liposomes are quickly absorbed by the epidermis and reach the deepest layers of the skin. In addition to introducing higher concentrations of a drug into the skin, hydration layers increase absorption by molecular mixing of the liposome bilayer with intracellular lipids in the stratum corneum.

A noninvasive macromolecule transdermal delivery system has been developed using protein encapsulation in liposomes. Many advanced liposomes have been used to increase macromolecule permeation through the stratum corneum, which limits drug delivery through the skin. For more efficient macromolecule delivery into the skin, liposomal formulations are integrated into a dissolving microneedle array.

Because of their enhanced and sustained drug release, biocompatibility, biodegradability, and non-immunogenicity, liposomes are the most effective pulmonary carrier for protein drugs. By changing the drug's physicochemical properties (making it hydrophobic) and decreasing mucociliary clearance, liposomes may improve drug permeation through the alveolar epithelium (due to their surface viscosity). Liposomal formulations for protein drug delivery to the lungs have resulted as a result.

In addition to liposomes, solid lipid nanoparticles (SLNs) have been tested as carriers of biopharmaceuticals to deliver through the lung epithelium; for example, spray-dried powders containing SLNs have solved the issue of low inertia, which can prevent nanoparticles from settling in the lungs.

4.5.4 Higher concentration formulations

High concentrations are frequently required to accommodate the lower volume when combined with their desirable applications for subcutaneous administration. Unfortunately, as protein concentrations rise the physical properties of such a material can change dramatically. The opalescence, viscosity, and protein aggregation/immunogenicity of the solution could all be significantly improved. Biotherapeutic products' manufacturing, administration, and marketability are all threatened by such altered properties.

Because subcutaneous routes have a limited volume (1.5 ml), formulations with large doses of more than 1 mg kg^{-1} or 100 mg per dose are formulated at concentrations greater than 100 mg ml^{-1}. Developing high-concentration formulations for proteins with the ability to aggregate is a difficult task. Protein interaction at higher concentrations can lead to reversible self-association, which can lead to the formation of insoluble aggregates. The likelihood of one molecule colliding with another increases, increasing the likelihood of reversible oligomers such as dimers, tetramers, and so on. Covalent linkages, for example, are one mechanism that causes aggregates to form (e.g. disulfide exchange). Even when the native structure undergoes minor conformational changes, aggregates can form. The chances of this happening are higher at higher concentrations.

A high-concentration solution is one in which the solutes take up a significant amount of space. Another definition of a high-concentration solution is when the molecular size and the distance between the van der Waals' surfaces are on the same scale. The molecular proximity intermolecular space is referred to as 'high concentration'. The main impediment to achieving a concentration formulation is the target protein's solubility. Solubility is controlled by molecular properties (sequence, charge distribution, etc) as well as solution state (pH, ionic strength and concentration, etc). The amount of protein that can be present in a solution without forming visible particles, precipitates, or other clumps is known as solubility. A more technical term is the maximum amount of protein that remains in solution after 30 min of centrifugation at 30 000 g in the presence of co-solute. Aside from solubility, there are a few other factors to consider when formulating mAbs. Examples include opacity, viscosity, and aggregation. In nephelometry, the turbidity unit is commonly used to express opalescence. Reversible protein–protein and liquid–liquid phase separation causes opacity. Protein–protein interaction is a significant factor that can affect opalescence and viscosity at high concentrations. Reversible self-association, viscosity increase, opalescence, and even aggregation can occur when molecules are close together.

The increase in viscosity has a direct effect on manufacturability and injectability at high concentrations. In large-scale manufacturing, tangential flow filtration (TFF) is a common technique for buffer exchange and protein concentration (clinical and commercial). Because rapid pumping and continuous circulation through the narrow tubing cause significant cavitation and shear stress, an increase in viscosity at high concentrations can cause back pressure that exceeds the pump's capability, putting the mAb under a lot of stress. Increased viscosity can lead to an increase in back pressure, making the protein more unstable. At the very least, this increases the manufacturing cost by lengthening the production time.

Increased viscosity has a significant impact on the administration of subcutaneous dosage forms. The glide force describes how easy it is to inject subcutaneously by describing the force used to drive the liquid through the syringe. The most important factor affecting the glide power of a subcutaneous solution is viscosity. As it rises, the pain at the injection site rises as well, resulting in lower patient compliance.

One of the most recent emerging concepts for addressing protein stability and high viscosity issues is the development of 'nanoclusters', tightly packed protein molecules developed in the presence of a crowder such as trehalose. At extremely high concentrations (up to 320 mg ml^{-1}), protein molecules are crammed into colloidally stable dispersions of distinct nanoclusters (35–80 nm) with hydrodynamic diameters of equilibrium. In nanoclusters, the distances between protein molecules may be smaller than in bulk solution. Although a shorter distance between proteins may improve protein–protein interactions, it may also compromise protein stability. The nanocluster concept is new and needs to be developed further.

4.6 Examples of formulation

Tables 4.15–4.27 list a few commercial product compositions.

This item is packaged separately as 0.54% solution (2 ml diluent for lyophilized product).

Table 4.15. Oprelvekin injection (interleukin IL-11).

Bill of materials (batch size 1 l)					
Scale/mL		Item	Material	Quantity	UOM
1.00	mg	1	Oprelvekin (interleukin IL-11)	1.00	g
4.60	mg	2	Glycine	4.60	g
0.32	mg	3	Dibasic sodium phosphate heptahydrate	0.32	g
0.11	mg	4	Monobasic sodium phosphate monohydrate	0.11	g
qs (quantity sufficient)	mL	5	Water for injection, qs to	1.00	L

Table 4.16. Interleukin injection (IL-2).

Bill of materials (batch size 1 l)					
Scale/mL		Item	Material	Quantity	UOM
0.25	mg	1	IL-2	0.25	g
0.70	mg	2	Sodium laurate	0.70	g
10.00	mM	3	Disodium hydrogen phosphate	10.00	M
50.00	mg	4	Mannitol	50.00	g
qs	mL	5	Hydrochloric acid for pH adjustment 1 M	qs	mL
qs	mL	6	Water for injection, qs to	1.00	L

Table 4.17. Interferon alfa-2a injection.

Bill of materials (batch size 1 l)					
Scale/mL		Item	Material	Quantity	UOM
3MM	IU	1	Interferon alfa-2a	3B	IU
7.21	mg	2	Sodium chloride	7.21	g
0.20	mg	3	Polysorbate 80	0.20	g
10.00	mg	4	Benzyl alcohol	10.00	g
0.77	mg	5	Ammonium acetate	0.77	g
qs	mL	6	Water for injection, qs to	1.00	L

Table 4.18. Interferon beta-1b.

Bill of materials (batch size 1 l)					
Scale/mL		Item	Material	Quantity	UOM
0.30	mg	1	Interferon beta-1b	0.30	g
15.00	mg	2	Albumin human	15.00	g
15.00	mg	3	Dextrose	15.00	g
5.40	mg	4*	Sodium chloride	5.40	g
qs	mL	5	Water for injection, qs to	1.00	L

Table 4.19. Interferon beta-1a injection.

Bill of materials (batch size 1 l)					
Scale/mL		Item	Material	Quantity	UOM
33.00	mcg	1	Interferon beta-1a	33.00	mg
15.00	mg	2	Albumin (human)	15.00	g
5.80	mg	3	Sodium chloride	5.80	g
5.70	mg	4	Dibasic sodium phosphate	5.70	g
1.20	mg	5	Monobasic sodium phosphate	1.20	g
qs	mL	6	Water for injection, qs to	1.00	L

Table 4.20. Interferon alfa-n3 injection.

Bill of materials (batch size 1 l)					
Scale/mL		Item	Material	Quantity	UOM
5 MM	U	1	Interferon alpha-n3	5B	U
3.30	mg	2	Liquefied phenol	3.30	g
1.00	mg	3	Albumin (human)	1.00	g
8.00	mg	4	Sodium chloride	8.00	g
1.74	mg	5	Sodium phosphate dibasic	1.74	g
0.20	mg	6	Potassium phosphate monobasic	0.20	g
0.20	mg	7	Potassium chloride	0.20	g
qs	mL	8	Water for injection, qs to	1.00	L

Table 4.21. Interferon alfacon-1 injection.

Bill of materials (batch size 1 l)					
Scale/mL		Item	Material	Quantity	UOM
0.03	mg	1	Interferon alfacon-1	0.03	g
5.90	mg	2	Sodium chloride	5.90	g
3.80	mg	3	Sodium phosphate	3.80	g
qs	mL	4	Water for injection, qs to	1.00	L

Table 4.22. Interferon gamma-1b injection.

Bill of materials (batch size 1 l)					
Scale/mL		Item	Material	Quantity	UOM
200.00	mcg	1	Interferon gamma-1b	200.00	mg
40.00	mg	2	Mannitol	40.00	g
0.72	mg	3	Sodium succinate	0.72	g
0.10	mg	4	Polysorbate 20	0.10	g
qs	mL	5	Water for injection, qs to	1.00	L

Table 4.23. Infliximab for injection.

Bill of materials (batch size 1 l)					
Scale/mL		Item	Material	Quantity	UOM
10.00	mg	1	Infliximab	10.00	g
50.00	mg	2	Sucrose	50.00	g
0.05	mg	3	Polysorbate 80	0.05	g
0.22	mg	4	Monobasic sodium phosphate monohydrate	0.22	g
0.61	mg	5	Dibasic sodium phosphate dihydrate	0.61	mg
qs	mL	6	Water for injection, qs to	1.00	L

Table 4.24. Daclizumab for injection.

Bill of materials (batch size 1 l)					
Scale/mL		Item	Material	Quantity	UOM
5.00	mg	1	Daclizumab	5.00	g
3.60	mg	2	Sodium phosphate monobasic monohydrate	3.60	g
11.00	mg	3	Sodium phosphate dibasic heptahydrate	11.00	g
4.60	mg	4	Sodium chloride	4.60	g
0.20	mg	5	Polysorbate 80 (Tween®)	0.20	G
qs	mL	6	Water for injection, qs to	1.00	L
Qs	mL	7	Sodium hydroxide for pH adjustment	qs	
Qs	mL	8	Hydrochloric acid for pH adjustment	qs	
Qs	Cu ft	9	Nitrogen gas	qs	

Table 4.25. Coagulation factor VIIa (recombinant) injection.

Bill of materials (batch size 1000 vials)					
Scale/vial		Item	Material	Quantity	UOM
1.20	mg	1	rFVIIa	1.20	g
5.84	mg	2	Sodium chloride	5.84	g
2.94	mg	3	Calcium chloride dihydrate	2.94	g
2.64	mg	4	Glyclyglycine	2.64	g
0.14	mg	5	Polysorbate 80	0.14	g
60.00	mg	6	Mannitol	60.00	g

Table 4.26. Reteplase recombinant for injection.

Bill of materials (batch size 1000 vials)					
Scale/vial		Item	Material	Quantity	UOM
18.10	mg	1	Reteplase	18.10	g
8.32	mg	2	Tranexamic acid	8.32	g
136.24	mg	3	Dipotassium hydrogen phosphate	136.24	g
51.27	mg	4	Phosphoric acid	51.27	g
364.00	mg	5	Sucrose	364.00	g
5.20	mg	6	Polysorbate 80	5.20	g

Table 4.27. Alteplase recombinant injection.

Bill of materials (batch size 1000 vials)					
Scale/vial		Item	Material	Quantity	UOM
58MM	IU	1	Alteplase	100.00	g
3.50	g	2	L-arginine	3.50	kg
1.00	g	3	Phosphoric acid	1.00	kg
11.00	mg	4	Polysorbate 80	11.00	g
qs	mL	5	Water for injection, qs to	1.00	L

4.7 Conclusion

The delivery of drugs requires a dosage form that can deliver the active molecule to the site of action at a pre-determined rate and concentration. Biopharmaceutical products are large molecules that are inherently unstable in the environment of many routes of administration, leaving the parenteral route as the only possibility. Unlike the chemical products, the safety and efficacy of biopharmaceuticals is significantly altered based on the formulation and manufacturing technology applied. Many novel routes of administration are studied to create new applications of biopharmaceuticals. This chapter provides several model formulations for various class of biopharmaceuticals. Also provided in the chapter is a list of the physicochemical properties of the proteins and peptides that are approved by the FDA. This listing should help understanding the relationship between the structure and its properties (table 4.A1).

Appendix A: Physicochemical properties of proteins and peptides approved by FDA

Table 4.A1. Physicochemical properties of proteins and peptides approved by FDA.

Name	MOA	MW	Formula	IEP	Hydrophobicity	MP	Half-life
Abarelix	IIIc	1416	C72H95ClN14O14	NA	NA	NA	13.2 ± 3.2 d
Abatacept	IIa	92 300	C3498H5458N922O1090S32	NA	NA	NA	12–23 d
Abciximab	IIa	145 651	C6462H9964N1690O2049S48	6.16	−0.424	71	0.5 h
Adalimumab	Ic	144 190	C6428H9912N1694O1987S46	8.25	−0.441	NA	240–480 h
Aflibercept	Ib	115 000	C4318H6788N1164O1304S32	NA	NA	NA	7.13 d
Agalsidase beta	Ia	45 352	C2029H3080N544O587S27	5.17	−0.307	NA	45–102 min
Albiglutide	Ib	72 970	C3232H5032N864O979S41	NA	NA	NA	4–7 d
Aldesleukin	Ib	15 315	C690H1115N177O202S6	7.31	−0.192	NA	0.22–1.42 h
Alefacept	IIa	51 801	C2306H3594N610O694S26	7.86	−0.432	NA	270 h
Alemtuzumab	IIa	145 454	C6468H10066N1732O2005S40	8.76	−0.431		288 h
Alglucerase	Ia	55 597	C2532H3854N672O711S16	7.41	−0.168	NA	3.6–10.4 min
Alglucosidase alfa	Ic	105 271	C4435H6739N1175O1279S32	NA	NA	NA	2.3 ± 0.4 h
Alirocumab	Ic	146 000	C6472H9996N1736O2032S42	NA	NA	NA	17–20 d
Aliskiren	IIa	552	C30H53N3O6	NA	NA	NA	24 h
Alpha-1-proteinase inhibitor	IIa	44 325	C2001H3130N514O601S10	5.37	−0.302	59	NA
Alteplase	Ib	59 042	C2569H3928N746O781S40	7.61	−0.516	60	NA
Anakinra	IIa	17 258	C759H1186N208O232S10	5.46	−0.412	NA	4–6 h
Ancestim	Ib	18 500	NA	NA	NA	NA	NA
Anistreplase	Ic	59 042	C2569H3928N746O781S40	7.61	−0.516	60	NA
Anthrax immune globulin human	IIa	NA	NA	NA	NA	NA	24.3 d
Anti-inhibitor coagulant complex	Ia	NA	NA	NA	NA	NA	4–7 h
Anti-thymocyte globulin (equine)	IIIb	NA	NA	NA	NA	NA	1.5–13 d
Anti-thymocyte globulin (rabbit)	IIIb	NA	NA	NA	NA	61	2–3 d
Antihemophilic factor	Ia	264 726	C11794H18314N3220O3553S83	6.97	−0.533	NA	8.4–19.3 h
Antithrombin alfa	Ia	57 215	C2191H3457N583O656S18	NA	NA	NA	11.6–17.7 h

Antithrombin III human	Ia	58 000	NA	NA	NA	NA	2.5–4.8 d
Anti-thymocyte globulin	IIa	NA	NA	NA	NA	61	2–3 d
Aprotinin	IIa	6511	C284H432N84O79S7	NA	NA	>100	10 h
Arcitumomab	IIb	144 483	C6398H9900N1714O1995S54	8.26	−0.423	61	1 h
Asfotase alfa	Ia	180 000	C7108H11008N1968O2206S56	NA	NA	NA	5 d
Asparaginase	Ic	31 732	C1377H2208N382O442S17	4.67	0.059	NA	8–30 h
Asparaginase erwinia chrysanthemi	Ic	140 000	C1546H2510N432O476S9	NA	NA	NA	16 h
Atezolizumab	IIa	145 000	NA	NA	NA	NA	27 d
Autologous cultured chondrocytes	Ia	NA	NA	NA	NA	NA	NA
Basiliximab	IIa	143 801	C6378H9844N1698O1997S48	8.68	−0.473	61	7.2 ± 3.2
Becaplermin	Ib	12 294	C532H892N162O153S9	9.38	−0.16	NA	NA
Belatacept	IIa	92 300	C3508H5440N922O1096S32	NA	NA	NA	9.8 d.
Belimumab	IIa	147 000	C 6358 H 9904 N 1728 O 2010 S 44	NA	NA	NA	19.4 d
Beractant	Ia	NA	NA	NA	NA	NA	20–30 h
Bevacizumab	IIa	149 000	C6538H10034N1716O2033S44	NA	NA	61	NA
Bivalirudin	Ia	2180	C98H138N24O33	3.91	−0.985	NA	0.42 h
Blinatumomab	IIIc	54 100	C2367H3577N649O772S19	NA	NA	NA	2.11 h
Botulinum toxin type A	Ic	149 323	C6760H10447N1743O2010S32	6.06	−0.368	NA	NA
Botulinum toxin type B	Ic	150 804	C690H1115N177O202S6	NA	NA	NA	NA
Brentuximab vedotin	IIb	149 200–151 800	C6476H9930N1690O2030S40	NA	NA	NA	4–6 d
Brodalumab	IIa	144 000	C6372H9840N1712O1988S52	NA	NA	NA	NA
Buserelin	IIIc	NA	C62H90N16O15	NA	NA	NA	50–80 min
C1 esterase inhibitor (human)	Ia	105 000	NA	NA	NA	NA	56 h
C1 esterase inhibitor (recombinant)	IIa	67 000	NA	NA	NA	NA	2.4–2.7 h
Canakinumab	IIIb	145 200 (deglycosylated)	C6452H9958N1722O2010S42	NA	NA	NA	26 d
Canakinumab	IIa	145 200	C6452H9958N1722O2010S42	NA	NA	NA	26 d
Capromab	IV	NA	NA	NA	NA	NA	NA

(*Continued*)

Table 4.A1. (*Continued*)

Name	MOA	MW	Formula	IEP	Hydrophobicity	MP	Half-life
Certolizumab pegol	IIb	91 000	C2115H3252N556O673S16	NA	NA	NA	14 d
Cetuximab	IIIc	145 782	C6484H10042N1732O2023S36	8.48	−0.413	71	114 h
Choriogonadotropin alfa	Ib	25 720	C1105H1770N318O336S26	8.61	−0.258	55	29 ± 6 h
Chorionic gonadotropin (human)	Ia	25 719	C1105H1770N318O336S26	NA	NA	NA	NA
Chorionic gonadotropin (recombinant)	Ia	25 720	C1105H1770N318O336S26	8.61	−0.258	55	4.5 ± 1 h
Coagulation factor IX	Ia	46 548	C2041H3136N558O641S25	5.2	−0.431	54	19.4 h
Coagulation factor VIIa	Ib	45 079	C1972H3076N560O597S28	6.09	−0.311	58	NA
Coagulation factor X human	Ib	NA	NA	NA	NA	NA	NA
Coagulation factor XIII A-subunit (recombinant)	Ia	NA	NA	NA	NA	NA	5.1 d
Collagenase	Ic	112 023	C5028H7666N1300O1564S21	5.58	−0.714	NA	NA
Conestat alfa	Ia	NA	NA	NA	NA	NA	2.4–2.7 h
Corticotropin	IV	4541	C207H308N56O58S	NA	NA	NA	15 min
Cosyntropin	IV	2933	C136H210N40O31S	NA	NA	NA	15 min
Daclizumab	Iia	142 612	C6332H9808N1678O1989S42	8.46	−0.437	61	11–38 d
Daptomycin	IIa	1621	C72H101N17O26	NA	NA	NA	7 d
Daratumumab	IIIc	148 000	NA	NA	NA	NA	18 d
Darbepoetin alfa	Ib	18 396	C815H1317N233O241S5	8.75	−0.188	53	NA
Defibrotide	NA	NA	NA	NA	NA	NA	a few hours
Denileukin diftitox	IIb	57 647	C2560H4042N678O799S17	5.45	−0.301	NA	1.16–1.3 h
Denosumab	IIIc	144 700	C6404H9912N1724O2004S50	NA	NA	NA	25.4 d
Desirudin	Ib	6964	C287H440N80O110S6	NA	NA	NA	2–3 h
Digoxin immune fab (ovine)	IIa	47 302	C2085H3223N553O672S16	8.01	−0.343		15–20 h
Dinutuximab	IIIc	145 000	C6422H9982N1722O2008S48	NA	NA	NA	10 d
Dornase alfa	Ib	29 254	C1321H1999N339O396S9	4.58	−0.083	67	NA
Drotrecogin alfa	Ib	55 000	C1786H2779N509O519S29	6.78	−0.291	NA	5.5 h
Dulaglutide	Ib	59 670	C2646H4044N704O836S18	NA	NA	NA	5 d

Eculizumab	Ia	148 000	NA	NA	NA	NA	272 h
Efalizumab	IIa	150 000	NA	NA	NA		5 d
Efmoroctocog alfa	Ib	NA	NA	NA	NA	NA	NA
Elosulfase alfa	Ia	110 800	C5020H7588N1364O1418S34	NA	NA	NA	7.52–35.9 min
Elotuzumab	IIIc	148 100	C6476H9982N1714O2016S42	NA	NA	NA	NA
Enfuvirtide	IIa	4492	C204H301N51O64	4.3	−0.875	NA	3.8 h
Epoetin alfa	Ib	18 396	C815H1317N233O241S5	8.75	NA	53	NA
Epoetin zeta	Ib	18 200	C809H1301N229O240S5	NA	NA	NA	7.37 h
Eptifibatide	NA	832	C35H49N11O9S2	NA	−2.3	NA	29 ± 6 h
Etanercept	IIa	51 235	C2224H3475N621O698S36	7.89	−0.529	71	102 ± 30 h
Evolocumab	Iia	141 800	C6242H9648N1668O1996S56	NA	NA	NA	NA
Exenatide	Ib	4187	C184H282N50O60S	NA	NA	NA	2.4 h
Factor IX complex (human)	Ib	NA	NA	NA	NA	NA	11–28
Fibrinogen concentrate (human)	Ia	340 000	NA	NA	NA	NA	78.7 ± 18.13 h
Fibrinolysin aka plasmin	Ic	88 400	C3848H5912N1096O1185S60	NA	NA	NA	24 h
Filgrastim	1b	18 800	C845H1343N223O243S9	5.65	0.209	60	3.5 h
Filgrastim-sndz	IIIc	NA	NA	NA	NA	NA	3.5 h
Follitropin alpha	Ib	NA	NA	NA	NA	NA	24–53 h in females
Follitropin beta	Ib	22 673	C975H1513N267O304S26	7.5	−0.33	55	35–40 h
Galsulfase	Ia	56 013	C2534H3851N691O719S16	NA	NA	NA	6–40 min
Gastric intrinsic factor	Ib	NA	NA	NA	NA	NA	NA
Gemtuzumab ozogamicin	IIb	151 000–153 000	NA	NA	NA	61	NA
Glatiramer acetate	IIa	5000–9000	C254H422N70O72	NA	NA	NA	NA
Glucagon recombinant	IV	3767	C165H249N49O51S1	9.52	−1.197	NA	NA
Glucarpidase	Ic	44 017	C1950H3157N543O599S7	NA	NA	NA	5.6 h
Golimumab	IIb	146 943	C6530H10068N1752O2026S44	NA	NA	NA	2 weeks

(Continued)

Table 4.A1. (*Continued*)

Name	MOA	MW	Formula	IEP	Hydrophobicity	MP	Half-life
Gramicidin D	IIa	1882	C99H140N20O17	NA	NA	229	NA
Hepatitis A vaccine	IIIa	NA	NA	NA	NA	NA	
Hepatitis B immune globulin	IIIa	NA	NA	NA	NA	NA	22–25 d
Human calcitonin	Ib	NA	NA	NA	NA	NA	NA
Human clostridium tetani toxoid immune globulin	IIIb	NA	NA	NA	NA	NA	NA
Human rabies virus immune globulin	IIIa	NA	NA	NA	NA	NA	NA
Human Rho(D) immune globulin	IIIb	NA	NA	NA	NA	NA	24–30.9 d
Human serum albumin	Ia	66 472	C2936H4624N786O889S41	5.67	−0.395	62	NA
Human varicella-zoster immune globulin	IIIa	NA	NA	NA	NA	NA	26.2 d
Hyaluronidase	Ic	53 871	C2455H3775N617O704S21	5.73	−0.117	NA	NA
Hyaluronidase (human recombinant)	Ib	61 000	NA	NA	NA	NA	NA
Ibritumomab	IIb	143 376	C6382H9830N1672O1979S54	7.91	−0.359		0.8 h
Ibritumomab tiuxetan	IIIc	143 376	C6382H9830N1672O1979S54	7.91	−0.359	61	0.8 h
Idarucizumab	Ib	47 766	C2131H3299N555O671S11	NA	NA	NA	4.5–10.8 h
Idursulfase	Ia	76 000	C2654H4000N688O774S14	NA	NA	NA	44 ± 19 min
Imiglucerase	Ia	55 597	C2532H3854N672O711S16	7.41	−0.168	NA	0.06–0.173 h
Immune globulin human	IIIb	142 682	C6332H9826N1692O1980S42	NA	NA	NA	>20 h
Infliximab	IIa	144 190	C6428H9912N1694O1987S46	8.25	−0.441		9.5 d
Insulin aspart	Ia	582 580	C256H381N65O79S6	NA	NA	NA	81 min
Insulin beef	Ia	5734	C254H377N65O75S6	NA	NA	NA	NA
Insulin degludec	Ia	6104	C274H411N65O81S6	NA	NA	NA	25 h
Insulin detemir	Ia	5917	C267H402N64O76S6	NA	NA	NA	425 ± 78 min
Insulin glargine	Ia	6063	C267H404N72O78S6	6.88	0.098	81	30 h
Insulin glulisine	Ib	5823	C258H384N64O78S6	NA	NA	NA	42 min

Insulin lispro	Ia	5808	C257H387N65O76S6	5.39	0.218	81	1 h
Insulin pork	Ia	5796	C257H387N65O76S6	5.39	0.218	NA	NA
Insulin regular	Ia	5808	C257H383N65O77S6	5.39	0.218	81	NA
Insulin, porcine	Ia	5796	C257H387N65O76S6	5.39	0.298	NA	NA
Insulin, isophane	Ia	5808	C257H383N65O77S6	9	NA	NA	NA
Interferon alfa-2a, recombinant	Ib	19 241	C860H1353N227O255S9	5.99	−0.336	NA	6–8 h
Interferon alfa-2b	Ib	19 271	C860H1353N229O255S9	5.99	−0.339	61	2–3 h
Interferon alfa-n1	Ib	19 241	C860H1353N227O255S9	5.99	−0.336	61	1.2 h
Interferon alfa-n3	Ib	NA	NA	5.99	NA	61	NA
Interferon alfacon-1	Ib	19 271	C860H1353N229O255S9	5.99	0.339	61	2–3 h
Interferon beta-1a	Ib	20 027	C908H1408N246O252S7	8.93	−0.427	NA	10 h
Interferon beta-1b	Ib	20 011	C908H1408N246O253S6	9.02	−0.447	NA	10–20 min
Interferon gamma-1b	Ib	17 146	C761H1206N214O225S6	9.54	−0.823	61	NA
Intravenous immunoglobulin	Ia	142 682	C6332H9826N1692O1980S42	8.13	−0.331	61	20 h
Ipilimumab	IIIc	148 000	C6572H10126N1734O2080S40	NA	NA	NA	14.7–15.4 d
Ixekizumab	Ic	146 158	NA	NA	NA	NA	13 d
Laronidase	Ia	69 899	C3160H4848N898O881S12	9.09	−0.3	NA	1.5–3.6 h
Lenograstim	Ib	18 668	C840-H1330-N222-O242-S8	NA	NA	NA	2.3–3.3 h
Lepirudin	Ia	6963	C287H440N80O110S6	4.04	−0.777	65	1.3 h
Leuprolide	IIa	1209	C59H84N16O12	NA	0.1	NA	3 h
Liraglutide	Ib	3751	C172H265N43O51	NA	NA	NA	13 h
Lucinactant	Ib	2470	C126H238N26O22	NA	NA	NA	NA
Lutropin alfa	Ib	30 000	C1014H1609N287O294S27	8.44	−0.063	55	4.4 h
Mecasermin	Ib	7649	C331H518N94O101S7	NA	NA	NA	2 h
Menotropins	Ib	23 390	C1014H1609N287O294S27	8.44	−0.063	55	NA
Mepolizumab	Ib	149 000		NA	NA	NA	16–22 d
Methoxy polyethylene glycol-epoetin beta	Ib	60 000	NA	NA	NA	NA	134 ± 65 h
Metreleptin	Ib	16 155	C714H1167N191O221S6	NA	NA	NA	3.8–4.7 h
Muromonab	IIa	146 190	C6460H9946N1720O2043S56	8.31	−0.513	61	0.8 h

(*Continued*)

Table 4.A1. (*Continued*)

Name	MOA	MW	Formula	IEP	Hydrophobicity	MP	Half-life
Natalizumab	IIa	NA	NA	NA	NA	61	11 ± 4 d
Natural alpha interferon OR multiferon	IIa	19 300–22 100	NA	NA	NA	NA	NA
Necitumumab	IIIc	144 800	NA	NA	NA	NA	14 d
Nesiritide	Ib	3464	NA	NA	NA	NA	18 min
Nivolumab	IIIc	143 597	C6362H9862N1712O1995S42	NA	NA	NA	26.7 d
Obiltoxaximab	IIIa	148 000	NA	NA	NA	NA	NA
Obinutuzumab	IIb	146 100	C6512H10060N1712O2020S44	NA	NA	NA	28.4 d
Ocriplasmin	Ic	272 500	C1214H1890N338O348S14	NA	NA	NA	NA
Ofatumumab	IIIc	146 100	C6480H10022N1742O2020S44	NA	NA	NA	2.3–61.5 d
Omalizumab	IIa	145 058	C6450H9916N1714O2023S38	7.03	−0.432	61	624 h
Oprelvekin	Ib	19 047	C854H1411N253O235S2	11.16	−0.07	NA	6.9 h
OspA lipoprotein	IIIa	27 743	C1198H2012N322O422S2	6.72	−0.652	NA	1.2 h
Oxytocin	Ib	1007	C43H66N12O12S2	5.51	−2.7	NA	1–6 min
Palifermin	Ib	16 193	C721H1142N202O204S9	9.47	−0.65	NA	NA
Palivizumab	IIIa	NA	NA	NA	NA	61	18–20 d
Pancrelipase	Ia	131 126	C5850H8902N1606O1739S49	6.44	NA	48–50	NA
Panitumumab	IIa	NA	NA	NA	NA	NA	7.5 d
Pegademase bovine	Ia	40 788	C1821H2834N484O552S14	5.33	−0.428	NA	NA
Pegaptanib	IIa	NA	NA	NA	NA	NA	10 ± 4 d
Pegaspargase	Ic	31 732	C1377H2208N382O442S17	4.67	0.059	NA	NA
Pegfilgrastim	Ib	18 803	C845H1343N223O243S9	5.65	0.209	60	15–80 h
Peginterferon alfa-2a	Ib	60 000	NA	5.99	NA	61	80 h
Peginterferon alfa-2b	Ib	31 000	C130H219N43O42	5.99	NA	61	40 h
Peginterferon beta-1a	Ia	20 000	NA	NA	NA	NA	78 h
Pegloticase	Ib	34 193	C1549H2430N408O448S8	NA	NA	NA	14 d
Pegvisomant	IIa	22 129	C990H1532N262O300S7	5.27	−0.411	76	6 d
Pembrolizumab	IIIc	146 286	C6504H10004N1716O2036S46	NA	NA	NA	28 d

Pertuzumab	IIIc	148 000	NA	NA	NA	NA	18 d
Poractant alfa	Ia	NA	NA	NA	NA	NA	NA
Pramlintide	Ia	3949	C171H267N51O53S2	NA	NA	NA	48 min
Preotact	Ib	9420	C408H674N126O126S2	NA	NA	NA	1.5 h
Protamine sulfate	Ib	NA	NA	NA	NA	NA	4.5 min
Protein S human	Ib	69 000	NA	5.05	NA	NA	NA
Prothrombin complex concentrate	Ia	NA	NA	NA	NA	NA	48–60 h
Ragweed pollen extract	IIIa	NA	NA	NA	NA	NA	NA
Ramucirumab	IIIc	143 600	C6374H9864N1692O1996S46	NA	NA	NA	15 d
Ranibizumab	IIa	48 350	C2158H3282N562O681S12	NA	NA	NA	9 d
Rasburicase	Ic	34 110	C1521H2381N417O461S7	7.16	−0.465	NA	18 h
Raxibacumab	IIa	142 845	C6320H9794N1702O1998S42	NA	NA	NA	16–19 d
Reteplase	Ib	39 590	C1736H2671N499O522S22	6.86	−0.435	60	NA
Rilonacept	IIa	251 000	C9030H13932N2400O2670S74	NA	NA	NA	8.6 d
Rituximab	IIa	143 860	C6416H9874N1688O1987S44	8.68	−0.414	61	0.8 h
Romiplostim	Ib	59 000	C2634H4086N722O790S18	NA	NA	NA	3.5 d
Sacrosidase	Ia	100 000	NA	NA	NA	NA	NA
Salmon calcitonin	Ib	3432	C145H240N44O48S2	8.86	−0.537	NA	0.83–1.33 h
Sargramostim	Ib	14 435	C639H1006N168O196S8	5.05	NA	NA	NA
Satumomab pendetide	IV	141 479	C6268H9708N1666O1971S48	7.02	−0.427	61	0.80 h
Sebelipase alfa	Ia	55 000	C1968H2945N507O551S15	NA	NA	NA	5.4–6.6 min
Secretin	IV	3056	C130H219N43O42	9.45	−0.463	NA	NA
Secukinumab	IIa	147 940	C6584H10134N1754O2042S44	NA	NA	NA	NA
Sermorelin	Ia	3358	C149H246N44O42S	9.99	−0.33	NA	11–12 min
Serum albumin	IV	66 472	C2936H4624N786O889S41	5.67	−0.395	62	NA
Serum albumin iodonated	IV	66 472	C2936H4624N786O889S41	5.67	−0.395	62	NA
Siltuximab	Iia	145 000	C6450H9932N1688O2016S50	NA	NA	NA	20.6 d
Simoctocog alfa	Ia	170 000	NA	NA	NA	NA	14.7 h
Sipuleucel-T	IIIc	NA	NA	NA	NA	NA	NA

(*Continued*)

Table 4.A1. (*Continued*)

Name	MOA	MW	Formula	IEP	Hydrophobicity	MP	Half-life
Somatotropin recombinant	Ib	22 129	C990H1532N262O300S7	5.27	−0.411	76 at pH 3.5	NA
Somatropin recombinant	Ib	22 129	C990H1532N262O300S7	5.27	−0.411	76	NA
Streptokinase	Ic	47 287	C2100H3278N566O669S4	5.12	−0.728	NA	NA
Sulodexide	IIIc	5000–8000	NA	NA	NA	NA	11.7 ± 2.0 h
Susoctocog alfa	Ib	NA	NA	NA	NA	NA	~17 h
Taliglucerase alfa	Ia	56 638	C2580H3918N680O727S17	10.54	NA	NA	NA
Teduglutide	Ia	3752	C164H252N44O55S	NA	NA	NA	2 h
Teicoplanin	IIa	1880	C88H97Cl2N9O33	NA	NA	NA	70–100 h
Tenecteplase	Ib	58 951	C2561H3919N747O781S40	7.61	−0.528	60	1.9 h
Teriparatide	Ib	4118	C181H291N55O51S2	NA	NA	NA	NA
Tesamorelin	Ib	5136	C221H366N72O67S	NA	NA	NA	38 min
Thrombomodulin alfa	Ib	52 124	C2230-H3357-N633-O718-S50	NA	NA	NA	2–3 d
Thymalfasin	Ib	3108	C129H215N33O55	NA	NA	NA	2 h
Thyroglobulin	Ib	660	NA				65 h
Thyrotropin alfa	IV	22 673	C975H1513N267O304S26	7.5	−0.33	55	5 ± 10 h
Tocilizumab	IIa	148 000	C6428H9976N1720O2018S42	NA	NA	NA	11 d
Tositumomab	IIb	143 860	C6416H9874N1688O1987S44	8.68	−0.4144		0.8 h
Trastuzumab	IIa	145 532	C6470H10012N1726O2013S42	8.45	−0.415	61	28.5 d
Tuberculin purified protein derivative	IV	NA	NA	NA	NA	NA	NA
Turoctocog alfa	Ia	NA	NA	NA	NA	NA	NA
Urofollitropin	Ib	980	C42H65N11O12S2	7.5	−0.33	55	35–40 h
Urokinase	Ib	31 127	C1376H2145N383O406S18	8.66	−0.466	76	12 min
Ustekinumab	IIIb	14 690	NA	NA	NA	NA	NA
Vasopressin	Ib	1050	C43H67N15O12S2	NA	−4.9	NA	10–20 min
Vedolizumab	IIa	146 837	C6528H10072N1732O2042S42	7.6	NA	NA	336–362 h
Velaglucerase alfa	Ia	63 000	C2532H3850N672O711S16	NA	NA	NA	11–12 min

Mode of action (MOA): group Ia—replacing a protein that is deficient or abnormal; group Ib—augmenting an existing pathway; group Ic—providing a novel function or activity; group IIa—interfering with a molecule or organism; group IIb—delivering other compounds or proteins; group IIIa—protecting against a deleterious foreign agent; group IIIb—treating an a harmful disease; group IIIc—treating cancer; group IV—protein diagnostics. MW: molecular weight; IEP: isoelectric point; MP: melting point.

Chapter 5

Biopharmaceutical development cycle

Regulatory approval of biopharmaceutical products is subject to strict compliance demonstrated in an elaborate registration dossier. In this chapter we describe in sufficient detail the steps relating to the filing of a biological license application (BLA) in the US or a marketing authorization application (MAA) in the EU. The topics include early discovery documentation, use of pharmacopeia, preclinical research, and the steps of IND to phase 4. Expectations from regulatory inspections (audits) are presented based on the authors' long experience, including both the US and EU systems. This chapter also provides details of the stepwise approach taken in the US, EU, and Japan in evaluating a regulatory dossier. Biopharmaceutical manufacturing has changed significantly since the 1980s primarily because of regulatory authorities' stricter controls on biological therapeutic products; this chapter will provide a global view of how the agencies approve the products and the facilities to manufacture these products. A comprehensive approach requires a full understanding of these steps to make large-scale commercial manufacturing of biopharmaceuticals feasible.

5.1 Overview

Biopharmaceutical products fall under biological drug manufacturing regulatory compliance that had changed significantly since the 1980s when the regulatory agencies became aware of the contamination issues with viruses. Later, when recombinant manufacturing took root, many new requirements were introduced.

A biopharmaceutical must first go through significant testing to see if it is safe and effective in treating the ailment for which it was created, as well as the proper dosage and administration route. Regulatory agencies oversee and regulate all therapeutic testing. The medication development process is always tailored to meet the needs of regulatory bodies, and developers collaborate closely with them throughout the

doi:10.1088/978-0-7503-3175-3ch5

process. The US Food and Drug Administration (FDA), the European Medicines Agency (EMA), the Japan Pharmaceutical and Medical Device Agency (PMDA), and the Medicines and Healthcare Regulatory Agency (MHRA) in the United Kingdom are the key agencies responsible for the development of most new pharmaceuticals. More than 50% of all new drugs come from the US, France, UK, Switzerland, Germany, and Japan.

In this chapter we describe the overall process of regulatory approval of new drugs, more particularly biotherapeutic agents, along with the regulatory guidelines that are jointly developed under the umbrella of the International Council for Harmonization (ICH) that was founded by the US, EU, and Japan regulatory agencies.

5.2 The US drug development process

The stages of drug development are more other less the same within all agencies and comprise:

- Early drug discovery.
- Preclinical research.
- Investigational new drug application.
- Clinical research.
- Regulatory review, approval, and post-marketing safety surveillance.

5.2.1 Early drug discovery

The early drug discovery stage is always uncertain despite many advances in this field, such as artificial intelligence and high-throughput screening, topics that are presented elsewhere in this book (figure 5.1).

Typically, new drugs development follows a structured path:

- Developers may design a product to avoid or reverse the effects of a disease based on new insights into the disease process.
- Many molecular compound studies are conducted to discover potential therapeutic effects against a variety of diseases, including existing therapies with unexpected side effects.
- Drug repurposing is also widely considered using existing approved products.
- New technologies targeting specific body sites or genetic material are now explored widely.

At this phase in the process thousands of molecules could be possibilities for therapeutic development. However, only a limited number of molecules would appear promising enough to warrant further exploration following basic testing.

Several critical 'steps' occur throughout the drug development process. Scientists from academia and industry are collaborating to establish new drug targets for a variety of diseases. They are trying to find and optimize drug compounds that can impact disease-related biological targets in the hopes of curing them.

Key considerations include developing an 'optimal' drug, providing a good understanding of a disease's clinical nature, and determining the precise role that

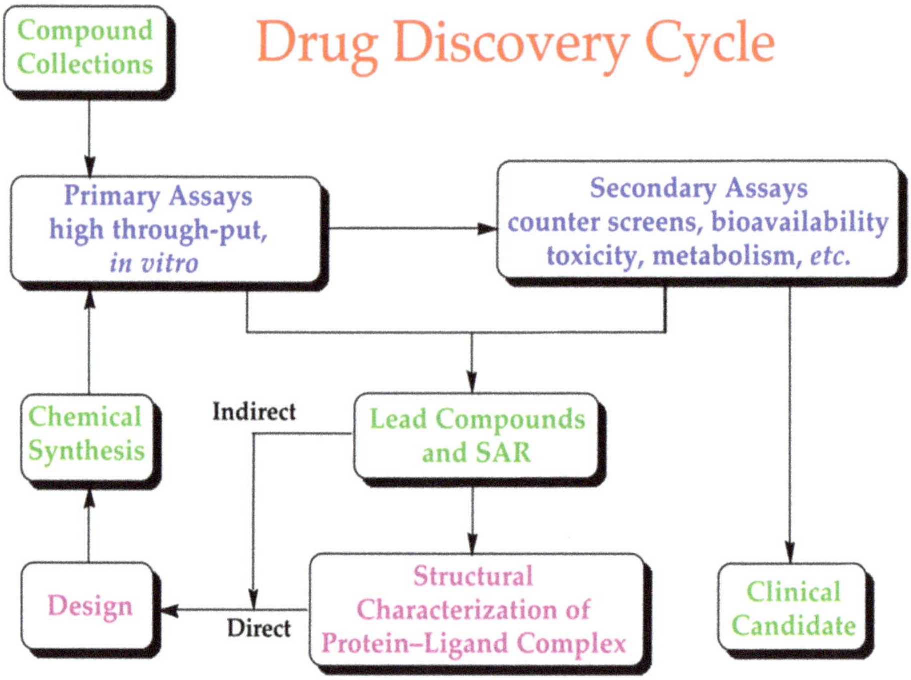

Figure 5.1. The drug discovery process. (Source: Wikipedia: Boghog https://commons.wikimedia.org/wiki/File:Drug_discovery_cycle.svg.)

a target plays in that disease. A biological target is said to be 'druggable' if a therapeutic agent, whether a small molecule or a biopharmaceutical, can change its activity. Drug targets that are 'focused' have remarkable benefits. They can be found in various ways, including reading current scientific literature, looking for databases, and using 'practical' methods such as target deconvolution and aim discovery. Once a goal has been established it will be validated to see if it is suitable for pharmaceutical production before a screening campaign to find possible 'hits' can begin.

To locate a 'hit' compound, various screening approaches might be applied. A chemical that interacts with the target of interest is referred to as a 'hit.' High-throughput screening, phenotypic screening, virtual screening, fragment-based screening, and structure-based design are some of the strategies used by researchers to uncover 'hits'.

Since the development strategy is based on detecting a phenotypic shift to determine if a compound exerts the desired effect, the drug target cannot be immediately apparent in phenotypic screens. While there is no requirement to know a drug's target if it exhibits good safety and efficacy properties, the target that underpins the observed phenotypic shift can be established later.

The key goal at this point is to refine some of the most promising 'hits' with 'optimized' pharmacokinetic features (how the body reacts to the drug) to develop more potent and selective candidates. The affinity of the 'first hits' for the biological target is usually low. Medicinal chemists attempt to boost affinity by many orders of magnitude. Advances in artificial intelligence (AI) have made it possible to collect large amounts of useful biological, structural, and chemical data quickly. Off-target interactions are also a major worry at this level, as they might have detrimental repercussions, thus researchers are looking into ways to improve molecular selectivity against other biological targets.

It is decided which aim will be classified as a clinical candidate from many prospective 'high-quality' leads. A drug candidate must preferentially bind to the target, elicit the intended functional response when interacting with the target molecule, and have appropriate bioavailability and biodistribution to elicit the desired reaction to be regarded suitable for preclinical and clinical testing. The molecule's toxicity profile would likewise be acceptable. Furthermore, one needs to consider that the drug's long-term performance would be heavily influenced by its manufacturing appropriateness and scale-up, market viability, and cost-effectiveness.

Once a potential molecule has been identified for development, researchers run studies to gather data on:

- Its methods of absorption, distribution, metabolization, and excretion.
- Its potential benefits and action methods in the future.
- The correct dosage.
- The best approach to administer the medication (such as by mouth or injection).
- Toxicity, a term used to describe side effects or bad outcomes.
- Its impact on several human groups (such as gender, race, or ethnicity).
- What other medicines and treatments it interacts with.
- Its efficacy compared to that of similar medications.

The production process starts once a lead compound has been found. If the medication is the first available cure for a disease or shows a major advantage over current medications, this step may be shortened. During the COVID-19 crisis in 2020, many new uses of existing drugs and vaccines were given fast-track approval. However, the process of development (figure 5.2) runs its slow course.

5.2.2 Pharmacopoeia

There are up to 40 pharmacopeias published around the world, with up to 60 active pharmacopeia commissions in charge of their establishment and upkeep. A timeline is included in table 5.1 which illustrates when many of the pharmacopeias were published. *De Materia Medica*, which first appeared in Greece and Rome in the first century CE, is thought to be the first example of a 'pharmacopeia', although the earliest work presenting medical information and herbal remedies may date back more than 3000 years to ancient Egypt. (In ancient Greek, the word 'pharmacopeia' meant 'drug-making.') This medical treatise listed herbal treatments and how they were prepared at the time. Several pharmacopeias providing medicinal prescriptions were published in the sixteenth century for apothecaries and physicians in major European cities such as Nuremberg, London, Edinburgh, and Dublin.

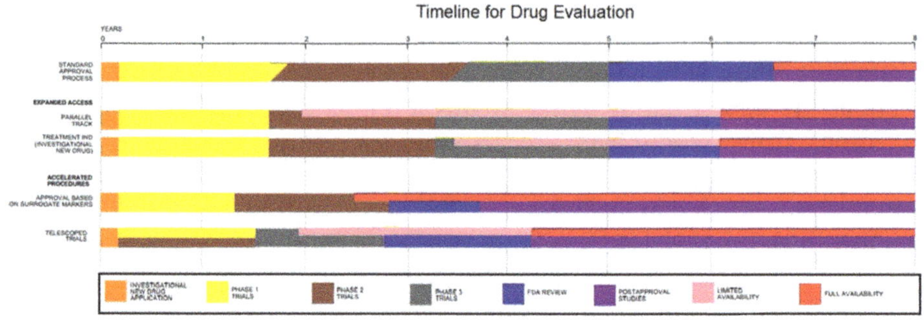

Figure 5.2. Drug development timeline. (Source: Wikipedia: Kernsters, graph created based on information provided in the *Scientific American* article, 'Faster evaluation of vital drugs', Creative Commons.)

Table 5.1. The introduction of pharmacopoeia across the globe.

Year	Country
1820	USA
1844	India
1846	Mexico
1858	United Kingdom
1866	Russia
1886	Japan
1898	Argentina
1929	Brazil
1951	International
1953	China
1955	Modern India
1958	Korea
1962	Indonesia
1964	Europe
1970	Vietnam
1980	USP-NF
1987	Thailand
2004	Philippines
2008	Kazakhstan
2020	EU
2021	South American

The development of drugs is highly dependent on the role of pharmacopeias in the regions of development. Where the pharmacopeias have legal status, these must be followed; only in the US, the pharmacopeia is not binding on the developers and the FDA.

The European Pharmacopoeia is a legally enforceable document. It is used as an official reference for public health purposes. It is one of the regulatory conditions for acquiring a marketing authorization (MA) for a pharmaceutical product (human or veterinary).

The BP has set regulatory standards for pharmaceutical products since 1864. The British Pharmacopoeia Commission and the British Pharmacopoeia were created under the Medicines Act of 1968. This act has been superseded by the Human Medicines Regulations of 2012.

The United States Pharmacopeia is the world's only independent, non-profit, non-governmental pharmacopeia. The USP establishes quality, purity, strength, and identity criteria for medications, food components, and dietary supplements that are enforceable by the US Food and Drug Administration (FDA) under US law. However, the FDA and the USP have no legal ties.

The Japanese Pharmacopoeia (JP) is a government-issued document that lays out the standards, procedures, and standard test methods that must be followed in order to verify that medications in Japan are of proper quality. According to paragraph 2 of article 41 of the Law on Securing Quality, Efficacy, and Safety of Products, including Pharmaceuticals and Medical Devices, full JP revisions are necessary every ten years.

5.2.3 Preclinical research

Before testing a medicine on humans, developers must establish whether it can cause toxicity. The FDA requires developers to follow good laboratory practices (GLP) for preclinical laboratory investigations, which are outlined in the medical product development regulation 21 CFR Part 58.1: Good Laboratory Practice for Nonclinical Laboratory Studies. These rules establish the bare minimal criteria for:

- equipment;
- facilities;
- operating procedures;
- personnel;
- study conduct;
- study reports;
- system of quality assurance oversight;
- written protocols.

To illustrate a candidate's biological effect, *in vitro* and *in vivo* models are often used. Before submitting an investigational new drug (IND) application, which is intended to accelerate clinical development, regulatory organizations such as the FDA and EMA require preclinical experiments. Several questions are addressed at this point:

- What is the drug doing to the body?
- What is the body doing to that drug?
- Is it powerful but also safe?

It is very crucial to pick the most appropriate animal model at this point and consider the gender of the animals to be utilized to avoid gender-specific bias.

A medication can cause a male animal to react differently from a female animal. In terms of metabolic pathways and genetics, the species-specific biology, and commonalities (for example, 99% of all mouse genes correspond to human genes) are demonstrated.

5.2.4 Investigational new drug application

Before clinical trials may begin, drug developers must apply an application to the FDA for an Investigational New Drug (IND).

Manufacturers may provide the following information in their IND application: Toxicity evidence from animal studies (side effects that cause significant harm), information data on manufacturing, and the testing protocols. Knowledge about the investigator from some prior human study are also included.

Drug developers are free to ask for help from FDA at any point in the drug development process, including:

- To review FDA guidance documents and get answers to questions that can help them improve their study, for this they must submit a pre-IND application.
- To seek advice on the design of major phase 3 studies after phase 2.
- To receive an evaluation of the IND application at any point during the process.

Despite the FDA's comprehensive technical assistance, drug developers need not follow the FDA's advice. Clinical trial design has a lot of leeway with the FDA as long as it is well-planned, represents what the developers know about a product, protects participants, and otherwise complies with Federal regulations.

A group of experts from various scientific fields makes up the review team. Each team member has a unique set of responsibilities:

- The *project manager* coordinates the team's efforts during the evaluation process and serves as the main point of contact for the sponsor.
- Before, during, and after the trial, the *medical officer* reviews all clinical research details and results.
- The *statistician* analyzes and interprets clinical trial designs and results, collaborating closely with the medical officer to assess procedures, as well as safety and efficacy data.
- A *pharmacologist* reviews preclinical trials.
- *Pharmacokinetics* is the study of a drug's absorption, delivery, metabolism, and excretion processes.

The initial IND application has 30 days to be reviewed by the FDA review team. Volunteers in clinical trials are protected from severe and undue risk by this mechanism. The FDA responds to IND applications in one of two ways:

1. The start of clinical studies is approved.
2. Clinical hold is used to postpone or end an inquiry. The FDA can put a clinical hold on a product for a variety of reasons, including:
   - i. Participants face unreasonably high or serious hazards.
   - ii. Investigators are inexperienced.

iii. The materials provided to volunteer participants are deceptive.
iv. The trial's risks are not adequately described in the IND application.

Instead of a clinical hold, the FDA frequently offers suggestions aimed to improve the uniformity of a clinical investigation. If the FDA is satisfied that the trial meets federal standards, the developers may proceed with the planned analyses in some situations.

During the trial the developer is responsible for informing the assessment team of any new methods or major side effects. Because of this, the team can keep a careful check on the trials for any symptoms or issues. After the testing is done, developers must submit study reports to the regulatory agency.

Before the developer stops clinical trials or submits the data for a marketing proposal, this cycle repeats itself. Before filing a marketing application, a developer must have enough evidence from at least two major, well-managed clinical trials.

The INDs are divided into three categories by the FDA:

- *Investigative*: This is sent by the doctor who oversees conducting and investigating the investigation. The investigational drug administration and dispensing would be done by the same practitioner. This form of application is typically used to investigate an unapproved prescription, an approved medication for use in an unlicensed indication, or a different category of patients.
- *Emergency use authorization*: The term 'emergency' refers to a situation in which you need to test a drug in an emergency. An IND requires the FDA to approve an investigational drug without the need to file an IND under 21 CFR, Sec. 312.23 or Sec. 312.20. This form of application is used in patients who do not meet the requirements for ongoing clinical studies or cases where a clinical procedure has not yet been accepted.
- *Treatment*: This form of IND application is used to obtain access to an experimental drug that has been shown in clinical trials to have the ability to cure life-threatening diseases. On the other hand, the FDA approves the new drug application after the final clinical work is done.

An IND may be classified as 'commercial' or 'research'. There are key areas to cover for an IND application: pharmacology and toxicology testing in animal models, information on manufacturing, recommendations for clinical studies, and information about investigators.

5.2.5 Clinical research

Until beginning a clinical trial, developers study prior knowledge about the drug to formulate research questions and objectives. Then they decide:

- Who is qualified to participate (selection criteria).
- How many people will be required to conduct the research.
- How long the research will last.

- Whether or not a control group would be used and other methods for limiting research bias.
- How the medication will be administered to patients, and at what dosage.
- What types of tests will be performed, when will they be conducted, and what data will be collected.

Clinical trials are intended to address basic research questions about a potential medicine under investigation. The studies must adhere to a research plan that details how the clinical trial will be carried out. To ensure the protection of participants and the accuracy of the data obtained during the report, the study specifies the key study goals, study design, and statistical considerations. A sequence of 'phases' accompanies the clinical stage of drug development:

- *Phase I*: The number of patients ranges from 20 to 100. It may be 'average' people or patients who need medication. A study time of several months is typical. The main goal is to develop safety and collect dosage information. Phase I trials are the best way to reduce toxicity while increasing therapeutic impact. Approximately 70% of prescriptions are progressed to phase II.
- *Phase II*: Several hundred people take part in this phase. Participants will be diagnosed with the disease or disorder for which they are developing a treatment. The study period will last anywhere from a few months to two years. The primary goal is to gather additional health data to determine the effectiveness and the side effects. The larger phase III study's design is optimized using this knowledge. Just about a third of drugs make it to phase III.
- *Phase III*: The number of participants ranges from 300–3000. Participants are given a diagnosis of the illness for which they are researching a cure. The length of the research ranges from one to four years, with the primary goal of determining the drug's efficacy and monitoring its adverse reactions. Owing to the increased number of participants in phase III, long-term or rarer side effects that could have gone undetected in phase I and phase II are normally observed. Phase III is where the majority of the safety data are obtained. After a drug has been accepted, about 25%–30% of drugs go on to phase IV trials.
- *Phase IV*: Thousands of people take part in this phase. The illness or disorder for which the medication is administered is diagnosed in the volunteers. A phase IV trial is designed to gather additional data on the long-term dangers and benefits of using a drug that is approved by the regulatory agencies. The 'real-world' data can also be used to assess if the drug has potential for expanding its use, such as other indications, other age groups, or other routes of administration.
- *Phase V*: These clinical trials for drugs on the market provide valuable information on effectiveness and safety on a larger scale. Despite the systematic steps in the drug development process, there are risks that can appear only when a drug is given to larger population. The FDA examines these data to decide if any warnings are required in the labeling.

5.2.6 Regulatory review, approval, and post-marketing safety surveillance

5.2.6.1 New drug applications

All biological products approved under the CBER's biological license application (BLA), but only for biological therapeutic pharmaceuticals, including (but not limited to) monoclonal antibodies (for *in vivo* use), cytokines, growth factors, enzymes, immunomodulators, proteins, and non-vaccine irradiation. The CDER, on the other hand, approves products through the New Drug Application (NDA) process. The identical method is known as a marketing authorization application (MAA) in the European Union and many other countries. The registration application's goal is to provide regulators with enough information—gathered during preclinical and clinical tests—to determine whether the medication is safe and effective as a treatment for the condition for which it was developed, whether the medication's therapeutic benefits outweigh the risks, whether the product's labeling is appropriate for use.

Various steps must be taken to prepare for the introduction of a drug once it has been approved by the regulatory authority. These include:

- production scale-up;
- serialization printing of final product label;
- packaging and artwork;
- product storage, shipping, and distribution arrangements;
- manufacturing personnel and quality team availability;
- post-marketing safety monitoring is the term used for tracking a drug after this has been approved and entered the market.

Supplemental applications are submitted to make substantive changes from the original BLA. Generally, the FDA approves any improvements in the strength of the formulation, labeling, or dosage.

Manufacturers must file an IND if they want to change an approved medication for a new indication, dosing frequency, dosage form, or type (such as injectable or oral liquid instead of tablet form), or if they want to conduct more clinical trials or post-market safety tests.

The Biologics Price Competition and Innovation Act (BPCI Act) of 2009 established an expedited licensure pathway for biological products that are proved to be biosimilar to or compatible with FDA-licensed biological products. This direction was established in order to broaden treatment alternatives, enhance access to life-saving treatments, and, in the long run, lower healthcare costs through competition. Chapter 7 contains more information on the growth of biosimilars.

Manufacturers, healthcare providers, and customers can all report licensed drugs through the FDA's various programs. MedWatch is a website that allows you to report medical items (drugs and devices) and get notified when new safety information becomes available. The Medical Product Protection Network (MedSun) keeps track of medical device safety and effectiveness. The FDA has enlisted the help of 350 healthcare professionals around the country to track any medical device faults that result in serious injury or death. The FDA publishes the

MedSun newsletter once a month. The newsletter informs readers on the importance of medical equipment safety.

The FDA's objective to protect public health includes guaranteeing the high quality of the products it regulates. Sentinel is the FDA's national computerized system for tracking the safety of FDA-regulated medical items like pharmaceuticals, vaccines, biologics, and medical devices. In May 2008, the FDA established the Sentinel Initiative in response to the FDA Amendments Act (FDAAA) of 2007.

5.3 Drug development in Europe

5.3.1 Research initiation

Every year, tens of thousands of compounds are studied by pharmaceutical and biotechnology companies, clinicians, and researchers for their potential to cure diseases. Only a small fraction of these products will show enough promise to be evaluated in people, and even fewer will ever have research findings that are acceptable enough to make it to market.

Every year researchers from the government and the business sector study a wide range of products for their potential as medications. Given below are the details related to approval of new medications in Europe.

5.3.2 Testing new medicines

Potential new medications are first examined in the lab, then in clinical trials using human volunteers. These assessments aid in the comprehension of how pharmaceuticals work as well as the assessment of their advantages and side effects.

Those who want to conduct clinical trials in the EU must apply to the national authorities of their respective countries. The European Medicines Agency (EMA) has no role in approving clinical trials in the European Union; state authorities oversee this.

Instead the European Medicines Agency (EMA) works closely with EU Member States to ensure that developers adhere to EU and international guidelines. Developers conducting research to support the marketing authorization of pharmaceuticals must adhere to strict guidelines, whether they are conducted within or outside the EU.

Good clinical practice (GCP) guidelines govern how studies are planned, data is recorded, and outcomes are reported. These standards are meant to ensure that research is both scientifically and ethically sound.

5.3.3 EMA influence in development

The EMA has no authority to require companies to conduct research into specific medicines for specific diseases, however, the EMA makes regions where new medicines are needed public in order to attract interested parties to research them.

The EMA does not have the authority to sponsor medications or fund clinical trials for specific medicines, nor can it compel companies to research certain medicines or therapies for specific ailments. As a medicines regulator the EMA must be unbiased and cannot have a commercial or other financial interest in any approved product.

When new medications, such as antibiotics, are needed, the EMA makes them public so that anyone interested can analyze them. EU law, on the other hand, includes provisions that allow companies to develop drugs for rare diseases. One example is fee reductions for seeking scientific assistance from the EMA.

The EU rule also establishes a system of pledges, rewards, and incentives to encourage pharmaceutical companies to invest in children's medical research and development.

5.3.4 Scientific advice

To be licensed, drug researchers must demonstrate that their product is trustworthy, secure, and of high quality.

The EMA will offer guidance and recommendations on the best methodologies and study designs for obtaining reliable data on a drug's performance and effectiveness during the manufacturing process.

When seeking marketing permission, the developer must submit all data generated on the medicine to the EMA. These data are analyzed by the FDA to see if the drug is safe and effective for patients.

5.3.5 Scientific advice for studies

For the benefit of patients the EMA provides scientific guidelines to assist the timely and sound manufacturing of high-quality, cost-effective, and safe medicines.

The EMA gives scientific guidance because studies with better design are more likely to produce solid and full results that demonstrate whether a drug performs and is safe. The more quickly a novel treatment can be proven to work and be beneficial, the sooner it can be made available to patients.

Patients are not denied access to life-saving treatments because of ill-conceived trials that failed to demonstrate that the drug is effective and safe. Patients are less likely to participate in trials that provide no useful information if the research design is better. More significant research comprises maximizing the few scientific tools available to assist patients. For the benefit of patients, the EMA provides scientific guidelines to assist the timely and sound manufacturing of high-quality, cost-effective, and safe medicines.

The EMA gives scientific guidance because studies with better design are more likely to produce solid and full results that demonstrate whether a drug performs and is safe. The more quickly a novel treatment can be proven to work and be beneficial, the sooner it can be made available to patients.

5.3.6 EMA providing other advice

The EMA issues technical advice to assist medication producers in determining the best way to conduct research on their medicines. These suggestions must address general situations and will not cover non-standard, unique methods as they emerge. As a result, scientific guidance complements and expands on current recommendations while also being adapted to the circumstances at hand. It could also be used to update or add new guidelines in the future.

Guidelines explain the best procedures and study strategies for developing certain pharmaceuticals, such as vaccines or antibiotics, as well as treatments for specific disorders, such as cancer. Guidelines, however, can only handle broad issues; they cannot address new and inventive ideas as they arise. They take a long time to mature.

In addition to recommendations, relevant scientific advice for the manufacture of certain medicines is available upon request. The recommendations are based on current clinical principles, however, they are adapted to the specific medicine and patient population in question.

As a result, scientific advice and experience with the evaluation of pharmaceuticals, particularly novel treatments, are factored into the development and revision of guidelines. For example, when a novel endpoint is offered in a number of recent scientific advice requests, the recommendations are updated to reflect the new endpoint's association. As a result, knowledge generated through scientific guidance benefits the scientific community.

5.3.7 How is scientific advice paid for?

For scientific advice, applicants must pay an administrative fee. The EU legislation mandates that EMA provide scientific advice and the administrative fees that must be paid to the applicant.

Certain forms of medicines and applicants are eligible for fee reductions: orphan medicines are eligible for a 75% fee reduction, whereas micro-, small-, and medium-sized businesses (SMEs) are eligible for a 90% fee reduction.

5.3.8 What happens during scientific advice?

- Experts point to specific scientific questions relating to the development of a certain treatment during scientific counseling.
- The drug's creator discusses the obstacles and possible solutions, as well as how the treatment will be manufactured. EMA then evaluates the developer's suggestions. During scientific advice, the EMA does not look at the outcomes of the trials or draw any conclusions about whether the treatment's benefits outweigh the risks.
- Clinical aspects (appropriateness of studies in patients or healthy volunteers, selection of endpoints, i.e. how best to measure effects in a study, post-authorization actions), nonclinical aspects (toxicological and pharmacological tests designed to demonstrate the activity of the medicine in the laboratory), and quality aspects (manufacturing, chemical analysis) are all topics that can be discussed (toxicological and pharmacological tests designed to show the activity of the medicine).
- Statistical tests that can be employed include data analysis, modeling, and simulation.

5.3.9 Who's involved in scientific advice?

Hundreds of specialists from various areas provide solutions to the questions.

The Committee for Medicinal Products for Human Use (CHMP) of the EMA is in charge of reviewing marketing authorization requests. One of its responsibilities is to offer scientific recommendations in order to promote research and development. The CHMP assigned this objective to the EMA's Scientific Advice Working Party (SAWP). The developer's concerns are addressed by the SAWP, and the CHMP then approves and publishes the final guidelines.

Some of the questions that are addressed during scientific advice are listed below:

- Is the patient population in a study sufficiently representative of the target population for the drug?
- Are the proposed methods for assessing the advantages of a drug accurate and relevant?
- Is the method for assessing the findings that has been suggested adequate?
- Is the study long enough and does it have a large enough sample size to give the necessary data for the risk–benefit analysis?
- Is there a viable replacement for the medicine that is being compared with?
- Are the monitoring arrangements for the product's long-term protection well-designed?

Experts from medical regulators across the EU, academia, and the EMA's committees for orphan medications, new treatments, medicines for girls, and pharmacovigilance and risk evaluation make up the SAWP, which has up to 36 members. CHMP members make up nearly a quarter of the overall membership. Because of this overlap, the CHMP can use the longer-term experience and expertise in medicine garnered through scientific advice in the later consideration of the marketing authorisation application.

Nonclinical protection, pharmacokinetics, technique and statistics, gene and cell therapy, and therapeutic areas where scientific aid is frequently sought, such as cardiology, oncology, diabetes, neurological disorders, and infectious diseases, are among the SAWP members' areas of expertise.

5.3.10 Is there any involvement of patients in scientific advice?

Patients are frequently invited to participate in scientific consultations. They are encouraged to give their personal thoughts and experiences with a particular medicine in their disease sector. This will aid drug developers and regulators in gaining a better understanding of how that patient group functions and what they value.

External specialists can be consulted as well, allowing the SAWP to access a bigger pool of expertise.

5.3.11 Is it true that providing scientific advice has an impact on the EMA's assessment of the medicine?

Regardless of how the medicine is tested to give robust proof, the evaluation at the time of marketing authorization looks at the actual data from the trials to decide if the treatment's advantages outweigh its hazards.

Concerns mentioned during scientific advice are not the same as those addressed during medical examination. The CHMP reviews the results of these trials during medicine review and determines whether the medicine's benefits outweigh its risks, allowing it to be approved for use in patients. Scientific advice addresses issues about the best way to test and investigate a drug; during medicine review, the CHMP reviews the results of these trials and determines whether the medicine's benefits outweigh its hazards, allowing it to be approved for use in patients.

Giving scientific advice can make medicine simpler and faster because the evidence to be supplied is more solid, relevant, and complete. This, however, has no affect on the regulator's thorough assessment of the drug's safety and effectiveness, nor does it imply that it would pass that test automatically. The benefit–risk balance is easier to determine with more evidence. It does not, however, ensure that the drug will be approved, it may just show that the medicine is hazardous or ineffective.

As a result, even if drug companies have received and implemented clinical advice, they may still be denied marketing approval. Drug makers who ignore the recommendations, on the other hand, may be awarded marketing approval.

While the scopes of these processes differ, scientific counsel provides long-term experience and expertise that is valuable in learning more about the drug and will be useful during the marketing application examination.

Both procedures produce decisions collaboratively, after lengthy deliberation and communication. A majority of SAWP or CHMP members must agree on a course of action before it can be executed.

5.3.12 What does the EMA say about the results of scientific advice?

The complete counsel of a company is not made public during the production and evaluation phases. This is since releasing knowledge now could jeopardize research and development efforts and stifle new medical research. However, information becomes available as soon as a drug is licensed for commercialization.

In June 2018 the EMA began publishing more extensive information on scientific guidance provided throughout the production of medicines in the evaluation reports of those who received EMA PRIME assistance. (Medicines that target illnesses for which there are no effective treatments and have showed promising early effects). This program has been implemented for all drugs assessed after 1 January 2019.

A list of the developer's questions appears at the beginning of the evaluation report. The report's linked sections comprise critical aspects of the suggestions offered. There are also details on the company's adherence to this advice.

As soon as the European Commission has agreed on marketing permission, the evaluation reports for drugs are published on the EMA website. On request, the entire advice can be made public.

One of the most important sources for amending the EMA's scientific recommendations on pharmaceutical manufacture is scientific advice. Certain disease recommendations are revised on a regular basis to reflect new knowledge and insights gained through scientific guidance and drug testing. In this way, the results of scientific guidance become available to the general public.

5.3.13 What safeguards are in place to protect experts' independence during scientific advice?

Before providing scientific advice, the EMA considers each expert's disclosure of interests, and if those interests are deemed to endanger impartiality, restrictions are imposed.

EMA has developed competitive interest policies to limit the participation of members, experts, and employees with potentially conflicting interests in the Agency's work while still allowing the Agency to access the best available expertise.

Prior to participation in EMA activities, members of the SAWP and any other experts interested in EMA activities must file a declaration of interest. The Agency assigns a risk rating to each statement of interests based on whether the expert has any direct or indirect financial or other interests that could influence their impartiality. Before beginning a new scientific advice procedure, the EMA reviews each member's or expert's declaration of interests. If a conflict of interest is discovered, the member's or expert's abilities will be limited.

Restriction examples include not being allowed to participate in a debate on a given subject or not being allowed to vote on the topic.

5.3.14 Details of the procedure based on scientific advice

With the support of independent teams, two experts conduct separate evaluations; other experts and stakeholders are frequently recruited.

A developer must first notify the EMA and submit a briefing paper if he or she wishes to seek scientific advice. It might be possible to host a gathering for first-time users of scientific advice or complex drugs. After that, the creator provides a list of basic science questions with possible responses. For scientific guidance the EMA determines whether the questions are true or not.

Two members of the SAWP with sound knowledge to address scientific concerns are designated as coordinators for each approved scientific advice procedure (or 'protocol assistance' procedure for orphan medicines).

Each coordinator assembles an evaluation team made up of assessors from their own national agency as well as other EU organizations. Each team writes a report in response to the scientific questions, generates a list of topics for discussion with the rest of the SAWP, and has the authority to request additional information or explanations from the applicant.

Assume the SAWP desires to discuss a specific issue with the developers. In this case, it will call a meeting, in particular if it opposes the planned plan and recommends alternative development options.

EMA committees and scientific working groups (such as the EMA's Committee for Advanced Therapies (CAT) and Committee for Orphan Medicines (COMP)) are consulted by the SAWP. External experts can also be consulted, allowing the SAWP to tap into a larger reservoir of knowledge.

Patients are consulted on a regular basis. Patients are asked to provide input if the EMA intends to react to the developers in writing; if the EMA prefers to meet with the developers, they are welcome.

The SAWP compiles a response in response to scientific inquiries. The CHMP debates and adopts the final advice before giving it to the pharmaceutical company.

5.3.15 What happens before a medicine evaluation begins?

To avoid unnecessary delays, the EMA provides developers with guidance a few months before the assessment begins to ensure that their marketing authorization applications meet legal and regulatory standards.

To acquire marketing permission, medication producers must submit thorough data about their product. The EMA then undertakes a thorough examination of these data to evaluate whether the drug is safe, dependable, and of high quality, and hence suitable for patients.

The EMA provides guidance on what information should be included in a marketing authorization application.

To ensure that their application complies with legal and regulatory standards, medicine developers should consult with the EMA roughly 6 to 7 months before applying. This means that the application required to meet all of the EU legislation's requirements and demonstrate that the drug worked as predicted.

These meetings will bring together EMA personnel responsible for various parts of the application during the review, such as consistency, protection, effectiveness, risk management, pediatric considerations, and others. Developers are encouraged to request such pre-submission meetings to improve the quality of their applications and minimize unnecessary delays, according to the EMA.

5.3.16 In a marketing authorisation application, what information must be provided?

Drug developers must utilize data that complies with EU legislation in their marketing permission applications. They must include, among other things, information about how the medicine is made, the results of laboratory tests, the benefits and side effects experienced by patients, risk management, and planned information to be sent to patients' physicians.

The data submitted in a marketing authorization application must include the patient population that the medication is intended to treat, as well as whether the medicine addresses an unmet medical need; the medicine's consistency, which includes chemical and physical properties such as stability, purity, and biological activity; and compliance with international laboratory research.

What data is expected to be collected from follow-up studies after approval?

The 'risk management plan' (RMP), a document that outlines any actual (known or possible) safety risks associated with the medication, how those risks will be addressed and tracked once the prescription is approved, and what information will be gathered in follow-up studies after approval. PRAC, the EMA's protection committee, evaluates the RMP to ensure its application.

The CHMP will examine and decide on the information that will be supplied to patients and healthcare professionals (for example, the description of product characteristics or SmPC, labeling, and package leaflet).

5.3.17 What sources do you have for medical data?

The majority of information on medicine gathered during its development comes from studies funded by the medicine's originator. Any further information about the applicant's medicine must be submitted and evaluated (for example, from recent medical literature study).

The studies that support the marketing authorization must meet stringent criteria and be conducted in a controlled environment. To guarantee that research is clinically sound and ethically done, international guidelines known as good clinical practice are used to study design, recording, and reporting. The type of evidence that must be used to assess the benefits and risks of a medicine is defined by EU law, and drug makers must follow it. EMA may conduct inspections to check that specified standards are being met.

Through programs such as Enpr-EMA and ENCePP, which bring together expertise from independent academic institutes across Europe, the EMA supports the conduct of high-quality investigations. Additional sources of evidence may be added to the evidence provided by developers because of these actions, notably in the area of ongoing safety monitoring of pharmaceuticals after they have been licensed.

5.3.18 What is the fundamental premise that underpins a medicine's evaluation?

The primary concept that guides a medicine's evaluation is the balance between its benefits and risks. Only if the benefits outweigh the risks will the medicine be approved.

Every drug has advantages and disadvantages. The EMA evaluates the evidence for a drug to see if the benefits of the medicine outweigh the risks in the patient population for which it is intended.

Because not all aspects of a medication's safety are known at the time of approval, how risks can be reduced, regulated, and monitored as the treatment becomes more widely used is typically a factor in the review and authorization decision.

While medicine approval is based on a favorable balance of benefits and risks across the population, each patient is unique. Before using medicine, physicians and their patients can determine if it is the best treatment option based on their knowledge and the patient's unique situation.

5.3.19 Who is responsible for evaluating marketing authorisation applications?

The applications are evaluated by a group of experts from the Committee for Medicinal Products for Human Use (CHMP). A team of assessors assists each of its members.

The CHMP of the EMA evaluates developers' applications. It makes recommendations on whether a medicine should be commercialized. Each EU member state, as well as Iceland and Norway, has one member and one alternate on the committee. The European Commission appoints up to five EU specialists in related fields such as statistics and pharmaceutical quality.

While conducting an evaluation, the CHMP members are assisted by a team of assessors from national agencies with a wide range of experience. They will consider the medicine's safety, efficacy, and method of action, among other factors.

During the evaluation, the CHMP collaborates with other EMA committees. The CAT is in charge of advanced treatment medicines (gene therapy, tissue engineering, and cell-based medicines), while the PRAC is in charge of risk management, the PDCO is in charge of pediatric medicine, and the COMP is in charge of orphan-designated medications.

5.3.20 How does the CHMP work?

The CHMP's reviews are based on peer review and collegial decision-making.

To lead the review of each new medication application, two committee members from different countries are named rapporteur and co-rapporteur (for generics, only one rapporteur is appointed). They are chosen based on objective criteria to ensure that the EU's knowledge is used to its full potential.

It is the responsibility of the rapporteur and co-job rapporteur to conduct independent scientific analyses of the medicine. They each put together a review panel that comprises assessors from their own national agencies, as well as assessors from other countries on occasion.

Each team summarizes the data from the application, makes judgments on the medicine's effects, and investigates any uncertainties or defects in the data in their evaluation reports. They also specify the types of questions that will be asked of the candidate. Both reviews take into account regulatory criteria, relevant research recommendations, and previous experience reviewing similar medications.

From among the CHMP members, the CHMP appoints one or two peer reviewers, as well as the rapporteur and co-rapporteur. Their job is to examine the two analyses and ensure that the scientific rationale is logical, consistent, and sound.

All CHMP members actively participate in the assessment process in cooperation with colleagues and professionals from their respective national agencies. They go over the rapporteurs' findings, make recommendations, and see if the applicant has any other questions. During a CHMP plenary meeting, the first assessment and input from peer reviewers and other committee members are discussed.

The scientific arguments are refined as a result of the disagreements and as new evidence becomes available throughout the examination, either from other experts or from the applicant's clarifications. The committee's analysis and assessment of the data resulted in a final recommendation. This could imply that the committee's views on the medicine's benefits and risks vary over time and differ from the rapporteurs' initial findings.

5.3.21 Is it possible for the CHMP to ask for extra information during the evaluation?

The CHMP asks the applicant questions regarding the facts in the application and requests answers or more analysis to address these concerns during the examination. Within a particular time range, responses must be received.

The CHMP has the authority to voice any issues or questions about medicine. Marketing permission will be rejected if major issues are not resolved.

The way the drug was tested, manufactured, or the results seen in patients, such as the severity of the benefits or the amount of side effects, are all major issues.

5.3.22 What other resources does the CHMP have?

Experts with advanced scientific competence or clinical experience are typically consulted during the assessment to add to the scientific discussion. During the examination, the CHMP can consult with other experts for advice on certain areas. The CHMP will seek for help and ask questions if they have experience in a certain field such as biostatistics or a therapeutic area such as cancer. Members of the EMA's working groups are up to date on the most recent scientific breakthroughs in their fields.

The committee may attract external experts through scientific advisory groups or ad hoc expert groups. Healthcare practitioners and patients are asked to reply to detailed questions on the medicine's future use and importance in clinical practice in these groups.

5.3.23 What role do patients and healthcare providers play?

Patients and healthcare professionals have firsthand knowledge of the problems. As a result, they are approached as experts and asked if the drug would be appropriate for their needs.

Scientific advisory committees or ad hoc expert groups are looking for patients and healthcare professionals to join as experts. Patients participate in talks by expressing their own personal experiences with the disease, as well as their needs and the risks they are willing to bear in exchange for the anticipated benefits. The success of prospective measures to reduce medication dangers in clinical practice, as well as unmet needs in groups of patients, should be examined.

Individual patients can be invited in person or through teleconference to CHMP plenary sessions, or they can be consulted in writing.

5.3.24 What safeguards are in place to protect the independence of experts?

If personal interests are thought to be jeopardizing impartiality, independence is protected through a high level of openness and the application of limits.

EMA has developed competitive interest policies to limit the participation of members, experts, and employees with potentially conflicting interests in the Agency's work while still allowing the Agency to access the best available expertise. Before engaging in EMA activities, members and specialists of commissions, working groups, scientific advisory councils, and ad hoc expert groups must file a statement of interest.

The Agency assigns a risk rating to each statement of interests based on whether the expert has any direct or indirect financial or other interests that could influence their impartiality. The EMA verifies the declaration of interests before allowing you to participate in an EMA activity. If a conflict of interest is discovered, the member's or expert's abilities will be limited.

Not being permitted to engage in a debate on a specific issue or vote on the subject are instances of restrictions. Members' and experts' declarations of interests, as well as details on the limits imposed during scientific committee meetings, are made public in the meeting minutes.

Members of scientific committees are subject to more severe rules than members of advisory boards or ad hoc expert groups. As a result, the EMA will rely on advisory groups' expertise to gather the most relevant and complete data while also imposing more severe decision-making procedures. Similarly, chairmen and members in positions of leadership, such as rapporteurs, have more stringent qualifications than regular committee members.

Members of commissions, working groups, scientific advisory groups (and experts attending these sessions), and EMA officials must all adhere to the EMA Code of Conduct.

5.3.25 How did the CHMP arrive at its final conclusion?

A formal vote determines the ultimate CHMP recommendation. In an ideal world, the CHMP will establish an agreement and recommend acceptance or denial of the marketing license unanimously; this will occur in 90% of situations. The committee's final recommendation will reflect the majority opinion if a final resolution cannot be reached by consensus.

5.3.26 What information is made public throughout the examination of new medicines and after a decision is reached?

The EMA publishes meeting agendas and minutes, as well as documents outlining how the medication was evaluated and clinical trial results provided by developers in their applications, all of which contribute to a high level of transparency in its medicine evaluation. A monthly updated list of new pharmaceuticals assessed by the CHMP can be found on the EMA website. All EMA committee meetings' agendas and minutes are also available, providing information on the evaluation level.

The EMA releases a comprehensive collection of documents known as the European public evaluation report after a decision on marketing authorization approval or rejection (EPAR). The public CHMP evaluation report explains the data analyzed and why the CHMP recommended approving or refusing authorization.

The EMA releases clinical trial results submitted by developers to aid their marketing authorization applications for applications received after 1 January 2015. You can receive clinical study results for previous applications by requesting access to the publication.

The guide to information on human medicines examined by EMA provides comprehensive information on what EMA publishes and when on human medicines, from early conception to initial evaluation and post-authorization changes.

5.3.27 A medicine's evaluation calendar

A new pharmaceutical application can take up to 210 working days to be evaluated. The time spent by EMA professionals assessing the applicant's evidence supporting a marketing authorization application is referred to as active assessment time. One or two 'clock-stops' break this session, during which the applicant prepares answers to any questions posed by the CHMP. The maximum length of a clock-stop is determined by the applicant's estimate of how long it will take to respond, but the CHMP must make the final decision. The initial clock-stop usually lasts 3–6 months, with a 1–3-month rest in between. A new drug is normally evaluated for a year.

Rapporteurs for advanced treatment medications will be chosen by the EMA's Committee for Advanced Therapies (CAT), and each will collaborate with a CHMP coordinator.

Meanwhile, the company's proposed risk management plan (RMP), which outlines how significant risks can be minimized or managed if the medicine is approved, as well as how more information about the medicine's risks and uncertainties can be obtained, has been assigned to two members of the EMA's Pharmacovigilance Risk Assessment Committee (PRAC) (e.g. through post-authorization safety studies). Both members of the PRAC are looking into this assessment.

The CHMP peer reviewers analyze the rapporteurs' assessment reports and submit their comments after examining the rapporteurs' assessment reports and sending their comments to ensure that the scientific argumentation is sound, consistent, and strong.

The CHMP plenary session then discusses the single evaluation paper. Any differences in opinion and difficulties may be resolved as a result of these discussions; nevertheless, new questions may arise, and the report will be updated accordingly. Following these deliberations, the CHMP accepts the research, which reflects a consensus view based on the facts and arguments presented thus far, as well as the applicant's list of questions.

All CHMP and PRAC participants receive a copy of the CHMP rapporteur's and co-assessment rapporteur's reports, as well as a list of questions based on their first evaluation that the applicant must respond. CHMP and PRAC participants are given a risk management plan evaluation that includes questions for the application.

During a 'peer-review' meeting, both participants, including the rapporteur and co-rapporteur teams, other CHMP members, PRAC members, and CHMP peer reviewers, discuss their points of view. This is a critical step in the appraisal of medicine since it is here that the preliminary viewpoints are integrated and consolidated. The result will be a single assessment report with a summary of the examination as well as a list of questions and concerns. Two members of the CAT are also working in advanced treatments.

It may take up to 120 days for this preliminary assessment to be completed. The evaluation is then put on pause (first clock-stop) while the applicant prepares responses to the CHMP's questions and updates the risk management strategy for the medicine.

5.3.28 Further assessment

The rapporteur and co-rapporteur study the applicant's information and include their conclusions in a fresh evaluation report in response to the CHMP's concerns.

CHMP members review and comment on the revised evaluation paper in the same way they did during the initial process.

By day 180 of the active appraisal phase, the opinions of CHMP and PRAC members have been compiled and incorporated into a new assessment report, which is examined and endorsed during a CHMP plenary meeting. Almost invariably, this report will include a new set of questions for the applicant, known as the list of unresolved issues. PRAC members also review and comment on the revised evaluation report, which is then addressed in a plenary meeting of the PRAC. After getting authorisation, the PRAC may mandate that the risk management approach include safety studies.

If a list of outstanding issues is established, the evaluation is postponed once more (second clock-stop) while the applicant prepares responses. After the second clock-stop, the applicant or the CHMP may request an oral clarification, in which the applicant speaks directly to the committee. When the CHMP still has major reservations about the application, it is usually coordinated. If this happens, the applicant will be asked to respond to the committee's unresolved concerns. The rapporteurs or any other CHMP member can now advise contacting a working group for specific questions or convening a scientific advisory group or ad hoc expert group meeting to bring in other experts, such as patients and healthcare professionals. This group will be asked several questions, most of which will revolve around the use of medicine in clinical practice. The group's chair will present the committee with the debate's conclusions.

5.3.29 Discussion and comment at the end

Once the solutions to the outstanding questions have been acquired and presumably addressed during an oral explanation with the organization, the CHMP rapporteur and co-rapporteur examine the applicant's new details. They include their review in a revised assessment report, just like the PRAC rapporteur and co-rapporteur for the risk management plan.

No later than day 210 of the active evaluation period, the CHMP will issue an opinion on the application. The committee would decide whether a medicine should be sold and, if so, under what circumstances. The committee would also agree on the phrasing of product details for healthcare practitioners and patients (i.e. the SmPC, labeling, and packaging leaflet) as well as any further documents that the manufacturer will be required to provide when the drug is approved.

The two members of the committee go over the updated evaluation report and discuss it during the CHMP meeting. The committee makes most of its decisions via consensus. If no such consensus can be reached, the committee's final proposal will reflect the views of the majority.

Divergent viewpoints are listed in the meeting minutes alongside the committee's opinion, along with the names of the members who expressed them. The public appraisal report is then accompanied by several points of view.

5.3.30 Re-examination

The applicant has 15 days from the date of receipt of the CHMP opinion notification to request a re-examination of the CHMP's opinion by stating the reasons for appeal. The scope of the re-examination is confined to the applicant's appeal arguments. In other words, it is solely based on the empirical facts available at the time the committee delivered its original conclusion. At this moment, the petitioner is unable to offer additional evidence. The applicant may suggest that the committee engage a scientific advisory board during the re-examination. If an expert panel was already contacted during the initial evaluation, different experts would be involved in the re-examination. Following that, the initial assessment is used to select a new rapporteur and co-rapporteur. After the re-examination, which can span up to 60 active days, the CHMP issues a final opinion.

5.3.31 Who grants EU-wide marketing authorization?

The EMA is a professional organization with the expertise to assess the benefits and dangers of medicines. However, under EU legislation, it lacks the competence to enable marketing in different EU countries. The EMA's job is to offer recommendations to the European Commission, which will then make a legally binding decision on whether the drug can be sold in the EU. This decision is made within 67 days of receiving the EMA's recommendation. As a result, the Commission is the approving body for centrally approved products.

Commission rulings are published in the Community Register of Medical Products for Human Use.

5.3.32 Who makes decisions on patient access to medicines?

Pricing and reimbursement choices are determined at the national and regional levels after a medicine has been granted EU-wide marketing authorization. Because pricing and payment decisions must be made within the context of each country's national health system, the EMA has no say in them.

To ensure that these processes run smoothly, the Agency collaborates with national agencies such as health technology assessment (HTA) bodies.

Medicines that have been granted marketing authorization by the European Commission are permitted to be sold within the European Union. On the other hand, the company that possesses the license oversees determining which EU nations the drug will be distributed in.

Furthermore, pricing and payment decisions are determined at the national and regional levels within the country's national health system before a drug is made available to patients in a specific EU country.

The EMA has no say in price or reimbursement decisions. To make these processes move more easily, the Agency collaborates with HTA bodies, which assess the new medicine's relative efficacy compared to existing treatments, and EU healthcare payers, who weigh in on the medicine's cost-effectiveness, impact on healthcare budgets, and disease severity. This alliance intends to create solutions for developers to address the data demands of medicines authorities, HTA organizations, and EU healthcare payers throughout the development of a medicine, rather than creating new data after a medicine is licensed. Assume that a single body of evidence can be created early in the development of a medicine that meets the demands of all of these groups. Pricing and reimbursement choices can be made more swiftly and easily at the national level in this case. To do this, the EMA and the European Network for Health Technology Assessment (EUnetHTA) offer developers the option to receive organized, concurrent guidance on their expansion plans. Patients' family members and friends attend these consultations daily to share their thoughts and experiences.

5.3.33 How is the safety of a medicine ensured once it has been put on the market?

After a medicine is approved for use in the EU, the EMA and EU Member States continuously evaluate its safety and act if new information indicates that it is no longer as safe and effective as originally thought.

Many routine activities are involved in the safety monitoring of medicines, including assessing how risks associated with the medicine will be managed and monitored once it is approved, continuously monitoring suspected side effects reported by patients and healthcare professionals identified in new clinical studies or reported in scientific publications, and regularly assessing reports submitted by the manufacturer.

On the request of a Member State or the European Commission, the EMA can also examine a medicine or a class of medicines. These procedures are known as EU referral procedures, and they are frequently initiated by concerns about a medicine's safety, the effectiveness of risk-mitigation methods, or the medicine's benefit–risk balance.

The Pharmacovigilance Risk Assessment Committee (PRAC) of the EMA is responsible for assessing and monitoring the safety of medicines. This means that, if an issue has been identified, the EMA and EU Member States can act rapidly to protect people by changing the information available to patients and healthcare professionals, restricting usage, or halting a medicine.

5.4 Japanese drug approval process

The classification of a new prescription drugs includes:

- Fresh active ingredients in prescription drugs.
- Prescription medications that come in a package.
- Prescription medications with new routes of administration.
- Fresh indications for prescription drugs.
- Fresh dosage formulations for prescription drugs.
- Fresh doses of prescription drugs.
- Items that are biosimilar.
- Prescription medications that come in different dosage types.
- Prescription medications in combination with identical formulations.

'Generic product' means products with the same active ingredients, strengths, dosage formulations, and dosage regimens as innovator's products.

Drugs and medical devices can be classed as orphan drugs or medical devices in Japan under Article 77-2* of the Act on Securing Quality, Efficacy, and Safety of Pharmaceuticals, Medical Devices, Regenerative, and Cellular Therapy Products, Gene Therapy Products, and Cosmetics. The Minister of Health, Labour, and Welfare (MHLW) may designate pharmaceuticals and medical devices that meet the following criteria as orphan drugs/medical devices after receiving orphan designation applications from applicants. An orphan drug is a medicine or medical equipment that has fewer than 50 000 potential users in Japan.

5.4.1 Overview

Before conducting clinical studies to collect data for submission with approval applications for new drug manufacturing, the MHLW must be notified of the study protocol and include various requirements, according to the Act on Securing Quality, Efficacy, and Safety of Pharmaceuticals, Medical Devices, Regenerative and Cellular Therapy Products, Gene Therapy Products, and Cosmetics, as well as the GCP.

Clinical trials on patients and phase I tests on healthy volunteers are all included under GCP, as are human bioequivalence studies, studies for new indications for authorized drugs, and post-marketing clinical trials.

A 'clinical trial consultation system', through which the PMDA aids and advises on research procedures, is now regulated by statute, as is a process through which the PMDA reviews the contents of the initial notice at the request of the MHLW.

5.4.1.1 Type of clinical trial notification

- Industry-initiated trial.
- *Investigator-initiated clinical trial*: The term 'sponsor-investigator' is used in the Ministerial Ordinance on Good Clinical Practice for Drugs to refer to an investigator who has submitted a clinical trial notification in accordance with the Act on Securing Quality, Efficacy, and Safety of Pharmaceuticals, Medical Devices, Regenerative and Cellular Therapy Products, and Gene Therapy Products (including a coordinating investigator who has submitted a clinical trial notification according to the Act, on behalf of all participating investigators, for a clinical trial conducted according to a single protocol but at more than one medical institution).

5.4.1.2 Products required for clinical trial notification

The following products are required to submit clinical trial (protocol) notifications:

- Pharmaceuticals with new active compounds.
- Drugs with unique administration routes (excluding bioequivalence studies).
- Drug combinations with novel indications, as well as new dose and delivery strategies, are all being developed (excluding bioequivalence studies).
- Pharmaceuticals that contain the same active ingredients as drugs that contain new active ingredients but have not yet gone through the re-examination process (excluding bioequivalence studies).

- Drugs considered biological products.
- Gene recombinant technology based drugs

When a clinical study is requested, the new GCP says that a contract for clinical trials can only be signed 30 days after the PMDA receives the research protocol's initial notice (at least two weeks have passed for subsequent notifications, as a rule).

Within 30 days of the study's start date, clinical research protocols for pharmaceuticals used in emergencies to prevent diseases that have a major impact on the patient's life or health, or to prevent other harm to the patient's health, must be filed (MHLW Ordinance No. 89, dated May 2003).

5.4.2 New drug approval (NDA) application

5.4.2.1 Procedure

The PMDA oversees the full approval review process, from inspections and clinical trial consulting to review works. The PMDA receives medication marketing authorization application forms. When the PMDA receives new medication marketing authorization application forms, review teams from the PMDA perform a compliance review of the application data (certification from source data), a GCP on-site inspection, and a thorough review.

To address important difficulties, the approved review process includes expert conversations with members of the review committee and experts. A general review conference involving team members, experts, and applicant representatives follows the expert meeting. Clinical trial participants must submit a 'list of persons participating in the collection of attached data' and a 'list of competitor products and businesses' immediately after the application is submitted, before the expert meeting, and before the Committee on Drugs meeting.

5.4.2.2 PMDA review process

The evaluation process followed by the PMDA is as follows:

- Interrogation (presentation, inquiries, and replies).
- Examination by the whole group.
- Requests and responses.
- Request for GMP inspection (about six months before the meeting of the Committee on Drugs).
- Report on the review.
- Meeting of experts.
- Report on the review.
- For consultation with PAFSC, send a report to MHLW.

5.4.2.3 PAFSC consultation

The PAFSC is consulted after the PMDA study for discussions by the relevant committees and the Pharmaceutical Affairs Committee, as needed. After obtaining the PAFSC report and confirming that the conditions are completed in a second

Table 5.2. ICH guidelines for efficacy, multidisciplinarity, quality, and safety compliance.

E1 The extent of population exposure to assess clinical safety for drugs intended for long-term treatment of non-life-threatening conditions	**E10** Choice of control group and related issues in clinical trials	**E11** (**R1**) Addendum: Clinical investigation of medicinal products in the pediatric population
E11A EWG Pediatric extrapolation	**E12** Principles for clinical evaluation of new antihypertensive drugs	**E14** (**R3**) Q&As: The clinical evaluation of QT/QTc interval prolongation and proarrhythmic potential for non-antiarrhythmic drugs
E14 The clinical evaluation of QT/QTc interval prolongation and proarrhythmic potential for non-antiarrhythmic drugs	**E14/S7B** IWG Questions and Answers: Clinical and nonclinical evaluation of QT/QTc interval prolongation and proarrhythmic potential	**E15** Definitions for genomic biomarkers, pharmacogenomics, pharmacogenetics, genomic data, and sample coding categories
E16 Biomarkers related to drug or biotechnology product development: context, structure, and format of qualification submissions	**E17** General principles for planning and design of multi-regional clinical trials	**E18** Genomic sampling and management of genomic data
E19 EWG Optimization of safety data collection	**E20** EWG Adaptive clinical trials	**E2A** Clinical safety data management: definitions and standards for expedited reporting
E2B (**R3**) Clinical safety data management: data elements for transmission of individual case safety reports (ICSRs)	**E2B** (**R3**) EWG/IWG Electronic transmission of individual case safety reports (ICSRs)	**E2B** (**R3**) Q&As: Clinical safety data management: data elements for transmission of individual case safety reports
E2C (**R2**) Periodic benefit–risk evaluation report	**E2C** (**R2**) Q&As: Periodic benefit–risk evaluation report	**E2D** Post-approval safety data management: definitions and standards for expedited reporting
E2D (**R1**) EWG Post-approval safety data management: definition and standards for expedited reporting	**E2E** Pharmacovigilance planning	**E2F** Development safety update report

E3 (**R1**) Q&As: Structure and content of clinical study reports	**E3** Structure and content of clinical study reports	**E4** Dose–response information to support drug registration
E5 (**R1**) Q&As: Ethnic factors in the acceptability of foreign clinical data	**E5** (**R1**) Ethnic factors in the acceptability of foreign clinical data	**E6** (**R2**) Good clinical practice (GCP)
E6 (**R3**) EWG good clinical practice (GCP)	**E7** Q&As: Studies in support of special populations: geriatrics	**E7** Studies in support of special populations: geriatrics
E8 General considerations for clinical trials	**E8** (**R1**) EWG Revision on general considerations for clinical studies	**E9** Statistical principles for clinical trials
E9 (**R1**) EWG Addendum: Statistical principles for clinical trials	**M1** MedDRA—Medical Dictionary for Regulatory Activities	**M1** PtC WG MedDRA points to consider
M10 EWG Bioanalytical method validation	**M11** EWG Clinical electronic structured harmonized protocol (CeSHarP)	**M12** EWG Drug interaction studies
M13 Informal WG bioequivalence for immediate-release solid oral dosage forms	**M2** EWG Electronic standards for the transfer of regulatory information	**M3** (**R2**) Guidance on nonclinical safety studies for the conduct of human clinical trials and marketing authorization for pharmaceuticals
M3 (**R2**) Q&As: Guidance on nonclinical safety studies for the conduct of human clinical trials and marketing authorization for pharmaceuticals	**M4** (**R3**) Q&As: Organization of the common technical document for the registration of pharmaceuticals for human use	**M4** (**R4**) Organization including the granularity document that provides guidance on document location and paginations
M4E (**R4**) Q&As: CTD on efficacy	**M4E** (**R2**) CTD on efficacy	**M4Q** (**R1**) Q&As: CTD on quality
M4Q (**R1**) CTD on quality	**M4Q** (**R1**) IWG quality	**M4S** (R2) Q&As: CTD on safety
M4S (**R2**) CTD on safety	**M5** Data elements and standards for drug dictionaries	**M6** Virus and gene therapy vector shedding and transmission
M7 Assessment and control of DNA reactive (mutagenic) impurities in pharmaceuticals to limit potential carcinogenic risk	**M7** (**R1**) Assessment and control of DNA reactive (mutagenic) impurities in pharmaceuticals to limit potential carcinogenic risk	**M7** (**R2**) Maintenance EWG/IWG assessment and control of DNA reactive (mutagenic) impurities in pharmaceuticals to limit potential carcinogenic risk
M8 Electronic common technical document (eCTD) v3.2.2	**M8** Electronic common technical document (eCTD) v4.0	**M8** EWG/IWG Electronic Common Technical Document (eCTD)

Table 5.2. (*Continued*)

M9 Biopharmaceutics classification system-based biowaivers	**M9** Q&As on biopharmaceutics classification system-based biowaivers	**Q10** Pharmaceutical quality system
Q11 Development and manufacture of drug substances (chemical entities and biotechnological/biological entities)	**Q11** Q&As: Selection and justification of starting materials for the manufacture of drug substances	**Q12** EWG Technical and regulatory considerations for pharmaceutical product lifecycle management
Q13 EWG Continuous manufacturing of drug substances and drug products	**Q1A (R2)** Stability testing of new drug substances and products	**Q1B** Stability testing: photostability testing of new drug substances and products
Q1C Stability testing for new dosage forms	**Q1D** Bracketing and matrixing designs for stability testing of new drug substances and products	**Q1E** Evaluation of stability data
Q1F Stability data package for registration applications in climatic zones III and IV	**Q2 (R1)** Validation of analytical procedures: text and methodology	**Q2 (R2)/Q14** EWG Analytical procedure development and revision of Q2 (R1) analytical validation
Q3A (R2) Impurities in new drug substances	**Q3B (R2)** Impurities in new drug products	**Q3C (R6)** Maintenance of the guideline for residual solvents
Q3C (R7) Maintenance of the guideline for residual solvents	**Q3C (R8)** Maintenance EWG maintenance of the guideline for residual solvents	**Q3D** Guideline for elemental impurities
Q3D Training implementation of guideline for elemental impurities	**Q3D (R1)** Guideline for elemental impurities	**Q3D (R2)** Maintenance EWG revision of Q3D (R1) for cutaneous and transdermal products
Q3E Informal WG impurity: assessment and control of extractables and leachable for pharmaceuticals and biologics	**Q4A** Pharmacopoeia harmonization	**Q4B (R1)** Annex 1: Residue on ignition/sulphated ash general chapter
Q4B (R1) Annex 10: Polyacrylamide gel electrophoresis general chapter	**Q4B** Annex 11: Capillary electrophoresis general chapter	**Q4B** Annex 12: Analytical sieving general chapter
Q4B Annex 13: Bulk density and tapped density of powders general chapter	**Q4B** Annex 14: Bacterial endotoxins test general chapter	**Q4B (R1)** Annex 2: Test for extractable volume of parenteral preparations general chapter
Q4B (R1) Annex 3: Test for particulate contamination: sub-visible particles general chapter	**Q4B (R1)** Annex 4A: Microbiological examination of non-sterile products: microbial enumeration tests general chapter	**Q4B (R1)** Annex 4B: Microbiological examination of non-sterile products: tests for specified micro-organisms general chapter

Q4B (R1) Annex 4C: Microbiological examination of non-sterile products: acceptance criteria for pharmaceutical preparations and substances for pharmaceutical use general chapter	**Q4B (R1)** Annex 5: Disintegration test general chapter	**Q4B** Annex 6: Uniformity of dosage units general chapter
Q4B (R2) Annex 7: Dissolution test general chapter	**Q4B (R1)** Annex 8: Sterility test general chapter	**Q4B (R1)** Annex 9: Tablet friability general chapter
Q4B Evaluation and recommendation of pharmacopoeia texts for use in the ICH regions	**Q4B** FAQs Frequently asked question	**Q5A (R1)** Viral safety evaluation of biotechnology products derived from cell lines of human or animal origin
Q5A (R2) EWG Viral safety evaluation of biotechnology products derived from cell lines of human or animal origin	**Q5B** Analysis of the expression construct in cells used for production of r-DNA derived protein products	**Q5C** Quality of biotechnological products: stability testing of biotechnological/ biological products
Q5D Derivation and characterization of cell substrates used for production of biotechnological/biological products	**Q5E** Comparability of biotechnological/ biological products subject to changes in their manufacturing process	**Q6A** Specifications: test procedures and acceptance criteria for new drug substances and new dug products: chemical substances
Q6B Specifications: test procedures and acceptance criteria for biotechnological/ biological products	**Q7** Good manufacturing practice guide for active pharmaceutical ingredients	**Q7** Q&As: Good manufacturing practice guide for active pharmaceutical ingredients
Q8 (R2) Pharmaceutical development	**Q8/9/10 (R4)** Q&As: Implementation	**Q9** Quality risk management
Q9 (R1) Informal WG quality risk management	**S1 (R1)** EWG Rodent carcinogenicity studies for human pharmaceuticals	**S10** Photo safety evaluation of pharmaceuticals
S11 EWG Nonclinical safety testing in support of development of pediatric medicines	**S12** EWG Nonclinical biodistribution studies for gene therapy products	**S1A** Need for carcinogenicity studies of pharmaceuticals
S1B Testing for carcinogenicity of pharmaceuticals	**S1C (R2)** Dose selection for carcinogenicity studies of pharmaceuticals	**S2 (R1)** Guidance on genotoxicity testing and data interpretation for pharmaceuticals intended for human use
S3A Note for guidance on toxicokinetics: the assessment of systemic exposure in toxicity studies	**S3A** Q&As: Note for guidance on toxicokinetics: the assessment of systemic exposure—focus on microsampling	**S3B** Pharmacokinetics: guidance for repeated dose tissue distribution studies

Table 5.2. (*Continued*)

S4 Duration of chronic toxicity testing in animals (rodent and non-rodent toxicity testing)	**S5** (**R2**) Detection of toxicity to reproduction for medicinal products and toxicity to male fertility	**S5** (**R3**) EWG Revision of S5 guideline on detection of toxicity to reproduction for human pharmaceuticals
S6 (**R1**) Preclinical safety evaluation of biotechnology-derived pharmaceuticals	**S7A** Safety pharmacology studies for human pharmaceuticals	**S7B** The nonclinical evaluation of the potential for delayed ventricular repolarization (QT interval prolongation) by human pharmaceuticals
S8 Immunotoxicity studies for human pharmaceuticals	**S9** Q&As: Nonclinical evaluation for anticancer pharmaceuticals	**S9** Nonclinical evaluation for anticancer pharmaceuticals

GMP enforcement examination, the Minister provides the new medicine manufacturing/marketing approval.

5.4.2.4 Review period

The applicant will be given one year in addition to the MHLW's one year standard approval review time for new drug approvals beginning 1 April 2000 (dated 28 March 2000) (excluding time required by applicants to prepare responses, etc), bringing the total time from submission to marketing approval to two years. If responding to inquiries or doing additional research will take a lengthy time, the MHLW has urged the applicant to withdraw the application.

In June 2010, the document 'Points to Remember in Applications for Shortening the PMDA Approval Time for New Drugs' was published. This document provides information for applicants who want to fulfill the PMDA's 2013 approval deadlines of 12 months for ordinary reviews and nine months for priority reviews.

The Ministry has also established that the regular review timeframe for new medication applications resulted in a 12 month median cumulative review period for new pharmaceuticals (the Minister showed in 2012 that the median, standard review timeline for new drug approval is 12 months).

The PMDA evaluates the consistency, effectiveness, and protection of medicines, medical devices, and cellular and tissue-based products during the review process, considering existing scientific and technological standards. PMDA's evaluations and related services also include 'consultations' to help with regulatory submissions, GLP/GCP/GPSP inspections to ensure that submitted data meets ethical and scientific requirements, and GMP/QMS/GCTP inspections to ensure that the manufacturing facility for the product under consideration meets quality control standards.

Biosimilar products are called follow-on products in Japan and follow essentially the same guidance as the EMA.

5.5 Conclusion

The development of biopharmaceuticals is a complex and tedious process that takes years and hundreds of millions of dollars to secure regulatory approval in the EU and US. The regulatory approval process in the US and EU needs understanding by the development teams and the manufacturing teams to assure that the regulatory requirements are continually met. This chapter detailed how the regulatory process difference between the FDA and the EU. While the emphasis in this chapter is to familiarize the scientific teams on approaching the two agencies seeking advice during the development, the point-by-point advice provided should help create master plans for faster approval of biopharmaceuticals.

Chapter 6

Biosimilar biopharmaceuticals

Biosimilars represent the fastest-growing category of biopharmaceuticals, where a copy of the originator's biological drug is developed as a low-cost comparator to a first biological product. The regulatory agencies are extremely cautious about safety and, as a result, the development process takes a long, stepwise approach that is fast evolving as more biosimilars are approved, giving the regulatory agencies, as well as the developers, a better understanding of safety and efficacy testing of biosimilars. While regulators across the globe have different submission requirements, most agree on standard methods of proving biosimilars' safety: analytical assessment, nonclinical testing, clinical pharmacology testing, and clinical efficacy testing as required. This chapter provides both an overall view and more specific details on developing biosimilars to enable faster regulatory approvals.

6.1 Overview

Biopharmaceuticals represent the highest cost category of drug therapy, given that the cost of securing approval for a new biological drug easily exceeds a billion dollars and takes almost ten years to ensure regulatory approval. As a result, in most regulatory territories biological drugs are given exclusivity for marketing despite the expiry of their patents; in the US, all new biological drugs get at least 12 years to recoup their investment, while other territories provide different types of exclusivities.

As discussed in earlier chapters, proteins and antibodies have a large structure that remains variable because of the nature of its production in a living species. To prove that the two products, both with a variable structure, have equivalent variability is challenging because of the inevitable lack of understanding of the relationship between a product's structural attributes and its clinical performance.

As a result, biosimilars' development takes a different approach and direction compared to the development of a new biopharmaceutical product. A proposed biosimilar product (it is called proposed until it is established as biosimilar) must demonstrate 'biosimilarity', meaning that it is highly similar to the reference product

doi:10.1088/978-0-7503-3175-3ch6

notwithstanding minor differences in clinically inactive components and that there are no clinically meaningful differences between the proposed biosimilar and the reference product in terms of the safety, purity, and potency of the product.

Biosimilarity is a judgment made by the regulatory agencies, not the developers, as a totality of the evidence based on:

- Analytical studies that demonstrate that the biological product is highly similar to the reference product notwithstanding minor differences in clinically inactive components.
- Animal studies (including the assessment of toxicity).
- A clinical study or studies (including the assessment of immunogenicity and pharmacokinetics or pharmacodynamics) that are sufficient to demonstrate safety, purity, and potency in one or more appropriate conditions of use for which the reference product is licensed and intended to be used and for which licensure is sought for the biological product.

In this chapter, we provide an overview of the current global guidance on the evaluation and approval of biosimilars, along with recommendations on making the regulatory filing more persuasive to the regulatory agencies.

6.2 Biosimilarity

A biosimilar is sometimes referred to as a follow-on biologic and is a therapeutic drug that is highly similar but not structurally identical to a brand-name biologic (i.e. the reference product). This contrasts with a generic chemical drug, an exact copy of a brand-name chemical drug (i.e. the reference listed drug). Because biologics are more complex than chemical drugs, both in composition and method of manufacture, biosimilars will not be exact replicas of the brand-name product; they may instead be shown to be highly similar. However, after decades of experience, both the developers and the regulatory agencies have agreed that the inherent variability in biological products, as evidenced in the batch-to-batch variability, occurs for both brand-name biologics and biosimilars, and can be assessed and managed effectively to declare products as biosimilars.

The regulatory guidelines and standard scientific literature relating to biosimilars have adopted a specific vocabulary, as described in table 6.1.

Table 6.1. Terminology related to biosimilars.

Biobetter	A biological product cannot claim to be a proposed biosimilar product if it has a higher efficacy, lower dose requirement, fewer side effects, more convenient drug administration, or any other difference considered an improvement over the reference product. Such products are considered new biological products and are not accepted for evaluation by Agencies under this guideline.
Comparability testing	Comparability guidance such as ICH Q5E and ICH Q6B applies to a change in the manufacturing process of an *approved* product. The testing is conducted using the final approved biological product as the

	reference product. The testing is limited to critical quality attributes that are well-known to the manufacturer. These guidelines may serve as overall guidance but do not apply to biosimilar development, any stage of development, including the final scale-up. However, the Agencies guidance may apply once the product has been approved. To avoid any confusion in referring to these guidelines, the current guidelines use the term 'comparative testing' rather than 'comparability testing'.
Product	When used without modifiers, this term is intended to refer to the intermediates, drug substance, and drug product, as appropriate. The use of the term *product* is consistent with the use of the term in ICH Q5E. This should not be confused with the regulatory consideration of a drug or a biological product's approval pathway. A drug product can be approved as a drug or as a biological product. During the development process, a proposed biologic is labeled by adding 'proposed' to differentiate it from a proposed biosimilar that will be an authorized product.
High similarity or highly similar	The similarity is a binomial attribute; a protocol for testing may fail or pass. If it fails, a close examination is conducted to determine the cause of failure if any of the failed attributes have clinical significance. Often the terms 'high similarity' or 'highly similar' are used to indicate that there is residual uncertainty remaining that is clinically meaningful.
No clinically meaningful differences	A proposed biosimilar product is not identical to the reference product. There are differences with the reference product; if the differences do not affect safety or efficacy, we can claim no clinically meaningful difference.
No residual uncertainty	Every test must meet a predetermined qualification; however, where the testing fails, it leaves uncertainty about the safety and efficacy as supported by the given test. When such delays are removed, we can say that a test left no residual uncertainty.
No one-size-fits-all	Every testing of biosimilars is specific to the product type and requires highly individualized development protocols.
Fingerprint-like similarity	At the highest possible level reaching this level of similarity will significantly reduce the burden of additional testing.
Totality of the evidence	The evidence of safety and clinical efficacy is accumulated through multiple studies and when we combine all the results we can establish biosimilarity.
Stepwise	Testing is conducted at predefined steps and only when the testing meets the criterion at one step can the testing move on to the next step; and a higher step does not resolve residual uncertainties of the lower steps.
Phase 1, 2, and 3 studies	Phase 1–3 studies are either stand-alone, as in phases 1 and 2, or placebo control, as in phase 3. None of these conditions apply to any testing of proposed biosimilar products. These terms are widely used by regulatory agencies and mostly by the developers. The correct designations are nonclinical pharmacology, clinical pharmacology, and comparative efficacy studies.

6.3 Regulatory guidance

6.3.1 Background

A product must have the same administration method, dose, potency, and mode of action as the reference medicine to be considered a proposed biosimilar. The application must include evidence that the biological product is comparable to the reference product, including modest changes in clinically inactive components (e.g. it may be possible, for example, for a proposed biosimilar product formulated without human serum albumin to demonstrate biosimilarity to the reference product formulated with human serum albumin). The developers would also have to show that there is no clinically substantial difference in terms of safety, purity, or potency between the proposed biosimilar and the reference product. Most regulatory agencies use a similar definition of biosimilars, with minor exceptions, primarily in the way biosimilarity is determined.

Biosimilars' field is relatively new, creating a new regulatory class genre requiring new regulatory guidance policies. Figure 6.1 shows a proposed biosimilar guidance introduction across the globe.

Since the EU was the first to issue proposed biosimilar guidance, many agencies followed these guidelines, mainly because of the references to the International Conference on Harmonization (ICH) guidelines incorporated in the EU guidance. The US brought out its guidelines in 2009 with several differences to the EU guidance, and even though there are efforts made to harmonize the global guidance, many differences remain among the regulatory agencies on the evaluation and marketing authorization of biosimilars.

An estimated 182 companies around the world are developing biosimilars including the companies that had brought in new biological products including Amgen, Pfizer, Merck, Hospira, Mylan, Sandoz, STADA Arzneimittel, Teva

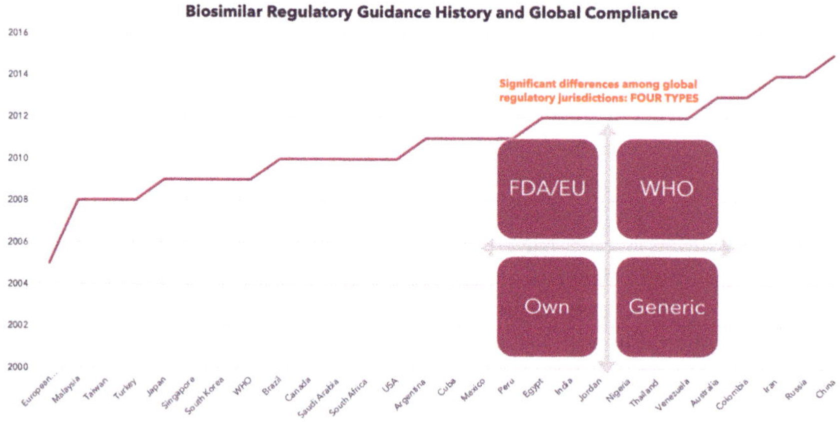

Figure 6.1. Dates of implementation of proposed biosimilar guidance based on four different approaches to creating guidance. Most developed countries follow either FDA or EU guidance; most developing countries follow the WHO guidance. Some have developed their guidance (Latin America, China, and India), while some treat biosimilars only as chemical generics.

Pharmaceutical Industries, 3SBio, Beijing ShuangLu Pharmaceuticals, Qilu Pharmaceutical, Shanghai Fosun Pharmaceuticals, Tonghua Dongbao, Biocon, Dr Reddy's Laboratories, Intas Biopharmaceuticals, Ranbaxy, Reliance Life Sciences, Wockhardt, Zydus Cadila, Probiomed, Biosidus, Amega Biotech, Celltrion, LG Life Sciences, Dong-A, Bioton, and Biocad, to name a few.

6.3.2 Latin America

Most Latin American countries have set out their own rules based on WHO recommendations, but there is no standardization in the evidence required for biosimilar certification.

6.3.3 Canada

The regulatory body for the approval of biologicals is the Biologics and Genetic Therapies Directorate (BGTD) of the Health Products and Food Branch (HPFB) of Health Canada that regulates biologicals (products derived from living sources) and radiopharmaceuticals for human use in Canada, whether manufactured in Canada or elsewhere. Some of the products regulated by the BGTD include blood and blood products, hemostatic agents, bacterial and viral vaccines, hormones, enzymes, cytokines, monoclonal antibodies, allergenic extracts, gene and cell therapies, tissues, and organs. There is no automatic extrapolation of indications allowed in Canada.

In March 2010, Health Canada finalized biosimilars guidelines previously called subsequent entry biologics in Canada, realizing that the correct definition is biosimilars.

6.3.4 The World Health Organization (WHO)

The WHO labels biosimilars as similar biotherapeutic products (SBPs) and the following three documents provide up-to-date guidance on the evaluation of SBPs:

- WHO 2009 Guidelines on evaluation of similar biotherapeutic products (SBPs), Annex 2 *Technical Report* Series No. 977, https://www.who.int/biologicals/publications/trs/areas/biological_therapeutics/TRS_977_Annex_2.pdf?ua=1.
- WHO 2018 WHO Questions and Answers: similar biotherapeutic products, https://www.who.int/biologicals/expert_committee/QA_for_SBPs_ECBS_2018.pdf?ua=1.
- WHO 2016 Guidelines on evaluation of monoclonal antibodies as similar biotherapeutic products (SBPs), Annex 2 *Technical Report* Series No. 1004, https://www.who.int/biologicals/biotherapeutics/WHO_TRS_1004_web_Annex_2.pdf?ua=1.

The WHO guidance is followed widely by developing countries. There remains a significant scientific gap in the understanding of biosimilars in how the WHO advises compared to the EU or US:

- The selection of the label 'similar biotherapeutics' instead of 'biosimilar' assumes that a target product is 'similar' that happens to be a biotherapeutic product. 'Biosimilar' means that the product is biologically similar, not necessarily similar in the first place. While this difference may seem mostly a jargon, it has caused much confusion leading to some countries calling biosimilars generics.

- The WHO also suggests that before a proposed biosimilar product can be approved, the reference product should have been marketed for a 'suitable duration' of time in 'a jurisdiction with a well-established regulatory framework and principals'. This suggestion creates doubts about the safety of an approved product. The WHO guidance does not indicate how long this waiting period should be. There is no rationale for this requirement.
- The WHO recommends that regulators use their prior knowledge, based on the evaluation of the reference product's manufacturing process, to judge the safety and efficacy of a proposed biosimilar product. In addition to ethical considerations, such comparisons are scientifically flawed because it is impossible to draw a parallel between products manufactured using different processes.
- The WHO analytical similarity testing is deeply flawed. The WHO allows the use of production lots from different manufacturing sites to establish analytical and biological similarity with the reference product, an allowance that violates the basic principles of similarity testing, even if each of the manufacturing sources is a good-manufacturing-process compliant. No developed country regulatory agency allows this practice.
- On a more specific scientific consideration, the WHO states that 'It may not be possible to set a definite number of lots to be analyzed in the comprehensive comparability exercise.' Nothing is farther from the truth. The number of lots tested is well-defined in the standard equivalence testing models, where at least 6–10 lots must be analyzed to establish an acceptable statistical basis. The WHO considers that statistical models should not be applied to analytical similarity assessment. This misconception has caused many countries to accept analytical data based on a small number of lots, leaving the evaluation of biosimilars' safety at high risk.
- In another significant scientific mistake, the WHO considers comparative stress stability testing of a proposed biosimilar and unnecessary reference product. However, this is one of the most sensitive methods to test structural similarity, and all developed country agencies require it.
- The WHO also states that stability testing studies only help establish appropriate conditions for shipping and storage. The fact is that a proposed biosimilar must have the same storage condition as the reference product, and there is no need to justify a new storage condition. The shipping stress testing is conducted under a different protocol based on the prevalent practices of shipping.
- One of the most divisive suggestions by the WHO is that toxicology studies in monkeys should be avoided. This has led to the approval of many biosimilars across the globe based on valueless testing in rats, which do not have receptors to show toxicity to monoclonal antibodies.
- The WHO suggestion to study pharmacodynamics and immunogenicity in animal species does not consider that animals do not replicate human responses.

- The WHO's animal testing protocols based on human dose multiples are also flawed because this dosing is not in the linear toxicity range to detect any difference between a proposed biosimilar and the reference product.
- For clinical efficacy testing, the WHO suggests using a non-inferiority testing model, allowing acceptance of higher efficacy, without realizing that higher efficacy may also lead to higher adverse events. The correct protocol for efficacy testing is the equivalence model.
- The WHO does not recommend evaluating the most frequent adverse effect of local reactions at the injection site caused by the agent's formulation and not by the active molecules. Because biosimilars can have a different formulation composition, evaluation of local reactions is pivotal in assuring the products' safety.

The WHO guidance is widely adopted in many countries that have added their arbitrary and unjustified changes. For example, in China, India, and Russia, local clinical trials are required, but not a suitable statistical protocol. Iran, Peru, Nigeria, and Venezuela have removed all clinical testing requirements.

6.3.5 India

As of April 2020 there were 98 approved biosimilars in India, with at least 50 on the market, the most of any country in the world. The Indian CDSCO and DBT guidelines have enabled manufacturers to bypass what they call 'phase 3' clinical trials in circumstances where sufficient pharmacodynamic (PD) and pharmacokinetic (PK) data are available, opening the door for faster product approvals. The safety and efficacy of trial requirements can also be waived for such products if PD and PK data are available. However, there is no description of what is considered 'sufficient' of any correlation between the PK/PD data for waivers of safety and efficacy studies. Additionally, CDSCO and DBT include a requirement for 'phase 4' post-marketing trials, including at least 200 study participants and be conducted within two years of market approval. There is no statistical rationale for this requirement of the number of subjects, and none of the products has ever failed in these token studies. The statistical validity of 'phase 3' trials is also questionable because many trials in India recruit only a limited number of participants and use a non-inferiority range that is much larger than what is used in the studies submitted to the FDA or EMA, making such studies of little value.

One of the most conflicting aspects of the Indian and WHO guidelines is that they allow testing monoclonal antibodies in rats because of religious restrictions on using monkeys in India; these studies are useless in resolving any critical quality attribute—antibodies have no toxicity in rodent species because rodents do not have relevant receptors. As a result, none of the monoclonal antibodies approved in India meet the US or EU definition of biosimilars. The Indian guidance further suggests using animal dosing based on human dosing and accepts testing immunogenicity in animals, a useless exercise because animals and humans have different immune systems.

Like the WHO, the Indian guidance does not recognize the statistical treatment of analytical testing, allowing testing a smaller number of batches, most commonly three, to establish analytical similarity.

India released its official guidelines in June 2012, before which 20 biosimilars were approved under an ad hoc basis.

6.3.6 Australia

Australia has adopted the EU guidelines without any changes, while Singapore and Malaysia amended their guidelines while mainly following the EMA guidelines. Brazil and Cuba chose the WHO and Canadian guidelines to develop regulations for their respective countries. However, there are considerable variations in definitions, terminology, reference product choice for comparative testing, the extent of data requirements, and other aspects.

6.3.7 Japan

In Japan, recombinant plasma proteins, recombinant vaccines, PEGylated recombinant proteins, and nonrecombinant proteins that are highly purified and characterized are defined in the guideline's scope. However, some EU inclusions, such as polyglycans (low molecular weight heparin) and synthetic peptides, have been excluded since the extensive characterization is not required for simple molecules without higher-order structures. Comparative stability studies with reference biologics are not mandatory in Japan. This is the case with toxicology studies, wherein impurities need not be evaluated through nonclinical studies. Somatropin and epoetin alfa have been authorized for marketing in Japan following the abovementioned route.

6.3.8 Korea

Korea guidelines (KFDA) follow the WHO regulatory guidelines, except for the clinical evaluation to demonstrate similarity. Equivalence testing, where the upper and lower margins need to be specified with justified preset limits, is mandatory with KFDA. In contrast, in the WHO guidelines the noninferior mode of testing where only the lower margin is specified is preferred.

6.3.9 Iran

In Iran a proposed biosimilar product can serve as the reference product if the original reference product is not registered in Iran or if a proposed biosimilar has been approved in the EU or USA. Also, Iran specifically recommends that specifications for a proposed biosimilar product should be the same as for the reference product. Still, it does not consider a side-by-side comparative accelerated stability study as a mandatory requirement. The Iranian NRA does not insist on comprehensive clinical studies; it relies more on national post-marketing surveillance data for drug safety. The biosimilars approved in Iran do not qualify as biosimilars anywhere else in the world.

6.3.10 EU

6.3.10.1 Background

The EMA had approved 59 products while the FDA has approved 30, including the biologics in table 6.2.

It is worth noting that the products developed in the EU largely match those in the US, probably because of the confidence it gives to the developers to file in both jurisdictions after it is approved in one.

The CHMP recommendations for designing and conducting proposed biosimilar comparability studies should always be read in conjunction with relevant scientific guidelines and European Union regulatory provisions.

The EMA does not make recommendations on whether a proposed biosimilar should be used interchangeably with its reference drug during its evaluation of proposed biosimilar medicines for authorization purposes. The EU member states are responsible for substitution policies.

The following are relevant guidelines:

- EMA 2014 Guideline on similar biological medicinal products containing biotechnology-derived proteins as active substance: quality issues, EMA/CHMP/BWP/247713/2012
- EMA 2014 Guideline on similar biological medicinal products containing biotechnology-derived proteins as active substance: non-clinical and clinical issues, EMEA/CHMP/BMWP/42832/2005 Rev1.

Table 6.2. The number of biosimilars approved in EU and US.

Product	EU	US
Adalimumab	10	5
Bevacizumab	2	2
Enoxaparin sodium	2	0
Epoetin alfa/zeta	5	1
Etanercept	2	2
Filgrastim	7	2
Follitropin alfa	2	0
Insulin glargine	2	2
Insulin	1	0
Infliximab	4	4
Insulin lispro	1	1
Pegfilgrastim	7	3
Rituximab	6	2
Somatropin	1	0
Teriparatide	2	1
Trastuzumab	5	5
Total	59	30

Specific product-related guidelines can be found on the EMA website:
https://www.ema.europa.eu/en/human-regulatory/research-development/scientific-guidelines/multidisciplinary/multidisciplinary-biosimilar#-product-specific-biosimilar-guidelines-section

The scientific principles of a proposed biosimilar comparability exercise are based on those used to evaluate the impact of changes in a biopharmaceutical product's manufacturing process (as outlined in ICH Q5E). However, this assertion creates a conflict because the ICH Q5E does not use a reference product, rather, it compares an approved product before and after the changes.

Extrapolation to other reference products' indications could be permissible with proper scientific justification if biosimilarity has been shown in one indication.

Once the marketing authorization has been obtained, there is no regulatory necessity to repeat the demonstration of biosimilarity against the reference product, e.g. in the context of a change in the manufacturing method.

6.3.10.2 Reference product

The reference product must be a product authorized in the EEA based on a complete dossier.

It may be possible for an applicant to compare a proposed biosimilar to a non-EEA-authorized comparator (i.e. a non-EEA-authorized version of the reference product) in certain clinical studies and *in vivo* nonclinical studies (where needed) to facilitate global development of biosimilars and avoid unnecessary repetition of clinical trials (e.g. ICH countries). It will also be the applicant's responsibility to show that the comparator approved outside the EEA is equivalent to the reference product approved inside the EEA.

If certain development program clinical and *in vivo* nonclinical studies are conducted with a non-EEA-authorized comparator, the applicant must provide sufficient data or information to scientifically justify the comparative data's relevance and establish an acceptable bridge to the EEA-authorized reference product. Data from analytical studies (e.g. structural, and functional data) that compare all three products will always be needed as bridging data in a scientific sense (a proposed biosimilar, the EEA-authorized reference product, and the non-EEA-authorized comparator). Data from clinical PK and PD bridging studies for all three drugs may also be included. The overall appropriateness of such an approach, as well as the sort of bridging data required, will be determined on a case-by-case/product-type basis, and should be reviewed with the regulatory authorities ahead of time. The conclusion of the scientific justification and bridge sufficiency, on the other hand, will be made only during the application's evaluation.

6.3.10.3 Nonclinical studies

To support biosimilarity, necessary nonclinical investigations should be done before initiating clinical trials.

Data from several comparative *in vitro* tests, some of which may already be available from quality-related assays, should generally be provided to examine any potential variation in biological activity between a proposed biosimilar and the reference product. These studies should include relevant assays on: Binding to target(s) known to be involved in the pharmaco-toxicological effects and pharmacokinetics of the reference product (e.g. receptors, antigens, enzymes). Signal transduction and cell functional activity/viability influence the pharmaco-toxicological consequences of the reference product.

Rather than focusing simply on the answer, the investigations should be comparative. In order to achieve precise results, the processes used must be scientifically valid and appropriate for their purpose.

The studies must be sensitive, specific, and selective enough to show that reported quality differences are not clinically relevant. At the pharmacological target(s), the research should investigate the concentration–activity/binding relationship of a suspected biosimilar and the reference product, covering a concentration range where potential differences can be noticed most sensitively.

Biotechnology-derived proteins are known to have *in vivo* consequences that *in vitro* studies cannot fully explain. As a result, if an acceptable *in vivo* model for the species or design is available, nonclinical evaluation *in vivo* study may be required to provide further data.

When determining the need for *in vivo* nonclinical trials, factors to examine include, but are not limited to:

- The inclusion of potentially crucial quality traits that are not present in the benchmark product (e.g. new post-translational modification structures.
- Significant quantitative differences in quality attributes between a proposed biosimilar and the reference product.
- Relevant formulation adjustments, such as the use of excipients that are not frequently used for biotechnology-derived proteins.

Although *in vivo* testing is not always required for each of the reasons given above, they should be considered together to assess the level of concern and if *in vivo* testing is necessary. If the physicochemical and biological properties of a suspected biosimilar, as well as nonclinical *in vitro* tests, are deemed good, and no problems are uncovered later that would impede direct human use, an *in vivo* animal study is usually not considered necessary.

If product-specific factors impacting PK and biodistribution, such as significant glycosylation, cannot be examined adequately on a quality and *in vitro* level, *in vivo* testing may be necessary. The applicant should then decide whether these tests should be performed on animals or in humans as part of clinical trials. If more *in vivo* data are needed, consider whether a suitable animal species or other relevant models (e.g. transgenic animals, transplant models) are available. In the absence of a relevant *in vivo* animal model, the applicant may choose to perform human trials while taking risk-mitigation measures into account.

As animal immunogenicity testing is not necessarily indicative of human immunogenicity, *in vivo* animal findings may need to be interpreted. As a result, blood samples should be acquired and stored for any future pharmacokinetic/toxicokinetic data analysis. It is unknown whether or not this procedure will be beneficial.

Biosimilars do not require safety pharmacology, reproduction toxicology, or carcinogenicity investigations in nonclinical testing.

In most cases, local tolerance studies are not required. If excipients are introduced for the first time or with little experience with the predicted therapeutic route of administration, local tolerance may need to be determined. If more *in vivo* experiments are carried out, local tolerance testing could be incorporated into that study's design rather than being done independently. Even if the inactive compounds are comparable, we believe that local tolerance studies are required since the sources and levels of the inactive substances may differ.

6.3.10.4 Clinical studies

A proposed biosimilar development program must include comparative pharmacokinetics (PK) studies to show that the proposed biosimilar and the reference product have similar PK patterns for key PK parameters.

Despite the necessity of comparing target-mediated clearance in the biosimilarity exercise, it may not be possible in patients due to high heterogeneity in target expression, including variability over time. However, because *in vitro* studies are expected to show similar interactions between a proposed biosimilar and its target(s) (including FcRn for a mAb), the absence of a pivotal PK study in the target population is acceptable if additional PK data are collected during the efficacy, safety, and PD studies to allow evaluation of the clinical impact of variable pharmacokinetics and possible changes in the target population. The PK profile in a subgroup of patients or population pharmacokinetics can be used to make these linkages.

It is desirable to carry out single-dose crossover research with complete PK profile characterization, including the late elimination phase. With drugs with a lengthy half-life and a high risk of immunogenicity, a parallel-group design may be required (avoiding second dosing of the same drug substance). The doses used in the single-dose PK suggested biosimilar comparison research in healthy volunteers could be lower than those used in clinical practice. Healthy subjects are not usually available for PK research. If a single-dose trial is not possible, the PK must be examined in patients as part of a multiple-dose study. A sensitive model/population should be investigated, with fewer elements that generate considerable inter-individual or time-dependent variation.

If the reference product can be supplied both intravenously and subcutaneously, the subcutaneous administration evaluation will usually suffice to address both absorption and elimination. If suggested biosimilar equivalence in both absorption and elimination has been proven for the subcutaneous route, the intravenous administration evaluation may be waived. When the chemical has an absorption constant that is substantially slower than the elimination constant, an intravenous PK study is required (flip flop kinetics).

The $AUC_{0\text{-inf}}$ for intravenous administration and $AUC_{0\text{-inf}}$ and usually C_{max} for subcutaneous injection are the major parameters in a single-dose PK investigation. Secondary parameters including t_{max}, distribution volume, and half-life should also be calculated. The shortened AUC after the first administration until the second administration ($AUC_{0\text{-}t}$) and AUC throughout a dosing interval at a steady state should be the major metrics in a multiple-dose trial. At a steady state, secondary parameters are C_{max} and C_{trough}.

Anti-drug antibodies should be assessed in tandem with PK parameters using suitable sampling time points in any PK investigation.

6.3.10.5 Pharmacodynamic studies

When it is possible, pharmacodynamic (PD) markers should be included in pharmacokinetic investigations. The relevance of the PD indicators to the clinical outcome should be considered while choosing them. Comparative PK/PD studies may be adequate in some cases to demonstrate clinical comparability between a proposed biosimilar and the reference product if the following conditions are met:

- The PD marker/biomarker used is a recognized surrogate marker. It can be linked to patient outcomes in the sense that a similar effect on the PD marker would assure a similar influence on the clinical outcome. An absolute neutrophil count to evaluate the efficacy of granulocyte-colony-stimulating factor (G-CSF), early viral load decrease in chronic hepatitis C to evaluate the effect of alpha interferons, and a euglycemic clamp test to compare two insulins are all examples of relevant tests.
- There may be PD-markers that are not established surrogates for efficacy but are relevant for the pharmacological action of the active substance, and a clear dose–response or concentration–response relationship has been demonstrated.
- Magnetic resonance imaging of disease lesions can be used to compare two events. A single or multiple dosage–exposure–response study at two or more dose levels may be enough to avoid a clinical efficacy trial in this scenario. This study design ensures that a proposed biosimilar and a reference can be evaluated in the steep section of the dose–response curve.
- If physicochemical, structural, and *in vitro* biological analyses, as well as human PK studies and a combination of PD markers that reflect the pharmacological action and concentration of the active substance, can provide robust evidence for proposed biosimilar comparability, the confirmatory clinical trial may be waived in exceptional cases.

It is recommended to discuss such a ('fingerprinting') strategy with regulatory authorities when data to prove clinical proposed biosimilar comparability is provided from PK investigations backed by studies using non-surrogate PD/biomarkers. A suggestion for the magnitude of the equivalency margin(s) with clinical reasoning and steps to demonstrate a comparable safety profile should be included in the plan.

6.3.10.6 Efficacy trials

In the absence of surrogate effectiveness markers, it is normally necessary to show that a proposed biosimilar and the reference product have comparable clinical efficacy in appropriately powered, randomized, parallel-group comparative clinical trials with efficacy outcomes, ideally double-blind. The research population should be sensitive enough to detect any discrepancies between the proposed biosimilar and the reference product's approved therapeutic indications. Changes in clinical practice, such as the use of a concomitant medicine in combination treatment, the line of therapy, or the severity of the disease, may occasionally necessitate a departure from the approved therapeutic indication. Deviations from the standard must be justified and discussed with the appropriate regulatory organizations.

6.3.10.6.1 Study designs

An equivalency design should be employed in most cases. If justified by a compelling scientific reason and considering the characteristics of the reference product, such as safety profile/tolerability, dose range, and dose–response relationship, the implementation of a non-inferiority design may be acceptable. Only when the prospect of a large and clinically relevant increase in efficacy can be ruled out on scientific and mechanistic grounds can a non-inferiority trial be considered. As with equivalency trials, however, assay sensitivity must be considered.

Because efficacy has already been shown with the reference medicine, efficacy trials of prospective biosimilar products do not demonstrate efficacy per se. The goal of efficacy trials is to confirm that a proposed biosimilar and the reference product perform similarly in clinical trials.

The selection of clinical endpoints and time points for endpoint analysis in the development of a proposed biosimilar product may differ from the guidance for novel active substances. For the development of novel products, the CHMP has published disease-specific guidelines. Comparability should be established in relevant, sensitive clinical models and study conditions in the absence of such a guideline. The applicant must demonstrate that the model used is appropriate and sensitive enough to detect any variations in efficacy and safety. Distinction from objectives indicated in disease-specific guidelines, on the other hand, must be scientifically justified. Significant disparities in quality features cannot be justified using clinical data.

Previous clinical trials using the reference product may have proven a link between the 'hard' clinical endpoints suggested by the recommendations for new active substances and other clinical/pharmacodynamic endpoints that are more sensitive to identify clinically relevant differences. It is not necessary to employ the same primary efficacy objectives as those used in the marketing authorization application for the reference product in this circumstance. However, some common endpoints (e.g. secondary endpoints) should be included to allow for comparisons to clinical trials using the reference product.

Comparability margins should be pre-specified and justified using the reference product's data on the choice of the non-inferiority margin on both statistical and clinical grounds.

6.3.9.10.2 Clinical safety

Clinical safety is a critical component of the clinical development process, and it is assessed throughout the initial PK and PD evaluations as well as the pivotal clinical efficacy study. Pre-authorization comparative safety data should be collected in most cases, with the amount of data collected dependent on the nature and severity of known safety risks for the reference product. The length of pre-authorization safety follow-up should be justified.

Immunogenicity testing of a prospective biosimilar and the reference product should be carried out as part of a proposed biosimilar comparability exercise, using the same assay format and sample schedule that must meet all existing criteria. Analytical assays should be performed in parallel (blinded) using both the reference and suggested biosimilar molecules to assess the immunological response to the product received by each patient. Antibodies to both the proposed biosimilar and the reference molecule should be detected in the analytical assays. Nonetheless, they should be able to detect all antibodies generated against a potential biosimilar. Antibodies and antibody titers should usually be quantified and displayed according to their frequency and nature (e.g. cross-reactivity, target epitopes, and neutralizing activity). They must be evaluated and interpreted in terms of their possible impact on clinical efficacy and safety.

The length of the immunogenicity research should be justified on a case-by-case basis, taking into account the length of the therapy, the product's disappearance from circulation (to avoid antigen interference in the assays), and the time it takes for the humoral immune response to arise (at least four weeks when an immunosuppressive agent is used). The time course and characteristics of undesired immune responses described for the reference product, such as a low risk of clinically significant immunogenicity or no significant tendency for growing immunogenicity over time, should be used to justify the duration of follow-up. In the case of chronic administration, pre-authorization usually requires one year follow-up data. Based on the immunogenicity profile of the reference product, a shorter follow-up data pre-authorization period (e.g. six months) may be appropriate. Post-authorization, immunogenicity data for an extended period of up to one year could be submitted if necessary. Refer to product-specific proposed biosimilar advice for specific products.

Increased immunogenicity compared to the reference product could skew the benefit–risk analysis and cast doubt on biosimilarity. Reduced immunogenicity for a proposed biosimilar, on the other hand, is a possibility that does not rule out biosimilar approval. The efficacy analysis of the overall study population may falsely suggest that the proposed biosimilar is more efficacious than the reference product if the biosimilar reduces the production of neutralizing antibodies. As a result, a subgroup analysis of efficacy and safety in those who did not develop an anti-drug antibody response throughout the clinical trial is recommended. If an immune response does not affect the efficacy of a proposed biosimilar and the reference drug, this subgroup study could help illustrate that they are, in principle, equivalent.

6.3.10.6.3 Extrapolation of efficacy and safety from one therapeutic indication to another

There could be multiple therapeutic indications for the reference product. Extrapolation of clinical data to other indications of the reference product may be permissible if proposed biosimilar comparability has been proven in one indication, but it must be scientifically supported. Additional evidence will be necessary if it is uncertain whether the safety and efficacy confirmed in one indication will be relevant in another. When proposed biosimilar comparability has been demonstrated by thorough physicochemical and structural analyses, as well as *in vitro* functional tests, and clinical data (efficacy and safety, as well as PK/PD data) in one therapeutic indication, it is expected that safety and efficacy can be extrapolated. Additional data are required in certain situations, such as:

- The active ingredient in the reference product interacts with numerous receptors, which may have varied effects on therapeutic indications that have been investigated and others that have not been tested.
- There are multiple active sites in the active chemical, and the sites may have varied effects on different therapeutic purposes.
- In terms of efficacy and safety, the researched therapeutic indication is not important for the others, i.e. it is not sensitive to differences in all relevant characteristics of efficacy and safety.

Multiple factors influence immunogenicity, including administration route, dose regimen, patient-related factors, and disease-related factors (e.g. co-medication, type of disease, immune status). As a result, immunogenicity may vary by indication. It should be justifiable to extrapolate immunogenicity from the studied indication/route of administration to additional applications of the reference product.

6.3.10.6.4 Pharmacovigilance

Pre-authorization clinical trials frequently do not provide enough information to detect unusual side effects. As a result, during the post-approval phase, biosimilars' clinical safety must be continuously maintained, including ongoing benefit–risk assessments.

During the approval process, the applicant must provide a description of the pharmacovigilance system as well as a risk management plan that adheres to current EU legislation and pharmacovigilance principles. The risk management plan should include a description of how the identified and potential risks associated with the usage of the reference product will be handled during post-marketing follow-up. Immunogenicity should be addressed specifically in this context. The biosimilar pharmacovigilance plan should effectively handle any unique safety monitoring placed on the reference product or product class. Applicants are encouraged to participate in the reference product's existing pharmaco-epidemiological investigations. New studies, on the other hand, may be required. Risk-mitigation activities that are in place for the reference product should, in theory, be incorporated in the biosimilar's risk management program as well.

When it comes to suspected adverse reactions involving biopharmaceutical drugs, a precise identification of the product and its manufacturing is critical. As a result, depending on the brand name and batch number of any biopharmaceutical product that is the subject of a suspected adverse reaction report, all necessary steps should be taken to identify it.

6.3.10.7 Quality issues

Comparability cannot be determined only by comparing a proposed biosimilar to a publicly available benchmark, such as a pharmacopeial monograph. A proposed biosimilar must be shown to be similar to a reference product that has been approved in the community and chosen by the biosimilar developer. As a result, a thorough comparison with the chosen reference product will be necessary to show that a proposed biosimilar product has a similar quality, safety, and efficacy profile to the reference product.

It is understood that the maker of a proposed biosimilar product would not generally have access to all information necessary to compare the product to the reference product comprehensively, particularly in terms of the manufacturing process. Nonetheless, the analytical data should show that a strong decision on the physicochemical and biological similarities between the reference product and the proposed biosimilar may be reached.

A proposed biosimilar comparison exercise at the quality level, which includes examination of important quality features using suitably sensitive analytical instruments, could allow for the submission of a marketing authorization application if it is properly carried out. In this case, the applicant would be required to complete relevant nonclinical and clinical comparability programs in order to complete a proposed biosimilar development, as specified by legislation and technical recommendations.

6.3.10.7.1 Manufacturing process

The development and documentation for proposed biosimilars should cover two distinct aspects:

- The target product profile's molecular features and quality attributes (QA) should be equivalent to the reference product.
- On its own, the performance and consistency of a prospective biosimilar's production process.

A proposed biosimilar's quality target product profile (QTPP) should be based on data acquired on the reference product, including publicly available information and data gathered during the reference product's comprehensive characterization. The QTPP should serve as the foundation for developing a biosimilar product and its manufacturing process. This QTPP should be viewed as a work in progress, with some goal ranges changing as more information about the reference product becomes available.

A prospective biosimilar is created and controlled in accordance with its development, considering the most up-to-date information on manufacturing procedures and their implications for product attributes. A proposed biosimilar product is defined by the molecular composition of the active substance because of its manufacturing process, which may introduce molecular variants, isoforms, or other product-related substances as well as process-related impurities, just like any other biopharmaceutical product. As a result, the production process must be properly designed to attain QTPP. When compared to the reference product, the expression system should be carefully chosen, taking into account expression system differences that may result in undesirable outcomes, such as atypical glycosylation patterns, increased variability, or a different impurity profile.

A proposed biosimilar's formulation does not have to be identical to that of the reference product. The adequacy of a suggested formulation in terms of stability, compatibility (i.e. interaction with excipients, diluents, and packaging materials), integrity, activity, and strength of the active component should be proved regardless of the formulation chosen. Assume you choose a formulation and container/closure system that differs from the reference product (including any material in contact with the product). In that instance, its possible impact on the efficacy and safety of a proposed biosimilar should be well justified.

According to ICH Q5C, the stability of a proposed biosimilar product should be assessed. Any statements about stability and compatibility must be backed up by facts and cannot be inferred from the reference product.

The existence of a putative biosimilar's lifespan is acknowledged. A comparability assessment (as stated in ICH Q5E) should be done when changes to the manufacturing process (active substance and finished product) are introduced during development. Any comparability exercises done to demonstrate biosimilarity versus the reference product should be identified in the dossier and addressed separately from the comparability exercise performed to demonstrate biosimilarity versus the reference product for clarity. During the development of a proposed biosimilar product, process changes may occur. However, it is strongly recommended that the required quality, safety, and efficacy data be generated using the product manufactured using the commercial manufacturing process and thus represent the quality profile of the batches to be commercialized in order to demonstrate biosimilarity against the reference product.

6.3.10.7.2 Comparability: quality aspects

The use of publicly available reference standards (e.g. *The European Pharmacopoeia*) as the reference product to demonstrate biosimilarity is not permitted. The use of these standards, on the other hand, is critical for method qualification and standardization.

To show that a proposed biosimilar has a highly similar quality profile to the reference product, a comprehensive comparability study will be necessary. This should comprise thorough assessments of a proposed biosimilar and a reference product utilizing sensitive and orthogonal methodologies to discover quality attribute similarities and differences. Unless otherwise warranted, these assessments should contain side-by-side comparison research. Any discrepancies in quality qualities that are discovered will need to be properly justified in terms of their possible influence on safety and efficacy.

Assume that relevant quality differences are discovered (for which the absence of a clinically relevant impact will be difficult to justify). In that instance, claiming similarities to the reference product may be difficult, hence a full marketing authorization application may be preferable. Alternatively, the applicant could explore making appropriate changes to the manufacturing process to reduce or eliminate these discrepancies.

All quality attributes of a proposed biosimilar product are unlikely to be identical to those of the reference product. Where qualitative and quantitative variations are found, they must be justified and, where applicable, proved to have no effect on the product's clinical performance. Additional nonclinical and clinical data, as stated in the 'Guideline on similar biological medicinal products containing biotechnology-derived proteins as active substance: quality issues', as well as the 'Guideline on similar biological medical products including biotechnology-derived proteins as active substance: non-clinical and clinical concerns', may be included. Quality factors that may affect immunogenicity or potency but were not recognized in the reference product should be given special consideration.

The applicant must show that the desired product (including product-related chemicals) present in a proposed biosimilar's completed product is identical to the reference product. Process-related impurities, on the other hand, may differ between the originator and suggested biosimilar products, however, they should be kept to a minimum. It is preferable to rely on purifying techniques to remove impurities rather than setting up a nonclinical testing program to qualify them. Differences that may confer a safety advantage (for example, lower levels of contaminants) should be explained, although biosimilarity is unlikely to be prevented.

Wherever possible, quantitative ranges should be established for a proposed biosimilar comparison exercise. These ranges should be based mostly on the reference product's measured quality attribute ranges. Unless otherwise warranted, they should not exceed the range of variability of sample reference product batches. The ranges' significance should be evaluated in light of the number of reference product lots tested, the quality attribute explored, the age of the batches at the time of testing, and the test method utilized. If properly justified, a descriptive statistical technique to establishing ranges for quality attributes could be utilized.

It should be emphasized that acceptable ranges for a proposed biosimilar comparability exercise versus the reference product should be addressed separately from release specifications. It is understood that the production process of the reference product evolves over time, which may result in visible deviations in some quality parameters. Such occurrences could occur throughout the development of a proposed biosimilar product, resulting in QTPP development that is no longer fully representative of the reference product on the market. Before and after the observed shift in quality profile, the ranges identified might normally support a suggested biosimilar comparability exercise at the quality level. The reference product can be found in any range. Values for quality attributes that are outside or between the ranges indicated for a quality characteristic of the reference product should be justified in terms of their potential influence on safety and efficacy. Once a marketing authorization has been obtained, there is no regulatory necessity to re-demonstrate biosimilarity.

6.3.10.7.3 Analytical considerations

It is the applicant's responsibility to show that the procedures chosen for a proposed biosimilar comparison exercise would detect minor changes in all elements important to quality assessment (e.g. the ability to detect relevant variants with high sensitivity).

The methods employed in characterization studies are an important component of the quality data package and should be validated for comparability. For method qualification and standardization, standards, and reference materials (e.g. from *The European Pharmacopoeia* and WHO) should be employed if applicable.

Direct or side-by-side comparisons of a proposed biosimilar and a reference product may not be possible or provide limited information for analytical procedures (e.g. due to the low concentration of the active substance and the presence of interfering excipients such as albumin). As a result, samples might be made from the final product (e.g. extraction, concentration, and other suitable techniques). The methods utilized to prepare the samples should be described in such circumstances. Their impact on the samples should be properly documented and described (for example, comparing active ingredients before and after formulation/reformulation preparation).

6.3.10.7.4 Physicochemical properties

The composition, physical characteristics, and primary and higher-order structures of a potential biosimilar should all be determined using approved approaches as part of a physicochemical characterization effort. A suggested biosimilar's target amino acid sequence should be validated and is intended to be identical to that of the reference product. The amino acid sequences at the N- and C-termini, as well as free SH groups and disulfide bridges, should all be compared. Any changes or truncations should be quantified, as well as any intrinsic or expression system-related variability. Any discrepancies between a proposed biosimilar and the reference product should be substantiated in terms of the reference product's micro-heterogeneous pattern (e.g. C-terminal lysine variability).

After glycosylation, oxidation, deamidation, and truncation, the presence and amount of post-translational changes (e.g. glycosylation, oxidation, deamidation, and truncation) should be determined. Carbohydrate structures, including the overall glycan profile, site-specific glycosylation patterns, and site occupancy, should be thoroughly compared if they are present. The appearance of glycosylation structures or variants not seen in the reference product may cause worry and necessitate justification, with non-human structures receiving special attention (non-human linkages, sequences, or sugars).

6.3.10.7.5 Biological activity

As part of generating a thorough characterization profile, a proposed biosimilar comparability exercise should include assessing the biological features of the proposed biosimilar and the reference product. The biological activity of a product refers to its ability or capability to produce a specific biological impact. As needed, biological assays utilizing various and complementary techniques to detect biological activity should be explored. Different test formats (e.g. ligand or receptor binding assays, enzymatic assays, cell-based assays, functional assays) might be utilized depending on the biological features of the product, considering their limitations.

To account for limitations in the validation characteristics of single bioassays, complementary or orthogonal techniques should be used. Separate assays should be used to investigate receptor binding and activation, if applicable. Cross-references to nonclinical and clinical sections of the dossier may be established as needed. It must be shown that biological assays are sensitive, specific, and selective enough. When available and suitable, the results of the relevant biological assay(s) should be reported and represented in units of activity calibrated against an international or national reference standard. If relevant, these assays should meet the standards of the *European Pharmacopoeia* for biological assays.

6.3.10.7.6 Immunochemical properties

Monoclonal antibodies and related compounds (e.g. fusion proteins based on IgG Fc) should have their immunological functions compared thoroughly. This would usually entail a comparison of the items' affinity for the desired target. Unless otherwise justified, the Fc binding affinity to relevant receptors (e.g. FcR, C1q, FcRn) should also be compared. To compare the ability to produce Fab- and Fc-associated effector functions, appropriate approaches should be used.

6.3.10.7.7 Purity and impurities

By combining analytical methodologies, the purity and impurity profiles of a proposed biosimilar and the reference product should be assessed qualitatively and quantitatively. To identify and compare product-related chemicals and contaminants, appropriate orthogonal and state-of-the-art methodologies should be utilized. Specific degradation processes (e.g. oxidation, deamidation, aggregation) of a suggested biosimilar product, as well as potential post-translational alterations of the proteins, should be considered in this comparison. The reference product's age/shelf-life at the time of testing should be indicated, as well as its potential impact on the quality profile, if applicable. Comparing the relevant quantitative features measured at different time points and storage circumstances (e.g. accelerated or stress conditions) could help to prove that the degradation pathways of the reference product and the biosimilar are identical.

Process-related impurities (e.g. host cell proteins, host cell DNA, reagents, downstream contaminants, etc) are likely to vary qualitatively. As a result, a suggested biosimilar comparability exercise may not require a qualitative evaluation of these criteria. Nonetheless, state-of-the-art analytical technologies should be used in accordance with existing rules and compendial requirements. The possible dangers associated with these detected impurities (for example, immunogenicity) must be well documented and justified.

6.3.10.7.8 Quantity

Quantity should be determined using the same assay as the reference product and represented in the same units. A proposed biosimilar's strength should be compared to that of the reference product.

6.3.10.7.9 Specifications

The selection of tests to be included in the specifications (or control plan) for both drugs and drug products, as with any biotechnology-derived product, is product-specific and should be defined as described in ICH Q6B. It's important to explain why you came up with a proposed set of acceptance criteria for routine testing.

Full stability data collected with a proposed biosimilar product should be used to justify the product's claimed shelf-life. There are no requirements for comparative real-time, real-condition stability tests between a proposed biosimilar and a reference product.

6.3.11 United States

6.3.11.1 Background

Both biologics and chemical medications are regulated by the FDA. The FDA must license (approve) a biologic or putative biosimilar before it can be commercialized in the United States. The Biologics Price Competition and Innovation Act (BPCIA) introduced a streamlined approval process for biosimilar biological products or biosimilars.

To strike a balance between competition and innovation, the BPCIA created two periods of exclusivity for brand-name biologics (i.e. the reference product), one of four years and the other of twelve years. If certain statutory requirements are met, periods of regulatory exclusivity attach upon approval or licensure of a drug or biologic, respectively, limiting competitors' ability to reference data provided by brand-name drug producers.

A BLA for a proposed biosimilar or interchangeable product referencing the brand-name biologic may not be filed to FDA during the four-year exclusivity period. A BLA approval for a proposed biosimilar or interchangeable medicine referencing the brand-name biologic may not be made effective during the 12 year exclusivity period. This means that a BLA for a proposed biosimilar or interchangeable product cannot be approved by FDA until 12 years after the reference product was first licensed, and a BLA for a proposed biosimilar or interchangeable product cannot be submitted to FDA until four years after the reference product was licensed.

Certain biologics, for example, if an application is for a small change to a previously licensed biologic, are not eligible for reference product exclusivity. If the developer conducts pediatric trials in response to a written request from the FDA, a new biologic may be eligible for an extra six month period of exclusivity in addition to the 12 and 4 year periods. Furthermore, a biologic approved to treat a rare disease or condition may be granted seven years of orphan drug exclusivity for the protected indication, after which FDA may not license another biologic for the protected orphan indication until the 7 year or 12 year exclusivity period has expired, whichever comes first. The first interchangeable product is eligible for exclusivity, while the first biosimilar for a brand name is not. This means that until the exclusivity period expires, FDA will not issue an interchangeability determination for a later biologic based on the same reference product for any condition of use. The exclusivity periods available under the PHSA for biological products are generally longer than those available under the FFDCA for chemical medications (i.e. five year new chemical entity exclusivity, three year new clinical study exclusivity).

The FDA is not compelled by law to disseminate information regarding authorized biologics and biosimilars, but it does so voluntarily through the publishing of the *Purple Book*. Unlike the *Orange Book*, which is available in both paper and electronic form, the *Purple Book* is divided into two lists: one for CDER-licensed biological products (including potential biosimilar and interchangeable products) and another for CBER-licensed biological products. The list identifies the date the biologic was licensed for brand-name biologics (i.e. approved). The exclusivity would expire if the FDA reviewed the product for reference product exclusivity.

6.4 Development master plan

Biosimilars play a significant role in enabling a lower-cost alternative to an expensive biological product. Being a new genre of products, proposed biosimilar development has been marred with many conflicts of understanding that we wish to clearup in a description based on our hands-on experience in developing biosimilars at the lowest possible cost.

Every proposed biosimilar developer must, first, create a detailed and comprehensive development master plan, as described below.

6.4.1 Choice of product

Most proposed biosimilar developers face a dilemma in choosing the product to develop because of cost and time constraints for taking the product through regulatory approval. The traditional business development teams follow the hard rules of the market, competitors, now and by the time the product is approved, and the cost of products, to decide which product to manufacture. While these considerations have survived the test of time, there are many reasons why these do not always apply to select a proposed biosimilar product to develop.

6.4.1.1 Competition

First, unlike chemical generics, the competitors' field will always be much smaller, not so much for a financial reason but for the need for deep science that is not available to many. Thus, regardless of the nature of the product, the competition will always be limited.

Given that almost every proposed biosimilar product can be a blockbuster, a different type of projection is required to qualify a product regardless of the competitors. Figure 6.2 shows the frequency of approval of biosimilars in the US and EU.

6.4.1.2 Regulatory profiling

The FDA has approved 28 proposed biosimilar products as of September 2020, including 19 monoclonal antibodies, six filgrastim cytokines products, two etanercept fusion products, and one erythropoietin cytokine.

The common practice of developers is to examine the public domain data, particularly the BLA documents available (https://www.accessdata.fda.gov/scripts/cder/daf/). However, a critical analysis of the developers' studies shows that they vary widely in number and details. Table 6.2 summarizes the studies submitted to secure the licensing of

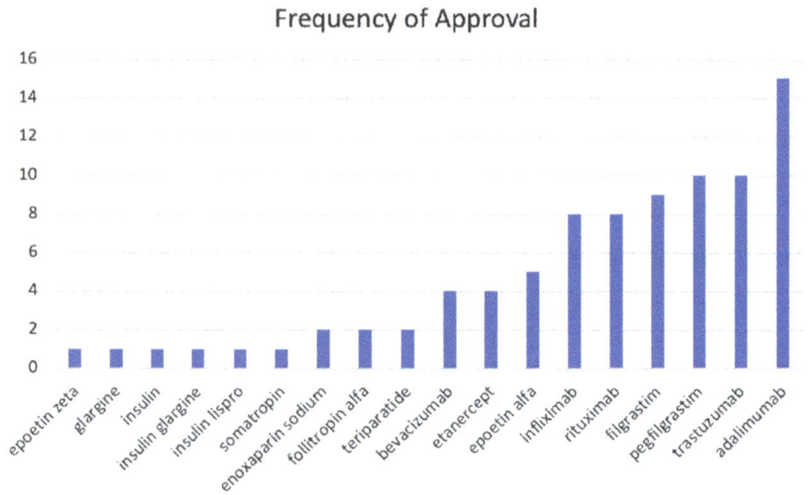

Figure 6.2. Frequency of approval of biosimilars in the US and EU, as of May 2020 (based on the definition of a proposed biosimilar in each jurisdiction).

27 proposed biosimilar products (as of writing). The numbers indicated in parenthesis show the studies that were not reviewed by the FDA as it was unnecessary.

A detailed analysis of the regulatory submissions that led to these products' approval shows high diversity, frequent redundancy, and reliance on studies that may not assure biosimilars' safety and efficacy. The paradigm of stepwise development and evaluation suggested by the FDA has not worked as well. We have sufficient data to indicate that a significant change in the biosimilars approval guidance is required to remove redundant testing and reduce the risk of approval of unsafe biosimilars.

Table 6.2 lists over 1100 analytical similarity, 96 animal pharmacology, 42 *in vitro*/*ex vitro* pharmacology, 52 clinical pharmacology, and 32 clinical efficacy studies. Here are the highlights of the compiled data:

- Twenty-seven animal pharmacology studies were not reviewed by the FDA, labeling them redundant or unnecessary.
- No animal pharmacology or *in vitro*/*ex vitro* study failed.
- A few clinical pharmacology studies had to be repeated to meet acceptance criteria due to the wrong choice of the study population. None failed.
- No clinical efficacy studies failed, where the primary endpoints did not meet, post hoc analysis, and additional scientific justification allowed approval. In two cases, higher immunogenicity was overcome by making minor changes to the manufacturing process. No product was rejected based on a failed efficacy study.
- There was no correlation between submission for the same molecule; trastuzumab had 48–111 total studies by different developers.

In summary, all analytical similarity testing met the acceptance criteria, the animal pharmacology studies added little to the knowledge, all clinical pharmacology studies reached the criteria, and even when there was a difference in clinical efficacy studies, these were overcome by a discussion of data and allowed marketing authorization. Given these observations, for a proposed biosimilar the developers have an opportunity to present to FDA testing protocols that may not be as extensive as used in the approval of all current products.

The analysis presented in table 6.3 is for the developers to meet with the FDA, first in a Biosimilars Advisory Meeting that requires having expressed the biological entity at a small scale with initial analytical similarity testing. This meeting should be followed by type 2 meetings to secure an agreement with the FDA on the minimal studies required.

Table 6.3. Studies submitted for licensing of biosimilars approved by the FDA.

Licensed product	Analytical	Animal pharmacology	*In vitro*/*ex vitro* pharmacology	Clinical pharmacology	Clinical efficacy	Total
Humira						
Adalimumab-atto	41	2	0	1	2	46
Adalimumab-adaz	52	5	1	4	1	63
Adalimumab-adbm	70	6 (2)	26 (10)	2	1	105
Adalimumab-afzb	25	1	0	3	2	31
Adalimumab-bwwd	38	2	0	2	2	44
Avastin						
Bevacizumab-awwb	56	7 (2)	0	1	1	65
Bevacizumab-bvzr	42	2	4	1	1	50
Epogen						
Epoetin alfa-epbx	32	15 (13)	0	4	0	79
Enbrel						
Etanercept-szzs	53	5	0	4	1	88
Etanercept-ykro	52	3	0	1	1	57
Neupogen						
Filgrastim-aafi	38	1	0	3	0	42
Filgrastim-sndz	41	5	0	5	1	52
Infliximab-abda	52	3	0	1	1	57
Remicade						
Infliximab-axxq	61	1	2	1	1	66
Infliximab-dyyb	33	4 (2)	2	4	5	48
Infliximab-qbtx	51	2	0	2	1	80
Neulasta						
Pegfilgrastim-bmez	NA	13 (8)	0	2	2	17
Pegfilgrastim-cbqv	31	1	2	2	0	36
Pegfilgrastim-jmdb	31	2	0	2	1	36
Rituxan						
Rituximab-abbs	50	1	1	1	1	171
Rituximab-pvvr	40	2	0	1	2	98

(*Continued*)

Table 6.3. (*Continued*)

Licensed product	Analytical	Animal pharmacology	*In vitro/ex vitro* pharmacology	Clinical pharmacology	Clinical efficacy	Total
Herceptin						
Trastuzumab-anns	27	5	2	1	1	111
Trastuzumab-dkst	37	2	2	1	1	84
Trastuzumab-dttb	48	2	0	1	1	99
Trastuzumab-pkrb	44	2	0	1	1	48
Trastuzumab-qyyp	44	2	0	1	1	48

6.4.1.3 Market

While it is plausible for the developers to seek out the largest revenue products such as adalimumab, the most valuable opportunities reside in the products that have not been popular. However, the frequency of approvals is not proportional to the market of the product. Nine filgrastim products with total global sale of about $1B compared to four bevacizumab with total sales of more than $10B compared to adalimumab with a market approaching $20B leaves much to discuss. It is noteworthy that only one monoclonal antibody, trastuzumab is included in the WHO list of essential drugs.

6.4.1.4 Intellectual property

Given that it will take 3–4 years to get to a regulatory filing if you were to start a project today, the total number of choices available to you is well over 100, so why not choose a product with the following attributes, which are highly valuable (table 6.4):

- *Less complex in structure*: fewer analytical testing required; a product with fewer PTMs is less complicated.
- *Easier to express*: less batch-to-batch variability; a fusion protein or mAb is more difficult.
- *Has an established market known for its safety and efficacy*: reducing redundant testing; molecules such as hormones and cytokines are here to stay forever; MAbs will change with time?
- *Does not require large CAPEX*: mAbs will take many times more CAPEX than cytokines
- *Has already been approved as a proposed biosimilar in the EU or USA*: being number two or three is an advantage to many.

6.4.1.5 Cost of products

The cost of products (COGs) can be controlled. Most products will cost the same within a small variation if you follow the development cycle to reduce your future COGs. The COGs is often considered a selection criterion, but this is a poor indicator since the production cost of biological drugs is relatively uniform, such as $150–$300 per gram for monoclonal antibodies; cytokines vary not that much in their category. Most of the cost goes into the cost of media since the production

Table 6.4. Patent expiry of biological products licensed by the FDA as of May 2020.

Expired as of 2021 Albumin (human) (1990-11-16); Zoster vaccine live (1991-11-08); Collagenase (2004-01-16); Hyaluronidase (2004-02-24); Urokinase (2004-02-24); Fibrinogen (human) (2004-02-27); Aprotinin (2004-03-17); Insulin human (2004-03-24); Interferon beta-1b (2004-03-28); Immunoglobulin g (2004-04-21); Parathyroid hormone (2004-04-21); Dermatophagoides farinae (2004-07-21); Dermatophagoides pteronyssinus (2004-07-21); Anti-inhibitor coagulant complex (2004-08-18); Histamine phosphate (2004-11-10); Interferon alfa-2b (2004-12-28); Plasma protein fraction (human) (2005-01-19); Natalizumab (2005-01-30); Asparaginase (2005-03-08); Filgrastim (2005-08-23); Factor IX complex (2005-10-25); Tuberculin purified protein derivative (2006-03-06); Rho (d) immune globulin (2006-11-25); Gonadotropin, chorionic (2007-01-30); Sebelipase alfa (2007-04-17); Human immunoglobulin g (2007-07-31); Antithrombin iii (human) (2008-07-25); Immune globulin (human) (2008-09-24); Anakinra (2008-12-24); Alemtuzumab (2010-05-11); Abciximab (2010-09-14); Interferon alfa-n3 (2011-03-01); Bacillus calmette-guerin (2011-05-03); Hepatitis a vaccine (2011-06-06); Antihemophilic factor/von Willebrand factor complex (human) (2011-07-19); Pancrelipase (amylase/lipase/protease) (2012-01-16); Aldesleukin (2012-02-03); Etanercept (2012-03-07); Epoetin alfa (2012-08-15); Thrombin human (2012-09-09); Becaplermin (2012-10-10); Ibritumomab tiuxetan (2012-11-13); Pegfilgrastim (2013-12-03); Bevacizumab (2014-01-25); Reslizumab (2014-06-17); Cetuximab (2014-08-10); Trastuzumab (2014-08-10); Trastuzumab/hyaluronidase-oysk (2014-08-10); Pegaspargase (2014-10-20); Von Willebrand factor (recombinant) (2014-11-14); Denileukin diftitox (2015-02-01); Rituximab (2015-02-01); Rituximab/hyaluronidase (human recombinant) (2015-02-01); Sargramostim (2015-02-01); Imiglucerase (2015-03-24); Menotropins (fsh/lh) (2015-09-29); Palifermin (2015-09-29); Urofollitropin (2015-09-29); Peginterferon alfa-2b (2015-11-02); Basiliximab (2016-05-16); Daclizumab (2016-05-16); Latrodectus mactans (2016-05-18); Alteplase (2016-05-22); Equine thymocyte immune globulin (2016-09-05); Ambrosia artemisiifolia (2016-11-26); Denosumab (2016-12-23); Insulin lispro (2017-01-10); Infliximab (2017-04-04); Panitumumab (2017-05-05); Adalimumab (2017-09-26); Reteplase (2017-09-26); Blinatumomab (2018-04-21); Insulin aspart (2018-05-19); Rilonacept (2018-09-25); Romiplostim (2018-10-23); Human plasma proteins (2018-12-10); Interferon beta-1a (2020-01-14); Interferon gamma-1b (2020-01-14); Palivizumab (2020-05-03); Capromab pendetide (2020-08-21); Ranibizumab (2020-08-24); Gemtuzumab ozogamicin (2020-11-28); Ocriplasmin (2020-12-21); Abatacept (2021-02-15); Golimumab (2021-03-07); Rasburicase (2021-05-01); Dornase alfa (2021-09-04); Tenecteplase (2021-10-30); Insulin detemir (2021-11-19); Laronidase (2021-11-30).

2021–2026 Follitropin alfa/beta (2022-01-22); Pertuzumab (2022-05-17); Peginterferon alfa-2a (2022-08-01); Corticorelin ovine triflutate (2022-08-05); Omalizumab (2022-08-14); Darbepoetin alfa (2022-08-29); Pegvisomant (2023-01-09); Imciromab pentetate (2023-03-05); Thyrotropin alfa (2023-06-24); Agalsidase beta (2023-10-01); Selumetinib (2023-12-12); Albiglutide (2025-01-04); Antihemophilic factor (human) (2025-01-04); Insulin aspart recombinant (2025-01-04); Lixisenatide (2025-01-04); Methoxy polyethylene glycol-epoetin beta (2025-01-04); Aflibercept (2025-03-15); Ramucirumab (2025-03-15); Belimumab (2025-05-20); Insulin glulisine recombinant (2025-05-23); Certolizumab pegol (2025-11-08); Ipilimumab (2025-11-08); Fremanezumab-vfrm (2025-11-14); Calfactant (2026-01-10); Tocilizumab (2026-04-13); Eculizumab (2026-04-27); Inotuzumab ozogamicin (2026-06-23); Mepolizumab (2026-06-23); Ocrelizumab (2026-06-23); Ofatumumab (2026-06-23); Raxibacumab (2026-06-23); Coagulation factor ix (recombinant) (2026-07-13); Desirudin recombinant (2026-08-02); Poractant alfa (2026-11-02).

Table 6.4. (*Continued*)

2027–2032	Benralizumab (2027-01-11); Galsulfase (2027-06-13); Evolocumab (2027-08-23); Ustekinumab (2027-11-30); Capmatinib (2027-12-12); Human fibrinogen, human thrombin (2028-05-22); Mecasermin rinfabate recombinant (2028-06-19); Ado-trastuzumab emtansine (2028-10-22); Belatacept (2028-12-05); Osilodrostat (2029-03); Opicapone (2029-12-12); Ozanimod (2029-12-12); Canakinumab (2030-03-29); Abobotulinumtoxina (2030-03-30); Asparaginase erwinia chrysanthemi (2030-03-30); Incobotulinumtoxina (2030-03-30); Dulaglutide (2030-05-05); Metreleptin (2030-05-05); Brentuximab vedotin (2030-05-26); Insulin degludec (2030-06-24); Sipuleucel-t (2030-07-19); Necitumumab (2030-09-08); Brodalumab (2030-10-08); Secukinumab (2030-10-08); Influenza vaccine, adjuvanted (2030-10-28); Nivolumab (2031-02-04); Pembrolizumab (2031-02-04); Insulin detemir recombinant (2031-04-01); Vedolizumab (2031-05-02); Idursulfase (2031-06-08); Ecallantide (2031-06-10); Human c1-esterase inhibitor (2031-06-10); Elotuzumab (2031-08-05); Olaratumab (2031-08-05); Siltuximab (2031-08-05); Sarilumab (2031-10-11); Alirocumab (2031-10-25); Alirocumab (2031-10-25); Asfotase alfa (2031-10-25); Elosulfase alfa (2031-10-25); Insulin aspart/insulin degludec (2031-10-25); Insulin degludec/liraglutide (2031-10-25); Insulin glargine recombinant (2031-10-25); Insulin lispro recombinant (2031-10-25); Ixekizumab (2031-10-25); Obinutuzumab (2031-10-25); Peginterferon beta-1a (2031-10-25); Daratumumab (2031-10-28); Somatropin (2031-12-05); Eptinezumab-jjmr (2031-12-12); Selpercatinib (2031-12-12); Selpercatinib (2031-12-12); Atezolizumab (2032-08-13); Avelumab (2032-08-14); Durvalumab (2032-08-14); Talimogene laherparepvec (2032-08-30); Ziv-aflibercept (2032-10-31); Ripretinib (2032-12-12).
2033–2039	Choriogonadotropin alfa (2033-02-06); Alglucosidase alfa (2033-03-11); Hepatitis a and hepatitis b (recombinant) vaccine (2033-03-11); Immune globulin infusion (human) (2033-03-11); Insulin recombinant human/insulin susp isophane recombinant human (2033-03-11); Sacrosidase (2033-03-11); Dupilumab (2033-03-14); Coagulation factor viia (recombinant) (2033-04-24); Meningococcal group b vaccine (2033-04-24); Pneumococcal 13-valent conjugate vaccine (2033-04-24); Bezlotoxumab (2033-09-09); Hepatitis b vaccine (recombinant) (2033-11-14); Pemigatinib (2033-12-12); Sacituzumab govitecan-hziy (2033-12-12); Tucatinib (2033-12-12); Ravulizumab-cwvz (2034-03-07); Idarucizumab (2034-07-31); Dinutuximab (2034-10-06); Guselkumab (2035-02-24); Obiltoxaximab (2035-02-24); Isatuximab-irfc (2035-10-05); Taliglucerase alfa (2036-02-11); Velaglucerase alfa (2036-02-11); Polatuzumab vedotin-piiq (2039-10-22).

process is producing a carbon-based entity. More details are provided below in the selection of the manufacturing process.

6.4.2 Qualification of product

A product must have the same administration method, identical dose, potency, and mode of action as the reference medication to qualify as a proposed biosimilar product. The application must include evidence that the biological product is comparable to the reference product, including modest changes in clinically inactive components (e.g. it may be possible, for example, for a proposed biosimilar product formulated without human serum albumin to demonstrate biosimilarity to the reference product formulated with human serum albumin). A product in development is referred to as a 'proposed biosimilar product', and any data supplied do not claim to be 'very comparable' or 'biosimilar', as those designations are only issued by the regulatory body.

A marketing authorization application must include information demonstrating that a proposed biosimilar product:

- Offers a qualified reference product.
- For a suggested condition(s) of use, claims the same mechanism(s) of action—but only to the extent that the mechanism(s) are known for the reference product.
- The proposed labeling condition(s) of use have previously been approved for the reference product.
- Has the same mode of administration, dose form, and strength, as well as a suitable, if not identical, formulation and delivery device.
- Is made, processed, packed, or stored at a facility that complies with industry standards to ensure that a proposed biosimilar product remains safe, pure, and potent.
- Is available in logical presentation options.
- Has graded extrapolation of indicators.
- Has available QT/QTc and pediatric waivers.

6.4.2.1 Qualified reference product

To obtain marketing authorization, developers must demonstrate that a proposed biosimilar product is biosimilar to a single reference product that has been previously approved in one of the ICH+ regions: the European Economic Area (EEA) in accordance with Article 8 of Directive 2001/83/EC, as amended; and the United Kingdom, Switzerland, and Canada. It should be good enough for the government of the country where it was made to approve it.

Only one reference product can be used as the comparator throughout the comparative testing program for quality, safety, and efficacy during the development of the proposed biosimilar product to make it easier to examine the entirety of the data. The comparative study's reference product should have a distinct trade name, pharmaceutical form, formulation, manufacture date, and expiration date.

The reference product cannot be one of the pharmacopeia's reference standards. If the reference product has different strengths, the proposed biosimilar product to be compared should have the same strength. Multiple reference items may be used to create a global dossier, but they cannot come from different production locations, regardless of their cGMP compliance level.

6.4.2.2 Mechanism of action

In several instances, the mechanism of action is known for the reference product. This may involve multiple mechanisms that may have led to multiple approved indications. A proposed biosimilar product may not claim or demonstrate any mechanism of action that is not established for the reference product. This restriction comes from the correlation between the mechanism of action and the scope of the side effects of the product.

6.4.2.3 Route of administration

An application for authorization must include information proving that a proposed biosimilar product's mode of administration is the same as the reference product's. The developers may not seek marketing clearance for a mode of administration, a dosage form, or a strength that differs from the reference product in a proposed biosimilar application or a supplement to an authorized proposed biosimilar application. The developers of a proposed biosimilar medicine may acquire marketing clearance for all administration routes for which a specified reference product has been approved. However, the developers may not be required to conduct studies in all products for all routes of administration; in general, the route of administration that is more likely to show variability.

6.4.2.4 Dosage form

The developer must demonstrate that the dosage form of a proposed biosimilar product is identical to that of the reference product. For the purposes of enforcing this section, the Agencies consider the dosage form to be the physical manifestation containing the active and inactive ingredients that deliver a dose of the drug product. In the context of meaning to be injected, the Agencies consider 'injection' (e.g. a solution) to be a separate dosage form than 'for injection' (e.g. a lyophilized powder). Even if the developers can show that the biosimilar, when constituted or reconstituted, can meet the other requirements for a proposed biosimilar product application, if the reference product's dosage form is 'injection', they will be unable to obtain marketing authorization for a proposed biosimilar product with a dosage form of 'for injection'. The Agencies also regard emulsions and suspensions of drugs for injection to be separate dosage forms. Liposomes, lipid complexes, and products with extended-release features present new scenarios due to their unique composition and perspective. For more information, the developer should contact the Agencies. This interpretation of the identical dose form, however, only applies to biosimilars, not generic chemical medicines.

6.4.2.5 Strength

The producers must show that a proposed biosimilar product's 'strength' is equivalent to that of the reference product. The data and information gathered during the analytical evaluation could help determine whether a proposed biosimilar product is as strong as its reference product. In general, developers of a proposed biosimilar product with a 'injection' dosage form (e.g. a solution) can show that their product is equivalent to the reference product by demonstrating that both products have the same total content of drug substance (in mass or units of activity) and the same concentration of drug substance (in mass or units of activity per unit volume). In general, developers can demonstrate that a proposed biosimilar product that is a dry solid (e.g. a lyophilized powder) from which a constituted or reconstituted solution is prepared has the same strength as the reference product by demonstrating that both products have the same total content of drug substance (in mass or units of activity). If a proposed biosimilar product is a dry solid (e.g. a lyophilized powder) from which a constituted or reconstituted solution is prepared, the application should generally include information that the biosimilar's concentration when constituted or reconstituted is the same as that of the reference product when constituted or reconstituted. Using the same procedure, the developers should determine the content of drug substances in both the reference product and the proposed biosimilar product.

6.4.2.6 Formulation

Different inactive substances that have been demonstrated to be safe and do not affect the efficacy of a proposed biosimilar product may be used in a proposed biosimilar product.

6.4.2.7 Drug delivery device

A combination product is a planned biosimilar product in a delivery device. The distribution method or container closing system of a proposed biosimilar product may differ from that of the reference product. Some design variations in a proposed biosimilar product's delivery method or container sealing system may be permissible. Even if the reference product is in a vial presentation for the same dosage form, it may be possible for the developers to obtain marketing authorization for a proposed biosimilar product in a prefilled syringe or an auto-injector device (which are considered the same dosage form), provided that the proposed biosimilar product meets the standard for biosimilarity and adequate perforation. The delivery device or container closure system for a proposed biosimilar product in a different delivery device or container closure system must be shown to be compatible with the final formulation of the biological product through appropriate studies, such as extractable/leachable studies and stability studies. Performance testing and a human factor research may also be required for design changes in the delivery device or container sealing mechanism. However, when a design variation in the delivery device or container closing system results in any of the following, the developers will be unable to get marketing authorisation for a planned biosimilar product:

- A clinically relevant variation in terms of safety, purity, and potency between a proposed biosimilar product and the reference product.
- A different dose type or mode of administration.
- A condition of use (e.g. indication, dosage regimen) for which the reference product has not previously been licensed; or does not otherwise meet the biosimilarity requirement.

6.4.2.8 Presentations

The developers of a proposed biosimilar product are not required to obtain marketing authorization for all presentations for which the reference product is licensed. If the developer is requesting marketing approval for numerous presentations, the testing will be done on just one of them.

6.4.2.9 cGMP compliance

To ensure the safety and efficacy of a proposed biosimilar product, it must be manufactured in accordance with cGMP compliance. Clinical batches of a proposed biosimilar product must be manufactured at-scale under cGMP since no changes to the manufacturing process can be made until the proposed biosimilar product is approved, without the need to repeat any testing. Once a product is approved, the developers can make whatever changes to the manufacturing process they want utilizing the ICH Q5E guidance, where the reference product becomes their product and the future product becomes their product. This is because the developer now understands the in-process controls and how they may affect the product's safety and efficacy.

6.4.2.10 Extrapolation and substitution

Extrapolation of all indications permitted for the reference product as of the date is permissible for a proposed biosimilar product, barring any intellectual property in the marketing authorization. Because of the idiosyncrasies identified in a proposed biosimilar product, the developer should present a justification document to claim extrapolation considering the observed similarity; there may be scenarios when extrapolation is not allowed. Assume that the reference product wins marketing permission for additional indications without undergoing any changes. Because this is a problem settled between the developer and the innovator company, the developers may request a revision to their label claim, regardless of the intellectual property involved.

The European Medicines Agency (EMA) does not allow a proposed biosimilar medicine to be automatically substituted for the reference product. However, different jurisdictions within the EU make their own decisions; in the US, substitution necessitates further testing to establish that switching and alternating poses no additional safety risk or reduces efficacy—the FDA has yet to approve any medicine with this classification.

6.4.2.11 Study waivers

The earlier determination of the safety, purity, and potency of the reference product, including any clinical QT/QTc interval prolongation and proarrhythmic risk, as well as drug–drug interactions, may be used in a marketing authorization application for a proposed biosimilar medicine. If such studies were not required for the reference product, these data would not be required for the marketing authorization of a proposed biosimilar product.

A proposed biosimilar drug is not considered to have a 'novel active component' if the reference product has received a waiver, and a pediatric assessment is generally not required. The developer of a proposed biosimilar product that has a different formulation must submit a statement explaining why the formulation poses no greater risk to children.

Biosimilars do not require specific research such as carcinogenicity, reproductive toxicity, and so on.

6.4.3 Public domain knowledge

Collect and analyze all publicly available reports, such as EPARs, BLA review documents, and other documents made available by the various agencies indicated above. When relying on published scientific literature, developers must ensure that there are no conflicts of interest, such as a manufacturer posting data submitted for marketing permission with conclusions that may differ from the regulatory authorities' judgment. While the US FDA makes BLA details available for many new products, developers can seek this information for a modest cost through the Freedom of Information Act. While the developer will have sufficient scientific expertise to make the most use of the available information, these documents will be redacted for confidential information.

Collect and analyze all available information on approved biosimilars to the developer's chosen reference product. Biosimilars have access to the same regulatory information as generics. The developers, in particular, can benefit from prescribing information that has been approved for these products. The developers may also want to compile an internal library of all accessible scientific studies on the reference product, its active ingredient, and the manufacturing and testing technologies used in biological products. It is possible that some commercial vendors have already obtained this information and are selling it to businesses. Regulatory information should only be regarded if it comes from Western Europe, the United States, Japan, Canada, or Australia. Some regulatory agencies in emerging markets have taken a chemical-generic strategy, allowed research waivers and not ensuring safety and efficacy.

6.4.4 Manufacturing plan and facility

Create a manufacturing strategy that uses a cell line that is like the reference product, even if it is not essential. The developers should be aware that employing unique cell lines will invariably raise the burden of proof required to prove the safety and efficacy of a proposed biosimilar product. The developers should be aware that the cell lines in the reference product are frequently decades old, from a time when productivity was low; nonetheless, manufacturers do not update cell lines to ensure safety and performance. In view of the availability of much higher yielding cell lines, using a similar cell line can be difficult. There are two important factors to consider. First, due to the decreased pressure, cell lines with lesser yield create more consistent products. Second, the real cost savings from adopting a high-volume production line may not be significant. The majority of the production cost is spent on media that is used in proportion to the protein output; however, a higher-producing cell line reduces the bioreactor's size. Before employing newer cell lines with extremely high production, developers should conduct a thorough analysis.

To reduce the possibility of viral contamination, developers must persuade authorities that there are no cross-contamination possibilities between the facilities utilized for bacterial and mammalian cell processing. The upstream manufacturing area should be rated 100 000 (class 8) and the downstream region should be rated 10 000 (class 7). Personnel and products should travel in one direction only. To achieve the OSHA standard, innovative clean area engineering concepts based on a single-pass system with only minimum air replaced could minimize the CAPEX and OPEX of HVAC systems by 50%–70%.

Most large pharmaceutical companies will have fixed-pipe stainless steel systems in place and will be less likely to adopt single-use systems. However, single-use should be the only option for all newcomers because it eliminates the need for cleaning validation, which adds more than just cost—the risk of contamination changing the molecular structure is a serious consideration.

6.4.4.1 Expression system

Generally, some quality attributes are strictly related to proteins' expression and related to the cell line's selection. When a new biological product is developed, the selected cell line product is characterized and taken through all three development phases. In biosimilars, the cell line must produce a similar product, and that can be challenging if the product has many post-translational modifications, higher molecular weight, and highly complex structure. In creating monoclonal antibodies, a single cell (monoclonal) is used to create a uniform cell line, which may not necessarily be the best choice. The developers should first conduct analytical testing of the reference product to establish the quality attributes required before selecting a cell line colony, to avoid extensive testing later to justify any analytical differences in the expression of proteins from the selected cell line. It is important to understand that a high titer cell line reduces the size of the bioreactor only; for low dose products such as cytokines, the titer's differences do not affect the size or cost as much as is touted by the suppliers of new cell lines.

6.4.4.2 Batch size

The first clinical investigation, PK/PD, must employ a commercial-scale lot since a proposed biosimilar product may be licensed without substantial comparative clinical safety and efficacy studies. If single-use upstream processing is chosen, the size of bioreactors available is currently limited. The developers could use a smaller bioreactor to make clinical lots and then scale up using ICH Q5E once the product has been authorized. The developers could potentially combine many smaller size upstream lots to eliminate scale-up difficulties and reduce ICH Q5E compliance costs. In general, the downstream phase is less likely to result in major product alterations than the upstream process, which frequently involves post-translational modifications.

Even though the lot may not have gone through the PPQ qualification at this point, it is recommended to include at least one commercial-scale lot when undertaking analytical similarity testing. Starting with multiple developing lots and at least one certified lot will provide several advantages in future studies.

6.4.5 Analytical testing

Because analytical similarity evaluation is so important in determining biosimilarity, you will probably spend the most money either setting up relevant testing in your labs or outsourcing it. Even if you are outsourcing, you will need to conduct in-house testing to confirm that the process changes are fair and compliant. It is not necessary to validate test techniques for analyzing similarity. The test techniques for comparing the primary, secondary, and tertiary structure elements, for example, must be acceptable and sensitive. Validation of product release test techniques is required. The difference in validation requirements stems from side-by-side testing of the reference product and a proposed biosimilar product, where method variance will be the same for both. Furthermore, some procedures are difficult to validate, such as mass spectrometry, NMR, DLS/SLS, CD, isothermal titration, and so on.

Regulatory authorities pay special attention to compliance with 21 CFR Part 11, which is one area where many businesses fall short. It is worth noting that the agencies do not always demand an IT foundation; in many cases, a manual record would suffice. However, the developer must always be able to demonstrate that the original data were not tampered with.

6.4.5.1 Critical quality attributes

While most qualities are common to numerous items, the criticality of each depends on the production process and its robustness in most cases. Some quality attributes are compared at the analytical evaluation level, while others are reviewed at the release level, of which developers should be aware. Post-translational changes, contaminants, aggregates, subvisible particles, and physical qualities are all part of the manufacturing process. Impurities are an important part of establishing biosimilarity since any unknown impurity, whether product-related (where it could be active) or process-related (where it is unlikely to be active), must be investigated for its safety potential. Because the quantity of undiscovered impurities is not always related to safety concerns, even a minor detectable impurity should be thoroughly investigated. Otherwise, contaminants may only make a minor difference in the product's bioactivity. The inventors are recommended to change the production process to eliminate any undisclosed impurities, which would necessitate additional nonclinical and clinical testing. Other characteristics, such as aggregates and subvisible particles, can only have a negative influence and must be reduced through process modifications. Shaking protein products can cause additional aggregates, as seen by the cautions on some products, such as erythropoietin, to avoid shaking the product.

6.4.5.2 Release specification

When a novel biological product is developed, the specification is based on the characteristics of many lots. Despite acknowledged constraints common to the dosage form, there is rarely any doubt as to why a given standard is utilized to release a product. Even though the expression systems are invariably different, the borders of the specifications are already established to match when a prospective biosimilar product is developed. The same can be said for the upstream and downstream processes. It is a fine line to walk to ensure that any deviations from the reference product are not clinically meaningful. Biosimilar companies must have a deep understanding of the product, technology, testing methodologies, and creative minds to assure regulatory agencies that there is no clinically meaningful difference between a proposed biosimilar and the reference product.

To define the specifications needed for drug and drug product qualification, many reference products should be employed. The number of reference product batches is determined by the variability of the reference product's quality parameters. Since the reference product manufacturer confirmed the manufacturing process over time, there is unlikely to be significant fluctuation in protein content, bioassay, post-translational modifications, contaminants, and other physical features of the reference product. Because the developer of a proposed biosimilar product does not have access to in-process controls, older lots of proposed biosimilar products may have more variance, necessitating the collection of substantial data on attribute variability in reference product lots. Assay, bioassay, physical properties, subvisible particles, total impurities, individual impurities, aggregates, and post-translational changes should all be included in the release specifications.

6.4.6 Nonclinical pharmacology

If nonclinical pharmacology studies are required once a product has been progressed to the development stage, the developers may explore them. Agencies encourage developers to justify nonclinical study waivers based on the analytical assessment performed, prior public knowledge about the product's toxicity, and the relevance of animal evidence to the product's safety and efficacy. Even for complicated chemicals, the authorities have recently begun to allow waivers. If pharmacological research on animals is required, the protocol must justify the species and dose chosen. For monoclonal antibodies, no rodent species testing should be done. Instead, the most ideal studies can be PK studies in monkeys with a small number of animals and a dose that is on the linear dose–response range in humans. The developers should be aware that the goal of animal PK research is to eliminate any structural similarity, since the product's disposition characteristics may reveal changes in the structure as well as any immunogenic response that could effect clearance from the body.

6.4.7 Clinical pharmacology

The PK/PD investigations are critical for determining analytical similarity, immunogenicity similarity, and, in some cases, bioavailability. In general, research in healthy participants will be more useful for two reasons: first, to recruit subjects with narrow demographics to minimize inter- and intra-subject variability; and second, to lessen the impact of disease and its therapies on the outcome profile. However, in some cases where the chance of anti-drug antibodies formation is quite high, such as in the case of an endogenous-related product, it may be unethical to expose healthy participants when a patient population is required. Regardless of the population chosen, the goal should be to reduce the number of subjects in the study; one approach that the developers could take is to conduct a two-arm (two doses) parallel two-phase (dose 1, dose 2) study with a follow-up suitable for immunogenicity evaluation; there may be situations where the use of this model is limited due to PD determination. Should a study fail to fulfill the preset acceptance criteria, the developers may undertake a post hoc analysis to assess whether the predetermined criteria may have been widened. The goal of the post hoc analysis is to see if the failed study creates sufficient residual uncertainty, not to change the admission criteria retroactively. As erroneously recommended in certain guidelines by other agencies, post hoc analysis may not contain additional characterization of results that are closer to the midpoint of the acceptability range.

6.4.8 Clinical immunogenicity

Immunogenicity testing can and should be coupled with PK/PD profiling. However, in some cases, if a specific population or procedure is required, a separate study must be undertaken, which may prevent the study from being mixed with PK/PD investigations. Before beginning clinical testing, the developers must ensure that there is no remaining doubt about the components that cause an immunogenicity response. The evaluation of anti-drug antibodies, which should be estimated from public domain data, is a more important aspect of these studies. For example, filgrastim has a very low immunogenic response and pegylation of the molecules makes it even less immunogenic. Therefore, any studies for filgrastim must include clinically meaningful acceptance criteria. Immunogenicity studies for simpler, low molecular weight medications where the immunogenic response is unlikely to alter clinical efficacy, such as insulin products, have recently been shown to be unnecessary. Based on similar or other innovative reasoning, the developers may propose barriers to undertaking immunogenicity testing.

The FDA has issued advice reducing the necessity to assess the clinical immunogenicity of insulin if it passes analytical similarity testing. The variables responsible for this waiver include a smaller, simpler molecule. This form of waiver is likely to become more common in the future.

6.4.9 Residual uncertainty

Following completion of all the above-mentioned testing, the developer should prepare a detailed rationale to show the agency that a proposed biosimilar product meets the required biosimilarity for license (in the US) or authorization (in other countries) (in other countries). This is the stage where a developer can save 24–36 months and tens of millions of dollars by obtaining permission. Based on their outdated notion of selling biopharmaceuticals products through prescribers, most suggested biosimilar companies are far too ready to jump to safety and effectiveness studies to back their marketing efforts. Additional clinical testing is unlikely to be necessary if the developer can establish that there is no residual ambiguity in a convincing argument, and even if it is, it may be restricted to additional clinical pharmacology studies.

The developers should make the case that the totality of the evidence is sufficient to decide whether a proposed biosimilar product is biosimilar to the reference product. The justifications should include a summary of any differences in the analytical, nonclinical, and clinical pharmacology assessments in this specific order, with the understanding that there must be some residual uncertainty (either proving it nonconsequential or demonstrating that it does not impact clinical safety and efficacy). One argument in the developer's side is that extra testing may not always eliminate any remaining doubt.

6.4.10 Clinical safety and efficacy

The developers should realize that if a PK/PD study has failed and any residual uncertainty related to the failure is not resolved, then a clinical safety and efficacy will not be allowed by the agencies to provide additional proof of biosimilarity, regardless of the size of the clinical efficacy study proposed by the developers.

However, where an efficacy study is conducted, its design must include justification of the indicators chosen to test the product where multiple indications are allowed through extrapolation. The study size should first present the effect size analysis (M1) based on public domain data. An equivalence interval (M2) was decided based on clinical judgment, and a rational argument justifying the M2 value. The choice of study model, equivalence margin versus non-inferiority, should be explained. A critical element of these studies is the population demographic that is often difficult to be a practical choice, particularly in anticancer drugs. While the option of naïve patients is always desirable, it is often not possible to achieve these criteria. These are some of the complications of safety and efficacy studies that make them less reliable than the PK/PD/immunogenicity testing. When conducted, the developers are encouraged to use clinical markers rather than hard efficacy results where possible, realizing that the study's purpose is not to demonstrate that a proposed biosimilar product is effective. Still, it is equally effective when compared to the reference product. Clinical markers that are relatively easier to evaluate provide greater robustness to the study than the hard efficacy results. Finally, the purpose of a safety and efficacy study is removing any remaining residual uncertainty and not provide proof of biosimilarity based on the study results alone.

6.4.11 Post-market surveillance

While a pharmacovigilance program will not satisfy any residual uncertainty in biosimilarity, in some cases, pharmacovigilance may provide additional confidence to regulatory agencies. For example, if the formation of anti-drug antibodies is dependent on the demographic aspects, only a large study can provide a reliable data, and such information can be collected in a pharmacovigilance plan, among other routine and common attributes.

6.5 Interchangeability and substitution

A generic medicine and its reference (brand-name) drug, as well as other generic products that use the same reference drug, are often interchangeable. Because a proposed biosimilar and its brand-name biologic are not structurally identical, determining interchangeability is a different process. The FDA oversees drug products in the United States, but states govern pharmacies and pharmacy practice. As of 22 October 2018 'at least 49 states have explored legislation setting state standards for substitution of a proposed biosimilar prescription pharmaceutical to replace an original biologic product', according to the National Conference of State Legislatures (NCSL). According to the NCSL, 45 states and Puerto Rico have passed laws, with varying parameters.

To assist sponsors in showing interchangeability of a proposed therapeutic protein product with its reference product, the FDA has published final advice. The author (Niazi) had two citizen petitions with comments that were mostly included into the final guidance.

It is worth repeating that the FDA recommendations are not legally binding, and that they do not bar a sponsor from submitting a different proposal to the FDA, even if most sponsors would be hesitant to do so.

FDA is yet to approve the first interchangeable product. As the FDA gains more confidence in the evaluation of interchangeability, after approving a few products, the guidance will change substantially.

6.6 Analytical assessment

6.6.1 Introduction

Proteins are classified in at least three ways: by their primary amino acid sequence, by their modification (glycosylation), and by their higher-order structure (protein folding and protein–protein interactions). Amino acid alterations can cause heterogeneity, which can be difficult to control. Protein modification and higher-order structure can be influenced by formulation and environmental elements such as light, temperature, moisture, packaging materials, container closing systems, and delivery device materials. Impurities in the manufacturing process and the final product can potentially increase the likelihood and severity of an immune reaction to a protein. Excipients might make it difficult to describe a protein product.

Analytical advances have allowed scientists to define protein products in terms of their physicochemical and biological properties, such as higher-order structures and functional capabilities. These analytical methods have increased the ability to identify and describe the active ingredient in a protein product, as well as excipients and product- and process-related impurities.

The regulatory guidance outlines how to evaluate analytical features to determine whether or not a biosimilar product meets the requirements for filing a marketing application. While the Agencies' proposals are focused on therapeutic protein products, comparable scientific ideas might be used to generate other protein products, such as *in vivo* protein diagnostics.

The developers will not be allowed to submit a marketing permission application if the reference product cannot be accurately characterized for relevant analytical qualities.

An application must include analytical studies demonstrating that the biosimilar product is equivalent to the reference product as part of the comprehensive CMC data submission. The rationale for the comparative analytical evaluation should be clearly described, taking into account the reference product's characteristics, known modes of action(s), and function.

Important quality criteria such as product identification, amount, safety, purity, and potency should be determined by physicochemical and functional characterization studies. The developers may have a reasonable scientific basis for a selective and targeted approach to subsequent animal and clinical studies to support a demonstration of biosimilarity based on the results of analytical studies assessing functional and physicochemical characteristics, such as higher-order structure, post-translational modifications, and impurity and degradation profiles. It can be beneficial to compare the differences in quality attributes between a biosimilar product and the reference product using a meaningful fingerprint-like analysis algorithm that covers many additional product attributes and their combinations with high sensitivity via orthogonal methods.

As stated in ICH Q8(R2), improved manufacturing science methodologies may make it easier to develop production procedures that are more closely related to the qualities of a reference product (RP). For further information on enhanced manufacturing processes, see the ICH advise for industry Q8(R2) Pharmaceutical Development (November 2009), Q9 Quality Risk Management (June 2006), Q10 Pharmaceutical Quality System (April 2009), and Q11 Development and Manufacture of Drug Substances (November 2012).

This method would allow for a more accurate evaluation of a molecule's overall similarity. It could open the way for future animal and clinical experiments that are more selective and targeted.

Any differences between a biosimilar product and the reference product, whether intentional or discovered after a thorough investigation of numerous production batches, should be described and investigated thoroughly. The debate should include the identification and comparison of essential product characterization quality features. To evaluate and support the potential clinical implications of discovered structural and functional differences between a biosimilar product and the reference product, animal or clinical studies should be used.

6.6.2 Sources of variation

The analytical evaluation of critical quality criteria is based on a detailed understanding of the sources of variation between a biosimilar and its reference equivalent. Changes to the expression system, upstream and downstream procedures, formulation, and manufacturing processes are some of the efforts made in biosimilars development to minimize any discrepancies between a biosimilar product and the RP.

6.6.2.1 Expression system

Therapeutic protein products can be made from prokaryotic or eukaryotic microbial cells, cell lines (e.g. mammalian, avian, insect, or plant), or tissues derived from animals or plants. The biosimilar product's expression constructs should encode the same primary amino acid sequence as the reference product. Minor truncations, such as the variability of a monoclonal antibody's C-terminal lysine, may be acceptable and should be described by the developer. Differences between a biosimilar product's chosen expression system (i.e. host cell and expression construct) and that of the reference product should be carefully considered because the type of expression system affects the types of process- and product-related substances, impurities, and contaminants (including possible adventitious agents) that may be present in the biosimilar product. The expression system, for example, will influence the types and degree of translational and post-translational modifications imparted to a biosimilar product, adding to the complexity of proving that the biosimilar product is biosimilar to the reference product.

The possibility of developing a biosimilar protein product can be raised by minimizing the discrepancies between a biosimilar product and the reference product expression systems. The application of various expression systems will be evaluated on an individual basis. However, the developers should consider the additional burden of testing required to qualify an alternate expression system. In the market, surveillance requires more emphasis on the demonstration of the safety of the product.

6.6.2.2 Manufacturing process

A thorough understanding of the manufacturing process for a biosimilar product should be developed throughout the product development phase. As a matter of science, the characterization tests, process controls, and standards derived from process development data must be specific to the biosimilar product and manufacturing process. Improved pharmaceutical development processes, in combination with comprehensive quality risk management and quality systems, will allow for constant high-quality product production. For further information on improved manufacturing science procedures, see the ICH advise for industry Q8(R2) Pharmaceutical Development (November 2009), Q9 Quality Risk Management (June 2006), Q10 Pharmaceutical Quality System (April 2009), and Q11 Development and Manufacture of Drug Substances (November 2012).

After completing the initial comparative analytical assessment or clinical studies required to support an application, developers considering manufacturing changes will need to demonstrate comparability between the pre- and post-change biosimilar product. They may also need to conduct additional studies after completing the initial comparative analytical assessment or clinical studies required to support an application. The type and magnitude of the alterations may influence the scope of this new research. Comparative analytical studies should include a considerable number of biosimilar products used in clinical studies as well as a suggested commercial process if the procedure used to manufacture the clinical study material differs.

The production procedure of a protein product may have an impact on its safety or effectiveness. Variations in the biological systems used to manufacture a protein product, for example, could result in various post-translational alterations, putting the product's safety and effectiveness in jeopardy. When a marketed protein product's manufacturing process is changed, the application holder must assess the effects of the change and demonstrate that it does not impair the product's identity, strength, quality, purity, or potency as they relate to the product's identity, strength, quality, purity, or potency as determined by analytical testing, functional assays, and, in some cases, *in vitro* testing.

The International Conference on Harmonization (ICH) has made suggestions to the industry. 'Comparability of Biotechnological and Biological Products' (Q5E) and 'The Evaluation of Comparability Principles of Science Manufacturing Process Changes' (ICH Q5E) specify the scientific criteria for assessing comparability of manufacturing process changes. It is often more difficult to prove that a biosimilar product is biosimilar to the reference product than it is to compare a product's comparability before and after a manufacturer's production adjustments. This requirement stems from a manufacturer's thorough grasp that the manufacturing process, including existing controls and acceptance criteria, may be changed. A biosimilar product's manufacturer, on the other hand, is likely to use a different manufacturing process than the reference product's manufacturer (e.g. a different cell line, raw materials, equipment, processes, process controls, and acceptance criteria) and is unlikely to have direct knowledge of the reference product's manufacturing process. While some of the scientific principles described in ICH Q5E may apply to establishing biosimilarity, Agencies anticipate that establishing biosimilarity will require more data and information than establishing that a manufacturer's post-manufacturing change product is comparable to the pre-manufacturing change product. Furthermore, because ICH Q5E does not provide a reference product, it is ineffective for showing biosimilarity right away.

6.6.2.3 Structural attributes

The structural features, such as primary, secondary, and tertiary structures, are dictated by the nature of the recombinant expression engine. Although the amino acid sequence of a protein is determined by its gene sequence, post-translational changes such as quality control (QC) in the endoplasmic reticulum (ER) and passage through the Golgi apparatus affect the protein's ultimate structure and function. The biopharmaceutical sector faces a barrier in developing a platform for the manufacturing of recombinant biologic therapies since these processes are species- and cell-specific. Both *in vivo* and *in vitro*, chemical changes to proteins and glycoproteins (P/GPs) occur. Non-self-variations can provoke an immunological response in the form of anti-drug antibodies (ADA); aggregated forms can be more immunogenic, and efforts are undertaken to avoid or remove them. The objective of monoclonal antibody therapies (mAbs) is to bind the target, resulting in the creation of immune complexes (ICs), a specialized aggregate form. Phagocytic cells with antigen-presenting capabilities can remove these ICs. These concerns may imperil efforts to reduce mAb immunogenicity by ensuring that aggregates are not present in medicinal products.

The pharmacokinetic profile of therapeutic antibodies is affected by FcRn binding and related structures, which is one of the quality attributes of therapeutic antibodies. Other quality attributes such as antigen binding, glycan structure, and isoelectric point are considered to have a potential impact on the product's pharmacokinetic profile.

- Compare the structural variants from validation lots representing the commercial process; if the primary structure does not match, discard the cell line and start over again. Chances are secondary and tertiary structure will follow if the primary structure is met. Do not use any public information data on structural attributes; only what is observed in a side-by-side comparison with the reference product. At this stage, the analytical methods need not be validated, only suitable and sensitive. The fact that the testing is conducted at one time and side-by-side, any impact of the test method is prevented. Testing does not require many lots since the variations are small, and acceptance criteria are very tight.

- Where post-translational modifications and other modifications are found to be variable in the testing above, the developers should return to upstream and downstream process refinement to match the profile as close as possible, realizing that it may not be possible to match all variants as they may move in opposite directions when one is modified. How much a match is required depends on the nature of the product. This is more critical for a monoclonal antibody; the developers can either justify any differences with additional studies that may require clinical efficacy testing or work backward towards a better matching. The post-translational modifications are release specification attributes, requiring multiple lots to establish a reliable quality range.

6.6.2.4 Functional attributes

Among the quality attributes of therapeutic antibodies, the FcRn binding and related structures are well known to affect the product's pharmacokinetic profile. Antigen binding, glycan structure, and isoelectric point are all known to have a potential effect on the pharmacokinetic profile of the product. However, these quality attributes are not part of the release specification and are tested once to establish similarity. The test methods also need not be validated since the testing is done side-by-side with the reference product. The equivalence margin approach is suggested to establish similarity, requiring 6–10 lots (see below).

6.6.2.5 Physicochemical properties

Developers should consider the notion of the desired product (and its variants) while developing and conducting characterization studies, as stated in ICH Q6B. Understanding the heterogeneity between the biosimilar and the reference product (e.g. glycosylation type, position, and extent) as well as the diversity of distinct isoforms, particularly those emerging from post-translational modifications, will be crucial. For industry Q6B Specifications, 'Test Procedures and Acceptance Criteria for Biotechnological/Biological Products', consult the ICH advice (August 1999).

The physicochemical properties of proteins can be characterized using analytical techniques. These approaches are described in scientific journals, regulatory recommendations, and pharmacopeia compendia, among other places. Various ways give information on a wide range of properties. Appropriate analytical test methods should be selected based on the type of the protein to be described, as well as information about the structure and heterogeneity of the reference product, a biosimilar product, and critical product performance characteristics.

6.6.2.6 Aggregates

While aggregates are generally considered to be immunogenic, and the goal is to have the lowest possible aggregates, such is not the case with monoclonal antibodies; the developers must develop a range within which all its lots must fall. An equivalence margin approach is suggested. However, like the post-translational modifications, this is also a release specification attribute, requiring multiple lots to establish a reliable quality range. A biosimilar product has lower aggregates so the equivalence margin testing may fail at the lower end; this is acceptable. However, it must be understood that the reference product lots have undergone a time-test and a transportation test, both of which can add to total aggregates. A biosimilar product will not be considered a biobetter with lower aggregates.

6.6.2.7 Impurities

There are two types of impurities: one type is *product-related* and the other is *process-related*. The product-related impurities can be active or inactive, and these must be so classified based on literature data. Any impurity, regardless of the source that is not found in the reference product, should be fully characterized. Its impact on safety and efficacy is determined; this applies irrespective of the impurity (if it is detectable) because of the immunogenicity potential. It is more appropriate for the developers to remove any impurity not found in the reference product through process changes rather than justify their safety, leading to inevitable additional nonclinical or clinical studies.

To the extent that it is possible, developers should specify, detect, and quantify product-related pollutants in both biosimilar and reference products. If a comparative physicochemical study indicates that a biosimilar product and the reference product have comparable quantities of product-related impurities, pharmacological and toxicological research to determine the potential biological implications of certain impurities may be unnecessary. If the manufacturing process for a biosimilar product contains contaminants that are not present in the reference product or at higher levels than those discovered in the reference product, additional pharmacological, toxicological, or other research may be required. (In ICH Q6B, the terms 'product- and process-related impurities' and 'process-related impurities' are interchangeable and mean the same thing.)

On page 2 of the ICH industry advice S6(R1) 'Preclinical Safety Evaluation of Biotechnology-Derived Pharmaceuticals' (May 2012), it is stated that 'It is preferable to rely on purifying techniques to eliminate contaminants rather than establishing a preclinical testing program to qualify them.'

Impurities in the cell substrate (e.g. host cell DNA, host cell proteins), cell culture components (e.g. antibiotics, media components), and downstream processing stages (e.g. reagents, residual solvents, leachable, endotoxin, and bioburden) all need to be considered.

Because manufacturing impurities in a biosimilar product are unlikely to be identical to those found in the reference product, they are excluded from the comparative analytical examination. Analytical methods should be able to detect, identify, and precisely quantify pollutants at biologically relevant concentrations. (See the ICH guidance from May 1997 for industry 'Q2B Validation of Analytical Procedures: Methodology'.) In immunological techniques for detecting host cell proteins, the assay reagents and cell substrate used are crucial. To ensure accuracy and sensitivity, such tests should be verified using the product cell substrate and orthogonal approaches.

As with any biological product, the safety of a biosimilar product against adventitious agents or endogenous viral contamination should be verified by screening critical raw materials and verifying virus removal and inactivation during the manufacturing process. (Further information can be found in the ICH guidance for industry 'Q5A Viral Safety Evaluation of Biotechnology Products Derived From Human or Animal Cell Lines' (September 1998).)

6.6.2.8 HCP and residual DNA

When a new product is developed, the allowable limits of HCP and residual DNA are decided by demonstrating the feasibility of lowering this negative attribute, the relative safety risk of the specific products, and monitoring these attributes. Both HCP and residual DNA are organism strain-dependent, so even if the developers are using the same expression system, there is no possibility of matching the originator's exact strain. Since the HCP and residual DNA are specific to the strain, a direct comparison of a biosimilar product's relative safety is not possible, short of conducting complete safety studies evaluating these attributes. Agencies require a critical analysis of HCP and residual DNA and accept that these may not be like the HCP and residual DNA in the originator product. A 2D SDS is also recommended to resolve these components for both the test and reference product. How does one claim a similarity between the HCP and residual DNA between the test and the originator product? While it is accurate that the HCP and residual DNA in a biosimilar product and the reference product may not be the same, however, if the same expression system is used (e.g. *E. coli*), then there is a good likelihood that the safety will be proportional to the quantity of these components, notwithstanding any specific activity associated with any particular component. Therefore, it is possible for the test product to demonstrate a non-inferiority test evaluation using an equivalence range approach.

6.6.2.9 Lot-to-lot variability

The observed heterogeneity between lots could be explained by manufacturing circumstances or analytical assay variability. The supply of certain raw components (e.g. growth medium, resins, or separation materials) as well as different manufacturing sites all contribute to lot-to-lot variability in the manufacture of a protein product. As a result, it is required to adequately describe the lot-to-lot variability of the reference product and the biosimilar product during the comparative analytical assessment.

In some cases, adjustments to a potential biosimilar product's production process may be required to reconcile inconsistencies discovered during the comparative analytical examination. Data indicating that any production improvements resolved the identified disparities and had no visible effect on other quality attributes should be given. If the manufacturing change has an influence on other attributes, data should show that the impact of the modification was assessed and remedied.

6.6.3 Product- or process-related substances

Advances in analytical sciences (physicochemical and biological) have enabled extensive evaluation of specific protein products' physicochemical and biological properties. These analytical procedures have increased the ability to detect and describe the targeted product and its associated molecules, as well as product and process-related impurities.

Product-related compounds and product- and process-related impurities are interchangeably used and have the same meaning in the ICH Q6B. A product-related substance is a derivative of the target substance that contains at least 80% of the active component in the target drug substance. At the same time, an impurity can be active or inactive.

Product-related impurities and substances should be identified, characterized, quantified, and compared using multiple lots of a biosimilar product and multiple lots of the reference product, to the extent feasible and relevant, as part of an assessment of the potential impact on the product's safety, purity, and potency.

6.6.4 Method sensitivity

Advances in analytical sciences (physicochemical and biological) have enabled extensive evaluation of specific protein products' physicochemical and biological properties. The ability to identify and characterize the targeted product and its related components, as well as impurities from the product and process, has improved thanks to these analytical methodologies.

In the ICH Q6B, the terms product-related compounds and product- and process-related impurities are interchangeable and signify the same thing. A product-related substance is a derivative of the target substance that contains at least 80% of the active component in the target drug substance. There are two types of impurities: active and inactive.

To the extent that it is feasible and relevant, product-related impurities and substances should be identified, characterized, quantified, and compared using multiple lots of a biosimilar product and multiple lots of a reference product to assess the potential impact on the product's safety, purity, and potency.

6.6.5 Comparative testing

The development of a new drug involves stand-alone testing of the product to establish its characteristics. The development of biosimilars does not require stand-alone testing, including establishing the safety of efficacy. All testing is conducted in a comparative mode for comparison, not characterization, a pivotal difference that establishes test methods, testing protocols, and each test's scope.

Comparative analytical data are used to develop a biosimilar product for application submission, and they may influence the type and quantity of animal and clinical data required to demonstrate biosimilarity. To compare a biosimilar product to the reference product, comprehensive, strong comparative physicochemical and functional testing (which may include biological assays, binding assays, and enzyme kinetics) should be performed.

For a meaningful comparative analytical assessment, available state-of-the-art analytical assays must be able to determine, among other things, the molecular weight of the protein, its complexity (higher-order structure and post-translational modifications), its degree of heterogeneity, its functional properties, impurity profiles, and degradation profiles denoting stability. The developer should describe the analytical methodologies used in these analyses, including their capabilities and limitations. Alternative analytical study methods offer a distinct viewpoint on a quality characteristic and do not necessitate recurrent testing or investigation of the same attribute using the same method. Orthogonal testing alone cannot be used to dismiss a failed test.

In the comparative analytical assessment, a variety of features are assessed utilizing risk ranking and data analysis, with numerous orthogonal assays often used. When analyzing analytical data, agencies consider the entirety of the evidence; even if the results of a particular assay do not match set standards, this does not prohibit a showing of similarity. If differences between products are discovered during the comparative analytical assessment (including components of the assessment not included in the risk ranking), the developers may provide additional scientific information (risk assessment and additional data) as well as a justification for why these differences do not prevent the products from being demonstrated to be identical.

6.6.6 Side-by-side testing

The emphasis on side-by-testing lies in resolving variability in test methods, particularly the methods that cannot be fully validated. Testing side-by-side resolves most, if not all, differences in the test method reliability. For this reason, any published data about reference product quality attributes cannot be used to compare with a biosimilar product, and neither can one use a reference standard to match the attributes of a biosimilar product with the reference product. These limitations apply even to the most established attributes, such as the molecular weight, the sequence, or other fixed attributes.

6.6.7 Heterogeneity

The developer of a biosimilar product should conduct comparative testing by comparing an appropriate number of lots of the biosimilar product and the reference product side by side, as well as comparing the biosimilar product to an internal reference standard for appropriate attributes, where available and appropriate (e.g. potency). Several lots of the reference product and multiple lots of the biosimilar product are evaluated to determine product variability across lots. The number of lots required to fully appreciate the lot-to-lot variability of the reference and biosimilar products may vary depending on the circumstances. It is the developers' responsibility to make a scientific justification for it.

6.6.8 Structure confirmation

A protein's three-dimensional structure is crucial for its biological action. Because of their enormous size and the rotational properties of their alpha carbons, proteins usually have complex three-dimensional conformations (tertiary and, in some cases, quaternary structure). As a result of the flexibility, complex yet subtle changes in protein structure may occur over time, some of which may be required for functional activity. Low-energy interactions such as hydrogen bonds and van der Waals forces are typically used in these rotations, and they can be quite sensitive to environmental influences. Many proteins can now be evaluated in three dimensions thanks to modern analytical technologies. Researchers may be able to better detect tertiary and, to a lesser extent, quaternary protein structure using a range of cutting-edge methodologies, while also adding to the body of evidence showing biosimilarity. Simultaneously, determining a protein's three-dimensional conformation with conventional physicochemical analytical methods may be problematic. Any variations in higher-order structure between a biosimilar and the reference product should be assessed for their impact on protein function and stability. The stability of higher-order structures can also be determined through functional experiments.

6.6.9 Acceptance criteria

When a biosimilar product and a reference product are compared in a formal study that requires a specific number of lots based on statistical calculation, the acceptance criteria for each test are established by first testing the reference product and biosimilar product in an exploratory mode. This leads to the establishment of specifications and acceptance criteria when a biosimilar product and a reference product are compared in a formal study that requires a specific number of lots based on statistical calculation. Developers should consult with agencies to ensure that enough of lots are examined. It is necessary to identify specific lots of reference products used in comparative analytical studies, as well as their expiration dates and time frames, as well as when the lots were evaluated and used in other types of studies (nonclinical or clinical studies). This understanding can be utilized to explain approval requirements aimed at ensuring product quality, as well as aid in the comparative analytical evaluation of a biosimilar and a reference product.

Acceptance criteria, on the other hand, should be based on the whole of the evidence, not just on the observable range of product attributes of the reference product. Because many product attributes interact to modify a product's safety, purity, and potency profile, their prospective interaction should be considered when conducting comparative analytical evaluations and setting requirements. For example, the number and distribution of tetra-antennary and N-acetylglucosamine repeats glycoproteins can affect *in vivo* potency and should not be evaluated independently.

Furthermore, data from lots used in nonclinical and clinical studies, as well as pertinent information about the relationship between an attribute and the medicinal product's performance, can be used to assist in the development of acceptance criteria. (See the ICH rules for industry 'Q8(R2) Pharmaceutical Development' (November 2009).)

6.6.10 Orthogonal testing

It is frequently essential to analyze the same quality attribute using various analytical approaches to effectively cover the full spectrum of physicochemical attributes or biological activities. Methods that analyze the same characteristic using different physicochemical or biological principles are very useful since they provide independent data to back up the attribute (e.g. orthogonal methods to assess aggregation). To give an useful and sensitive tool for comparing medications, complementary analytical procedures such as peptide mapping or capillary electrophoresis should be utilized in conjunction with mass spectrometry of the separated molecules.

A thorough analytical characterization may reveal discrepancies between the reference product and the biosimilar product when analytical techniques capable of identifying qualitative or quantitative differences in product attributes are used. The focus should be on creating orthogonal quantitative approaches for determining any variations in product qualities convincingly. However, an orthogonal test must present an alternative view of the quality attribute or where the validity of a test can be questioned. An example of the latter choice may be using both the UV absorbance and HPLC to determine protein content. However, the specification must always be definitive without leaving a choice of testing.

6.6.11 Accountability of lots

The developer must account for all obtained and characterized reference product lots. The application must include all reference products and biosimilar product lots that have been studied in any fashion, including the specific physicochemical, functional, animal, and clinical studies for which a lot was used. The justification should be provided when a lot is specifically chosen to be included or omitted from analytical tests. The date of the analytical tests, as well as the product's expiration date, should be included in the application. Expired reference product lots should not be used in comparative analytical evaluations in general because lots investigated after their expiration date may produce results outside of the range commonly observed in unexpired lots, resulting in an overestimation of reference product variability. Testing lots beyond their expiration date may be acceptable if samples are kept for a lengthy period (e.g. frozen at −80 °C). The developer provides statistics and data to demonstrate that storage has no effect on the product's quality.

The same information and data collection for each manufactured drug substance and drug product lot of a biosimilar product should be given as for reference product lots.

In the comparative analytical assessment, the reference product and biosimilar product batches utilized in clinical research should be included (e.g. PK and PD, if applicable, similarity, and comparative clinical study).

It is not authorized to integrate data from two different reference products in any test. It could lead to a greater variety of similarity criteria and acceptability criteria than those produced by relying exclusively on data from a single reference product lot.

To conduct analytical investigations, suppose the drug substance was extracted from the reference product. In that situation, the developers should describe the extraction technique and show that it had no effect on the product's quality qualities. Modifying or eliminating desired products and impurities, as well as relevant product-related chemicals, would be part of this undertaking. It should have adequate controls in place to guarantee that the extraction technique does not dramatically affect the protein's important properties.

6.6.12 Critical quality attributes

When developing a comparative analytical assessment to support a biosimilarity demonstration, developers should consider all factors that could affect the safety and efficacy of a biosimilar product.

The critical quality attributes that are the subject of analytical similarity can be classified into inherent and legacy attributes. Inherent attributes are those variable properties inherent in the manufacturing of the product, resulting in lot-to-lot variability. This classification of features includes both process- and product-related factors. A good example will be all types of post-translational modifications for both cytokines as well as antibodies. Legacy attributes are those characteristics that are not subject to a lot-to-lot variation and must be met essentially in their entirety. A good example of a legacy attribute is the total mass of a protein (if it is fixed as in some cytokines but not for monoclonal antibodies), its amino acid sequence, and other declared properties as the position of disulfide bonds, etc, that are reported in several protein databanks, the patents, and publications, as well as the pharmacopeia. Another category of legacy attributes is the labeled specification of inactive ingredients, product characteristics such as pH, osmolality, etc.

While the legacy attribute specifications provide a significant characterization of the reference product, a biosimilar product must still be tested against the reference product lots because the reference product is not required to comply with any legacy attribute, and additionally, testing side-by-side is one way to reduce the impact of test method variability in determining clinically meaningful differences between a biosimilar product and the reference product. However, the release specification for several quality attributes can be established independently, based on well-established principles of assuring the product's efficacy. Examples of such specifications may include protein content (e.g. ±3%) or bioactivity (e.g. ±15%), post-translational modifications (e.g. ±10%), subvisible particles (USP specification), physical properties (e.g. pH, osmolality, density, etc, ±10%), aggregates (e.g. ±10% of the average from multiple lots of RP), fill volume (USP specification), impurities (e.g. no more than 3% and no single impurity more than 1% and no unidentified impurity). The developers may suggest other limits based on multiple reference product lot analyses. The fundamental principle behind establishing analytical similarity and release specification of drug substance and drug product is that there are no residual uncertainties that might first affect the safety and then the product's efficacy.

The impact of critical quality qualities on the product's safety and efficacy is determined. Table 6.5 is a partial, but not exhaustive, list of numerous attributes and their classifications. The test method selected is determined by the attribute's criticality.

6.6.13 Reference standard

Comparing a biosimilar product to a publicly available standard, e.g. a pharmacopeia monograph, or using a reference standard are *not* allowed for comparative

Table 6.5. Common critical quality attributes.

Quality attribute	Criticality	Potential impact	Suggested analytical methods
Amino acid sequence	Very high	Efficacy, safety	Peptide mapping, MS, Edelman degradation
Biological activity	Very high	Efficacy, safety	Bioassay
Deamidation, oxidation	Low	Efficacy	Chromatography
Glycan structure and content	Very high	Efficacy, safety	Glycan analysis
High molecular weight aggregates	High	Safety	SE HPLC, AUC, SDS PAGE
High-order structure	High	Efficacy, safety	Spectrophotometric, thermodynamic methods
Host cell proteins	High	Safety	SPR, cell-based assay
Immunochemical identity	Very high	Efficacy, safety	SDS PAGE+ immunoblotting, immunoassay
Insoluble aggregates	High	Safety	Light obscuration
Isoform distribution	High	Efficacy	Isoelectric focusing
Protein content	High	Efficacy	UV; use HPLC as an orthogonal method
Receptor binding	High	Efficacy	SPR, cell-based assay
Truncated forms	Low	Efficacy	Reference product HPLC, other chromatography

testing to establish biosimilarity. It is emphasized that any pharmacopeia monograph specifications or a reference standard provide by third parties are not considered appropriate to establish biosimilarity.

Pharmacopoeia reference standards are not suitable to conduct any comparative study, and their use is limited only to the qualification of test methods. A qualified reference standard is developed internally by the developer based on its own fully characterized batch.

If a relevant, publicly available, and well-established reference standard for the protein exists, comparing the physicochemical and functional properties of a biosimilar product to it may give useful information. While studies that use such a reference standard can be beneficial, they are insufficient to show biosimilarity between the biosimilar and the reference product. If a global standard for calibrating potency is available, for example, a biosimilar product's relative potency should be compared to it. According to ICH Q6B, the production process and product shall always be governed by qualified in-house reference standard(s).

Typically, lots from the early stages of research or lots utilized in a clinical study are used to create an in-house reference standard. Additional reference standards may be qualified and submitted later for submission. In an ideal scenario, the developers would have created and qualified primary and working reference standards for the biosimilar product lots utilized in the application's clinical investigations.

When developing a biosimilar product, the reference product lot is typically used as a first reference standard. Each clinical batch of a potential biosimilar product must be appropriately qualified for use as a release and stability reference standard, as well as comparable analytical testing after production (including bridging to prior reference standards). Once a biosimilar product has been suitably validated, enough of an in-house reference standard should be available for use throughout the development process. All lots of reference standards utilized in the development of a biosimilar product should be qualified adequately. All analytical procedures that report results comparable to the reference standard, as well as methods for release testing, should be included in the qualifying methodology for reference standards.

For all methods that report findings relative to a reference standard, a potency assignment of 100% should include a narrow-permitted potency range and ensure that product drift is limited. For example, developers should consider using a fixed two-sided confidence interval (CI) of the mean of the replicates, where both the mean relative potency and the 95% CI are within a narrow range (e.g. 90%–110%). To identify potential drift, an examination of the history of various reference standard qualifications should be conducted.

To adjust for discrepancies in reference standards' potency or biological activity, the developers should avoid applying a correction factor.

The use of unqualified reference standards for analytical techniques that compare results to the reference standard is likely to raise concerns about comparative analytical evaluation. If possible, holding reference product lots in long-term stable conditions could be one solution to these problems. Before applying, the prospective developer should review the biosimilar and reference product lots using the same reference standard for all processes that provide results in comparison to the reference standard.

6.6.14 The finished drug product

A biosimilar product's characterization should be done on the most downstream intermediate that is most suited for the analytical techniques. During any future processing steps, the qualities examined should keep their stability. Characterization studies on the drug material are frequently undertaken for these reasons. The impact of these alterations should be evaluated if a drug substance is reformulated, and the completed dosage form is exposed to new components. Developers should assess the finished medication product whenever possible to see if it is best suited for a certain analysis. Assume that an analytical approach is more sensitive in detecting specific properties of the drug material. Nonetheless, the qualities it assesses are crucial and may change during the final medicine product's manufacture. Comparative characterization of both the extracted protein and the final medicinal product may be necessary in this instance.

Proteins are particularly sensitive to their surroundings. As a result, changes in excipients or main packaging may have an impact on the product's stability and clinical efficacy. Dissimilarities in a biosimilar product's formulation and primary packaging compared to the reference product may determine whether subsequent clinical studies are selective and targeted. (See the ICH guidelines for industry 'Q8 (R2) Pharmaceutical development' (November 2009).) The excipients used in a biosimilar product should be distinguished from those utilized in the reference product by the developers. Any variations between the finished biosimilar product and the finished reference product should be assessed and supported by suitable evidence and rationale. Furthermore, existing toxicological data for the excipient or further toxicity studies to develop the biosimilar product should be backed up by the different excipients utilized in a biosimilar product.

It is worth noting that once a drug substance has been purified through the downstream process, this is considered a chemical entity, not a biological entity. Therefore, it is subject to all cGMP requirements pertinent to any other chemical drug fill and finish process. However, extraordinary considerations such as the desorption of proteins to the filling line, aggregate formation during the filling process, and changes in the concentration during the filling process need to be considered. Cleaning validation of the fill and finish equipment is also a bigger challenge. In contrast, a developer may present a protocol and the results to show the validation process, the risk of any chemical entity, albeit at a very low concentration, contaminating the protein solution may trigger structural changes. To avoid this, a preferred method is to use a dedicate filling head for the filling operations. The use of single-use heads further reduces the risk of contamination.

6.6.15 Excipients

Biosimilar products can have different inactive ingredients from what is disclosed in the reference product prescribing information. This allowance helps the developers overcome any patent protection of formulations that may extend beyond the biological entity's gene patents. However, this possible change creates additional challenges to demonstrate that the choice of inactive ingredients is not producing any adverse response in terms of toxicity, pharmacokinetics (including absorption), or efficacy of the product. Such incidences of toxicities are rare, but a prominent example is an incidence of pure red cell aplasia (PRCA) in the use of erythropoietin caused by changes in the formulation to without albumin and subcutaneous administration in a packaging that used uncoated rubber stoppers. (See McKay J *et al* 2008 Epoetin-associated pure red cell aplasia: past, present, and future considerations *Transfusion* **48** 1754–62.) This product was one of earlier biosimilars when the understanding of biosimilars was still in its infancy. None of these changes would be allowed today without a thorough investigation that would readily identify these risks.

The developers are advised to consider the reference product's formulation as the choice formulation based on the reference product's analysis. The developers should establish release limits of any inactive substance based only on the profile obtained for the RP. For example, surfactants are widely used in biological drug formulations and it is not uncommon for the surfactant concentration to vary by 30%–50% in the formulation. The purity of surfactants is also a critical measure. In some instances, it is advised to discard any remaining surfactant left in a fresh container to avoid safety risks to the product.

The developer's second choice is to use a different formulation based on ingredients that are generally known to be safe for parenteral products and not try any unusual components.

The last choice of using unique inactive ingredients will require more extensive safety studies that may leave residual uncertainty about the formulation. It is often not possible to extrapolate the toxicology of proteins from any analytical assessment. It is important to think about excipient interactions as well as direct toxicity.

6.6.16 Stability

It is recommended that forced deterioration experiments, as well as accelerated and stress stability studies, be carried out. These studies should be carried out under a number of stress conditions (e.g. high temperature, freeze–thaw, light exposure, and agitation), each of which has the potential to cause incremental product degradation over time. These studies' findings may show product variations that necessitate further investigation, as well as manufacturing and storage settings that necessitate extra controls. ICH 'Q5C Quality of biotechnological products: stability testing of biotechnological/biological products' (July 1996) and 'Q1A(R2) Stability testing of new drug substances and products' (July 1996) provide industry guidelines. The predicted shelf-life should be supported by sufficient data from a biosimilar product's real-time, real-condition stability.

Agencies expect a biosimilar product to be tested side-by-side both qualitatively and quantitatively for the type of degradants as well as their levels. Two questions arise. First, it is impossible to source lots of reference products with similar production dates as a biosimilar product. Second, how much deviation is allowable from theoretical degradation rates based on the products' declared shelf-life?

The developers must provide data on at least three lots that may include development lots placed on stability testing. The stability study should be conducted for at least six months, and linear regression coefficients of degradation are obtained with the expectation of high statistical significance (r^2). The slopes of degradation rates are then compared with theoretical rates. For example, if the product has a shelf-life of 36 months and the stability limit is no more than 3% degradation, then the regression line's theoretical slope can be calculated. A biosimilar product may not have a higher degradation rate than predicted theoretically, even if the reference product demonstrates a higher rate resulting in the reference product not meeting a projected expiration dating.

The developers should realize that biosimilar products' stability data are used to determine structural differences in the protein structure, in addition to other known factors that enhance degradation. Forced degradation studies are particularly important as they point to the finer structures' stability within the molecule.

6.6.17 ICH

The regulatory guidance also covers the CMC data that must be taken into account when determining if a biosimilar product is biosimilar to the reference product.

To demonstrate that the manufacturing process consistently produces a product with the desired quality characteristics, all product applications must include a comprehensive CMC section with all necessary and appropriate information (e.g. characterization, adventitious agent safety, process controls, and specifications). Several ICH guidelines (table 6.6) are applicable to the presenting of this advice. Agencies, on the other hand, are not required to follow any of these guidelines' recommendations.

Table 6.6. Pertinent ICH guidelines for biosimilars.

ICH guidance for industry M4: the CTD—quality (ICH M4Q) (August 2001)	ICH guidance for industry Q1A(R2) stability testing of new drug substances and products (ICH Q1A(R2)) (November 2003)
ICH guidance for industry Q2(R1) validation of analytical procedures: text and methodology (ICH Q2(R1)) (November 2005)	ICH guidance for industry Q2B validation of analytical procedures: methodology (ICH Q2B) (May 1997)
ICH guidance for industry Q3A(R) impurities in new drug substances (ICH Q3A(R)) (June 2008)	ICH guidance for industry Q5A viral safety evaluation of biotechnology products derived from cell lines of human or animal origin (ICH Q5A) (September 1998)
ICH guidance for industry Q5B quality of biotechnological products: analysis of the expression construct in cells used for production of r-DNA derived protein products (ICH Q5B) (February 1996)	ICH guidance for industry Q5C quality of biotechnological products: stability testing of biotechnological/biological products (ICH Q5C) (July 1996)
ICH guidance for industry Q5D quality of biotechnological/biological products: derivation and characterization of cell substrates used for production of biotechnological/biological products (ICH Q5D) (September 1998)	ICH guidance for industry Q5E comparability of biotechnological/biological products subject to changes in their manufacturing process (ICH Q5E) (June 2005)
ICH guidance for industry Q6B specifications: test procedures and acceptance criteria for biotechnological/biological products (ICH Q6B) (August 1999)	ICH guidance for industry Q7 good manufacturing practice guidance for active pharmaceutical ingredients (ICH Q7) (September 2016)
ICH guidance for industry Q8(R2) pharmaceutical development (ICH Q8(R2)) (November 2009)	ICH guidance for industry Q9 quality risk management (ICH Q9) (June 2006)
ICH guidance for industry Q10 pharmaceutical quality system (ICH Q10) (April 2009)	ICH guidance for industry Q11 development and manufacture of drug substances (ICH Q11) (November 2012)
ICH guidance for industry S6(R1) preclinical safety evaluation of biotechnology-derived pharmaceuticals (ICH S6(R1)) (May 2012)	

6.6.18 Examples of test methods

The developers are advised to create a comprehensive protocol for testing based on the nature of the product. Given below is a suggested list of tests that may be used for physicochemical and biological assessment of TNF alpha-blockers (table 6.7) and oncology antibodies (table 6.8)

Table 6.7. Quality attribute testing for monoclonal antibodies (TNF-α blocker).

Category	Product quality attributes	Analytical attributes
Physicochemical characterization		
Primary structure	Molecular mass The sequence of amino acids The sequence of the terminal Oxidation of methionine Deamidation of C-terminal and N-terminal variants Mapping disulfide linkages	Intact mass in reducing/non-reducing conditions Peptide mapping by LC-ESI-MS/MS using a combination of digestion enzymes Peptide mapping in non-reducing conditions
High-order structure	Protein secondary and tertiary structure	Far- and near-UV CD spectroscopy, ITF HDX-MS, antibody conformational array DSC
Glycosylation	N-linked glycosylation site determination N-glycan identification N-glycan profile analysis	LC-ESI-MS/MS Procainamide labeling and LC-ESI-MS/MS 2-AB labeling and HILIC-UPLC
Aggregation	Soluble aggregates	SEC-UV, SEC-MALS/RI SV-AUC
Fragmentation	Low molecular weight	Non-reduced CE-SDSReduced CE-SDS
Charge heterogeneity	Acidic variants Basic variants	CEX-HPLC and icIEF
Biological characterization		
Fab-related biological activity	TNF-α neutralization activity TNF-α binding activity Apoptosis activity Transmembrane TNF-α binding assay	TNF-α neutralization assay by nuclear factor-κB reporter gene assay FRET Cell-based assay FACS
Fc-related biological activity	FcRn binding FcγRIIIa (V/V type) binding ADCC using healthy donor PBMC CDC C1q binding FcγRIa binding	AlphaScreen® SPR Cell-based assay Cell-based assay ELISA FRET

FcγRIIa binding	SPR
FcγRIIb binding	SPR
FcγRIIIb binding	SPR

ADCC: antibody-dependent cell-mediated cytotoxicity; AlphaScreen® and AlphaLISA® (amplified luminescent proximity homogeneous assay) are bead-based assay technologies used to study biomolecular interactions in a microplate format; CD: circular dichroism; CDC: complement-dependent cytotoxicity; CE-SDS: capillary electrophoresis–sodium dodecyl sulphate; CEX-HPLC: cation exchange–high-performance liquid chromatography; DSC: differential scanning calorimetry; FcRn: neonatal Fc receptors; FRET: fluorescence resonance energy transfer; Gal: galactosylated glycans; HDX-MS: hydrogen–deuterium mass spectrometry; HILIC-UPLC: hydrophilic interaction liquid chromatography–ultra-performance liquid chromatography; HMW: high molecular weight; icIEF: imaging capillary isoelectric focusing; ITF: intrinsic fluorescence spectroscopy; LC-ESIMS: liquid chromatography–electrospray ionization–mass spectrometry; LC/MS: liquid chromatography–mass spectrometry; LC-ESI-MS/MS: liquid chromatography–electrospray ionization–tandem mass spectrometry; PBMC: peripheral blood mononuclear cells; SEC: size exclusion chromatography; SEC-MALLS/RI: size exclusion chromatography–multi-angle laser light scattering/refractive index; SPR: surface plasmon resonance; SV-AUC: sedimentation velocity analytical ultracentrifugation; UV: ultraviolet; UV/VIS: ultraviolet visible.

6.6.19 Risk assessment

The developer should design a risk assessment tool to rate and evaluate the quality attributes of reference products in terms of their potential impact on the product's mechanism(s) of action and function. Certain quality attributes of the reference product should be excluded from the risk ranking in general (e.g. degradation rates obtained by stability or forced degradation studies). These assessments, on the other hand, should be included in a biosimilar and reference product's comparative analytical examination.

The risk assessment tool should be built with the following qualities in mind:

- *The influence of a characteristic on clinical performance*: Specifically, developers should consider how a characteristic affects activity, pharmacokinetic/pharmacodynamic characteristics, safety, efficacy, and immunogenicity, among other things. When examining the likely influence of a characteristic on clinical performance, developers should examine publicly available information as well as their appraisal of the reference product.
- *Uncertainty regarding a particular qualitative attribute*: When there is a lack of knowledge of the relationship between the degree of change in an attribute and the clinical outcome, it is proposed that the attribute be ranked as having a higher risk of increasing ambiguity.
- Any trait that poses a high risk for any performance category (activity, PK/PD, safety, efficacy, and immunogenicity, for example) should be classed as high risk. In an ideal world, the risk assessment tool would generate a list of qualities that were graded based on the patient's risk level. As a result, the proportionality of attribute risk scores to the patient's risk is critical. The risk assessment scoring criteria should be clearly established and supported. Each attribute's risk ranking should be backed up by appropriate references to the literature and data.

Table 6.8. Testing methods for oncology monoclonal antibodies.

Category	Attribute (method or methods to query that attribute)
Primary structure	Primary sequence (e.g. UPLC peptide map, LC-MS/MS, amino acid analysis, Edman degradation, carboxypeptidase sequencing) Disulfide structure (e.g. LC-MS of non-reduced protein digest) Intact mass (e.g. LC-MS) Isoelectric point (e.g. IEF gels, cIEF, iCE) Extinction coefficient (e.g. UV/AAA or UV/RI)
Secondary and tertiary structure	Low-resolution secondary structure or indirect tertiary structure measurements (e.g. CD, DSC, FTIR, fluorescence) High-resolution measurements of higher-order structure (e.g. 2D-NMR, HDX-MS, x-ray crystallography)
Glycosylation	Glycosylation (e.g. HILIC, MS (MALDI, ESI), exoglycosidase sequencing, HPLC-FLD, HPAEC-PAD, CE-LIF) Glycosylation site mapping/site occupancy (e.g. peptide mapping by LC-MS)
Dose	Protein content (e.g. UV A280, RP-HPLC) Deliverable volume (extractable volume)
Particulates	Subvisible particles (e.g. light obscuration, MFI, NTA)
Function	Biological activity (e.g. for mAb: proliferative bioassay, cytotoxicity assay, ADCC, and CDC; other assays may be appropriate for other proteins, e.g. enzyme kinetics for biosimilar enzyme) Receptor and ligand binding (e.g. SPR, ELISA)
Product variants (product-related substances and impurities)	High molecular weight species (e.g. SEC-MALS, AF4/HF5, AUC, DLS) Covalent dimers (e.g. SDS PAGE, CE-SDS) Purity and impurities (oxidation, deamidation, glycation, isomerization, fragmentation, disulfide reduction, e.g. RP-HPLC, CEX, SEC, IEX, IEF, cIEF, LC-MS) Amino acid misincorporations (e.g. LC-MS/MS) Micro sequence heterogeneity (e.g. LC-MS) C- and N-terminal modifications (e.g. LC-MS, Edman degradation)

6.6.20 Statistical considerations

The assessment of biosimilarity is primarily based on statistical considerations; for the sake of clarity and to describe how Agencies interpret data, table 6.9 lists statistical concepts as applied to biosimilarity testing.

6.6.20.1 Quantitative and qualitative data analysis

To prove that a biosimilar product is comparable to the reference product, appropriate studies of comparative analytical data are required, even slight variations in clinically inactive components. Analyzing the relevance of analytical data can be done in a variety of ways. Agencies want the developer to be well-versed in statistical approaches and to be able to defend the statistical modeling that was employed. The variability of the important quality features is proportionate to any documented difference in clinical response between the reference product and a biosimilar product. This assumption is the foundation for determining key quality attributes (CQAs), as biological products are known to have a wide dose–response relationship. As a result, any modeling must account for these possibilities.

The use of descriptive quality ranges to determine the quantitative quality attributes of high and moderate risk, as well as raw data/graphical comparisons for quality attributes with the lowest risk ranking or those that cannot be quantified, is one method of data analysis (e.g. primary sequence).

Acceptance criteria for the quality range (QR) approach used in the comparative analytical evaluation should be generated using the developer's examination of the reference product for a specific quality feature. The QR stands for the sample standard deviation, which is calculated using reference product lot numbers and should be interpreted as the sample mean. The multiplier (X) for that attribute should be scientifically validated and agreed with Agencies. Based on our previous experience, we do not recommend using tolerance intervals to construct similarity acceptance criteria.

When deciding what type of quantitative data analysis to apply to a feature or test, other considerations should be examined in addition to risk ranking. The following are some additional factors to consider while choosing on the optimal strategy for data analysis and interpretation:

- Over unknown but potentially high-risk features, the nature of the attribute should be stressed (i.e. attributes with a high risk ranking due to uncertainty).
- *The distribution of the attribute* : In general, the developer should build the manufacturing process as nearly as feasible to the distribution centers of the reference product's quality qualities. As a result, it is acceptable to use a QR code to demonstrate that a biosimilar product is identical to the reference product. It implies that the mean and standard deviation of the population are comparable. More information or tests may be required to support the QR technique or to support a new research methodology if there are issues about the distribution. Because of the nature of the attribute and its participation in the product's mechanism of action, the distribution of an attribute in a biosimilar product that is skewed towards one side of the

Table 6.9. Statistical concepts applied to biosimilarity determination.

<table>
<tr><th>Subject</th><th>Definition and concepts applicable to biosimilarity testing</th></tr>
<tr><td>A stepwise analytical assessment</td><td>Assess the quality attributes following the ICH Quality Guidelines Q8, Q9, Q10, and Q11's risk management principles.
Consider the criticality risk ranking of quality attributes concerning their potential impact on activity, PK/PD, safety, and immunogenicity.
Use a selective approach for assessment.
Equivalence interval testing for some critical attributes ($K \times \sigma R + \Delta$ and the default value of Δ is zero). The equivalence margin can be determined by a proposed sample size/power-adjusted method.
Quality ranges (mean $\pm$ $K \times$ SD) for other less critical attributes; $X \leqslant 3$ unless otherwise justified to be higher. The quality range can be determined by mean $\pm$ $K \times$ SD, where $K = 2$~3 based on the targeted coverage; quality range tests will also assess all biosimilar lots used in equivalence tests; 90% of biosimilar lots need to be within the quality range.
Raw/graphical comparisons for other least critical attributes. Graphic data displays are a useful tool to identify the potential issues with statistical methods listed above.</td></tr>
<tr><td>Significance level (also called size of a test)</td><td>The probability of making a Type I error is determined by the magnitude of a test, often known as the significance level. While a null hypothesis is rejected when it is true, this is known as a Type I mistake. The test size is indicated by the symbol α (alpha). The confidence level is a number that is used in the $(1-\alpha) \times 100\%$ confidence interval of a parameter.

<table><tr><th>Reality of Attribute</th><th>Conclusions about Attribute</th><th>Description</th></tr><tr><td>Not Similar</td><td>Similar</td><td>Type I error: patient (or regulatory) risk (α)</td></tr><tr><td>Similar</td><td>Not Similar</td><td>Type II error: the developer's risk (β)</td></tr><tr><td>Similar</td><td>Similar</td><td>Power ($1-\beta$)</td></tr></table></td></tr>
<tr><td>Confidence level</td><td>Self-assurance level. The confidence level relates to the likelihood of discovering the true population parameter in a specific proportion of all potential samples. Assume that all available samples were selected from the same population and that each sample was given a confidence interval. A confidence level of 90% indicates that the genuine population parameter is used in 90% of confidence intervals. The confidence level can be reduced in</td></tr>
</table>

	some circumstances, and the research suggests that the lowest confidence level for evaluating analytical similarity is around 80%.
Confidence interval	The confidence interval expresses the degree of uncertainty associated with a sample statistic. A confidence interval is a combination of an interval calculation and a probability assertion. For example, suppose an analysis allows the computation of an interval estimate; a confidence level will describe the uncertainty associated with the interval estimate. One may describe the interval estimate as a '90% confidence interval'. This means that we can expect the true population parameter to fall within the interval estimates 90% of the time if we use the same sampling approach to pick different samples and calculate an interval estimate for each sample.
Statistical power	The statistical power is a function of the sample size, the effect size, and the probability level chosen:

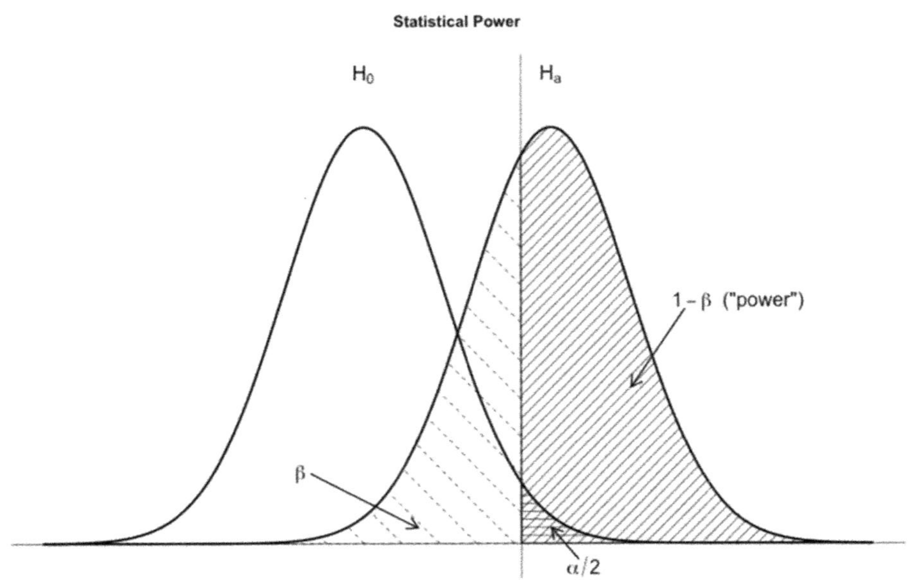

(*Continued*)

Table 6.9. (*Continued*)

Subject	Definition and concepts applicable to biosimilarity testing
Equivalence interval model	Two one-sided *t*-tests (TOST) should be used to test if the two-sided 90% confidence interval (CI) of the mean differences between a biosimilar product and the reference product falls entirely within $K \times R$ (R is the population standard deviation of the lots of the reference product that is estimated by the sample s) to meet the 'equivalence interval' most critical quality attributes. Enough lots must be used to reach at least 80% statistical power. The TOST for testing the hypotheses are represented as follows: H_0: $\mu_B - \mu_R \leqslant -\delta$ or $\mu_B - \mu_R \geqslant \delta$ $H_{a:}$ $-\delta <$ $\mu_B - \mu_R < -\delta$ where μ_B and μ_R are the mean responses of a biosimilar product and reference product lots, respectively, and $\delta > 0$ is the equivalence margin. If the null hypothesis is accepted, the equivalence will be rejected.

$$H_0 \quad \mu_B - \mu_R \leq -\delta \text{ or } \mu_B - \mu_R \geq \delta$$

$$H_a \; -\delta < \mu_B - \mu_R < \delta$$

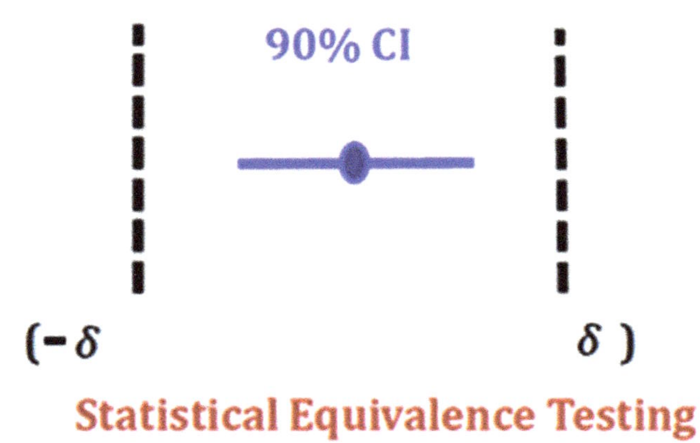

Quality range model

The reference product data define the quality range for a specific quality attribute as ($\hat{\mu}_R - K^*\hat{\sigma}_R$, $\hat{\mu}_R$ $\hat{\sigma}_R$) where $\hat{\mu}_R$ is the sample mean, and $\hat{\sigma}_R$ is the sample standard deviation based on the reference product lots. The standard deviation multiplier (K) should be scientifically justified for that attribute. The statistical analysis will generally support analytical similarity for the quality attribute if a sufficient percentage of test lot values (e.g., 90 percent) fall within the quality range. For establishing the scientific justification for the standard deviation multiplier (K), the two-sided tolerance intervals of the univariate normal distribution is used to calculate the multiplier K. the following equation estimates the K with given confidence ($1 - \alpha$) and a targeted portion (P) of the univariate normal distribution:

$$Pr_{\overline{X},\, S}\left\{Pr_X(\overline{X} - \mathrm{kS} < \mathrm{X} < \overline{X} + \mathrm{kS} \mid \overline{X}, \mathrm{s}) > \mathrm{P}\right\} = 1 - \alpha.$$

For a sample size $N = 10$ per product, the two-sided tolerance interval of the 90% coverage for the distribution is 2.86 at the 95% confidence level. If the significant digit for attribute reported at one decimal, set the tier multiplier as 2.9 (generally 3.0) in this study to cover the 90% of reference product distribution at $N = 10$ per group. That is, the quality range is $\hat{\mu}_R - 2.9_*\hat{\sigma}_R$, $\hat{\mu}_R$ $\hat{\sigma}_R$) (if 90* lot values will fall within the quality range.
The errors in the Mean ± 3SD Method for Quality Ranges.

(μ-3σ, μ+3σ) = (-3, 3)	N = 6	N = 30	N = 100
Simulation 1	(-4.3, 4.4) too wide!	(-3.4, 3.6)	(-3.1, 3.3)
Simulation 2	(-1.4, 2.0) too tight!	(-3.2, 3.1)	(-3.0, 2.9)

It is possible that either the equivalence margin and/or equivalence range may fail or vice versa. Products with similar standard deviation but different means may fail the equivalence range test but not equivalence margin. For products where the means are the same, but the standard deviations are not, they may fail the equivalence margin but not equivalence testing.

(*Continued*)

Table 6.9. (*Continued*)

Subject	Definition and concepts applicable to biosimilarity testing
Fixed margin approach	The fixed margin approach used for bioequivalence testing used the fixed margin $(\delta_1, \delta_2) = (0.80, 1.25)$ or (0.90, 1.11) for the ratio of means regardless of the variability of the tested CQAs. The testing requires a different number of lots for different CQA. However, this approach may lead to equivalence with the out-of-specification product. The margins are based on scientists' prior knowledge, which may require excessively large EAC by sponsors to assess the equivalence of a small number of the lot (e.g. a few lots) that may be highly variable. Note that after decades of scientific discussions, the 80%–125% limits of bioequivalence testing were adopted, admitting that there is not any real scientific rationale, except that it has been shown to work. Lately, there is a focus on the scaled average bioequivalence, wherein the standard deviation of the reference product determines the acceptance criteria.
Sample size and variance approach	The sample size and variance come from selecting an equivalence testing hypothesis: H_0: $\mu_T - \mu_R \leq -\delta$ or $\mu_T - \mu_R > \delta$ H_a: $-\delta < \mu_T - \mu_R < \delta$ Now we give lower power γ^* (for target equivalence) to small sample sizes and solve for the (symmetric) margin so that power (margin, sample sizes, true mean diff, variability) = γ^*. The rationale for this approach is that it captures the product and quality attribute variability, allows an acceptable shift $\delta_0 \geqslant 0$, determined by scientists for each CQA and rewards large sample sizes by controlling the power of small samples.
Minimum to maximum approach	The quality criterion regarding coverage cannot be defined using this approach as it is unstable for small sample sizes and too wide for large sample sizes. For these reasons, Agencies reject this approach.
Raw and graphical comparisons	To evaluate which qualities have the lowest risk ranking, this testing approach uses raw data/graphical comparisons. The evaluation of similarity for QAs using this method is not in any way less rigorous, which is acceptable given that they have the least impact on clinical outcomes in the sense that significant dissimilarity has no effect on clinical results. Depending on the nature of the testing output, if a numerical output is a result of graphical output, this method of testing must always be made; a fluorescence spectrum is tested on this basis, and the maximum emission wavelength or its ratio with 310 nm can be subjected to an equivalence range level testing. There is no guarantee that a given QA, which passes the equivalence margin or equivalence range test, will pass this testing test and vice versa. Evaluation is based on raw data and graphical presentation; it is somewhat subjective and biased.

Number of lots	The number of lots tested side-by-side selected will depend on three factors: • The power of testing is accepted; generally, like the bioequivalence testing, where 80% (0.80) power will be acceptance. • Confidence level: generally, it is taken to be 5% giving CL of 0.90. • Standard deviation: Large standard deviations can be reduced either by increasing the number of lots and adjusting the analytical methods that are less variable. • To ensure (i) adequate characterization of the biosimilar product and understanding of manufacturing variability; and (ii) adequate comparison to the reference product, which should include lots manufactured using both investigational and commercial-scale processes, developers should include at least six to ten lots of a biosimilar product in the comparative analytical assessment. These batches should be representative of the commercial production procedure. If a manufacturing method is changed before authorisation, it may be possible to use data from lots made using a different process if a solid scientific explanation is presented. However, data supporting the synthesis of drug substances and drug products using various techniques and scales should be included in the authorization application. The extent of process development design (as defined in ICH 'Q8 (R2) Pharmaceutical development' and 'Q11 Development and manufacture of drug substances') and process understanding should be used to justify the number of biosimilar product lots included in the application's comparative analytical assessment. In this scenario, the ICH Q5 standard is irrelevant. • The number of lots used is more critical when statistical modeling is used to compare a biosimilar product with the RP; the statistical model may include using an equivalence interval or an equivalence range where applicable; in such instances, the power of the test is important as derived from the following equation: $$Power(n_B) = 1 - \exp(- 0.53948 - 0.14694n_B + 0.00205n_B^2$$ (Niazi S 2018 *Biosimilarity—The FDA Perspective* (Orlando, FL: CRC Press).) • Using four lots will give a power of 65%, ten lots to give 85% power, and 20 lots allow a power of 92%. Generally, a power of 80% is recommended. Since not all testing is conducted using statistical approaches, fewer lots may be sufficient for some testing.
Types of lots	Biosimilar lots should be derived from various drug substance lots to the extent possible in order to effectively portray the diversity of features inherent in the drug substance production process. Except for specific drug product attributes for which variability is primarily dependant on the manufacturing process, drug product lots formed from the same drug substance batch(es) are not judged to be sufficiently indicative of such variability (e.g. protein

(Continued)

Table 6.9. (*Continued*)

Subject	Definition and concepts applicable to biosimilarity testing
	concentration). While it may be preferable to compare biosimilar product lots to reference product lots, it may be necessary to incorporate independent drug substance lots in order to obtain enough lots for the comparative analytical assessment (if the drug substance was not utilized to create the drug product). A large number of biosimilar products utilized in primary clinical research, as well as a potential commercial product, should be included in the comparative analytical assessment submitted with the marketing application to demonstrate biosimilarity between the biosimilar and the reference product. After completing the initial comparative analytical review or clinical studies supporting an application, developers considering manufacturing changes may need to conduct additional comparable analytical tests comparing the biosimilar product to the reference product. The nature and magnitude of the alterations may define the scope of these further analytical studies.
Replicates	Each biosimilar lot should have the same number of replicates as each reference product lot, according to the guidelines. For equivalency testing, quality range testing, and visual assessment of graphical displays, the same lots are used.
Equal number side-by-side	When comparing lots side by side, there should be an equal number of them. Assume that additional lots are available for reference. In that situation, an unbiased selection should be made to identify the equal number, with the remaining being utilized to independently construct range acceptance criteria. This recommendation is debatable until we demonstrate that the test has sufficient power. If the number of samples of the reference product is higher, this will lead to a smaller standard deviation and a larger difference range that may cause the test to fail because of β error. The critical minimum number of lots tested side-by-side is approximately 8; however, if there is an unbalanced selection between test and reference, a higher number may be required.
Unbalanced samples	Samples of reference and test products are blinded before the testing of analytical similarity; if the reference samples to test are out of specification, these are to be removed from the statistical calculation, creating an unbalanced design since Agencies require an equal number of samples of test and reference product to be tested side by side. How can this be resolved? The developers should remove an equivalent number of test samples through a randomization process; this process must be presented in the analytical testing SOP and properly documented. It is important that if the reference product fails one attribute and is declared out of specification, other values drawn from those samples cannot be

	used, even if they are not affected by the out-of-specification reading for one attribute. This exercise will create a situation where the test's power is reduced; to overcome this possibility, the developers should try to start with a larger number of samples, such as 11 instead of 10, if the target is ten samples. The developers may also adopt a procedure wherein the reference samples are tested before being blinded to avoid this situation. However, the selection and blinding of the reference product must still be made on a random basis.
Blinding	Before conducting the side-by-side analytical similarity testing, the developers would create a protocol that will include all acceptance criteria and blind the test and reference samples; where a larger number of samples are available, this will be preceded by a random selection of lots. However, when establishing EAC or other acceptance attributes using a separate set of lots, there is no need for blinding the reference samples.
Combining lots	The biosimilar developer conducts several biosimilarity studies at different stages and accordingly conducts analytical and functional similarity testing; can these lots be combined to provide a composite description instead of one study?
Mixed graphic and numerical data	Some testing provides a graphical output, but the peak height and area under the peak curve can be quantitated. Whenever there is a graphical output, it should demonstrate no extraordinary peak, no extraordinary heights of the peaks, and no extraordinary baseline—this is an overall evaluation of the graphical output; however, where the peaks have known significance relating to potency, purity, and safety, and as a result, the quantifiable graphical attributes have clinical meaningfulness, these can be compared for equivalence margin and equivalence range analysis; for visual testing, there is no further need to perform any quantitative evaluation. However, if the numerical comparisons do not have any clinical relevance, a graphical representation alone would be sufficient, regardless of the attribute.
Significant figures	The data must be rounded off before statistical analysis to balance the analytical method's sensitivity, instrument sensitivity, and other variables. For example, if a balance is sensitive to 0.1 g, the weight, all weight values will be rounded off to one decimal point: 1.09 becomes 1.1 and 1.02 becomes 1.0; another approach is to express significance to a percentage of the value. For example, a method sensitive to giving ±1% will have the following values: 1.00; 10.0; 100.

reference product distribution, for example, may cause complications. If such a distribution is discovered, sufficient empirical support for comparative analytical product evaluation will be necessary. Developers can reach out to Agencies if an attribute in the reference product is not normally distributed.

- *The attribute's abundance*: Because of the inherent variability of protein products, a feature of the reference product that may pose a high risk when present in high abundance (e.g. percent aggregation or percent oxidation) may pose a significantly lower risk (or no risk) when present in low abundance. Both the reference product (as indicated by a biosimilar product, the developer's assessment of the RP) and the biosimilar product should have a lot of this attribute. Limit assays do not require QR evaluation, but the chosen limits for an attribute's total must be defined and justified. The explanation should also take into account how the property's value changes over time.
- *The sensitivity of an assay is used to evaluate an attribute*: While using a variety of orthogonal assays to evaluate an attribute is advised, not all tests must be evaluated in the same way. Although QR should be used to choose the most sensitive assay for detecting product deviations, a graphical comparison of the results of various assays for the same characteristic may be sufficient. For each sort of assay, an explanation for the technique of evaluation should be supplied.
- *Many different qualities and judgments*: It is likely that some characteristics are not well-suited to quantitative analysis (e.g. protein sequence or certain assays used for higher-order structure evaluation or only qualitative assays). Which assays will not be exposed to quantitative data analyses, and why, should be stated in the comparative analytical evaluation plan.
- *Data that are available to the general public*: Publicly available data may influence the type of data analysis and acceptance criteria used in the comparative analytical assessment. If any publicly available data should be included in the comparative analytical assessment, the developers may request additional input from agencies.

Side-by-side data presentation (e.g. spectra, thermograms, graphical portrayal of data) is recommended for qualitative studies of lower risk features to allow for a visual comparison of a biosimilar product to the reference product. With side-by-side testing, NMR, mass spectrometry, biological activity, and other difficult-to-verify assays can only be used for analytical purposes rather than product release.

6.6.20.2 Risk ranking

The risk rating of features, the type of data measurement used for each attribute/assay, and the final data analysis approach should all be included in the final comparative analytical evaluation plan. For each attribute/assay, the proposal should describe why a specific number of lots was chosen for the evaluation, as well as the projected availability of biosimilar and reference product lots. Agencies should be informed about the comparative analytical evaluation plan early in a possible biosimilar product development program to reach a consensus on which attributes/assays should be investigated. Before beginning the final analytical evaluations, the final comparative analytical evaluation plan should be submitted to Agencies; this is usually done in a meeting with Agencies. The developers must create and sign off on an empirical assessment strategy, which includes approval and rejection criteria, before starting the testing.

6.6.20.3 Assay variability

A large R would not be justified if the assay variability was high. The assay would need to be streamlined to minimize uncertainty, and the number of replicates would need to be increased. In addition, if the equivalence margins or consistency ranges are too broad, it may be scientifically justified and reasonable to limit them. The sponsor may use a non-parametric tolerance interval if the data does not follow a normal distribution, but a broad sample size is usually expected.

What is considered a high variability acceptable is difficult to answer, but several guiding principles can be applied. First, it involves instrumentation variability—can we use better equipment? Second, it requires understanding the limit of the CQA variability—for example, many pharmacodynamic responses are highly variable by nature involving biological testing systems and the same applies to bioassays or binding assays (which will likely be equivalence range tested). For example, if the literature confirms, in particular the originator data reports, that the variability is ±30%, then that could be taken as the limit to achieve. The developers may also want to adopt tests other than those prescribed in the compendia (if so) or other routine testing methods and instead adopt higher sensitivity and repeatability testing. Agencies are not recommended on which method to be used, if it can be justified and demonstrated to be appropriate and suitable. Later, it can be validated if it is used for release purposes. One way to reduce the variability is to perform replicate analysis of the samples, this takes care of assay variability; Agencies do not allow multiple samples from the same lot.

6.7 Functional assessment

Most biological products have a specific means of interacting with the body, such as binding to a receptor, ligand, or substrate. At the molecular or cellular level, the functional consequences are typically obvious. Human cells or human receptors can be employed in *in vitro* experiments to detect target binding and functional effects, and these assays are extensively used in structure and function characterization. The complementarity determining region (CDR) of monoclonal antibodies binds to the principal target of the antibody. Nonetheless, the Fc region of the molecule contains receptor binding sites that can cause a range of effector activities, such as complement activation, complement-dependent cytotoxicity (CDC), and antibody-dependent cell-mediated cytotoxicity (ADCC). Fc-related binding characteristics and effector functions can also be tested *in vitro*.

6.7.1 Orthogonal studies

- To establish the potential difference in biological activity between a biosimilar product and the reference product, data from several comparative *in vitro* experiments, some of which may already be accessible from quality-related assays, should normally be presented.
- Relevant assays on binding to target(s) known to be involved in the pharmaco-toxicological effects and pharmacokinetics of the reference product should be included in these studies (e.g. receptors, antigens, enzymes).
- Cellular signal transmission and functional activity/viability studies should be carried out.

The studies should be comparative and not solely focus on the answer. The methods used should be acceptable and suitable for their function to obtain unambiguous results. Validation is not needed for these test methods.

To establish that observed quality attribute disparities are not clinically significant, the research must be responsive, comprehensive, and selected. The studies should compare the concentration–activity/binding relationship between a possible biosimilar and the reference product at the pharmacological target(s), using a concentration range that allows for the most sensitive detection of potential changes.

6.7.2 Lots tested

A sufficient number of lots of the reference product and a potential biosimilar product that is indicative of the substance intended for clinical use should be included in the study. The number required will be determined by the assay and batch-to-lot variability. For both the biosimilar and the reference product, the number of samples evaluated should be adequate to draw firm judgments about the variability of each parameter as well as their similarity. While it is generally recommended to use 6–10 lots to achieve adequate test power, assays that do not yield quantitative data suitable for statistical calculations may be performed on a smaller number of lots; however, developers should be aware that using a smaller number of lots increases the risk of lots failing.

These tests should cover the entire spectrum of pharmacological/toxicological properties regarded clinically important for the reference product and the product class when the reference product and the product class are used together. *In vitro* assays are often more specific and sensitive than animal studies when it comes to detecting differences between a biosimilar and the reference product, hence they could be considered crucial for nonclinical biosimilar comparisons.

For a variety of reasons, functional tests are employed to characterize protein products. These assays are used in conjunction with physicochemical evaluations to provide a quantitative assessment of the function of the protein component.

If the reference product has many functions, the developers should conduct a series of relevant tests to determine the full scope of the product's capabilities. Developers of proteins with many functional domains that express enzymatic and receptor-mediated activities, for example, should think about how important these activities are to the final product. The comparative characterisation of each parameter between products should be evaluated for things whose functional activity can be determined by more than one parameter (e.g. enzyme kinetics or interactions with blood clotting factors).

Developers should be aware of the limitations of various functional tests, such as excessive variability, which can make tiny but important differences between a biosimilar and the reference product difficult to detect. Because a highly variable test may be unable to determine whether a biosimilar product is identical to the reference product, assays that are less variable and responsive to changes in the product's functional activities are encouraged. In addition, *in vitro* bioactivity studies do not accurately reflect a protein's clinical activity. For example, these tests rarely predict a product's bioavailability (PK and biodistribution), which might affect PD and clinical outcomes. Furthermore, even minor changes in glycoform distribution or other post-translational modifications might significantly affect bioavailability. As a result, while evaluating the consistency of data showing biosimilarity and the need for additional knowledge to settle residual uncertainties, these limits should be considered. Finally, in both nonclinical and clinical trials, functional tests are critical for detecting the presence of neutralizing antibodies.

When binding is part of the role assigned to the protein component, analytical tests should be performed to identify a biosimilar product in terms of its specific binding properties (e.g. if binding to a receptor is inherent to protein function, this property should be measured and used in comparative studies).

Surface plasmon resonance, microcalorimetry, and basic Scatchard analysis are some of the methods used to determine the kinetics and thermodynamics of binding. This knowledge can be linked to the higher-order structure of a biosimilar product's functional behavior and characterization.

6.8 *In vivo* assessment

The fact that biotechnology-derived proteins can mediate *in vivo* effects that *in vitro* studies could not fully explain is a well-known scientific fact. As a result, if a specific *in vivo* model for the species or design is not available, nonclinical *in vivo* evaluation studies may be required to provide additional details.

6.8.1 Determination of the needs for *in vivo* studies

When determining the need for *in vivo* nonclinical studies, factors to consider include, but are not limited to:

- Quality attributes that are not included in the reference product (e.g. new post-translational modification structures).
- Quantitative differences in quality characteristics between a biosimilar product and the reference product that are theoretically significant.
- There are several formulation differences to consider, such as the use of excipients that are not often used in biotechnology-derived proteins.

Although *in vivo* testing is not always required, the following factors should be considered together to determine the level of concern and if *in vivo* testing is required.

Assume that a putative biosimilar product's physical and biological characteristics, as well as nonclinical *in vitro* testing, are sufficient to ensure the safety of human study. In that case, an *in vivo* animal study may not be necessary.

In vivo studies may be required if product-inherent factors that affect PK and biodistribution, such as extensive glycosylation, cannot be properly established on a quality and *in vitro* level. Following that, the developers must decide if these studies should be carried out on animals or on humans as part of clinical trials.

If more *in vivo* data are needed, the availability of a specific animal species or other relevant models (e.g. transgenic animals, transplant models) should be investigated. If no adequate *in vivo* animal model is available, the developers can use risk-mitigation principles to conduct human experiments.

6.8.2 *In vivo* animal studies

If an *in vivo* evaluation is required, the goal of any analysis (PK, PD, and safety) will be defined by the need for further information. To acquire the most information possible, animal experiments should be meticulously planned. When constructing any *in vivo* analysis, the 3Rs (replacement, refinement, and reduction) should be taken into account. It may not be necessary to kill the animals at the end of the trial, depending on the endpoints used. The length of the analysis (including the observation time) should be justified in light of the reference product's PK profile and therapeutic relevance.

The PK and PD of a putative biosimilar product and the reference product should be quantitatively examined where the model allows and if justified, including, if feasible, a dose–concentration–response evaluation considering the estimated exposure in humans.

Regardless of the species chosen, a versatile solution should be considered for safety investigations. The use of non-human primates in typical repeating dosage toxicity investigations is generally discouraged. An in-life assessment of safety criteria (such as clinical signs, body weight, and vital functions) or a repeated dose toxicity study with refined design (e.g. using only one dose level of biosimilar product and reference product, only one gender, and no recovery animals) can be considered if it is sufficiently justified. In repeated dose toxicity studies, when only one dose is studied, the dose is frequently selected at the high end of the dosing range and justified using the reference product's expected toxicity.

Conducting toxicity tests on organisms that are not relevant is not a good idea (i.e. to determine unspecific toxicity solely based on impurities). Qualitative discrepancies in process-related contaminants can develop because to the various production procedures used by biosimilar and reference product manufacturers (e.g. host cell proteins). Limiting pollutants to a bare minimum is the greatest method to reduce any associated threat.

Qualitative or quantitative alterations in product-related polymorphisms can affect the biological activity of biotechnology-derived proteins (e.g. glycosylation patterns, charge variants). To evaluate them, appropriate *in vitro* assays should be used. These variations and pollutants may have an impact on immunogenicity and hypersensitivity potential. Clinical trials should be done to further examine these findings, which are difficult to predict based on animal tests.

While animal immunogenicity is rarely predictive of human immunogenicity, it may be necessary in *in vivo* animal investigations. As a result, blood samples should be collected and processed for any pharmacokinetic/toxicokinetic data analysis that may be required.

Local tolerance studies are only required in rare cases. Local tolerance may need to be investigated if excipients are introduced for the first time or if the anticipated therapeutic route of administration is new. Local tolerance evaluation could be incorporated into the design of those studies rather than being done separately if further *in vivo* investigations are conducted.

Animal experiments could be used to help evaluate the safety of a biosimilar product as well as demonstrate biosimilarity between a biosimilar and a reference product.

6.8.2.1 Animal toxicity studies

An animal toxicology study may be required for the following reasons:

- If there are any remaining questions regarding the safety of a potential biosimilar product before clinical trials begin, animal toxicity data can help. The publicly accessible data and information on the reference product and the biosimilar product, as well as the level of discovered similarities and differences between the two, as submitted in a biosimilar application, will define the nature and scope of animal toxicity studies.
- Animal toxicity data are significant in the scientific community because, based on the results of detailed structural and functional characterization, there are still concerns regarding the safety of a potential biosimilar product that must be addressed before human clinical trials can commence (assuming results from animal studies can meaningfully address the remaining uncertainties).

Information about the reference product and information about a potential biosimilar product can be used to define the scope and duration of any animal toxicity studies based on the degree to which a biosimilar and a reference product are known to be alike or dissimilar. Agencies encourage the developer to give arguments for any toxicological testing in animals, including presenting adequate scientific grounds for any other research to address any remaining ambiguities from analytical assessment studies.

Animal toxicity studies are usually not useful if there are no animal species that can give pharmacologically suitable data for the substance (i.e. no species in which the product's biologic behavior mirrors the human reaction).

Regulatory agencies frequently refer to the ICH guidance for industry 'S6(R1) Preclinical safety evaluation of biotechnology-derived pharmaceuticals' (ICH S6 (R1)) to plan animal toxicology studies; however, the Agencies' guidance is intended for the evaluation of new biological products where the mode of action is unknown and a previous evaluation leading to marketing authorization is not required. As a result, ICH S6(R1) should be used as a guideline but not as the only foundation for developing a toxicological procedure. Multiple species animal toxicity testing is not advised.

A general toxicity investigation, which includes detailed animal pathology, histopathology, PD, PK, and immunogenicity analyses, may be required if the structural and functional data are restricted in scope or there are concerns about the safety of a proposed biosimilar product. Toxicology testing on animals are carried out in comparison to a standard product (i.e. comparative bridging toxicology studies). The dosage, regimen, time, and test species used in these tests should be able to distinguish the biosimilar from the reference product in terms of toxicology. It is crucial to recognize the limits of animal studies when interpreting findings comparing a possible biosimilar product to the reference product (e.g. limited sample size, intra-species variations). The dosing is a critical consideration given that there may not be any similarity in the dose–response in an animal species to the human dose–response. Since the purpose of these studies is to compare toxicologic symptoms, a dose too low to show a toxic response will not be useful, just as a dose that provides an overt toxic response. The developers should not base the dosing on human dose or devise any multipliers of human dose; instead, the developers will establish the dosing based on public information about toxicity to selected animal species. In such cases where these data are not available, a small-scale animal toxicology study at multiple doses may be required to establish an effective dosing range.

Some regulatory agencies make specific recommendations regarding using primate species and frequently discourage such species, in particular when the animals are eventually euthanized. Agencies maintain that unnecessary animal toxicity studies should be avoided, yet where such studies are essential, Agencies do not recognize any differential treatment of any animal species.

6.8.2.2 Animal PK and PD measures

A single-dose animal study comparing a biosimilar to the reference product using PK and PD data, if available, could supplement the totality of the evidence that supports a biosimilarity demonstration in some situations. Animal PK and PD testing is used to identify structural differences between a potential biosimilar and the reference product. Animal PK and PD assessments would not negate the need for human PK and PD studies. In most cases, only a few animals are required for PK/PD research in primate animal species. The major purpose is to indicate an overall similarity among the studied animals rather than any statistical comparability.

6.8.2.3 Animal immunogenicity testing

Animal immunogenicity assays, for the most part, do not predict human immune responses to protein products. Even though some regulatory agencies maintain that animal immunogenicity assessments represent possible structural or functional differences between a biosimilar product and the reference product that are not captured by other analytical methods, some regulatory agencies have approved immunogenicity testing in animals that would rarely fail to support biosimilarity. A limited number of studies has been conducted and given below are a few examples:

- Detection of anti-etanercept antibodies in *Cynomolgus* monkey serum of preclinical study 1 (under general toxicology) by an ECL bridging immunogenicity assay. Subcutaneous, 1 or 15 mg kg^{-1} administered twice per week.
- Epoetin showed a higher incidence of ADA development to the reference product compared to Epoetin in rats. The ADA is likely due to the route of administration (SC is more immunogenic than IV administration) and the presence of HAS in the reference product.
- Filgrastim was tested in rats with evaluation of potentially associated immunogenicity. Sprague Dawley rats: 10 rats/sex/group + 5 rats/sex/group in control and HD for biosimilar product or RP; TK control: 6 rats/sex/group; TK: 12 rats/sex/group; Doses tested: 0, 20, and 500 μg/kg; SC daily for 29 days,
- Infliximab efficacy, pharmacokinetics, and immunogenicity in the Tg197 transgenic mouse arthritis model. Ten treatment groups, 4/sex/group for efficacy study + 3 males/group for TK; Doses: 1, 3, or 10 mg/kg of a biosimilar product, RP; twice weekly by IP injection for seven weeks beginning at three weeks of age. A supplemental control group of 2 animal per sex was sacrificed at three weeks of age to initiate treatment.

6.9 Clinical pharmacology assessment

6.9.1 Introduction

In contrast to generic chemical drugs, where chemical equivalence is established, clinical pharmacology studies what the body does to the drug (PK) and what the drug does to the body (PD), profiling of which often cannot be adequately predicted from analytical assessment, functional assays, and nonclinical studies alone. Clinical pharmacology studies are used to determine any changes in the structure of the molecule that may effect the product's safety and efficacy due to the intricacy of the molecular structure.

The developers should understand that the aim of clinical studies is to compare a biosimilar product to the reference product, not to characterize the product (which is already done in the development of RP). This concept necessitates a unique strategy.

A traditional 'phase 1' study, as defined by the developers, is a stand-alone study designed to evaluate the disposition characteristics of a novel medicine. A typical 'phase 3' study is a placebo-controlled trial to determine safety and efficacy; neither of these criteria apply to biosimilars, hence the word 'phase' should be avoided. Comparative clinical pharmacology and comparative efficacy testing are the correct terms.

Clinical pharmacology investigations are, in essence, an extension of the analytical assessment, a biological tool for identifying structural differences that no other testing method can detect.

In essence, clinical pharmacology studies are an extension of the analytical assessment, a biological tool to identify the structural difference that is not possible to study by any other testing method.

6.9.2 Scope of studies

Agencies anticipate that the developers will perform comparable human PK studies (if the route of administration allows for such a study to be meaningful). To remove any lingering question about the safety and efficacy of a biosimilar product after all priors, PD studies (if appropriate PD measure(s)) as well as a clinical immunogenicity assessment are required. Clinical pharmacology trial results, such as those mentioned below, may be adequate evidence to support the claim of no clinically significant difference between a biosimilar and the reference medicine, allowing the biosimilar to be approved for marketing. If a biological product is administered by a route that does not result in a definable blood concentration profile, the developers may justify the waiver of PK studies. Less complex biological products that are well-characterized and where it is established that variations in immunogenicity do not affect clinical efficacy, the developers may justify a waiver of clinical immunogenicity studies, more particularly in the case of insulin products.

- A human PK study demonstrates that a biosimilar product and the reference product have similar pharmacokinetic properties. Exposure may be used to help a biosimilarity demonstration (e.g. serum concentration over time).
- Human PD study demonstrates a similar influence on crucial PD metrics associated to efficacy or major safety hazards, adding to the evidence for biosimilarity (except for immunogenicity, which is studied separately). Biosimilarity uncertainty can be decreased by analyzing responsive PD endpoints, even if adequate PD measures are not available. Combining PK and PD studies is a possibility.

In analyzing the similarity of a biosimilar product and the reference product, the PK and PD indicators are often more sensitive than clinical efficacy endpoints. For example, a comparison of two thyroxine products based on thyroid-stimulating hormone (TSH) levels would be more sensitive than a comparison based on clinical symptoms of euthyroidism. The same holds for evaluating filgrastim and erythropoietin products, where the endpoints of change in white blood cells and red blood cells respectively serve as a more objective tool to evaluate these products' efficacy.

Even if there is still some doubt regarding biosimilarity, the evidence from PK and PD studies can be used to provide a theoretical foundation for a selective and focused approach to comparative protection and efficacy research in the future.

6.9.3 Study plan

To detect variations in PK and PD profiles, comparative human PK and PD studies should use a sensitive population, dose(s), and route of administration (if available). However, there is a substantial difference between evaluating a biosimilar and a stand-alone biological product. Because the goal of a biosimilar product evaluation is to compare it to the reference product rather than to characterize the PK/PD profiles, the developers may choose a population that is least likely to produce highly variable results in order to reduce the population size and improve the study's ability to distinguish the biosimilar from the reference. This concept is distinct from the considerations that must be made when developing a new product.

The developer must justify the human PK and PD study population (e.g. patients versus healthy subjects) and parameters with scientific evidence, taking into account the population's relevance and sensitivity, the population and parameters studied for the reference product's marketing authorization, and the interdisciplinary team's current knowledge.

When developing study techniques, it is critical to think about how long it takes for a PD measure to change as well as the possibility of nonlinear PK.

Agencies also promote the use of modeling and simulation in the design of comparative human PK and PD studies. *In silico* PK/PD modeling is a valuable tool for reducing the need for clinical efficacy studies.

The clinical backdrop, protection, and PK characteristics of the reference product all influence the design of a PK analysis (target-mediated disposition, linear or nonlinear PK, time-dependence, half-life, and so on). Bioanalytical assays must also be appropriate for their intended purpose and adequately verified.

- *Patient versus healthy subject*: Clinical PK and PD experiments can be conducted on healthy subjects or patients if the chemical can be given safely. If the drug can be delivered safely, clinical PK and PD tests should be undertaken in healthy volunteers. Because it is predicted to provide less PK and PD variability, research in healthy participants is likely to be more sensitive in determining product similarity than a study in patients with possible confounding factors such as underlying and concomitant disease and concomitant drugs. Assume that for some items (e.g. immunogenicity or known toxicity from the RP), safety or ethical concerns prevent healthy people from participation in human PK and PD trials, or that PD biomarkers can only be useful in patients with the condition or disease in question, clinical pharmacology investigations on these patients should be conducted in this situation. A population that is representative of the patient group for whom the medicine is intended is ideal, but it is not required if the chosen population can demonstrate the likely difference between the biosimilar and the reference product.
- *The scope of the research project is as follows*: The overall number of participants tested should provide sufficient statistical power for PK and, if necessary, PD similarity analyses. Any post hoc statistical analysis should only be exploratory, and the data should be examined in accordance with the previously specified study plan. Given the scientific reasoning that both investigations will provide a better assessment of structural changes, the developers may prefer a well-selected sample to limit intra- and inter-subject variability. Several failed PK studies in the past required retesting in a restricted population, such as patients with a specified BMI, age, gender, and race.
- *Crossover studies*: A single-dose, randomized crossover analysis is commonly recommended for PK similarity analyses. For a product with a short half-life (less than five days), a rapid PD response (e.g. the period of beginning, maximal impact, and absence in combination with drug exposure), and a low

projected incidence of immunogenicity, a crossover study is indicated. This is the most sensitive approach for measuring PK similarity with a limited number of patients, and it can provide precise estimates of exposure disparities. The PD effect is delayed or not parallel to a single-dose medication's PK profile. A multiple-dose technique may be acceptable for PD similarity tests, and the presence and absence of immunogenicity, as well as their link to the washout phase, should be examined in crossover studies.

- Because biological products have a long half-life and trigger immunogenic reactions, several of them must be produced at the same time. A parallel-group strategy is useful for items with a long half-life or those that may evoke a stronger immune response, which could affect PK and PD testing. This technique is also appropriate for illnesses with treatment response changes over time.
- *Several studies*: In addition, the developers may undertake a smaller study known as a pilot study to assess the population size needed to obtain at least 80% power. In most cases, choosing a population with a specific demographic, such as age, gender, BMI, race, and other criteria available, will lower the population size requirement. Because the reference product would have been tested in a more diverse population, the PK/PD data are unlikely to be available in the public domain. Let us say follow-up research is needed, and the developer can match the study population, using the same lots of biosimilars and the reference product. The results from the supplementary study(ies) can then be merged to conduct the final bioequivalence analysis in this scenario. Before undertaking a follow-up study, the developers should obtain consent from the Agency.
- To reduce test subject exposure, the developers may adopt a study design that integrates the PK and PD studies with clinical immunogenicity testing. For example, to evaluate immunogenic qualities, a two-arm parallel research with two dose levels and a follow-up after the second (higher) dose may suffice. However, the developers must justify the research size, subject selection criteria, and dose selection in order to establish appropriate power.
- All clinical pharmacology investigations should be undertaken using biosimilar product components from the final manufacturing process that would be used for the marketed product if marketing permission is granted. All biosimilar products must be manufactured in one location. There may be more than one comparator in a study, but each comparator must be used throughout the investigation. During the creation of a biosimilar product, developers should be aware that the ICH Q5E concerns do not apply to biosimilars.

6.9.4 Dose selection

When assessing PK/PD or selecting healthy volunteers for testing, a lower dose on the steep section of the exposure–response curve is usually acceptable. Nonlinearity can make it difficult to establish true changes in PK/PD parameters, hence doses higher than clinical dosing are not recommended. Assume you are working with a group of patients on a research project. In that case, the reference product's permissible dosage may be the better option because it best describes the pharmacological effects in a clinical scenario. Model based simulations can be used to explain the dose chosen for the PK and PD study or studies using publicly available data for the dose– (or exposure)–response relationship of the reference product.

For determining clinical PK and PD similarities, a dose chosen from a range of doses may be useful in some cases. If the concentration–effect relationship for the reference product is regarded as extremely variable or nonlinear, different doses may be used to determine dose–response. When many doses are measured, a lower dose is utilized, at least one dose, possibly one-half of the therapeutic amount.

Assume that only patients have access to the product. A modified dosing strategy, such as a single dose for a chronic indication or a lower dose than the permissible dose, may boost sensitivity in identifying differences if the allowed dose induces nonlinear PK or exceeds the dose required for maximum PD impact. Several factors will determine if an alternative dosage regimen is appropriate, including whether the lower dose has been shown to have the same effect as the established dose and whether it is ethical to provide lower doses despite differences in impact. There should be a compelling justification for moving to a different dosing regimen.

6.9.5 Route of administration

A biosimilar product's administration route(s) must be the same as the reference products. Assume the reference product is approved for multiple administration routes (e.g. intravenous, and subcutaneous). In that situation, the approach for determining PK and PD similarity that is more sensitive to clinical inconsistencies should be used. Extravascular pathways may reveal PK differences that occur throughout the absorption, delivery, and removal processes. Subcutaneous or other extravascular modes of administration are likely to be the most vulnerable in most cases. Additionally, extravascular administration methods may allow for a more sensitive examination of immunogenicity variations.

6.9.6 Pharmacokinetic parameters

Bioequivalence studies are generally not required for generic chemical drug products administered by the subcutaneous or intravenous route since, by definition, they are considered 100% bioavailable. The purpose of a bioequivalence test for generic drugs is to compare the dosage form's property to release the drug at the action site. Once a drug has been released, the rest of the drug's time course from a generic product will be the same as expected for the molecules coming from the reference product. A biosimilar product and the reference product are chemically identical. The study design of a bioequivalence testing of a generic product is based on universal acceptance of variation (0.8, 1.25) because this testing is *not* intended to compare the product's safety or efficacy—both of which are already established.

The testing of the PK profile of biosimilar product against the reference product uses the same premise as the testing for generic chemical drugs because, just like the chemical generics, the purpose is to establish efficacy based on overall exposure that depends on the disposition profile that may vary because of subtle differences in the chemical structure of the molecule, differences in clearance induced by differences in immunogenic response, and other factors that might alter the safety and efficacy of the biosimilar product.

In those instances where the products are administered by routes other than intravenous, there may be differences in the absorption rates, such as the subcutaneous route. Still, in most cases, these will not be significant. However, where a delayed or prolonged absorption is anticipated, the PK profile will be appropriate for determining the differences; examples include long-acting insulin products.

The developers should contest the equivalence criterion (0.8, 1.25). The PK is used to identify potential variations in the way the reference product and a biosimilar product interact with the body, which may require change within a pre-specified acceptable range. When examining similarity, the location as well as the width of the confidence range should be taken into account. For example, statistically significant discrepancies in 90% confidence intervals within the justified acceptance range for critical PK parameters must be addressed and justified to avoid excluding biosimilarity. If the 90% CI, on the other hand, exceeds the established boundaries, the developers must explain the deviations and pinpoint the root causes. On a case-by-case basis, protein content modification may be acceptable provided it is pre-specified and sufficiently justified. The procedure includes the results of the test and reference product assays.

For both the biosimilar and the reference product, all PK measures should be taken. $AUC_{(0\text{-inf})}$ for intravenous administration and $AUC_{(0\text{-inf})}$ and typically C_{max} for subcutaneous injection are the important parameters in a single-dose PK sample. Secondary characteristics such as T_{max}, distribution volume, and half-life should also be estimated. In a multiple-dose trial, the shortened AUC from the first to the second administration ($AUC_{0\text{-t}}$) and the AUC across a steady-state dosage cycle ($AUC_{0\text{-ss}}$) should be the most important metrics. At steady state, C_{max} and C_{trough} are secondary parameters. The formula $AUC_{0\text{-}} = AUC_{0\text{-t}} + C_t/K_{el}$ (or C_t (concentration at the last observable time point) divided by K_{el}) is a good way to calculate (elimination rate constant). Cmax should be calculated straight from the data without the use of interpolation.

The primary endpoint in intravenous trials will be $AUC_{0\text{-}}$, and C_{max} and AUC will be considered co-primary study outcomes in subcutaneous studies. In multiple-dose studies, the primary endpoint is the region beneath the concentration–time profile from time zero through the end of the dosing period at steady-state ($AUC_{0\text{-tau}}$). Secondary endpoints include the concentration before the next dose during multiple doses ($C_{trough\text{-ss}}$) and C_{max}. To estimate PK similarity, using population PK data will not suffice.

Anti-drug antibodies should be assessed concurrently with PK assessment using acceptable sampling time points in every PK sample.

Table 6.10 lists the PK parameters and their significance.

Table 6.10. Relevant pharmacokinetic parameters.

Parameter	Significance
C_{max}	Rate of absorption, reaching minimum effective levels (dose and drug delivery system).
T_{max}	Rate of absorption with minimal impact of disposition because of faster absorption from parenteral administration (dose and delivery system).
$AUC_{0\text{-inf},\ 0\text{-ss},\ 0\text{-t}}$	Absorption, distribution, and elimination (structural difference, dose dependence).
V_{dss}	Distribution, receptor binding (structural differences).
dV_{dss}/dt	The thermodynamic potential of molecules to interact with tissue receptors, distribute and bind due to structural differences.
$AUCV_d$	Distribution, receptor binding (structural differences, dose).
K_{el}	Metabolism (structural differences, dose dependence).

6.9.7 New approach to PK analysis

The PK parameters mentioned above are the most widely used in PK profile comparisons. However, a new method involves using the diffusion volume to calculate a drug's time-course thermodynamics in the body. V_d is a pharmacokinetic measure that connects drug plasma concentrations to the amount of drug in the body. It is important for determining drug loading and maintenance doses because the amount of drug delivery following an intravenous bolus change over time. During this redistribution, an avascular phase and a washout or dilution phase with slightly distinct time periods occur. A medication is combined in the blood plasma volume second by second or minute by minute during the vascular process. During the dilution phase, hydrophilic medicines diffuse through the body's interstitial fluids over hours or days. Although most drugs actively redistribute from plasma into body tissues, plasma concentrations are primarily attributable to this recycling rather than actual drug removal. The following assumes first-order kinetics, which is the most common drug kinetic, in which drug removal is proportional to its concentration. The apparent volume of drug distribution at time zero (V_0), the apparent terminal volume of distribution following bolus intravenous administration (V_{area}), and the estimated volume of distribution, V_{ss}, which is the apparent terminal volume of distribution, similar to V_{area} in a bolus model, are the three most commonly measured volumes of distribution. The most commonly measured volumes of distribution are the apparent volume of drug distribution at time zero (V_0), the apparent terminal volume of distribution following bolus intravenous administration (V_{area}), and the estimated volume of distribution, V_{ss}, which is the apparent terminal volume of distribution, similar to V_{area} in a bolus model. V_{ss} is the predicted physical volume of the drug delivery in bolus experiments, and it is invariant between constant infusion and bolus experiments, unlike V_{ss}.

In a thermodynamic approach to employing the volume of distribution, a temporally variable apparent volume of the distribution model for the number of exponential functions based on mass conservation is utilized. This variable volume model argues that redistribution and the rate at which medication delivery expands in volume over time are inextricably linked. A model like this is employed when the medication volume surpasses a particular size around its apparent terminal delivery volume. A variable volume model like this could potentially reveal previously unknown information on the impact of time-based tissue drugs on tissue metabolism and eliminations. This version of the model also has consequences for medication impacts on body tissues, whether medical, detrimental, or radiation exposure.

The optimal instantaneous dose for obtaining the intended effector site concentration without overdosing the drug could be determined using a variable volume model. For practically any concentration washout fit characteristic, a time-dependent apparent volume of distribution model based on mass conservation could be applied. The elimination constant can be more sensitive to structural variants leading to differences in the degradation of the administered drug, and the distribution parameters can be sensitive to biological drugs that act by receptor binding. The appropriate parameters to include in a PK study are V_{dss} as a function of time, in addition to other V_d parameters.

The elimination rate constant, K_{el}, or half-life, is sensitive to how the body disposes of molecules; structural differences that are difficult to quantify using traditional methods may show differences in the metabolism of the administered drug, and this parameter should be considered a key consideration.

6.9.8 PK/PD waivers

Although comparing target-mediated clearance is critical in the biosimilarity process, it may not be feasible in patients because of the considerable variability in target expression, including fluctuation over time. Because *in vitro* studies are expected to show similar interactions between a biosimilar product and its target(s) (including FcRn in the case of a mAb), the absence of a pivotal PK study in the target population is acceptable if additional PK data are collected during the efficacy, protection, and PD studies, allowing for more investigation of the clinical effect of variable pharmacokinetics. You might use a subset of patients' PK profiles or population pharmacokinetics to accomplish this.

Biosimilars require comparative pharmacokinetic studies, which are frequently more sensitive than clinical effectiveness trials in detecting potential product differences. This may explain why finding different pharmacokinetic profiles may not be enough to overcome evidence of similar efficacy.

Many product categories (insulin, low molecular mass heparins, and (peg) filgrastim) no longer require comparative efficacy trials since crucial evidence for similarity can be obtained through physicochemical, functional, PK, and pharmacodynamic (PD) comparisons. Complex, multifunctional biologicals are an exception, necessitating biosimilar development to include comparative efficacy and safety clinical trials in humans.

6.9.8.1 Route of administration

If the reference product may be administered intravenously and subcutaneously, the subcutaneous administration evaluation would normally cover both absorption and removal.

If a biosimilar product is comparable in both absorption and removal for the subcutaneous path, the assessment of intravenous administration may be waived. When the molecule has an absorption constant that is much slower than the removal constant, the omission of the PK analysis of intravenous administration must be explained (flip flop kinetics).

While most biological drugs are administered systemically, a special case situation arises for drugs such as ranibizumab, bevacizumab, and aflibercept which are administered intravitreally. Since the site of action is aqueous humor, this is the most desirable sampling site; systemic testing shows that ranibizumab appears in the smallest concentration. It is cleared fast, followed by aflibercept and then bevacizumab. The developers are encouraged to develop an appropriate animal testing model using an *in silico* approach to suggest a waiver of systemic PK/PD studies. Additionally, since ocular immune privilege protects the eye, significantly lowering immunogenic response, the direct immune response cannot be tested. However, as the drug is cleared through systemic fluids, it is important to establish the ADA response.

6.9.9 Pharmacodynamic parameters

A well-designed clinical PK and PD analysis compares a biosimilar product's PK and PD characteristics to those of the reference product in a biosimilar development program. A well-designed clinical PK and PD trial should contain information on biological product exposure and, if possible, exposure–response, which is critical for determining if the two drugs vary clinically. Using a wider panel of PD biomarkers that capture the product's various pharmacological effects (e.g. performing a protein or mRNA microarray analysis) may be beneficial. The following five characteristics should be considered when deciding which biomarkers to use to test response:

- The initiation of a shift in the PD biomarker after dosing and when dosing is stopped the biomarker returns to baseline.
- The PD biomarker's dynamic spectrum over the biological product's exposure range.
- The PD biomarker's sensitivity to discrepancies between a biosimilar product and the reference product.
- The PD biomarker's significance in the drug's mode of action (to the extent that the mechanism of action is known for the RP).
- The PD biomarker assay's analytical validity.

In other cases, the qualities of PD biomarkers are uncertain. However, because these PD biomarkers could be employed as orthogonal tests to support similarity, it is recommended that developers use PD biomarkers with a high dynamic range spanning the concentration range of the PK assessment.

When PD biomarkers are not sensitive or specific enough to identify clinically significant variations, the PK parameters that arise should be employed to compare clinical pharmacology. Biomarkers for Parkinson's disease can be utilized to enhance the PK data. A combination of PK and PD similarities can be useful in proving that a biosimilar and the reference product have no clinically relevant differences.

Where it is possible, pharmacodynamic (PD) markers should be used in pharmacokinetic studies.

Comparative PK/PD trials may be required in some cases, as specified in the testing requirements below, to demonstrate clinical comparability between a proposed biosimilar product and the reference product:

- The specified PD marker/biomarker has been accepted as a surrogate marker. In the sense that a similar impact on the PD marker ensures a similar clinical outcome, it is linked to patient outcome. An absolute neutrophil count can be used to determine the effect of granulocyte-colony-stimulating factor (G-CSF). An early viral load reduction in chronic hepatitis C can be used to evaluate the effect of alpha interferons, and a euglycemic clamp test can be used to compare two insulins. In multiple sclerosis, magnetic resonance imaging of disease lesions can be used to study interferons.
- There may be PD markers that are not approved as efficacy surrogates but are important to the active substance's pharmacological action and have a strong dose–response or concentration–response relationship. One or more dose–exposure–response trials at two or more dose levels may be enough to avoid a clinical effectiveness study in this case. This design allows a biosimilar and a reference product to be compared inside the steep part of the dose–response curve.
- Once a dose–response or systemic exposure–response relationship has been found, it is critical to explore a dose(s) on the steep region of the dose–response curve of a biosimilar product whenever possible (the response may be PD measurements or clinical endpoints). Analyzing doses around the dose–response curve's plateau is unlikely to reveal differences between a biosimilar and the reference product.
- When non-surrogate PD/biomarkers are used in PK trials to see if there is a clinically significant difference, the strategy should include a recommendation for the size of the equivalence margin(s), as well as steps to show a comparable safety profile.
- The features of the PD biomarkers (e.g. the PD response timing following administration of the product based on the product's half-life and the projected period of the product's action) would dictate the allowed time points and durations for assessing PD biomarkers.
- An examination of multiple-dose and steady-state settings may be crucial when a PD response occurs after the start of product administration, especially if a proposed therapy is intended for long-term use.
- A comparison should be made between the region under the effect curve for the PD biomarker(s) assessed for a biosimilar product and the reference product (AUEC). Due to the features of the PD biomarker, if only one PD

measurement is available, it should be connected to a simultaneous drug concentration measurement. The relationship between drug concentration and the PD biomarker can therefore be utilized to compare treatments.

- The use of a single, scientifically acceptable PD biomarker, such as the one mentioned above, or a composite of multiple specific PD biomarkers, such as the one mentioned above, will assist in resolving any remaining doubts about whether there are clinically meaningful differences between products and will significantly contribute to overall biosimilarity demonstration. It may also be advantageous to use a larger panel of biomarkers to capture the substance's various pharmacological effects (e.g. by performing a protein or mRNA microarray analysis).
- Crossover design is preferred for PD trials including drugs with a short half-life (e.g. less than five days), a rapid PD response, and a low incidence of immunogenicity. For products with a longer half-life, a parallel design is frequently necessary (e.g. more than five days). The developers should have a logical reason for the research dose (e.g. one dose or many doses) and administration route they choose.
- For PD measurements, the optimum sampling technique may differ from that for PK measurements. When it comes to PK sampling, the most effective way to characterize the concentration–time profile is to repeat sampling with decreasing frequency at early time points after product administration. The PD–time profile, on the other hand, may differ from the PK–time profile. In these instances, PD sampling should be justified. When collecting both PK and PD data in a clinical pharmacology sample, the sampling approach should be tailored for PK and PD data.

If physicochemical, structural, and *in vitro* biological tests, as well as human PK studies and a combination of PD markers that represent the pharmacological activity and concentration of the active drug, can provide robust evidence for biosimilar product protection and efficacy, the confirmatory clinical study may be waived in some cases.

6.9.10 Statistical treatment of PK and PD results

In PK and PD trials statistical analyses are used to determine the clinical pharmacology similarity of a putative biosimilar medicine and the reference product. The suggested clinical pharmacology similarity examination includes (i) comparison criteria, (ii) a trust interval for the criterion, and (iii) a biosimilarity assessment acceptable limit. Before performing the statistical analysis, it is recommended that the exposure measurements be log-transformed. To compare PK and PD parameters for replicates and non-replicate design studies, the developers can use an average equivalency statistical approach. (For more information, see the FDA's 'Statistical approaches to establishing bioequivalence' guidance.) Calculating a 90% confidence interval for the ratio between the geometric means of the parameters of a biosimilar product and the reference product is part of the average equivalence technique. The measured confidence interval must be within a suitable range to determine PK and PD similarities. The confidence interval and reasonable restrictions will differ depending on the product. If additional limits are offered, the developers must justify why the biosimilar product's limits were chosen. If other limits are proposed, the developers should justify the limits selected for a biosimilar product. A reasonable starting point for an acceptable limit for the confidence interval of the ratio is 80%–125%; if other limits are proposed, the developers should justify the limits selected for a biosimilar product. Sometimes the results of PK and PD tests diverge from the predetermined parameters. Agencies advise developers to analyze and clarify such findings before moving on to the next stage of the development process, since they may show the presence of underlying differences between a biosimilar product and the reference product.

The developers should be aware that differences in the PK/PD profiles of biosimilar products when compared to reference products are generally due to bioactivity, referring primarily to the disposition profile, which can be extremely important. As a result, PK/PD studies are more sensitive than comparative efficacy studies. The developers are not allowed to submit comparative clinical efficacy and safety testing until any remaining questions concerning failed PK/PD comparison studies are resolved.

One may question the relevance of using the same equivalence intervals for all types of products as a 'one-size-fits-all' approach. However, it should be realized that each stepwise study of a biosimilar product against the reference product is a means of showing a lack of dissimilarity rather than establishing absolute similarity. It is correct that the dose–response of biological products varies widely, and the equivalence interval may be too wide or too narrow concerning the clinical response. Yet the PK/PD profiling is not directly related to clinical efficacy, only one of many critical steps.

It is well known that differences in immunogenicity can cause significant changes in the clearance of therapeutic proteins—such differences may be associated with subtle structural differences that cannot detect using available analytical methodologies.

6.9.10.1 Utility tools

When preparing a PK and PD analysis, modeling and simulation tools can be beneficial. These methods could, for example, aid in the selection of the most informative dose or doses for assessing PD similarity. When employing a biomarker-based comparison, the chosen dose should be in the steepest part of the reference product's dose–response curve.

Assume that the reference product's exposure–response statistics are not available. In that case, the developers may decide to gather this information through a small study in order to determine the most beneficial dosage (e.g. a dose representing the effective dose to achieve the 50% maximal response (ED50) the RP). In this small study, the PK/PD relationship will be investigated at various dosage levels (e.g. the lowest, moderate, and highest permitted doses). Alternatively, if it is feasible, the developers should undertake a PK/PD similarity comparison between the reference medicine and the biosimilar product using the lowest, medium, and maximum authorized dosages with a strong dose–response. If many doses are being investigated, PK/PD parameters such as EC50,

E_{max}, and the slope of the concentration–effect relationship should be compared to see if they are comparable. When clinical pharmacology assessment is expected to be the main source of knowledge for determining clinically meaningful discrepancies, such investigations may be useful for showing PK, PK/PD, and PD similarities. Publicly available data on biomarker–clinical endpoint relationships, as well as modeling and simulation, can be used to determine acceptable PD similarity levels.

6.9.11 Assay considerations

6.9.11.1 Specific assays

To produce biosimilar products, three types of assays are particularly important: ligand binding assays, concentration and activity assays, and PD assays:

- It is crucial to use the right bioanalytical approaches to compare the PK and PD parameters of a biosimilar and the reference product when assessing clinical pharmacology similarity. Traditional analytical methods cannot be utilized to analyze biological products due to their complex chemical structure. In PK and PD research, bioanalytical techniques that are reliable, accurate, specific, responsive, and repeatable should be used.
- The developers should choose or build an assay based on a thorough understanding of the mechanism of action (to the degree that the reference product's mechanism of action is known) and structural properties of the biosimilar and reference products that are essential for operation. It is ideal to use an assay that generates concentration data that corresponds to pharmacological/PD findings. A biosimilar's and a reference product's concentrations should be assessed using the same assay that has been validated for both. Analytical tests should follow industry best practices in terms of design and performance parameters.
- The developer should use the most relevant assays and methodologies to obtain results that are meaningful and represent the PK, biological activity, and PD impact of a biosimilar product and the reference product. In addition, developers should explain why they chose an assay and why it is significant to drug activity in submissions to Agencies.

6.9.12 Reserve samples

Reserve samples provide the legitimacy of the items examined in the real study, allow for the validation and consistency of results, and make it easier to investigate any follow-up questions that arise after the tests are done. It is recommended that biosimilar product developers keep reserve samples for at least five years after the application is approved or for at least five years after the completion of a comparative clinical PK and PD study of the reference product and the biosimilar product if the application is not approved (or another clinical study). For a three-way PK similarity investigation, samples of both comparative and biosimilar commodities should be retained.

For most protein therapies, the following quantities of material and dosage units are recommended, as they are expected to be suitable for evaluation by state-of-the-art analytical methods:

- A minimum of ten dosage units of a biosimilar product, reference product, and, if applicable, another comparator product is required, depending on the amount of product in each unit. In general, a total product mass of equal to or more than 200 mg should be obtained in a volume of equal to or greater than 10 ml.
- In the following scenarios, it is recommended that developers contact Agencies to discuss the required amounts of reserve samples:
  1. Many dosage units are required when the product mass is equal to or greater than 200 mg in a volume of equal to or greater than 10 ml.
  2. Other biological products than medicinal protein products.

6.9.13 Examples of pharmacokinetic studies

Pharmacokinetics research has proven to be a big stumbling block in the clinical biosimilarity process, one that must be resolved. Before any further clinical evidence may be considered satisfactory, all findings must be completely justified. The developers may send *in silico* PK studies that could eliminate the need for larger efficacy studies.

According to EPARs and EMA product-specific criteria, PK studies in healthy volunteers are appropriate for the following molecules: teriparatide; low molecular mass heparin; insulin; interferon-; pegfilgrastim; somatropin; follitropin-; epoetin; etanercept; trastuzumab; bevacizumab; adalimumab; and teriparatide. For rituximab, the EMA only recommends a PK analysis in one therapeutic area and an efficacy/safety trial (including PK data) in the other.

The EPARs released a number of failed studies. Initial PK studies for Cyltezo (adalimumab), Hyrimoz (adalimumab), Ziextenzo (pegfilgrastim), Terrosa (teriparatide), Grastofil (filgrastim), and Efgratin (pegfilgrastim) failed completely.

6.9.13.1 Adalimumab

In healthy volunteers, two examples of a failed and successful PK study using adalimumab biosimilars have been published. Changes in glycan structures known to effect PK (high mannose content) were argued to be too minor in both cases to explain the initially reported PK differences, since only high mannose content of at least 20% may modify systemic exposure due to enhanced receptor-mediated clearance. In one of these examples, the initial failed PK research was conducted with a clinical trial formulation that differed from the commercial formulation in the buffering procedure. After controlling for confounders such as body weight and protein content, post hoc trials were unable to provide an adequate explanation for the observed disparities. The commercial version was employed in a second, enhanced PK study. A larger subject sample size, a consistent injection site, and the established factors of body weight and age were used to account for the considerable PK variability.

Researchers investigated batch collection, investigational medicinal product (IMP) storage and transport, IMP preparation, IMP administration, PK sampling, PK sample shipping and testing, the impact of body weight and anti-drug antibody (ADA) growth, and other subject characteristics in the other case. The developer, on the other hand, was unable to establish the fundamental cause of the PK study's failure.

As a result, a second PK research with a different study design was performed in order to diminish inter-subject variability (body mass index (BMI) restriction, the inclusion of only male subjects, and increased sample size). Prefilled syringes made IMP handling and dosing easier, and they did not require IMP compounding as in the previous trial. The similarity of PK may be illustrated with this upgraded design.

6.9.13.2 Pegfilgrastim

Pegfilgrastim is associated with a lot of PK variability. Based on normal comparability margins of 80%–125%, two of the six marketing authorization applications for the biosimilar pegfilgrastim failed PK studies.

The dose–exposure relationship for pegfilgrastim is highly disproportional, with a ten-fold increase in dose resulting in a 75-fold increase in exposure in healthy subjects, and linear models to correct for protein content are ineffective. Special attention should be paid to administering the same dose of test and reference product.

Surprisingly, a high level of PK variability does not imply a high level of PD variability. The exposure–response relationship, on the other hand, is flat, resulting in highly similar PD responses (i.e. ANCs) even in cases of high PK variability or failed PK similarity, making PD endpoints less sensitive than PK endpoints in detecting possible discrepancies between two pegfilgrastim-containing items.

One way to design pegfilgrastim pharmacokinetic studies is to select a population of highly selective volunteers; this is not phase 1 study. The purpose is to compare, not characterize, a projected patient population. All failed studies missed this valuable advice.

6.9.13.3 Trastuzumab

The upper bound of the predefined equivalence margins for the primary endpoint (pCR) for two trastuzumab biosimilars did not formally cross the upper bound of the predefined equivalence margins for the primary endpoint (pCR) in patients with human epidermal receptor-2 (HER2)-positive EBC/locally advanced breast cancer, confirming non-inferiority but not formally ruling out the possibility of superiority. The overall contribution of ADCC activity versus antiproliferative effects through suppression of ligand independent HER2 signaling to trastuzumab's therapeutic effectiveness is uncertain and has not been concluded as the cause. There were no clinically significant variations in the safety profile, such as cardiac toxicity as determined by left ventricular ejection fraction and symptom occurrence.

6.9.13.4 Rituximab

For the EU reference product and US Rituxan, quality attributes (such as charge variations, glycan structures, and ADCC) were observed. Despite this, because they were both on the market at the same time, both versions of the reference product were declared safe, dependable, and eligible for use in comparability studies.

6.10 Clinical immunogenicity assessment

6.10.1 Introduction

Immunogenicity or adverse immune response refers to a therapeutic biologic's ability to induce immune responses to itself and linked proteins, as well as to cause immunologically related nonclinical consequences or adverse clinical outcomes. In the development of therapeutic biologics, there are two types of immunogenicity: intrinsic and extrinsic. Intrinsic is based on the structure and extrinsic on impurities such as degradation products.

The desired immunogenicity of vaccines is frequently mentioned. An antigen (vaccine) is injected into the body to elicit an immune response to a pathogen to defend the organism (virus, bacteria, cancer cell, etc).

Immune responses to therapeutic biologics can also neutralize their biological activities and cause adverse events, not only by inhibiting their efficacy but also by cross-reacting with an endogenous protein counterpart, causing the endogenous protein's physiological function to be lost (e.g. neutralizing antibodies to therapeutic erythropoietin cause pure red cell aplasia by also neutralizing the endogenous protein). In this review, immunogenicity refers to the latter adverse immune response in therapeutic biologics research and development.

Antigenic processes are carried out by antigen presentation cells (APCs) such as DCs, macrophages, and B-cells. Antigen capture involves delivering antigens to the cellular antigen processing machinery, and antigen processing and presentation involves attaching antigenic peptides to MHC molecules and presenting them to adaptive immune cells. Extracellular antigens are collected by APCs through phagocytosis, macropinocytosis, and receptor-mediated endocytosis. In the acidic environment of the endosome or lysosome, antigens are degraded into many immunogenic peptides that carry T-cell epitopes. MHC class II molecules are produced in the endoplasmic reticulum and transported to the Golgi apparatus, where they combine with antigen peptides to form peptide–MHC II complexes that are delivered to the surfaces of APCs. TCR identifies peptide–MHC II complexes, which activates T-cells and causes an immunogenic response. The affinity of peptide–MHC complexes determines the quality of antigen presentation, and there is a direct relationship between peptide–MHC complex stability and the immunogenic response. APCs' ability to bind to CD4+ T-cells is also determined using the DC–T-cell assay.

Immune responses to therapeutic biologics can have a significant impact on their efficacy and safety. While many of these immunogenic responses have been reduced in humanized mAbs, there is still cause for concern. The first generation therapeutic mAbs were of murine origin, resulting in severe immunological reactions in patients due to the antibodies; while many of these immunogenic responses have been reduced in humanized mAbs, there is still cause for concern.

Anti-drug antibodies (ADAs) are produced in response to unfavorable immune responses and can result in several clinically significant outcomes (table 6.11), including anaphylaxis, cytokine release syndrome, and cross-reactive neutralization of endogenous proteins that mediate critical functions.

Table 6.11. Clinically relevant effects of immunogenicity.

Effects on bioavailability
Effect on safety and efficacy
Potential cross-reactivity with endogenous proteins affects PK
Inhibition of the function of endogenous protein
Injection site reactions
Mild or life-threatening systemic reactions
ADA (HAMA, HACA, HAHA) creation
Formation of neutralizing antibodies
Formation of immune complexes
Formation of anti-idiotypic antibodies

When ADAs have been identified, further characterization of the antibodies, such as immunoglobulin type in the case of acute hypersensitivity, can be beneficial. It may also be able to further classify clinically relevant ADAs or to determine a 'threshold' level of ADAs that has a significant impact on efficacy and safety.

The most important thing to remember when designing therapeutic biologics is that immunogenicity is a pharmacokinetic covariate. When immunogenicity develops against a therapeutic biologic, clearance increases and exposure to that medication decreases. The immunogenicity of therapeutic biologics can be influenced by both patient and product-related factors. These elements are critical in determining immunogenicity risk (table 6.12) and must be considered when establishing immunogenicity testing methods.

6.10.2 Immunogenicity investigation

Because animal research and *in vitro* models may not always predict human immune responses, immunogenicity should be investigated in the target population. Immunogenicity is crucial for demonstrating product comparability after manufacturing changes and for demonstrating similarities after a product is licensed in the context of manufacturing process changes. Nonetheless, for evaluating biosimilar products in the early stages of development, this methodology is ineffective. Minor differences, such as a biosimilar product's immunogenicity, can also affect bioactivity, effectiveness, or protection.

Impurities, heterogeneity, aggregation development, oxidation, and deamidation in therapeutic biologics can all be evaluated physiochemically or through formulation based features to better anticipate immunogenicity and design less immunogenic therapeutic agents. Furthermore, predicting possible immunogenic epitopes in therapeutic biologics is a crucial and successful technique for enhancing safety and efficacy. During the production of therapeutic biologics, several preclinical immunogenicity evaluation techniques are available, as shown in table 6.13.

Table 6.12. Immunogenicity associated factors.

Category	Example
Treatment-associated	Mechanism of action Route of administration Frequency of administration, duration of therapy
Patient-associated	Disease type Disease status Immune system function Genetic factors Concomitant disease Concomitant medications Prior exposure Prior sensitization
Drug property associated	Recombinant expression system Post-translational protein modifications Impurities Contaminants Aggregates

Table 6.13. Strategies for managing the immunogenicity of therapeutic biologics.

Prediction	Reduction
Physiochemical characterization	Deimmunization (epitope modifications)
In silico immunogenicity assessment	Humanization
T-cell epitope predictions	
B-cell epitope predictions	
In vitro immunogenicity assessment	Purity and formulations
Ex vivo immunogenicity assessment	Purity and formulations
T-cell response modifications	
HLA binding assays	Fusion proteins
In vivo immunogenicity assessment	Combination biologics or combination therapy

6.10.2.1 Assay

In clinical immunogenicity evaluation, the developers can test the following antibody parameters as a scientific matter:

- Titer, specificity, related isotype distribution, the time course of growth, longevity, absence, effects on PK, and clinical sequelae are all factors to consider.

- Product operation neutralization: power neutralization for all related functions (e.g. uptake and catalytic activity, neutralization for replacement enzyme therapeutics).

Before beginning any clinical immunogenicity evaluation, the developer should check with Agencies about the sufficiency of assays.

Even in circulating medicinal products, developers can create tests that can identify immune responses with high sensitivity (biosimilar product and RP).

A putative biosimilar product and the reference product should be tested in the same assay using the same patient sera when possible. Early in the development process, immunogenicity tests should be developed and validated, with the validation considering both the biosimilar and the reference product.

As part of a multi-tiered testing process, screening assays, confirmatory assays, titration assays, and neutralization tests should all be developed and validated. Antibodies that bind to therapeutic protein products are detected via screening assays, also known as binding antibody tests. According to confirmatory testing, ADAs are specific for the therapeutic protein. The degree of the ADA response is determined using titration tests. ADA is tested for neutralizing behavior in neutralization assays. In order to avoid false-negative results, it is vital to use the precise cut-point in assays.

Concurrently, binding ADA tests should be developed employing a biosimilar or reference product as the capture ligand. Each test should be able to detect ADA even when the drug is present. To demonstrate ADA cross-reactivity against a biosimilar product and the reference product, samples from both treatment arms should be evaluated for ADA using both assays. Any deviation from this method should be backed up with data.

- IgG and IgM ADA tests with a sensitivity of at least 100 nanograms per milliliter (ng ml^{-1}) are recommended for screening and confirmation. However, based on danger and recent experience, a sensitivity limit of more than 100 ng ml^{-1} may be justified.
- In an experiment, a lack of specificity could lead to false-positive results. This article discusses the challenges of confirming the specificity of antibody reactions to mAb, Fc-fusion proteins, and Ig-fusion proteins, as well as how to overcome them.
- The results can be skewed by interference from the matrix or onboard therapeutic protein products. A non-specific signal can be produced if selectivity is not maintained, obscuring a positive result.
- Assay accuracy is crucial in estimating ADAs since assay variability is the foundation for evaluating the cut-points and ensuring that low-positive samples are identified as positive.
- If a sample is being tested by many laboratories, the developers should ensure that the results are comparable in terms of assay sensitivity, drug tolerance, and precision.
- Because bioassays are sensitive to changes in assay settings, factors such as cell passage number, incubation times, and culture media components must

all be assessed and optimized. To retain antibody reactivity, a strategy for short- and long-term stability is required.
- Consider throughput, sensitivity, selectivity, dynamic range, the ability to detect numerous Ig isotypes, the ability to detect fast dissociating antibodies, and reagent availability when choosing an assay format. Washing frequency and intensity should be considered as well, as they can affect assay sensitivity and epitope exposure.
- If the assay is to be effective, positive control antibodies, negative controls, and device compatibility controls must all operate consistently.
- Reporting the findings: The appropriateness of approaches should be evaluated; qualitative methods are the most common.

Table 6.14 lists the common screening assays, and table 6.15 lists the assay methods for antibody testing.

6.10.2.2 Lifecycle management

In the absence of clinical consequences, differences in immune responses between a putative biosimilar product and the reference product can be problematic. It is possible that they are worth investigating further (e.g. an extended period of follow-up evaluation).

The time course for the generation of immune responses (such as the development of neutralizing antibodies, cell-mediated immune responses) and expected clinical sequelae (informed by experience with the RP) should be determined by (1) the time course for the disappearance of immune responses and clinical sequelae following the cessation of therapy, (2) the time course for the disappearance of irradiation, and (3) the time course for the disappearance of irradiation following the cessation of therapy. A one year follow-up period is recommended for chronically prescribed substances unless a shorter period can be scientifically justified based on the entire body of evidence confirming biosimilarity.

The developer can then create a lifecycle management report that includes these parts as the product progresses through the stages:

- *Risk evaluation for immunogenicity*: This should be tailored to the therapeutic protein and provide information on consistency and subject factors.
- *Summaries of the bioanalytical approach and assay validation for each clinical development process*: The developers can outline the evaluation strategies for each clinical development phase.
- *Clinical research design and comprehensive immunogenicity sampling plans*: For all clinical studies with an evaluation done, the developers should include the sampling plans.
- *Clinical immunogenicity data analysis*: For all clinical studies with an immunogenicity aspect, the developers should describe immunogenicity analyses.
- *Conclusions and risk assessment and mitigation strategies*: The developers should talk about how immunogenicity affects the protein product's protection and effectiveness in the subject population and how it would be controlled in the post-marketing process.

Table 6.14. Commonly used screening assays.

Type of assay	Advantages	Disadvantages
Direct/indirect ELISA	High throughput, low cost. It is simple to use and automate. There is a high therapeutic tolerance in the solution phase. Generic reagents and instruments	It is possible that it will bind in a non-specific way. High background potential. Antigen immobilization can change the structure of the antigen and mask epitopes. Low-affinity antibodies may go undetected. In solid-phase therapy, there is a low therapeutic tolerance. Requires a secondary reagent that is particular to the species.
Bridging ELISA	High throughput, low cost. It is simple to use and automate. Low-contrast background. Specificity is high (dual-arm binding). It is possible to employ it on a cross-species basis. Instruments and generic reagents.	Antigen labeling has the potential to change the antigen. Low-affinity antibodies may go undetected. Multivalent targets, very vulnerable to interference by therapeutic serum components, such as anti-human Ig molecules. IgG4 and IgM antibodies may not be detected.
Electrochemiluminescence (with direct/indirect bridging format)	Throughput is high. Extremely wide dynamic range. Matrix has a minor impact. Tolerance to treatment is high. Consistent detection signal over the life of the TAG conjugate.	Two antigen conjugates may be required (indirect). Antigen labeling has the potential to change the antigen. Therapeutic serum components, such as anti-human Ig molecules, can interfere with multivalent targets. Vendor-specific equipment and reagents may not detect IgG4.
Radioimmunoprecipitation assay	Throughput is moderate. Exceptional sensitivity. Specificity is possible. Inexpensive	It is possible that it is isotype-specific. Low-affinity antibodies may go undetected. A radiolabeled antigen is required. Antigen stability may be affected by radiolabel degradation.

(*Continued*)

Table 6.14. (*Continued*)

Type of assay	Advantages	Disadvantages
Surface plasmon resonance	Throughput is moderate. Specificity, isotype, and relative binding affinity are all determined. Allows both low-affinity and high-affinity antibodies to be detected. Tolerance to treatment is high. There is no need for a detection reagent.	Antigen immobilization may have an impact on treatment efficacy. The antigen may be degraded during the regeneration process. The sensitivity of the assay may be lower than that of the binding assay. Expensive. Equipment and reagents customized to each vendor.

Table 6.15. Methods for the detection of neutralizing antibodies.

Type of assay	Advantages	Disadvantages
Cell-based bioassay	Functional assay that reflects the treatment effect's mechanism of action. It is possible that it has something to do with clinical outcome.	It's a time-consuming process. It's possible to have a complicated protocol design. Frequently erratic. Interfering factors and serum (matrix) effects have an impact. Therapeutics may cause interference. Validation of cell lines, reagents, and other items might be difficult.
Competitive ligand binding assay	Rapid design of a simple assay. It is relatively simple to use. Does not necessitate the use of cell lines. Simple to create and test.	Antigen labeling may alter antigen susceptibility to interference by therapeutics. May not represent true functional read-out. May not correlate with clinical response. It is possible that clinical response has nothing to do with it.

6.10.2.3 Immunogenicity testing methods

Minor differences in immunogenicity that do not correspond at the quality level and have no detrimental impact on clinical effectiveness (lost or reduced efficacy) or safety may be acceptable. It can be difficult to determine the clinical impact of a reported difference in immunogenicity due to the small sample size and short follow-up period. Assume the clinical relevance of the detected difference is unknown, for example, because the potentially serious adverse impact is rare or the immune response evolves gradually. In that situation, post-marketing surveillance or investigations, which are accessible at EMA for monoclonal antibodies, may be required, as well as an update to the risk management plan.

Immunogenicity studies look for associations between ADAs and pharmacokinetics, pharmacodynamics, efficacy, and safety. As a result, immunogenicity should be considered part of the preparation of pivotal clinical trials, including synchronized sampling for ADAs and related biomarkers, if available, as well as an assessment of efficacy and safety.

6.10.2.4 Population

To compare immunogenicity, a suitable population should be chosen. Immunocompetence, previous or concurrent use of immunosuppressive drugs, and historical evidence on the immunogenicity of the reference product should all be considered when choosing an acceptable population. Since immunogenicity is usually studied as part of a pivotal comparative safety and efficacy study, the above considerations must be considered when designing the program's clinical component to demonstrate biosimilarity.

If immunogenicity studies are needed, the form of study should be justified based on the observed difference(s), route of administration, dose–response curve, therapeutic window, and possible clinical effects. The effectiveness, protection, and immunogenicity study's target population must be responsive to variations in immunogenicity and their effects and reflect the population(s) for whom the product is intended. The samples should be tested after approval in high-risk circumstances.

In several instances, a healthy population will be a preferred choice if the drug treatment, previous or concurrent, can suppress the immune response; a good case in point is evaluating filgrastim products preferred in a healthy population.

6.10.2.5 Side-by-side assessment

The goal of a comparative immunogenicity analysis is to rule out any immunogenic differences between a biosimilar and the reference product that are clinically meaningful. Antibodies that modify PK, cause anaphylaxis, or neutralize the product and its endogenous protein equivalent are of special concern. The frequency and amount of the ADA response, the time course of ADA growth, the persistence of ADA, and the influence of ADA on protection, efficacy, and PK should all be described in the comparative study(s) for each treatment arm.

To develop the biosimilar product, side-by-side comparative immunogenicity studies are needed. The molecules are strongly immunogenic, and there is a greater probability of detecting variations in immunogenicity if the initial physicochemical and *in vitro* research suggests a difference.

Immunogenicity testing should be done in conjunction with pharmacokinetics, safety, and efficacy testing. Unless there is a compelling reason not to, patient samples that test positive for binding ADA in confirmatory binding assays should be investigated for their ability to neutralize the drug. It is crucial to choose a neutralizing antibody testing method that takes the drug's mode of action into account. Competitive ligand binding assays or cell-based assays may be appropriate depending on the mechanism of action.

6.10.2.6 Endpoints

Immunogenicity endpoints or PD measures for immunological responses to therapeutic protein products (e.g. antibody production and cytokine levels) should account for any immunogenicity concerns that have arisen because of the usage of the reference product. Clinical immune response criteria (e.g. descriptions of severe clinical outcomes such as anaphylaxis) should be specified prospectively for each type of predicted immune response, using existing standards, when applicable. Before beginning the report, they should get consent from the agencies on these parameters.

6.10.2.7 Sampling schedule

The drug's risk assessment will dictate the frequency of sampling, as well as the time and scope of tests. Patients who are transiently positive should have a different sampling regimen than those who have a long-term antibody response. The post-treatment sampling period should be long enough to draw conclusions about the immune response elicited by the therapeutic protein's persistence, as well as to identify an immune response that the therapeutic protein blocks. The half-life of the protein and the drug tolerance of the ADA test define the time of the post-treatment sample(s).

More frequent sampling is required in the early stages of treatment, when patients are most at risk of developing ADA. Long-term immunogenicity surveillance with fewer samples contributes to our understanding of the progression and implications of immunogenicity. Immunogenicity proof for one year of treatment should be available prior to licensing in the case of continuous chronic treatment, but shorter follow-up is allowed with adequate reason.

Immunogenicity in patients should be examined on a regular basis with repeated sampling, as well as symptom-driven with an additional sample when an unfavorable immune response is identified.

A variety of product-related characteristics can alter the immune response to a therapeutic protein. As a result, the sampling schedule for detecting an immune response should be tailored to the pharmacokinetics (e.g. drug tolerance) and ADA-determined drug tolerance of each product. A baseline sample should be taken in all cases.

Considering the experience of similar products as well as relevant regulatory and scientific publications, the developer should characterize the kinetics of the ADA response and any immune-mediated adverse effects using readily recognized terms. Because residual levels of the active component in the sample can interfere with the assay, samples should be taken before delivering the product during treatment.

In the marketing authorization application, the developers must explain how they assessed immunogenicity for each channel if the product has several administration routes.

6.10.2.8 PK dependence

The ADAs have the potential to influence pharmacokinetics, particularly during the elimination phase. Non-neutralizing, 'binding' antibodies may alter rather than just reduce the effectiveness of a substance, for example, by extending the half-life. A change in pharmacokinetics could signal the start of antibody synthesis. As a result, all repeat dose studies should provide concomitant screening for both pharmacokinetics and immunogenicity, according to the developers.

6.10.3 Safety and efficacy correlation

Intermittent treatment's immunogenicity should be assessed using risk factors such as previous experience with similar products, potential immunogenicity hazards, boosting effect, and antibody persistence or appearance following exposure.

The presence of ADAs can be clinically significant or not. Clinical research should focus on identifying potential hazards as well as methods for detecting and minimizing them. A risk analysis, which incorporates prior experience with the product (class), potentially immunogenic structures in the protein, and the patient group, should be utilized to prepare for the evaluation of immune-mediated adverse effects. Patients with pre-existing ADAs may have a different efficacy and safety profile, thus if at all possible, they should be studied individually. The study plan should include a description of symptom complexes associated with acute or delayed hypersensitivity, autoimmunity, and effectiveness loss. Any potential immunological side effects should be considered in the risk management strategy.

6.10.4 Management of immunogenicity testing

Despite the greatest efforts of the inventors to select molecules with low immunogenic potential, a negative immune response to a therapeutic protein cannot always be avoided. If possible, developers should search for ways to mitigate the unfavorable impact of immunogenicity discovered during clinical development in such cases. In some circumstances, immunosuppressive or anti-inflammatory co-medication can assist to prevent or lessen immunological side effects. In certain cases, such as with coagulation factors, tolerization regimens, such as administering higher doses of a therapeutic protein, can be used to re-establish immunological tolerance. Such therapeutic regimens should be recorded in clinical trials.

6.10.5 Examples of studies

6.10.5.1 Infliximab

Although the biosimilar (Flixabi) and reference product had equivalent ACR20 response rates at week 30, ADA rates measured with a highly sensitive assay were around 5%–12% higher in the biosimilar cohort at the individual time points of determination (with about 50% of patients in a biosimilar cohort determined ADA positive at any time in the trial). There was no significant influence on any of the efficacy characteristics investigated because the primary endpoints were within the established comparability margins.

6.10.5.2 Etanercept

While the product (Benepali) passed all biosimilarity testing, the overall ADA formation at week 24 was significantly different. The clinical significance of the difference in ADAs appeared to be minor, and after eight weeks of treatment the difference had mostly dissipated.

The idea that PK investigations are more sensitive than clinical trials in detecting potential product-related variations could explain why a finding of equivalent efficacy could not overcome the PK discrepancies. This could be interpreted as being unduly strict. Efficacy trials, however, are determined not just by drug exposure but also by the biological substance's pharmacological effect *in vivo*. As a result, the goals of both types of investigations are different.

6.11 Clinical efficacy assessment

6.11.1 Introduction

Efficacy trials for a possible biosimilar product are not designed to prove efficacy in and of itself because this has already been established with the reference product. Efficacy trials are used to see if a biosimilar and a reference product have similar clinical outcomes.

Generally, 'additional clinical studies may be required' if there remains any residual uncertainty after completing the analytical assessment, nonclinical, and clinical pharmacology studies. Whereas in most cases a single study in one of several approved indications may be sufficient to establish biosimilarity, where multiple modes of action are involved a single study in one indication alone may or may not be sufficient to remove 'residual uncertainty'. The developers are strongly urged to consult with Agencies before conducting clinical efficacy testing to make sure that the selected protocol is suitable to remove any identified residual uncertainty.

6.11.2 Residual uncertainty

If there is any doubt about whether there are clinically significant differences between a biosimilar product and the reference product based on structural and functional characterization, animal testing, human PK and PD results, and clinical immunogenicity evaluation, comparative clinical research will be required to support a proof of biosimilarity. If the developers conclude that a comparative clinical study is not needed, they should include a scientific rationale.

A biosimilar product tested in a stepwise manner undergoes identification and removal of residual uncertainty at each step before moving forward to another testing in the following order: analytical, functional, *in vitro*, *in vivo*, clinical pharmacology, and immunogenicity. While most biosimilars developers routinely conduct clinical efficacy testing agencies consider unnecessary patient exposure not appropriate and encourage them to meet with Agencies to understand what the Agencies consider as the remaining residual uncertainty.

In most cases, at this stage of development, a failed PK/PD/immunogenicity testing will reject a biosimilar product for a biosimilar status. The developer may choose to refile the applications as a new biological drug. The residual uncertainties identified at this stage are structural, functional, or nonclinical testing, where the variation may not be evaluated fully for impact on safety and efficacy.

6.11.3 Waivers

PK/PD research can be adequate clinical study for marketing authorization if there are relevant PD endpoints and a good understanding of the mechanism of action. Comparative efficacy and safety clinical trials in humans are also considered crucial components of potential biosimilar production for complex, multifunctional biologicals. The need for patient testing stems from the complexity of interactions caused by the molecule's size, various moieties with various functions (e.g. Fab/Fc-parts), multiple mechanisms of action, the impact of glycosylation pattern, and the potential for immunogenicity and potentially life-threatening adverse effects (AEs).

As more biosimilar products are approved globally, and safety and efficacy data are becoming available to regulatory agencies, there is a trend developing to question any comparative efficacy testing relevance. Agencies expect the developers to adduce arguments in favor of avoiding clinical efficacy testing that does not result in a meaningful comparison, expose patients to avoidable risks, and do not establish a bridge between any multiple indications approved for the reference product

To rule out clinically significant variations in efficacy and safety between a biosimilar and the reference product, a comparative clinical trial is usually required. For example, a clinically meaningful PD endpoint may not necessitate a clinical efficacy trial. In such circumstances, a scientific reason for conducting efficacy research is required, but safety and comparative immunogenicity data are still needed.

6.11.4 Study types

The comparative safety and efficacy testing is conducted using three types of study designs.

In most cases, an equivalence design can be used. A non-inferiority design may be appropriate if it is supported by strong scientific evidence and takes into account the features of the reference product, such as its safety profile/tolerability, dosage range, and dose–response relationship. On empirical and mechanical grounds, a non-inferiority trial should only be considered if the potential of a major and clinically meaningful improvement in effectiveness can be ruled out. As with equivalence trials, however, assay sensitivity must be considered.

Previous research using the reference product could have shown a link between the 'hard' clinical endpoints provided by new active ingredient recommendations and other clinical/pharmacodynamic endpoints that are more sensitive to identify clinically significant differences. In this case, using the identical primary efficacy objectives as those used in the reference product's registration application is not a good idea. However, it is a good idea to add some common endpoints (as secondary endpoints, for example) to make comparisons with the reference product's clinical trials easier.

The developers are advised to consult the EPARs and BLA review documents relating to biosimilar products to establish the choice of the model for establishing clinical safety and efficacy. In the absence of such data, the developers are required to suggest to Agencies a model of testing for marketing authorization by Agencies before the conduct of the study. However, neither the Agencies nor the developers are bound or required to use any testing model published or submitted and accepted by any regulatory agency.

6.11.4.1 Traditional comparative (two-sided) study

The null hypothesis in a typical study notes that there is no difference between the biosimilar and the reference product. The study hypothesis of the disparity in efficacies bears the burden of evidence. If the proof for a distinction is insufficient, equality cannot be ruled out. This model is not appropriate for evaluating biosimilars.

6.11.4.2 Equivalence testing

Equivalence testing, in which the null hypothesis states that a biosimilar is not equal to the objective of proving equivalency, and the presumption of proof is placed there. Nonequivalence cannot be ruled out if the proof in favor of equivalence is insufficient. In measuring equivalence, the null and research assumptions are those of a conventional comparative study reversed. This is the most used model, except that the M_2 margin of difference is based on professional judgment, which can be disputed. The word 'equivalent' refers to the efficacy of two similar treatments that neither is considered superior or inferior to the other. This idea is formalized by the equivalence margin, a constant that defines a set of values for which the efficacies are 'similar enough' to be considered equal. In practice, the margin is the highest clinically permissible disparity that one is willing to consider in exchange for the secondary benefits of the new treatment.

An inherent disadvantage in using an equivalence model lies in the choice of difference (M_2) considered acceptable; this difference is established mainly on clinical judgment since objective studies will not be available to consult. The total response (M_1) is drawn from the reference product data.

In comparative efficacy trials, choosing equivalence margins is more difficult. The largest variations that would not be clinically significant are included in the predetermined margins. If dose-related toxicities occur or the dose used is above the plateau of the dose–response curve, the margins do not need to be symmetric. There is a slim chance of dose-related side effects. In most trials, such as the testing of Remsima for ACR20 at 30 weeks and Samsung Bioepis's infliximab biosimilar, a margin of 15% is indicated. However, the adalimumab 'similar biologic' approved in India (Exemptia) (Jani H *et al* 2016 A prospective, randomized, double-blind, multicentre, parallel-group, active controlled study to compare efficacy and safety of biosimilar adalimumab (Exemptia; ZRC-3197) and adalimumab (Humira) in patients with rheumatoid arthritis *Int. J. Rheum. Dis.* **19** 1157–68 https://www.ncbi.nlm.nih.gov/pmc/articles/PMC5215647/) used a 28.5% equivalence margin, allowing for much smaller sample size and a margin of 23% in India for an infliximab similar biologic. When comparing anticancer products, the margin option becomes more complicated because the result is binomial and therefore difficult to predict.

6.11.4.3 Non-inferiority testing

The null hypothesis of non-inferiority testing is that a biosimilar product is inferior to the reference product. According to the study hypothesis, the new therapy is either equal to or superior to the existing therapy. Only one margin (the lower or upper limit, depending on what is suitable for the study or endpoint) is used in a non-inferiority model. It usually necessitates a smaller sample size than an equivalence trial. The confidence interval represents the level of uncertainty in a statistical parameter, such as the difference between two treatment effects (such as risk difference) or the ratio. The risk ratio, for example, may be used as a primary efficacy parameter if the overall response rate (ORR) is the primary endpoint. The confidence interval for a population parameter differs from the point estimate. Point estimates of unknown population parameters include sample means, ORR, and median survival time. The non-inferiority trials do not rule out the possibility of a potential biosimilar product's increased activity correlated with further adverse events.

The impact size of the reference product and clinical judgment are utilized to establish comparability margins when statistical modeling is used. They should represent the most significant difference in effectiveness that would be unimportant in practice. Treatment discrepancies outside of this range would be tolerated because they have little clinical impact. Acceptable equivalence margins for the patient population, endpoints, backbone therapy, predicted treatment impact, and minor discrepancies found in comparison studies

A significant criticism of relying on non-inferiority studies to establish biosimilarity comes from the very nature of the need for such studies, when considered justified, on the rationality of the study protocols.

Second, when non-inferiority testing is conducted, it requires establishing a response (M_1) and an acceptable difference that can only be arbitrary, given the observed high variability of biological responses.

Agencies are open to suggestions on *in silico* pharmacokinetic studies and other modeling studies that may prevent the need for comparative efficacy studies. The developers are encouraged to minimize testing in patients to secure faster and low-cost marketing authorization biosimilars. In our opinion, such studies will be limited only to a few highly complex drugs with mixed modes of action and where other assessments cannot be matched well.

6.11.4.4 Extrapolation justification

Biosimilar products receive extrapolation of all indications approved for the reference product as of the marketing authorization date. However, suppose a biosimilar product is tested in a comparative efficacy trial. A single study is one of many available indications where the modes of action can differ among the indications. In that case, a question arises regarding the suitability of a single comparative efficacy testing to allow such extrapolation.

6.11.4.5 Ethics and practicality
In most cases, a suitable patient population is difficult to recruit, such as in the case of anticancer drug testing, where the patients have inevitably been exposed to many drugs, and it is unethical to put patients through treatment regimens of monotherapy that are not in the best interest of the patient. In several instances, the patients may not survive until the end of the protocol, creating a dilemma for both the developer and the patients.

6.11.5 Selection of study protocols

Within set comparability margins, the comparative clinical study should be sensitive enough to rule out clinically important changes. When designing a sufficiently responsive clinical sample, the developers should consider the following factors:

- The population(s) features under investigation (e.g. underlying disease, immune competence).
- Study length, route of administration, medication regimen, clinical endpoint (s), and evaluation time are all characteristics of clinical research.
- Immunogenicity's dangers and consequences.
- The influence of concurrent therapies (e.g. monotherapy versus combination therapy).
- Relevant comparability margins are used.

It may be appropriate to evaluate more than one sensitive group in some cases.

The length, mode of administration, pharmaceutical regimen, clinical endpoint (s), and evaluation period of clinical research are all factors to consider.

Immunogenicity's hazards and ramifications.

The effect of several treatments (e.g., monotherapy vs. combination therapy).

Relevant comparability margins are employed.

- The existence and complexity of the reference product, the depth of structural and functional characterization, and the results and limitations of comparative structural, functional, and nonclinical research, including the degree of discovered differences, are all things to think about.
- When combined with information of the reference product's MOA and disease pathology, the degree to which differences in structure, function, and nonclinical pharmacology and toxicity predict clinical outcomes.
- The ability of human PK or PD to predict clinical outcomes (e.g. PD measures known to be relevant to effectiveness or safety).
- The amount of clinical experience with the reference product and its therapeutic class, as well as the safety and risk–benefit profile (e.g. if off-target adverse effects are uncommon), as well as accepted safety and effectiveness endpoints and biomarkers, are all significant factors to examine (e.g. availability of established, sensitive clinical endpoints).

The creators should provide a scientific reason for how they plan to use these features to identify what type(s) of clinical research, as well as any study design, are required. Consider the following scenario: you need to conduct a comparative clinical trial. In this case, the developers might describe how the study's design was influenced by endpoint(s), population, similarity margin, and statistical analysis.

Furthermore, unique safety or effectiveness concerns about the reference product and its class (such as a history of manufacturing- or source-related adverse events) may necessitate a greater amount of comparative clinical evidence. Assume, on the other hand, that information about other biological products is available that may be used to determine biosimilarity (with marketing histories that demonstrate no apparent differences in clinical safety and effectiveness profiles). In that case, such data could be utilized to justify a more targeted and tailored treatment strategy.

To establish comparable clinical efficacy of a biosimilar and the reference product using efficacy endpoints in the absence of surrogate effectiveness markers, a suitably powered, randomized, parallel-group comparative clinical study(s), ideally double-blind, is often required. The study population should reflect the approved therapeutic indications for the reference product and be sensitive enough to detect differences between the biosimilar and the reference product. Clinical process changes, such as the use of a concomitant medicine in combination treatment, the line of therapy, or the severity of the disease, may result in a departure from the approved therapeutic indication. Deviations must be justified and reported to the appropriate regulatory bodies.

6.11.5.1 Study design

The trial's design, as well as the key effectiveness outcomes and clinical margins, should all be thoroughly examined. Each of these factors is critical and should be supported by clinical evidence. The analysis should be undertaken using a clinically acceptable and responsive endpoint to establish that there is no clinically significant difference between a biosimilar product and the reference product. The endpoint of the reference product differs from the study's original endpoint (e.g. a well-established surrogate or a more sensitive endpoint). In all circumstances, an acceptable comparability margin should be established based on publicly available historical data and the reference product's smallest effect size. When using multiple endpoints, the criteria outlined above should be followed.

Because they obey the similarity principle, equivalence experiments are often employed. Non-inferiority tests must be justified, and the developer should consult with regulatory agencies before proceeding with the study. Such trials show that a biosimilar product outperforms the reference product statistically, which should be recognized by the developers. In such cases, the observed superiority should be evaluated for clinical significance as well as any potential safety concerns. The product can no longer be considered a biosimilar if the superiority seen is clinically relevant and is associated with increased adverse drug responses when compared to the reference product. Furthermore, demonstrating non-inferiority of a biosimilar drug to the reference product may not provide good rationale for the approval of additional indications, particularly if the additional indications involve dosages that were not investigated in the clinical research.

6.11.5.2 Efficacy endpoints

In a comparative clinical study, endpoints can be used to assess clinically relevant differences between a proposed biosimilar and the reference product. If the results are statistically supported, they may differ from those used as primary objectives in the RP's clinical trials. Some endpoints are more sensitive than clinical endpoints, such as Parkinson's disease measures, allowing for more exact assessments of relevant therapy effects. In some cases, several PD measures in a comparative clinical trial may improve the study's sensitivity. The effectiveness of the endpoints is determined by the degree to which PD measures correspond with clinical result, the quantity of structural and functional evidence supporting biosimilarity, the interpretation of MOA, and the significance or severity of the outcome affected.

While the hard clinical outcome remains the most desirable endpoint, other clinical endpoints that require shorter study times have been used widely (table 6.16).

The creators must demonstrate that the model they chose is appropriate and sensitive enough to detect changes in efficacy and protection. Differences in efficacy between a potential biosimilar and a reference medicine should always be studied to see if they are clinically meaningful. Clinical data are used to correct small flaws discovered in previous steps and guarantee that a biosimilar and a reference product perform similarly in clinical studies. Significant disparities in quality attributes cannot be justified by clinical evidence.

(See ICH subject 'E9 Statistical principles for clinical research' and CHMP guideline CPMP/EWP/2158/99 for further information on the non-inferiority

Table 6.16. Clinical endpoints in biosimilar testing.

For a granulocyte-colony-stimulating factor, the absolute neutrophil count (ANC) is used (G-CSF).
Insulin-stimulated blood glucose concentrations in clamp studies.
Complete pathological response (pCR) in breast cancer.
Disease activity score (DAS)-28 versus American College of Rheumatology (ACR)-20 in rheumatoid arthritis (RA) disease.
Objective response rate (ORR) in solid tumors and lymphoma.
Factor X and anti-factor II activity, magnetic resonance imaging-related endpoints for interferon-β.
Use of serum calcium levels for teriparatide.
Bone mineral density (BMD) together with serum C-terminal crosslinks (CTX), a bone resorption marker, as co-primary efficacy endpoints for denosumab to treat and prevent osteoporosis.
α4-integrin receptor saturation for natalizumab as the binding is linked directly to clinical outcomes.
Serum lactate dehydrogenase levels and for eculizumab.

margin.) To pre-specify and justify comparable margins on both statistical and clinical grounds, the reference product's data, as well as any comparative clinical research designs and test sensitivity, should be employed.

6.11.5.3 Clinical safety

Clinical safety is a key issue throughout the clinical development process, and it is assessed during the initial PK and PD assessments, as well as any comparative clinical efficacy studies. Depending on the nature and level of acknowledged safety risks for the reference product, comparative safety data should be acquired before approval. The length of pre-authorization protection follow-up should be warranted. Comparing the form, intensity, and frequency of adverse events between a biosimilar product and the reference product, as documented in the literature, should be done with caution. In the application dossier, the developers should assess the pertinent hazards connected with a potential biosimilar product. This section summarizes any potential safety concerns that may arise because of a manufacturing process that differs from that of the reference products, such as infusion-related reactions and immunogenicity.

To comply with all existing standards, immunogenicity monitoring of a biosimilar and reference product should be performed as part of a comparative efficacy testing exercise for a biosimilar product, utilizing the same assay design and sample schedule. Analytical assays should be done in parallel (blinded) using both the reference product and the biosimilar product to measure the immune response to the product acquired by each patient. Antibodies to both the biosimilar and the reference product should be detectable in the analytical assays. They should, however, be able to recognize all antibodies generated against a biosimilar product's molecule. Antibodies and antibody titers should be evaluated and reported based on the frequency and type of antibodies (e.g. cross-reactivity, target epitopes, and neutralizing activity). They must be evaluated and understood in terms of their clinical efficacy and safety implications.

The length of the immunogenicity test should be justified on a case-by-case basis, taking into account the length of the therapy, the product's absence from circulation (to avoid antigen interference in the assays), and the time it takes for the humoral immune response to manifest (at least four weeks when an immunosuppressive agent is used). The duration of follow-up should be justified based on the time course and characteristics of unwanted immune responses identified for the reference product, such as a minimal risk of clinically relevant immunogenicity or no substantial tendency for immunogenicity to rise over time. In most cases, pre-authorization for chronic administration requires one year follow-up data. A shorter follow-up data pre-authorization time (e.g. six months) may be suitable based on the immunogenicity profile of the reference product. If necessary, immunogenicity data for up to a year can be submitted after approval. For the individual, consult product-specific biosimilar recommendations.

Increased immunogenicity versus the reference medicine could bias the benefit–risk analysis and call biosimilarity into question. Lower immunogenicity for a potential biosimilar medicine, on the other hand, is a possible situation that does not rule out biosimilar approval. Assume that the biosimilar prevents neutralizing antibodies from developing. In that case, the efficacy analysis of the entire sample group could lead to the incorrect conclusion that the biosimilar is more productive than the reference product. It is also a good idea to set aside time for an exploratory subgroup analysis of efficacy and safety in patients who did not develop an anti-drug antibody response during the clinical trial. Assume that the efficacy of a biosimilar and the reference product is unaffected by an immune response. This subgroup study could help determine whether they are theoretically identical in that situation.

6.11.5.4 Population research

Due to the research population's selection, a clinically suitable comparison of a putative biosimilar medicine to the reference product should be achievable. Frequently, the sample population shares the same characteristics as the group that was evaluated for the reference product's marketing authorization for the same indication. However, in some cases, the sample population used in clinical studies that led to the reference product's marketing clearance may differ, if, for example, a genetic predictor of response was developed after the reference product had been approved.

6.11.5.5 Study sample size and length

The sample size and duration of the comparative clinical trial should be sufficient to detect clinically significant differences between the biosimilar and the reference product. Some endpoints, such as Parkinson's disease measures, are more sensitive than clinical endpoints, allowing for a quicker study. When the size and length of a comparative clinical research are insufficient for finding meaningful safety signals, a separate evaluation of safety and immunogenicity may be required.

6.11.5.6 Analyses and design of the study

For a new biosimilar development program, a comparative clinical investigation should be planned to identify if there are any clinically relevant differences between the biosimilar and the reference product. The type and degree of residual uncertainty regarding biosimilarity should be included in the design based on evidence from comparative structural and functional characterization, animal research, human PK and PD investigations, and clinical immunogenicity assessment.

A clinical research or studies aimed to give statistical confirmation that a biosimilar product is neither inferior nor superior to the reference product by more than a (potentially different) predetermined margin are generally expected by Agencies. An equivalence design with symmetric inferiority and superiority margins will be the most prevalent. For example, where there are dose-related toxicities, symmetric margins might be appropriate.

An asymmetric interval with a higher upper bound to rule out supremacy than a lower bound to rule out inferiority would suffice in some instances. An asymmetry interval may be appropriate if the dose utilized in the clinical investigation is near the dose–response curve plateau. There is a remote possibility of dose-related adverse effects (e.g. toxicity). Utilizing an asymmetric interval provides for a smaller sample size than using symmetric margins in most circumstances. If strong superiority can be demonstrated, more thought should be given to whether a biosimilar product can be considered a biosimilar to the reference product.

Depending on the sample population and outcome, inferiority alone may be sufficient to demonstrate that there is no clinically significant difference between a biosimilar and the reference product(s). It would be unethical to utilize doses lower than those permitted by the FDA. A non-inferiority (NI) design will suffice if it is widely known that doses of the reference product pharmacodynamically saturate the target at the clinical dose level. Empirical evidence should be used to support the research design, study population, study endpoint(s), estimated effect size for the reference product, and margin(s) (how much difference to rule out). Before starting comparative clinical research, the developers should meet with the agencies to discuss their study proposal(s) and overall clinical development plan.

6.11.5.7 Clinical endpoints

In one or more of the indications for which the reference product is approved, a clinical study or studies are necessary to demonstrate the efficacy, purity, and potency of a biosimilar product. Immunogenicity, pharmacokinetics (PK), and, in certain instances, pharmacodynamics (PD) are usually assessed, and a comparative clinical efficacy analysis may be included. Clinical data should resolve any minor variations in analytical similarities, nonclinical pharmacology, and PK and PD (where possible in healthy subjects) to validate the clinical performance of a biosimilar and the reference product, according to EMA guidelines. On the other hand, medical evidence cannot be used to explain significant variations in quality attributes. The developers may choose to withdraw their biosimilar application and file it as a new biologic, a strategy that worked for Teva, for its filgrastim product that was ultimately approved as a new biologic (Granix). It is worth noting that in exploiting this path the developer is not required to provide any comparative data with the reference product, but the literature data can be used to justify safety and efficacy claims to some extent.

The success of efficacy trials is determined not just by drug exposure (PK profile) and by the biological substance's ability to perform pharmacologically *in vivo*. As a result, the goals of both types of studies are different. Efficacy trials are commonly configured as equivalence trials (or non-inferiority trials) to ensure that the efficacy of a biosimilar is lower or higher than the reference product. However, in rare cases, some residual ambiguity about a biosimilar's possible increased efficacy might be appropriate. Biosimilarity and protection were concluded based on data from other assessment exercises.

The clinical endpoints (table 6.16) used in biosimilar comparability studies should be able to identify potential clinically significant discrepancies between a biosimilar candidate and its reference product and should be able to quantify any unconfounded pharmacological effects. The clinical endpoints, on the other hand, do not have to be the same as those used in the formulation of the RP. If the endpoints are responsive enough to show any clinically significant difference, they should be appropriate.

Clinical efficacy comparability endpoints should ideally assess unconfounded pharmacological effects and be sensitive enough to detect clinically relevant variations between a biosimilar candidate and its reference product. In this regard, hard clinical objectives such as overall survival are rather insensitive, and disease

Table 6.17. Suggested comparative clinical study endpoints.

Pathological complete response (pCR) in breast cancer.
Disease activity score (DAS)-28 versus American College of Rheumatology (ACR)-20 in rheumatoid arthritis (RA) disease.
Objective response rate (ORR) in solid tumors and lymphoma.
Absolute neutrophil count (ANC) for granulocyte-colony-stimulating factor (G-CSF).
Blood glucose concentrations in clamp studies for insulins.
Low molecular weight heparins. Factor X and anti-factor II activity.
Magnetic resonance imaging-related endpoints for interferon-β.
Serum calcium levels for teriparatide.
Anti-factor X and anti-factor II activity.
BMD and serum C-terminal crosslinks (CTX), a bone resorption marker, were used as co-primary efficacy endpoints for denosumab, a monoclonal antibody used to treat and prevent osteoporosis.
For natalizumab, α4-integrin receptor saturation as the binding is linked directly to clinical outcomes.
For eculizumab, a potential PD marker to study biosimilarity is serum lactate dehydrogenase levels because of the sustained reduction observed in intravascular hemolysis for the treatment period in a reduced need for red blood cell transfusions and less fatigue.

and patient-related factors commonly influence them. Agencies encourage developers to offer novel, confirmed clinical markers rather than patients' responses as a better approach (table 6.17).

6.12 Extrapolation of clinical data across indications

Assume that a biosimilar product meets the biosimilarity and other regulatory criteria for biosimilar product marketing authorization, based on evidence from a clinical study or studies that show, among other things, protection, purity, and potency in an appropriate state of use. In that case, the developers would need to file for biosimilar product marketing authorization for one or more additional conditions of use not covered by the reference product.

The developers, on the other hand, must provide sufficient scientific support for extrapolating clinical evidence to assess biosimilarity for each condition of use for which marketing approval is sought.

For the tested and extrapolated conditions of use, such a scientific basis for extrapolation should address, for example, the following considerations:

- The MOA in each condition of use for which marketing authorization is requested; this may include:
    - Each related activity/function of the product's target/receptor(s).
    - When a target/receptor is engaged, the binding, dose/concentration-response, and pattern of molecular signaling change.

  - Relationships between the product structure and interactions between targets and receptors.
  - The expression of the target/position receptor(s).
  - The PK and biodistribution of the product in various patient populations (important PD measurements can also reveal useful details about the MOA).
  - Immunogenicity of the product in various patient populations.
  - Anticipated toxicity differences in each state of use and patient population (including whether expected toxicities are related to the pharmacological activity of the product or off-target activities).
  - Any other factor that may influence the product's protection or efficacy in each use condition and patient population for which marketing approval is sought.

Extrapolation is not always achievable due to differences in the conditions of use for the factors indicated above. A scientific justification should address these disparities in the context of the totality of the data to establish a proof of biosimilarity.

When deciding which condition of use to study, it is suggested that the developers consider choosing a condition of use that is sufficiently sensitive to detect clinically meaningful differences between a biosimilar product and the reference product, allowing clinical data to be extrapolated to other conditions of use.

The developers of a biosimilar product are required to obtain marketing authorization for all conditions of use that have been previously approved for the reference product at the time of applying. If the reference product receives marketing authorization of additional indications, the developers must add those indications before or after marketing authorization. However, if an indication is protected under IP laws, the developers may request fewer indications and then add more indications as the IP expiration allows them.

6.12.1 Extrapolation across indications

The reference product has a wide range of clinical applications. Extrapolation of clinical data to other indications of the reference product after biosimilarity has been proven in a comparative trial may be permissible, but it must be clinically justified. More evidence will be required if it is unknown whether the protection and efficacy exhibited in one indication will be applicable in another. When extrapolating, all data should be evaluated, including consistency, nonclinical, and clinical data. It is thought that the safety and efficacy can be extrapolated when rigorous physicochemical and structural studies have proven biosimilar comparative and *in vitro*, functional testing paired with clinical data (efficacy and safety, as well as PK/PD data) in one therapeutic indication. Additional data are required in certain situations, such as:

- The reference product's active ingredient interacts with many receptors, which can have differing impacts on clinical indications that have been investigated versus those that have not.
- There are several active sites in the active drug, and the sites can have different effects on different clinical indications.

- The researched therapeutic indication is not important for the others in terms of efficacy and safety, i.e. it is not subject to disparities in all relevant aspects of efficacy and safety.

Several factors influence immunogenicity, including administration route, dosing regimen, patient-related factors, and disease-related factors (e.g. co-medication, type of disease, immune status). As a result, immunogenicity differs from one indication to the next. It must be justifiable to extrapolate immunogenicity from the studied indication/route of administration to other applications of the reference product.

6.12.2 Additional conditions of use

After product-market authorization, agencies understand that a biosimilar product application holder may choose to seek marketing authorization for an additional condition(s) of use. While this option is generally available in many jurisdictions, Agencies allow a biosimilar product to add any new indications allowed to the reference product in the future, provided there were no changes made to the reference product. If an indication is protected under a patent, it is up to the developers to judge whether the patent is applicable in its region and solely responsible for litigation. Marketing authorization by Agencies does not constitute an opinion regarding intellectual property associated with the reference product.

6.13 Conclusion

Biosimilars present the fastest-growing new category of biopharmaceuticals where efficiency of manufacturing cost is pivotal. The cost of all pharmaceutical products is based on the development cost, the regulatory cost, and the cost of establishing a marketing plan. All these costs are amortized over a fixed time, adding to the depreciation. However, if the cost of products is higher, this makes it improbable for the companies to compete with reference products on the market. This chapter, the longest in the book, is complete teaching of every element that must be controlled to achieve a fast-to-market and the lowest possible cost. Given that most readers will be engaged in producing a competitively priced product, this chapter provides the most current teaching of science, technology, regulatory and novel approaches to enable biosimilars to help reduce the cost of biological drugs.

IOP Publishing

Biopharmaceutical Manufacturing, Volume 1
Regulatory processes
Sarfaraz Niazi and Sunitha Lokesh

Chapter 7

Intellectual property issues for scientists

The development of biopharmaceuticals is an expensive exercise. The burden of reducing the risk of litigation in infringement of other patents and protecting one's own intellectual property require close collaboration between scientists and legal teams. This chapter describes the process of protecting intellectual property and walks the reader through a strategy of how to avoid infringing on others. Fine points about the definition of the vocabulary used in a patent application and the legal language are described in simple words. All types of patents, differences in international patent laws, and a detailed description of patents related to biological drugs are provided. Details on writing freedom-to-operate documents are provided, along with a comprehensive listing of patent expiry of biological drugs and the approved BLAs and their patent expiry.

7.1 Overview

Intellectual property is the lifeline of every industry; it is a lot more than the patents (figure 7.1), but the focus pertinent to the reader of this book is the patenting process associated with new drug development.

The term 'patent' stems from 'letter of patent', which is an open letter in which a sovereign body bestows a specific privilege or right on a subject. In 1421, in Florence, Italy, Filippo Brunelleschi received the earliest recorded patent for an agricultural invention. Since then, countries have evolved their own patent rules, such as the amount of time a patent can be awarded, the categories of patents that can be granted, and the filing processes.

Patenting biopharmaceutical discoveries is an essential requirement for commercialization. The development of biopharmaceutical products undergoes a lengthy and expensive cycle adding to high cost, hovering over a billion-dollar mark, and such investment can only be justified if the developer can protect the exclusivity of the product to recover its investment. There are two types of exclusivities available.

doi:10.1088/978-0-7503-3175-3ch7

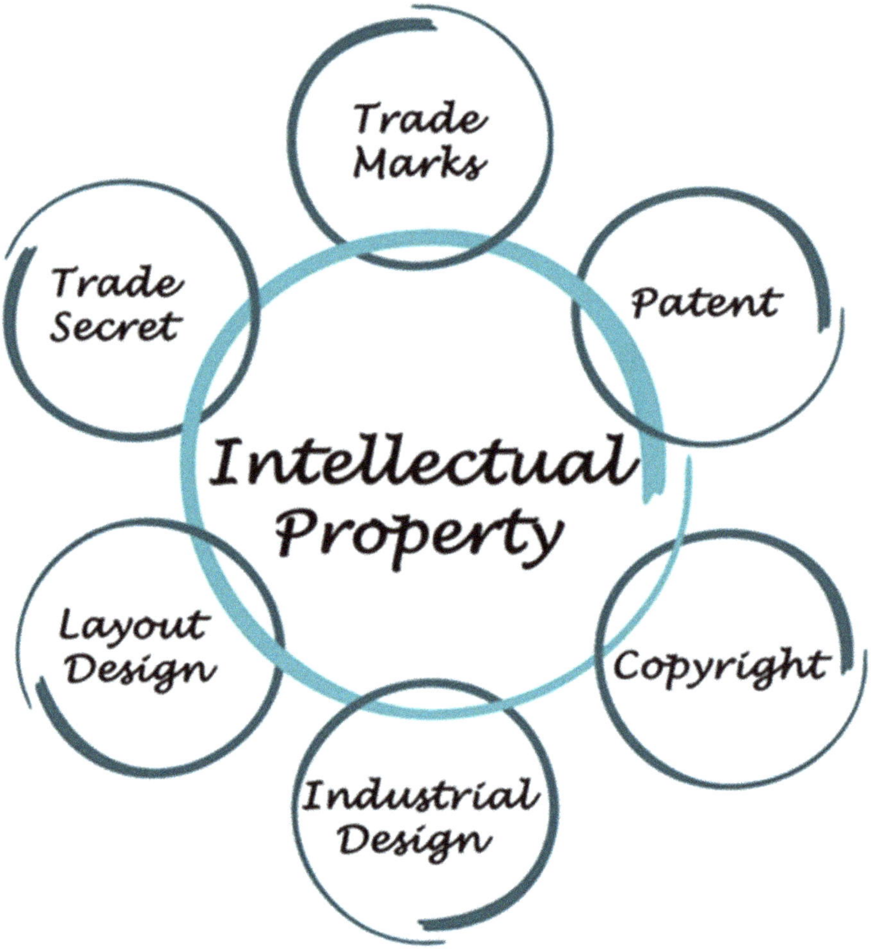

Figure 7.1. Types of intellectual properties.

One is from a patent granted. The other that applies in the US is the regulatory exclusivity of 12 years since the launch date, regardless of the patent protection.

On the other side, a manufacturer will not want to infringe any patent for the risk of being sued or being blocked from taking their product to the market. Most scientists and technicians who have not been trained in the concept of patenting may not realize that an inventor, an invention, and a patent are not the same as a scientific publication because these are legal documents that contain a great deal of detail and finesse that are important for new and subsequent biologics.

A new drug entity is generally protectable as a new chemical structure, a new synthesis or manufacture method, and new use. In chemical entities, the method of manufacture is generally not exclusive as other methods can be used to synthesize a product. Still, in biological drugs, the manufacturing process can be protected by various patents, including the upstream parameters, downstream processes, and further purification.

Table 7.1 lists the 51 566 patents protecting biological drugs in the order of the number of patents protecting each entity. This database is a starting point for deciding which product to develop and comes in handy when creating a freedom-to-operate document for the approved biological products.

Table 7.1. Patents protecting biopharmaceuticals.

Biological drug	Number of patents
Parathyroid hormone	3396
Interferon alfa-2b	3168
Peanut	3139
Gonadotropin, chorionic	3099
Somatropin	2605
Urokinase	2436
Botulism antitoxin heptavalent	1757
Bevacizumab	1712
Albumin human	1702
Hyaluronidase	1588
Rituximab	1390
Hyaluronidase	1330
Zoster vaccine live	1170
Histamine	1151
Alemtuzumab	972
Albumin human	847
Panitumumab	837
Trastuzumab	828
Filgrastim	816
Adalimumab	796
Collagenase	782
Trastuzumab	750
Hyaluronidase-oysk	720
Darbepoetin alfa	700
Etanercept	598
Insulin human	592
Asparaginase	583
Cetuximab	552
Epoetin alfa	518
Insulin recombinant human	504
Nivolumab	399
Bacillus anthracis	369
Infliximab	338
Pancrelipase amylase,	315

(*Continued*)

Table 7.1. (*Continued*)

Biological drug	Number of patents
Protease	315
Anakinra	303
Immunoglobulin g	282
Peginterferon alfa-2b	264
Aprotinin	254
Tocilizumab	253
Alteplase	243
Interferon beta-1a	238
Palivizumab	236
Sargramostim	230
Basiliximab	222
Insulin aspart	207
Insulin lispro	195
Onabotulinumtoxina	190
Ipilimumab	185
Denosumab	184
Abciximab	179
Insulin glargine	178
Pertuzumab	175
Golimumab	173
Human immunoglobulin g	165
Natalizumab	163
Ranibizumab	158
Daclizumab	154
Ibritumomab tiuxetan	137
Aldesleukin	134
Belimumab	129
Ustekinumab	129
Chymopapain	128
Hemin	113
Pembrolizumab	112
Ofatumumab	111
Gemtuzumab ozogamicin	108
Lixisenatide	108
Omalizumab	108
Bacillus calmette-guerin	101
Pegfilgrastim	94
Pegaspargase	89
Denileukin diftitox	84
Influenza virus vaccine	84
Aflibercept	80
Atezolizumab	80
Interferon beta-1b	79

Brentuximab vedotin	78
Eculizumab	75
Immune globulin	74
Peginterferon alfa-2a	74
Ramucirumab	72
Certolizumab pegol	70
Pancrelipase	70
Abatacept	68
Mepolizumab	66
Follitropin alfa/beta	60
Insulin degludec	60
Pegvisomant	60
Avelumab	56
Dulaglutide	56
Botulinum toxin type b	52
Albiglutide	51
Tenecteplase	49
Imiglucerase	48
Rasburicase	48
Sipuleucel-t	46
Ocrelizumab	45
Reteplase	44
Insulin detemir	42
Interferon gamma-1b	39
Palifermin	39
Elotuzumab	38
Menotropins fsh 1h	37
Fibrinogen human	33
Fibrinogen human	33
Hepatitis b vaccine recombinant	33
Durvalumab	32
Urofollitropin	32
Insulin lispro recombinant	31
Rabies vaccine	30
Ado-trastuzumab emtansine	28
Daratumumab	28
Obinutuzumab	28
Canakinumab	27
Agalsidase beta	26
Alirocumab	26
Thrombin human	26
Insulin aspart recombinant	25
Insulin detemir recombinant	25
Menotropins	25
Necitumumab	25

(*Continued*)

Table 7.1. (*Continued*)

Biological drug	Number of patents
Dermatophagoides farinae	24
Siltuximab	24
Thyrotropin alfa	24
Bacillus calmette-guerin substrain tice live antigen	23
Rilonacept	23
Dermatophagoides pteronyssinus	22
Reslizumab	21
Autologous cultured chondrocytes	20
Romiplostim	20
Becaplermin	19
Belatacept	19
Blinatumomab	19
Evolocumab	19
Vedolizumab	19
Antihemophilic factor recombinant	18
Laronidase	18
Sarilumab	18
Anti-inhibitor coagulant complex	16
Antihemophilic factor human	16
Inotuzumab ozogamicin	16
Interferon alfa-n3	16
Secukinumab	16
Benralizumab	15
Hepatitis a vaccine	15
Insulin glulisine recombinant	15
Metreleptin	15
Abobotulinumtoxina	14
Ambrosia artemisiifolia	14
Histamine phosphate	14
Human papillomavirus quadrivalent types 6, 11, 16, and 18 vaccine, recombinant	14
Incobotulinumtoxina	14
Dupilumab	13
Corticorelin ovine triflutate	12
Immune globulin intravenous human	11
Capromab pendetide	10
Collagenase clostridium histolyticum	10
Dornase alfa	10
Ixekizumab	10
Pneumococcal vaccine polyvalent	10
Tuberculin purified protein derivative	10
Ziv-aflibercept	10
Antihemophilic factor recombinant	9
Olaratumab	9

Raxibacumab	9
Antihemophilic factor human	8
Asfotase alfa	8
Beractant	8
Immune globulin human	8
Ocriplasmin	8
Salmonella typhi ty21a	8
Anti-thymocyte globulin rabbit	7
Antihemophilic factor/von willebrand factor complex human	7
Brodalumab	7
Cat hair	7
Factor ix complex	7
Galsulfase	7
Choriogonadotropin alfa	6
Dinutuximab	6
Human c1-esterase inhibitor	6
Latrodectus mactans	6
Sacrosidase	6
Varicella virus vaccine live	6
Velaglucerase alfa	6
Von willebrand factor recombinant	6
Antithrombin iii human	5
Somatropin [rdna origin]	5
Alglucosidase alfa	4
Antihemophilic factor, recombinant	4
Antihemophilic factor, recombinant	4
Calfactant	4
Desirudin recombinant	4
Ecallantide	4
Equine thymocyte immune globulin	4
Hepatitis a vaccine, inactivated	4
Idursulfase	4
Isatuximab-irfc	4
Methoxy polyethylene glycol-epoetin beta	4
Peginterferon beta-1a	4
Poractant alfa	4
Rho d immune globulin	4
Smallpox vaccinia vaccine, live	4
Talimogene laherparepvec	4
Coagulation factor ix recombinant	3
Fremanezumab-vfrm	3
House dust mite, dermatophagoides pteronyssinus	3
Human plasma proteins	3
Imciromab pentetate	3
Mecasermin rinfabate recombinant	3

(*Continued*)

Table 7.1. (*Continued*)

Biological drug	Number of patents
Plasma protein fraction human	3
Ravulizumab-cwvz	3
Sebelipase alfa	3
Taliglucerase alfa	3
Technetium tc-99m albumin colloid kit	3
Asparaginase erwinia chrysanthemi	2
Bezlotoxumab	2
Clostridium tetani toxoid antigen formaldehyde inactivated,	2
Diphtheria and tetanus toxoids and acellular pertussis adsorbed and inactivated poliovirus vaccine	2
Diphtheria and tetanus toxoids and acellular pertussis adsorbed, hepatitis b recombinant and inactivated poliovirus vaccine combined	2
Diphtheria and tetanus toxoids and acellular pertussis vaccine adsorbed	2
Guselkumab	2
Hepatitis a and hepatitis b recombinant vaccine	2
Idarucizumab	2
Immune globulin infusion human	2
Insulin glargine recombinant	2
Lixisenatide	2
Insulin recombinant human	
Insulin susp isophane recombinant human	2
Isatuximab	2
Meningococcal group b vaccine	2
Neisseria meningitidis group a capsular polysaccharide diphtheria toxoid conjugate antigen	2
Ovine digoxin immune fab	2
Pneumococcal 13-valent conjugate vaccine	2
Rotavirus vaccine, live, oral	2
Coagulation factor ix recombinant human	1
Coagulation factor viia recombinant	1
Elosulfase alfa	1
Human fibrinogen, human thrombin	1
Human papillomavirus 9-valent vaccine, recombinant	1
Influenza vaccine, adjuvanted	1
Insulin aspart	1
Insulin degludec	1
Liraglutide	1
Obiltoxaximab	1
Pegademase bovine	1
Polatuzumab vedotin-piiq	1
Poliovirus type 1 antigen	1

The various patents for many biological molecules make it difficult for scientists and technicians involved in the development and production of biopharmaceutical products to design manufacturing procedures for their goods. A freedom-to-operate document (as described later) defines the limits of technology used to develop a new process. Protecting patent rights also requires several formal exercises such as entering into a non-disclosure agreement, creating proof of inventions, and filing patents timely. Since scientists and technicians need to communicate with legal teams, a basic understanding of intellectual property protection is always needed.

7.1.1 Patent landscape

This chapter focuses on patents that form the most important part of intellectual property. A patent is awarded in various legal jurisdictions based on various considerations, but most commonly it defines the boundary lines of invention, novelty, non-obviousness, and utility. The concept of intellectual property includes additional elements (figure 7.1)

As of May 2020 more than ten million patents had been issued in the US. The first US patent was issued in 1790 (three that year), and it was not until 1836 that the first patent was issued to a foreigner. The rate of patent rejections is highly variable based upon the technology and, interestingly, upon whether the inventor is a US citizen (table 7.2).

> Everything that can be invented has been invented—Charles H Duell, United States Patent Office Director, telling President McKinley to abolish the office in 1899.

The global filing of patents is shown in figure 7.2 from 2004 to 2018, as reported from 160 patent offices worldwide.

Table 7.3 lists links to patent search resources around the world. There are over 160 patent offices around the world.

7.2 Patent law basics

Patents grant the patent holder the right to ban others from producing, using, selling, or importing a patented invention in the United States (or the patentee's nation) for a specified length of time (generally, twenty years from the date a patent

Table 7.2. Total patent applications and awarded patents at the USPTO from 1790 to 2019.

Type of application	Total filed	Percentage approved
Utility	19 774 364	53%
Design	870 770	69%
Plant	35 161	89%
All	21 064 418	54%
Foreigners	4 051 671	35%

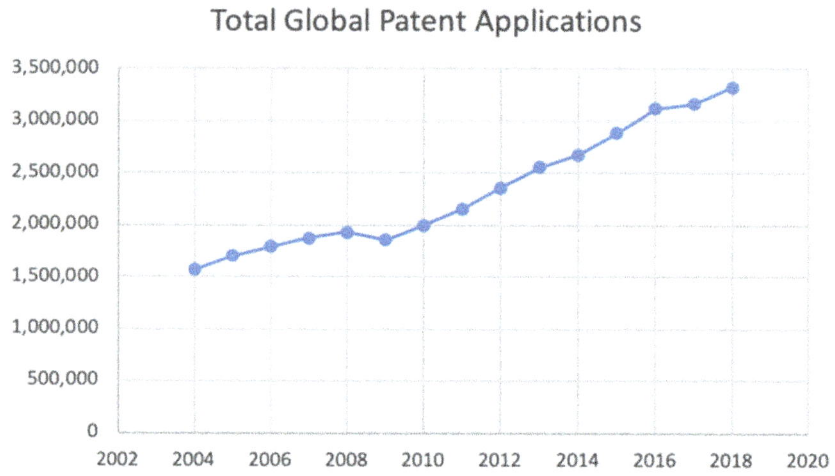

Figure 7.2. Yearly global patent filing 2004–2017.

Table 7.3. Patent search resources.

US	http://www.uspto.gov
India	http://www.ipindia.nic.in/
Europe	http://www.european-patent-office.org
Japan	http://www.jpo.go.jp/
Korea	http://www.kipo.go.kr
Italy	http://www.info-brevetti.org/
Canada	http://strategis.ic.gc.ca/sc_mrksv/cipo/
Australia	http://www.ipaustralia.gov.au/
African Region	http://www.aripo.wipo.net/
New Zealand	http://www.iponz.govt.nz/
Singapore	http://www.ipos.gov.sg/
UK	http://www.ukpats.org.uk
WIPO	http://www.wipo.int/patentscope/en/
Patent Cooperative Treaty (PCT)	http://www.wto.org/english/tratop_e/trips_e/trips_e.htm http://www.pctlearningcenter.org/
Public Patent Foundation	http://www.pubpat.org/index.html
Intellectual Property Owners Association	http://www.ipo.org/
Global: Espacenet	https://worldwide.espacenet.com/patent/
China	http://english.sipo.gov.cn
Trilateral: US, Japan, EU	https://www.trilateral.net/index
ASEAN	https://www.aseanip.org

application was filed). Individuals who do so without the consent of the patent owners breach the patent and may face monetary damages and other legal consequences. The purpose of patent exclusivity is to stimulate invention by allowing the patent holder to recoup any costs invested throughout the research and development process. Patentees can also benefit from exclusivity since it shields them from competition and allows them to charge higher-than-competitive prices for patent-protected goods. Patent protection is particularly vital for commodities such as pharmaceuticals, which are expensive to produce yet easy to copy once they are on the market.

Beyond the active ingredient, pharmaceutical patents can cover a wide range of aspects of a medication or biologic. Among the claims made by such 'secondary patents' are:

- Methods of employing the pharmaceutical (e.g. to treat a specific ailment); methods of manufacturing or technologies utilized to create the pharmacological product.
- Additional substances connected to the active ingredient, such as intermediates; or techniques or technologies for administering the pharmaceutical.

7.2.1 Pharmaceutical patenting practices

The behaviors mentioned here, in the opinion of patent holders, are legitimate uses of their patents' legal rights. However, critics view these practices as harmful strategies that exploit the patent system in ways that Congress did not intend:

- Evergreening, also known as patent 'layering' or 'lifecycle management', is a process, where some drug firms are allegedly attempting to extend their drug patent monopolies by obtaining new patents as old ones expire. A single pharmaceutical product can be covered by hundreds of patents since different pharmaceutical goods are patentable. According to critics of evergreening, secondary patents are typically for minor alterations or auxiliary pieces of a pharmaceutical product, and effectively extend patent protection of the primary product beyond Congress's term. Defenders contend that any additional patents must include significant new discoveries and improvements to existing products, and that so-called secondary patents must pass the same patentability and inspection procedures as any other patent.
- The process by which a brand manufacturer leverages its current dominant market position to urge doctors, pharmacists, and customers to 'jump' from one drug with soon-expiring patents to a newer version of the same (or similar) drug with later-expiring patents is known as 'product hopping'. An extended-release form, a new dosage, a changed mode of administration (e.g. capsules to tablets), or a slight chemical modification could all be part of the new edition of the medication. A marketing campaign, as well as discounts and rebates, may be used by the brand maker to encourage the shift. A 'hard switch', in which the brand removes the original product from the market, or a 'soft switch', in which the brand leaves the original product on the market, is the most common type of product hopping. Product hopping opponents

argue that the new product usually provides little or no clinical benefit. The adjustment is made solely to avoid generic competition by removing the original product's market before the predicted generic entrance date. Manufacturers have legitimate motivations to develop and patent new items, according to their defenders. Clinical advantages are frequently found in novel products (for example, fewer side effects or better patient compliance).

- 'Patent thickets', as the title suggests, refers to a brand manufacturer's strategy of acquiring many patents relating to a single product, preventing competitors from entering the market or making it too expensive and hazardous to do so. Manufacturers received an average of seventy-one patents on each drug, according to a recent survey of the top twelve drugs by gross US sales. When it comes to biologics versus small molecule chemical medications, concerns about patent thickets are common. This could be attributable, in part, to the difficulty of creating a drug from living cells, which provides many options for patenting novel techniques or patenting the use of a different medium for cell growth or a dosage change. Patent thickets, according to critics, are established by patenting minor or secondary innovations, significantly delaying competition because generics or biosimilars must challenge or design around each patent, which can be costly or complex. The patents on these products, according to their proponents, illustrate the types of advances that the patent rules were intended to encourage. During the patent examination procedure, each patent was found to be legitimate.
- 'Pay-for-delay' settlements. When generic (or biosimilar) manufacturers submit shortened applications for products covered by certain unexpired patents, brand manufacturers may commence patent litigation under the Hatch-Waxman and the BPCIA procedures. Some brand manufacturers have paid (or otherwise compensated) generic producers in exchange for the generic manufacturers agreeing to postpone market launch. The Supreme Court has ruled that this method, known as 'reverse payment' or 'pay-for-delay', may be a legal exercise of patent exclusivity in some cases but may violate antitrust laws in others. Pay-for-delay agreements, according to critics, are used by brand manufacturers to safeguard weak patents from invalidation; yet, because pay-for-delay agreements end the litigation, patent validity and infringement problems remain unanswered. As a result, critics argue that pay-for-delay harms competition by allowing the brand manufacturer to (i) avoid the danger of having its patents invalidated, (ii) postpone generic competition from entering the market, and (iii) extend the company's exclusive marketing rights for the specified medicine. Defenders argue that settlements are a reasonable means to decrease the cost and risk of litigation, pointing out that most claims are settled in all areas of law. Furthermore, defenders claim that the case might end with the brand maker winning, thereby barring competition until the patent period expires. Defenders argue that settling the case ensures generic entry before the patent period expires.

- Combinations of techniques. Despite the fact that these tactics are described separately above, opponents argue that they are sometimes utilized together. Some brand manufacturers, for example, may use a pay-for-delay settlement to postpone generic entry by combining product hopping with pay-for-delay settlements. The brand manufacturer, on the other hand, shifts the market to a new product that is protected by patent exclusivity.

7.3 US patent elements

A patent is a license to prevent others from practicing an invention. The inventor cannot practice his/her invention if it infringes in part on other inventions. If you patent a new application of an unexpected drug, then no one else can use that drug for the treatment you have invented. Still, if the drug molecules are protected under a chemical patent, then you can use the drug for the treatment you have invented. A patent is always negative support of invention.

If an invention is unique, non-obvious, and beneficial, it is eligible for a patent. The term ‘new and novel’ refers to the fact that the invention must have never been publicly revealed in any way, anywhere in the world, more than one year before the filing date of the patent application. The founder would not have a one year grace period in other countries. An innovation must require a creative process to be unique and non-obvious.

A utility invention (not to be confused with utility filing in the EU) can fulfill any of the following definitions:

- A kit for achieving a useful purpose.
- A method or process of synthesis or processing.
- A machine.
- An article of manufacture.
- A matter composition (such as a chemical compound).
- A means to enhance any of the above categories.

The specification (not specifications) is a complete description of the invention as well as instructions for making and using it. The specification should be stated in such a way that an expert in the field could make and use your innovation. The US Patent Office (https://mpep.uspto.gov/RDMS/MPEP/current) provides details on current methods for preparing a patent application, and this URL should be utilized to read the current regulations. The rules are current as of the time of publication of this book.

Patents are given to people in the United States, who may then delegate them to others. A business person may also file a patent in other jurisdictions. At least one of the patent’s arguments must have had an individual artistic contribution from each co-inventor. Co-authorship on a research paper is not included in this category.

A patent application constitutes a structured document that includes the following elements. This structure is followed by most patent offices around the globe, but they often allow different naming of headings and order of presentation.

7.3.1 Title of invention

The invention's title may have up to 500 characters and should be as short and specific as possible.

7.3.2 Cross-reference to related applications

Any non-provisional utility patent application claiming the benefit of one or more prior-filed co-pending non-provisional applications (or foreign applications) must provide a reference to each prior application in the first sentence of the specification after the title, identifying it by the application number or international application number, among other things, under laws 120, 121, or 365(c).

7.3.3 Statement regarding federally sponsored research or development

A declaration on rights to innovations developed under federally financed research grants or intramural programs should be included in the application if it is suitable.

7.3.4 Background of the invention

This part should include a statement about the invention's intended use. This part can also include a summary of the relevant US patent classification definitions or the asserted invention's subject matter. This section may have been known as 'field of invention' or 'technical field' in the past. This part should also offer a summary of the information.

7.3.5 Brief summary of the invention

This section should present the alleged invention's content, purpose, or general concept in its summarized form. The benefits of the invention and how it addresses previously known problems in the specification will be highlighted in the overview. The description is not the same as the abstract.

7.3.6 Background of the invention

The following elements can be included in the invention's background. (i) The invention's field, a description of the art form to which the invention belongs. This statement can include a paraphrase of the relevant patent classification definitions from the United States. The argument should be directed at the alleged invention's subject matter. (ii) A description of the relevant art, as well as a data disc.

7.3.7 Brief description of the drawing

Where there are drawings, a list of all figures by number (e.g. Picture 1A) and supporting explanations describing what each figure depicts.

7.3.8 Detailed description of the invention

Both the definition and the statements are included in the specification. In this section, the innovation and the process of creating and using it must be detailed in great detail, in absolute, simple, brief, and exact terms. This section should distinguish the invention from previous work and other inventions. In biomedical patent descriptions, experiments involving materials and procedures are frequently included.

7.3.9 Claim or claims

These arguments are the distinguishing characteristics of the invention and serve as the legal foundation for its defense. The argument or claims must specifically identify and claim the subject matter that you consider to be the invention. The arguments determine the extent of the patent's defense. The choice of claim wording determines whether a patent would be issued in large part.

The most crucial part of a claim include:

- *Scope*: Each argument can only have one sentence, which may be wide or narrow, but not all at once. A narrow argument, in general, specifies more specifics than a wide claim. You may have legal title to many parts of your invention by having several arguments, each with a different reach.
- *Characteristics of significant importance*: When drafting the claim, there are several things to keep in mind:

The patent examiner evaluates each argument. Based on its validity, each argument is either approved or rejected. As a result, the vocabulary of the argument is often repetitive in stating the invention's novelty.

The first phrase defines the innovation category and, in some cases, the intent, such as 'a diagnostic test kit' or 'a cancer-treating composition'.

- *Claim merit*: Each claim will be judged on its own merits by the patent examiner. To ensure that you get the most cover possible, you can make claims on all aspects of your invention. Writing an initial claim and referring to it in narrower scope claims is one way to ensure that specific inventive features are included in some or all claims.

7.3.10 Abstract of the disclosure

The abstract must include a summary of the disclosure found in the introduction, statements, and any drawings. The summary must indicate the technological area to which the invention relates and be written in a way that allows a clear understanding of the technical issue, the gist of the invention's solution to that problem, and the invention's primary application or use.

7.3.11 Drawings

If sketches are needed to understand the subject matter to be patented, they must be included in the patent application. Every feature of the invention must be depicted in the sketches, as stated in the claims. An application can be deemed incomplete if drawings are missing. In every patent drawing, every aspect of the invention listed in the claims must be shown.

7.3.12 Oath or declaration

An oath or declaration must include the following information: (i) the legal name of the inventor or joint inventor who is executing the oath or declaration; (ii) the application to which it is directed; and (iii) certify that the person administering the oath or announcing the true inventors is one of the declared or joint inventors. The declaration is a brief document that each inventor must sign to claim ownership of the invention.

7.3.13 Sequence listing (when necessary)

Because amino acid and nucleotide sequences are considered concepts, they must be used if they are part of your invention. This section must be prepared to disclose a nucleotide and amino acid sequence, with a sequence listing that complies with patent rules 1.821, 1.822, 1.823, 1.824, and 1.825 (37 CFR 1.821 'Nucleotide and amino acid sequence disclosures in patent applications' and WIPO Standard ST.25 (1998)).

7.4 Types of patents

Utility patents are patents that cover novel methods, formulations, or gadgets. A design patent protects a new decorative design for a manufactured item. Plant patents provide protection against infringement for any asexually reproduced distinct and novel type of plant.

Patents on utility and plant inventions typically last 20 years, starting on the date of issuance and ending 20 years after the application was filed. The appropriate maintenance payments must be paid on time. Design patents are granted for a period of 14 years from the date of issuance. There are no continuing maintenance expenses with feature patents.

The patent is personal property, which means the owner can sell, assign, or transfer it at any time. If there are any differences, the competent authorities or jurisdiction may arbitrate and prosecute the infraction. If the violation is identified, a decision concerning imposing sanctions and compensating the rightful owner must be made.

In the 1990s, the World Trade Organization was founded. It created a minimal set of rights that governments should provide to all patent owners, as well as a 20 year patent duration (from the date the application was filed).

7.4.1 Unpatentable inventions

Patents cannot be awarded for natural products that have not been altered. A natural chemical, gene, protein, or animal or plant species that has not been modified cannot be patented. A modified version of a natural object, on the other hand, could be copyrighted if the change is helpful. It is possible to patent the utilization of natural ingredients in useful devices, chemicals, or diagnostic tests.

To summarize, rather than attempting to patent a gene or protein as a composition of matter, patent claims should be tailored to the gene or protein's non-obvious functional use or altered form.

Nature's laws, physical facts, abstract notions, literary, dramatic, musical, and aesthetic works are all exempt from patent protection. They are not write-protected, but they can be copied. Patents would not be granted to technologies that are not helpful or technically impossible (for example, perpetual motion machines) or those are considered objectionable to public morality by the USPTO.

Any technique, machine, manufacturing, or composition of matter may be patented, according to Section 101 of the Patent Act (the 'congressional categories'). For decades, courts have understood that certain inventions do not fall into one of the four legislative categories. Such examples include abstract ideas, natural facts, and natural laws (the 'judicial exceptions').

Since 2012, the Supreme Court has issued three landmark rulings prohibiting patenting such forms of inventions. First, in Mayo versus Prometheus, patents on medical diagnosis and research were outlawed. Then, in Association with Molecular Pathology versus Myriad Genetics (aka Myriad Genetics), patents for artificial DNA characterized by a natural nucleic acid sequence were prohibited. Finally, the Alice ruling prohibited patents on computer hardware and software that are characterized by their use in financial transactions or other 'abstract concepts'. Although many anti-patent factions (notably Silicon Valley and the generic drug industry) have applauded these decisions (collectively known as the 'Alice' decisions), organizations that depend on creativity, such as research universities, solo inventors, and biotechnology firms, have slammed them. The Alice trifecta also overturns over a century of contrary legal precedent, brings the US out of line with international patenting standards, and breaches TRIPS section 5 Article 27 Part 3. Congress has finally introduced legislation to fix the Alice trifecta, but nothing has come to fruition as of this writing.

Molecular profiling and customized therapy provide fresh insights into illness management as well as novel tools and treatments. Even though genomic and proteomic research is not new, the availability of patentable knowledge derived from these molecular insights is constantly challenging existing health and patent rules.

Every country's economic health depends on software, medical methods, and business methods advancements, and it is commonly agreed that continuing investment in these fields demands appropriate recompense for the innovators. However, it is commonly known that granting patents for such ideas requires careful consideration, and numerous governments are still debating how to handle such requests.

7.4.2 Software patents

The basic theory in the United States is that a software invention is patentable if it meets two criteria:

1. It is one-of-a-kind, which means it is something different.
2. It is connected to a computer, in the sense that the type of hardware platform on which the program runs is defined, ensuring that a patent is not awarded

for the interpretation of an abstract process but rather for something that necessitates a particular type of physical hardware. (As we will see, this is a little more open-ended than the machine specifications in other countries.)

However, there are three forms of applications that are not patentable:
1. An algorithm is not patentable.
2. It is impossible to copyright a scientific law.
3. Patenting an abstract concept is impossible.

Software is not patentable on its own in the European Union. Only a 'computer-implemented invention' can be patented, which is defined as a software program that performs some innovative and beneficial function within a patented hardware system.

Countries such as Japan, India, and South Korea normally follow the EU's lead and only allow software to be patented as part of a physical invention. China has had a similar stance regarding software patents in the past. Newly announced patent review rules, on the other hand, show that the country is warming to the idea of patenting software as a separate entity. The State Intellectual Property Office (SIPO) of China has guidelines that allow both a (storage) medium and a computer program execution to be patented. According to some researchers, the two components—storage devices and software—may be patentable individually.

7.4.3 Medical method patents

In the United States, a medical procedure is patentable if it meets three criteria:
- Specificity—it is detailed enough to reveal its flaws.
- A way of treating a specific condition with a specific medication is referred to as a practical application.
- It has a primary transformative effect, meaning it fundamentally changes the goal's nature.

The EPO will approve a medical technique patent application if it is new and creative and does not involve surgery, therapy, or diagnostics. As previously stated, the purpose of denying patents for medical, pharmaceutical, and diagnostic processes is to alleviate clinicians of the fear of unwittingly infringing on a patent while treating a patient.

Japan, like the EPO, approves medical patents as long as they do not interfere with the work of practicing physicians. China has allowed pharmaceutical patents since 1992. Surprisingly, Chinese examiners do not look for a patent claim that claims to follow a set of approved medical technique standards. Instead, they keep track of all non-patentable medical operations in a database. Medical procedure patents are not accepted in South Korea. India does not qualify since it prohibits 'any method for the medical, surgical, curative, preventive, diagnostic, therapeutic, or other treatment of human beings or any process for a similar animal treatment that renders them disease-free'.

7.4.4 Business method patents

In the United States, business process patents have been licensed since 1988, but the Alice judgment has placed doubt on their future validity. A business method that generates a 'useful, measurable, and observable outcome' and has real-world value can be patented. It cannot just be a theory or a procedure under investigation. This includes running a certain hardware system or combination of devices in an obvious, current, and inventive manner.

A business process that is applied using a hardware can be patented in Japan and not China. Novel technologies that enhance the technological features of an automated system can be protected in South Korea. A business process cannot be patented in India.

7.4.5 Utility model in the EU

A utility model, like a patent, is a sort of intellectual property protection that protects inventions. Many nations have this sort of right, but the United States, the United Kingdom, and Canada, for example, do not. A utility model is like a patent, but it is usually less expensive to acquire and retain. It has a shorter period (usually 6 to 15 years), a faster grant time, and fewer patentability requirements. It is only eligible for technologies in specific technological fields and goods in specific countries. utility models are second-class patents.

Although no international treaty mandates the protection of utility models (unlike copyright, trademarks, or patents) and the TRIPS agreement does not cover them, the Paris Convention covers them for the Protection of Industrial Property, which means that countries that do protect utility models must adhere to rules such as national treatment and priority. Utility models are also available through the Patent Cooperation Treaty (PCT) scheme of foreign patent applications (in countries with a utility model system).

A utility model is a statutory exclusive right given for a limited time (the 'term') in return for an inventor providing enough teaching of his or her invention to allow a person of ordinary ability in the related art to perform it. Utility model laws grant privileges similar to those granted by patent laws, but they are more suited to 'incremental inventions'. A utility model is a 'right to prohibit anyone from commercially using a protected invention without the permission of the right holder(s) for a limited period'.

'Petty patent', 'innovation patent', 'short-term patents', 'minor patent', and 'tiny patent' are all terms that are often used to describe the utility model. The 'Gebrauchsmuster' is a German and Austrian utility model that inspired other countries such as Japan.

The technology must be new in most countries with utility model rules. Many patent or utility model offices, on the other hand, do not perform a substantive review and instead simply grant the utility model after verifying that utility model applications meet formal requirements. The granting procedure for a utility model is often referred to as 'utility model registration'. Furthermore, some countries exempt such subject matter from the security of utility models. Methods (i.e. processes), chemical substances, plants, and animals, for example, are not protected as utility models in some countries.

In the EU, an invention can also be protected under utility models that have:

- Registered territorial IP right.
- Available in a limited number of countries.
- No central filing in Europe.
- Protection for up to 10 years.
- Search report in some countries only.
- Registered and published after a few months.
- Generally, no substantive examination (novelty, inventiveness, industrial applicability).
- Reviewed only in invalidation or infringement proceedings.

The means of protecting the utility model include:

- Contractual:
    - Restrictive covenants in employment contracts.
    - Non-disclosure agreements.
- Practical:
    - Limited access to information.
    - 'Need to know.'
    - Non-disclosure agreements (NDAs).
    - Encryption of data.
    - Monitored entry to installations.

7.4.6 Provisional application

A provisional application for patent may be filed in the United States. The provisional application contains extensive details about the invention, but not to the same extent as the standard (or non-provisional) application. Within one year, one must file a routine patent application on the invention of one's provisional application.

The provisional application establishes a registration date for your innovation that is much ahead of the regular application's final date of patent issue. The provisional application has no resemblance to a regular utility patent because it expires in a year and does not start a 20-year patent term running.

Provisional applications are typically submitted to establish priority dates when there is a pressing need for them. However, there are a number of strong reasons to file a provisional application, including higher overall expenses and a lengthier wait time for a patent to be issued (prosecution will begin only on the utility application). Under US patent law, the provisional application can now be upgraded to a utility patent application.

Another way to demonstrate that you were the first to learn about an idea is to use a disclosure document. A disclosure document serves as a 'proof of concept' for a new idea or product. It is not meant to be used in place of a provisional or normal utility patent application. If the standard patent application is filed within two years of receipt of the disclosure document at the USPTO, the applicant can obtain a registered proof of date of conception for a charge of $10. Unlike a provisional submission, the date of the disclosure document cannot be used as an effective filing date. Because the provisional application allows for an earlier filing date, most intellectual property offices prefer to file the provisional application and forego the disclosure document.

7.5 Comparison of patent laws

7.5.1 Jurisdiction

In the United States a patent does not grant the patent holder an absolute right to utilize his or her creation. Instead, under 35 USC 271(a), the patent owner has the right to prevent anyone from making, using, selling, or importing the patented invention.

Patents are territorial in nature and must be filed in the country where protection is sought. Because a US patent only protects an inventor in the United States and not in other countries, an inventor who wishes to have his or her invention protected in other countries must apply for a patent in each of those countries or regional patent offices. Patent laws can be found in practically every country. A person who intends to obtain a patent in each country must apply for a patent in that country, following the country's regulations.

Patent rules in many nations differ from those in the United States in a variety of ways. In most countries if an innovation is publicized before the application's filing date the right to a patent is lost. Also in most nations, maintenance expenses are required. Most countries insist that the copyrighted technology be produced in their country after a specified amount of time, usually three years. If the patent is not manufactured within this time span it may become invalid in some countries. In most countries, however, providing obligatory licenses to anybody who applies for a patent is possible.

7.5.2 The Patent Cooperative Treaty

You must have a patent that has been awarded in that country to be covered in that jurisdiction. As a result, most new patent applications are filed under an international agreement that allows countries to pool patent applications. The Patent Cooperation Treaty (PCT) Learning Center (http://www.pctlearningcenter.org/) is a non-profit organization that educates people about the PCT. The PCT is an international treaty that controls the filing of patent applications in 117 nations. Although the PCT scheme does not allow for the issuance of a foreign patent, it is designed to (a) simplify the process of filing patent applications in different countries, (b) delay the costs of seeking patent protection in other countries, and (c) give the inventor more time to consider the commercial feasibility of his or her invention.

When a PCT application is filed, it does not mean that separate applications are filed in each of the countries covered by the treaty. The invention must, in addition, be filed independently in each jurisdiction, and the inventor must follow the regulations of that country. Although the standards for PCT patent applications are equal to those for US patent applications, each country has its own requirements. The filing fees for a PCT application are substantial, and the inventor must pay additional fees for each nation filing. Finally, some nations may grant the innovation while others do not, and the patent claims that are accepted in each country may differ.

The PCT was signed in June 1970 during a diplomatic summit in Washington, DC. The pact went into effect on 24 January 1978 and over 160 countries have ratified it, including the United States (2 May 2020). The treaty establishes centralized filing procedures and a common application format, among other things, to make it easier for member countries to file patent applications for the same invention. When an application is filed on time, the applicant receives an international filing date in each of the countries included in the international application, as well as (i) an invention search and (ii) a later deadline for submitting national patent applications. Most patent attorneys specialize in obtaining patents in other nations. If you seek for treaty immunity within one year of filing in the United States, you will have up to 30 months from the first filing date in the United States to file in each of the other signature nations.

In the case of inventions made in the United States, it is necessary to obtain permission from the director of the USPTO before applying for a patent in another nation. Unless a filing receipt with a license grant is provided earlier, a license grant is necessary if the overseas application is to be submitted before an application is filed in the United States, or before the six month period from filing an application in the United States expires. When a patent application is filed a request for a license is made and the granting or denial of that request is indicated on the filing receipt mailed to each applicant. A license is not necessary after six months from the filing date in the United States unless the innovation has been asked to be kept secret. If an order of secret has been granted for the invention, the director of the USPTO must consent to the filing being made outside of the United States while the order of secrecy is in place.

7.5.3 First to invent rule

In the United States, the first inventor who conceives and implements the innovation, such as a functional prototype or a well-written definition, is granted a patent. Other countries use the first to file rule, which provides a patent and all rights to the person who submits the first patent application for an invention.

Clause 101 of US Code 35 states, 'Whoever invents or discovers any new and useful technique, system, manufacturing, or composition of matter, or any new and useful improvement thereof, may acquire a patent therefore …'

On 1 January 1996, clause 104 of US Code 35 was changed to allow World Trade Organization member countries to rely on the 'first to invent' approach in assessing innovation precedence in the United States.

One of the proofs of the date of invention is the inventor's logbook. In an ideal world the inventor's logbook would be a distinct book or a collection of observed highlighted pages or entries in a continuous laboratory notebook. A logbook should contain detailed records of ideas, test findings, and other aspects of the invention process. Look for pre-printed numbered pages, fade-away backgrounds, slots for you to sign and date, and a witness to sign and date the document. Never use a loose-leaf notebook or three-ring binders as a logbook. Always avoid using a taped-together legal pad or notepad. Make use of a notebook with bound or sewn-together pages. The pages must be bound so that, in the event of a valid patent dispute, you may establish that you did not add the notebook record afterwards and backdate it.

7.5.4 First to file rule

When two people apply for a patent on the same invention, the patent will go to the first person who filed his or her application (assuming the invention is patentable, of course). This is true even though the invention was created by the second person first. The filing date is the only thing that matters. In 2013 the United States adopted the first to file system. The first plaintiff to apply has a *prima facie* claim to a patent. When two parties assert the same invention under the first to invent scheme, the USPTO will hold an intervention proceeding between them to examine proof of creation, reduction to practice, and diligence. Interference is a time-consuming and costly operation.

If the technology was publicly available in some manner before the patent application was filed, the application would be refused. 'Publicly accessible' means selling the idea, giving a talk about it, exhibiting it to an investor without a non-disclosure agreement (NDA), publishing it in a journal, and so on. It makes no difference whether the person releasing it is the inventor, a neutral third party, or someone else.

In the United States there is a one year grace period (35 US Code section 102). This means that the innovator can publish his innovation without fear of losing his patent. However, this is exclusively true in the United States. If an inventor does so, all future European patent rights are forfeited (as well as many other countries in the world). This grace time is effectively limited under the Leahy-Smith America Invents Act to publications by the inventor or those who directly got the inventor's knowledge. The uniqueness of the innovation would be jeopardized by a third-party publication. (However, this is arguable; resolving the matter may necessitate a court judgement.)

Article 30 of the Japan Patent Act provides a six month grace period for disclosures made by an experiment, publication, presentation at a study conference, or exhibition (a trade fair or the World's Fair), or if the invention becomes known to the public against the applicant's will. These kinds of disclosures are not counted as previous art. This is a significantly broader exemption than the one provided by European patent law (Article 55 EPC), but much narrower than the one offered by US patent law.

In Japan a person who is the first to apply for a patent for an invention may obtain that patent, rather than a different person who is the first to invent the same invention. Sharing the same rule as other jurisdictions, Japanese patent law does not grant exclusionary rights to existing technologies. Article 29(1) of the Patent Act stipulates this point; an inventor may not obtain a patent for inventions that were known to the public ('publicly known') (Item (i)), inventions that were publicly worked ('publicly used') (Item (ii)) or inventions that were described in a distributed publication or made publicly available through an electric telecommunication line (Item (iii)), in Japan or a foreign country before the filing of the patent application.

7.5.5 Best mode requirement

The inventor must include the best technique for executing the patent application, according to US patent law (35 US Code section 112). This prevents the inventor from obtaining a patent while concealing an essential or valuable feature. If the best mode is not included before the Leahy-Smith America Invents Act, the patent may be invalidated. This Act is no longer sanctioned, but it must still be formally included.

In European patent law, however, such a clause does not exist. Article 83 EPC mandates that the proposal include at least one method of carrying out the invention, although this method does not have to be the sole or even the best one.

7.5.6 Patent publication

Public disclosure (making copies available to the public) is required for most plant and utility patent applications. Patent applications are also published by the World Intellectual Property Organization (WIPO) and the United States Patent and Trademark Office (USPTO). When filing a US plant or utility application, an applicant may request that the application not be published if the invention has not been and will not be the subject of a foreign application that requires publication 18 months after filing (or earlier asserted priority date) or under the PCT.

Both the USPTO and the WIPO/PCT publish patent applications after 18 months have passed since the applicant's earliest successful filing date or priority date. After a patent application is released, the USPTO and the WIPO no longer hold it in trust. Any member of the public can seek access to the whole file history of the application.

Until 2001 patents in the United States were only issued once they had been obtained. Unless withdrawn or filed with a non-publication order asserting that the application is just for the United States, patent applications in the United States are now revealed 18 months after they are submitted.

With the exception of the order, this is similar to the European position, in which all patent applications are revealed 18 months after filing unless they are withdrawn. If the novelty search has been finished by that time, the search report is published with the publication.

The filing of a patent application has no bearing on whether or not the invention is patentable. Simply put, that indicates the application has been open for 18 months. People who are accustomed to the US system of exclusively publishing awarded patents may believe that anything published by the European Patent Office (EPO) is a granted patent.

The distinction between a patent application and one that has been granted can be expressed in two ways. The top-right corner number is an 'A' when the publication is an application and a 'B' when it is protected. Second, there is no abstract on the front cover of any European patents that have been issued.

As a result of the publishing, an applicant can claim temporary rights. These rights allow a patentee to sue a third party who infringes on a claim in a published application for a lawful royalties. As a result, anyone who breaches one or more of the invention's claims prior to the grant of a patent can now face penalties.

7.5.7 Rights conferred

A US patent is an enforceable property right in the United States. It allows the patent holder to prevent others from making, using, or selling the patented invention in the United States. This is the situation since the US patent law (35 US Code) is a federal statute.

In contrast, the European Patent Convention (EPC) is a treaty signed by 27 European countries: Austria, Belgium, Bulgaria, Switzerland, Cyprus, the Czech Republic, Germany, Denmark, Estonia, Spain, Finland, France, Greece, Hungary, Ireland, Italy, Liechtenstein, Luxembourg, Monaco, the Netherlands, Portugal, Romania, Slovenia, Slovakia, Sweden, Turkey, and the United Kingdom. EPC patents are granted by the EPO, which is based in Munich.

In the EPC nations chosen for an application, a European patent issued under the EPC offers its owner the same rights as a national patent. As a result, a European patent becomes essentially a 'bundle' of national patents. This means that after a European patent has been awarded, it may only be canceled in each chosen country. Anyone can file an opposition with the EPO for the first nine months after the patent is awarded to have the patent canceled in all of these countries at the same time.

7.5.8 Opposition after grant

Anyone can file an opposition with the EPO within nine months of the issuance of a European patent, arguing why the patent should not have been granted (of course, with arguments and evidence). After that, the patent holder and the opponent will debate each other. Finally, the EPO will decide based on all parties' evidence and arguments.

The patent holder and the opponent usually present their case during oral proceedings at the EPO in Munich after the parties have submitted their case to the EPO (by exchanging various letters). The EPO normally makes a final judgment after these hearings, though the proceedings are often continued in writing. Both parties can appeal the ruling, which would result in a letter exchange followed by oral proceedings.

Although the United States has a re-examination procedure, it is not as effective as an opposition. Anyone may question the validity of a granted patent in a re-examination by presenting reasons and facts to the USPTO. On the other hand, the patent holder engages in dialogue with the USPTO examiner to determine the validity of the reasons. The challenger is not involved in all of this.

The Leahy-Smith America Invents Act established an 'inter partes' test in which the challenger sits in on the hearing. It is uncertain if this investigation will proceed.

7.5.9 Inventive step

The two most important requirements of European patent law are that an invention be both patentable and novel, as well as inventive (Article 52 EPC). This is analogous to the criteria in the US that the innovation be novel and not obvious (35 US Code sections 102 and 103). The PCT stipulates that innovation must be both original and inventive, which simplifies the filing procedure in participating nations. Being non-obvious, on the other hand, is sufficient to necessitate an imaginative step.

The EPO, on the other hand, takes a more rigid stance on the subject. A European patent application is considered innovative if it addresses a technological challenge in a non-obvious way. It is important to note that this introduces two new requirements: it must come up with a solution to an issue (no problem solved means no inventive step) and the problem must be technical (solving economic problems means no inventive step).

The following is the technique for determining the technical issue. To begin, it is established whether the invention is novel. If it is, the closest prior art document with the most features in common with the invention is chosen (or that which most closely resembles the invention in some other way). After that, the differences are compared to find a problem that the invention solves.

For example, the technological challenge is to increase the driver's visibility in low-light settings if the innovation is for a bicycle with a reflector and the closest prior art is for a bicycle with no lights. The difficulty can be rectified by attaching a reflector on the bike, which allows other road users to see the driver in the dark due to the light reflected on it.

The next question is whether or not the solution to a technological problem will be obvious. Adding a headlight to advertise the object's existence, as in the example above, is a typical option for vehicles, airplanes, and other items such as signal towers. A professional will use the traditional procedure to install a headlight on the bike. As a result, the answer would be non-obvious, necessitating a creative phase in the innovation process.

It is worth mentioning that in patent law, the terms 'qualified individual' and 'obvious' have more distinct meanings than in their common use.

It is worth noting that the technical issue is not decided after the fact. The first notion that comes to mind when confronted with an invention is that it is clear because the solution appears to be so natural. This does not reflect the amount of effort required to bring the invention to fruition. Assume that everyone was satisfied with the mediocre answer (e.g. a bike with a headlight, which requires a dynamo charged by human effort). In that situation, recognizing that a particular method would be desirable could be called an innovation.

A patent must first be industrially relevant in Japan to be granted (proviso to Patent Act Article 29 (1)). In the case of innovations involving genes, chemical substances, or species, a concrete, appropriate usage must be identified. The manufacturing, agricultural, fishing, and forestry industries, as well as mining, commercial, and service industries, are all included in the phrase 'industry'. Medical enterprises, on the other hand, are not regarded as 'industries', therefore patents for medical care act discoveries are not accepted because they are not industrially significant. This prohibition is founded on the ethical concept that patent rights should not limit the diagnosis and therapies available to medical practitioners in the care of patients in any way. It is a legal view based on the fact that there is no clear constitutional provision prohibiting patent rights for medical care acts. Patent rights can be obtained for fields of medical practice, such as drugs and medical equipment, as well as their manufacture (note that pharmaceutical patent rights are subject to certain limits under Article 69(3) of the Patent Act).

7.5.10 Two-part claim

In European patents and applications, two-part claims are used frequently (almost always). In other words, an argument begins with a list of traits, followed by the phrase 'characterized in that' or 'with an augmentation containing', and one or more more attributes. The latter characteristics characterize the innovation (and so are often called the characterizing features). The prior art contains the first two qualities.

Assume you have submitted an application for a single-part allegation. In that instance, the examiner will propose that the argument be delimited from the closest prior art (the text that has the most features in common with the invention or is the most similar in some other way).

One-part claims, on the other hand, are virtually invariably present in US patent applications (and patents). If a patent in the United States includes a two-part argument, it is most likely held by a European corporation. If you employ two-part claims in the United States, anything before the characterizing portion is deemed prior art (also known as 'Jepson claims' after the first patent attorney to use them). If a novel feature is mistakenly included in the pre-characterizing part, it will be regarded a previous art and patentability may be jeopardized.

Consider the case when an applicant includes a feature in the pre-characterizing section of an argument that is not included in the closest prior art in Europe. In such instance, all the applicant needs to do is move the feature to the characterizing section. This is common because the applicant often starts with a document that has been designated as the closest prior art. During the review, however, it is determined that another text is the closest prior art, and the allegation must be amended. However, this has no discernible impact on patentability.

7.5.11 Patent assignment

Assigning your patent is like selling your house. You do not own it any longer. Licensing your patent is like renting your house; you can evict the renter if they violate the lease terms.

Patent law provides for the transfer or sale of a patent by a written agreement called an 'assignment' that can transfer the patent's entire interest. When the patent is assigned to him or her, the assignee becomes the owner of the patent and has the same rights that the original patentee had. Patent law also provides for the assignment of a part interest, that is, a half interest, a fourth interest, etc, in a patent. An assignment can be granted for a particular part of the invention or a specific field of use.

The US Patent Office records assignments, grants, and similar instruments for recording, and the recording serves as notice. Suppose an assignment, grant, or conveyance of a patent or an interest in a patent (or an application for a patent) is not recorded in the US Patent Office within three months from its date. In that case, there can be no subsequent purchaser(s).

Patent licensing and joint ownership—patents may be owned jointly by two or more persons, as in the case of a patent granted to joint inventors or the assignment of a part interest in a patent. Any joint owner of a patent, no matter how small the part interest, may make, use, offer for sale and sell and import the invention for his or her profit provided they do not infringe another's patent rights, without regard to the other owners, and may sell the interest or any part of it, or grant patent licensing to others, without regard to the other joint owner.

A patent licensing agreement promises the licensor not to sue the licensee for patent infringement. No particular form of license is required; a license is a written contract and may include whatever provisions the parties agree upon, including the payment of royalties, etc.

7.5.12 Patent infringement

Patent infringement consists of the 'unauthorized making, using, offering for sale or selling any patented invention within the United States or United States Territories, or importing into the United States of any patented invention during the term of the patent.' (35 United States Code 271; http://www.uspto.gov/web/offices/pac/mpep/consolidated_laws.pdf). When patent infringement happens, the patentee may sue for relief in the appropriate Federal court. The patentee may ask the court for an injunction to prevent patent infringement and may also ask the court for an award of damages because of the patent infringement.

The defendant usually challenges the validity of the patent, which is then decided by the court. For example, an invalid patent could be wrongly granted because the Examiner did not know about a prior art reference produced by the defendant that anticipated the invention. The defendant may also try to say that what is being done does not constitute infringement. Infringement is determined primarily by the specific language of the claims. The defendant will argue that the specific terms of the claims are not violated.

Suits for infringement of patents follow the rules of procedure of the Federal courts. From the district court's decision, there is an appeal to the Court of Appeals for the Federal Circuit. The Supreme Court may after that take a case by writ of certiorari. If the United States Government infringes on a patent, the patentee has a remedy for damages in the United States Court of Federal Claims. The Government may use any patented invention without the patentee's permission, but the patentee is entitled to obtain compensation for the use by or for the Government.

Anyone who sells patented articles must mark the articles with the word 'Patent' and the number of the patent. The penalty for failure to mark is that the patentee may not recover damages from an infringer unless the infringer was duly notified of the infringement and continued to infringe after the notice. The marking of an article as patented when it is not patented is against the law and subjects the offender to a penalty.

Patent Pending—articles can be sold with the terms 'Patent Applied For' or 'Patent Pending.' These phrases have no legal effect but only give information that a patent application has been filed in the US Patent and Trademark Office or WIPO. The protection afforded by a patent does not start until the actual grant of the patent. False use of these phrases or their equivalent is prohibited.

7.6 Biological patents

Gene patenting is a controversial topic in bioethics. There have been three different ways that people have communicated their concerns regarding genetic patenting. To begin with, some say that patenting genetic material is unethical since it treats life as a commodity. Others say that living elements cannot be patented because they are prevalent in nature. Finally, others worry that granting patents on genetic material will jeopardize human and other species' integrity by allowing outsiders to possess their genes. Agreements such as the Agreement on Trade-Related Aspects of Intellectual Property Rights (TRIPS) require the World Trade Organization (WTO) to have intellectual property security regulations in place for most biological inventions, making it unlikely that many governments would outright ban gene patents.

The usage of gene patents after they are issued is a major ethical concern. One major issue is that, due to the patent owner's constraints, employing proprietary products and processes would be prohibitively expensive, if not impossible. Furthermore, considering the vast markets for these goods, the innovators construct a wall around their composition patent in order to profit well beyond the initial exclusivity period; this goes against the basic ethical principle that patents are granted so that humanity can prosper until they expire.

In Australia, patents on naturally occurring DNA sequences are valid.

If natural biological compounds (together with any related method or usage) are sufficiently 'isolated' from their normally occurring states, they can be patented in the United States. Patents on adrenaline, insulin, vitamin B12, and other genes are just a few examples of previous patents. Natural biological products, including gene sequences, can be protected, according to the European Patent Organization, provided they are 'isolated from [their] natural environment or generated using a technological approach'. However, the United States Supreme Court concluded in a landmark ruling in June 2013 that naturally occurring DNA sequences are not patentable.

European patents cannot be granted for treatments that necessitate the killing of human embryos, according to the EPO.

The Supreme Court concluded in Diamond versu Chakrabarty (447 US 330; 1980) that discoveries involving live organisms transformed by man were patentable. The Court's interpretation of section 101's reach gave the embryonic biotechnology industry the incentive it needed to kickstart and propel a frenetic and exhilarating period of expansion.

Biotechnology patent claims, like all invention claims, define the scope of a patentee's enforceable rights. Failure to provide the largest claim breadth possible could be a serious hurdle for a patentee attempting to defend its rights. Given the importance of claims in understanding and effectively using patent rights, it is worth looking into how claims work in the field of biotechnology to defend biotechnology breakthroughs. Consider the situation in which a broad but badly worded patent claim is rejected due to formalistic defects; the patentee is left with a defense scope that is not much larger than the actual protein species developed. Rivals can then make slight adjustments to avoid the remarks' literal meaning. The goal of the claims, according to the Federal Circuit, is to identify the enforceable breadth of patent rights. Patents must continue to play this role if they are to continue to stimulate innovation.

The counterparts' theory, on the other hand, provides a reasonable foundation for discouraging the alleged infringement from making minor changes to escape the claim. In exceptional circumstances, it seems reasonable that a patentee with overly narrowed protein patent claims should be able to circumvent the restrictions imposed by the literal scope of patent claims to protect himself from 'the unscrupulous copyist' who makes 'insignificant and insubstantial changes and substitutions in the patent which, while adding nothing, would be enough' (Graver Tank, 339 US at 607).

Before 1995 the expiration dates of the various components of a biotech invention may be years apart. Unless they acquire an extension owing to a patent office or regulatory delay, or a pediatric exclusivity extension, most patents now expire 20 years after they are filed in the United States.

7.7 Monoclonal antibody technology

Biotechnology medications are derived from the immune system's systems for producing white blood cells, or lymphocytes. In the thymus gland, these cells begin as stem cells in the bone marrow and subsequently differentiate and grow into

B-lymphocytes (B-cells) or T-lymphocytes (T-cells). B-cells' main job is to make antibodies in response to foreign material exposure via interactions with B-cell surface receptors. After activation, the activated B-cell divides swiftly, generating a clone of identical plasma cells that release antibodies with the same antigen specificity as the original B-cell. Antibodies (also known as immunoglobulins or Ig molecules) are complex proteins with antigen-binding sites in their branches that resemble the letter Y. Antibodies attach to antigen molecules and produce a cross-linked, insoluble complex that keeps the antigen from spreading. Antigens are proteins that are located on the surface of invading cells such as bacteria. Antibodies bind to the cell's surface, making it exposed to macrophages and other immune system components (opsonization).

Antibodies isolated from human blood, particularly immunoglobulin-G (IgG or gamma-globulin), have long been used to treat viral infections; the effectiveness of the antibodies varies depending on how recently the donor was infected. Antibodies can also be used to diagnose disease and distinguish between biological species. Hybridoma lines were deposited as part of the patent application disclosure because early inventions were unable to characterize the amino acid sequence in antibody molecules. Later, as more sophisticated methods became available, antibody sequencing was submitted instead of or in addition to cell line deposits. Antibody amino acid sequencing was characterized, enabling for the generation of antibodies using recombinant DNA methods. Another concern with the clinical usage of monoclonal antibodies was that after repeated injection, mouse proteins interacted with the patient's immune system, reducing their potency or potentially inducing a deadly allergic reaction. Antibodies are produced by recombinant (rDNA) methods using chimeric mAbs in which the variable regions (the arms of the Y) remain murine, but the constant regions (the base of the Y) are replaced with the constant regions of a human antibody. Nonetheless, the constant portions (the Y's base) are human. Another technological advance was the replacement of all hypervariable regions with specificity, resulting in a humanized antibody.

More technological advances allowed segments of antibody genes to be produced on a carrier's surface, such as a bacteriophage, allowing hypervariable sections of the appropriate specificity to be selected and combined into genes that can be expressed to make completely human monoclonal antibodies. The technology for chimeric and humanized antibodies had become obsolete by the time the first of these medications hit the market. Phage display can also be used to find large or tiny compounds that bind to a given structure, such as a receptor or its ligand. Antibodies with special qualities are frequently used as catalysts, allowing the process to proceed by retaining two reagent molecules in the correct conformation.

Claims to the nucleic acid encoding the antibody protein, the vector constructs, the cell line harboring the vector that expresses the protein, the method of harvesting the antibody from the cell line, the method of purifying the protein from the cell line components, the formulation in which the antibody is administered, and the device in which the antibody is administered can all be used to protect a monoclonal antibody product.

7.8 Antisense technology

If the genetic code of a disease-causing gene is identified, it could be used to avoid the gene entirely. Genes are made up of double-helical DNA. When a gene is turned on the genetic code in that portion of DNA is transcribed out as a single strand of RNA called messenger RNA. Messenger RNA is referred to as a 'meaning' sequence since it may be translated into a string of amino acids to produce a protein. In a DNA double helix, the 'antisense' strand is the opposite strand (A opposite T, T opposite A, C opposite G, G opposite C). Because of the illness gene's antisense coding sequence, short antisense DNAs can be produced to behave as medicines by attaching to messenger RNAs from disease genes. Because the genetic code in the RNA cannot be read, the disease-causing protein cannot be produced.

A 20-base fragment usually only affects the expression of one gene and does not affect the expression of other genes. Antisense medications confront a number of obstacles, including single-stranded DNA instability *in vivo* and the development of adequate delivery vehicles. To alleviate the stability problem, chemical alterations to the DNA chain's backbone, such as replacing phosphate groups with less readily hydrolyzed groups, can be applied, and these modifications can be copyrighted.

7.9 Transgenic plants

Plant cells, unlike animal cells, have a tough cell wall that makes it difficult to incorporate genetic information. The cellular environment also makes it tough for the vector to move about in the cell. As a result, unique approaches are employed, such as directly blasting DNA molecules onto the surface of micronized glass beads. After they have been transformed, plants can be bred normally. Increased yields, improved nutritional quality, and cheaper production costs are all goals of transgenic plants.

7.10 Exclusivities for biological products

The three basic sources of market exclusivity for biological goods are regulatory exclusivities, patents, and trade secrets or private knowledge. Patents and regulatory exclusivities ensure that a product's market is protected for a specific length of time. Under the BPCIA, innovative biologic medicines that file a full BLA are granted 12 years of regulatory exclusivity. In general, a '20 year period' begins when the patent is awarded and ends twenty years after the application was filed in the United States. (The '20 year term' does not apply to patents that were in force on 8 June 1995, or that were issued from an application filed before that date.) The greater of the '20 year term' or seventeen years from the grant date determines the length of a patent in this category. (See 35 USC 154(c) for further information.) Pediatric exclusivities, patent term extensions, and patent term amendments, on the other hand, can prolong any of the legislative provision's baseline terms of exclusivity for extra periods.

Regulatory exclusivities provide market protection for creative goods regardless of whether they are patent protected. In the lack of patent protection, a generic medication may reach the market after the regulatory exclusivity period has elapsed. As a result, a biopharmaceutical company's economic strategy requires them to acquire and maintain a strong patent portfolio.

Patents can be granted at any point throughout the development of a medicine. Patents claiming the drug product itself, for example, may be issued prior to or simultaneously with the NDA or BLA submission. Other patents, such as those claiming a commercial formulation, the use of a tailored delivery system, or a comprehensive treatment regimen, would almost certainly be granted when human clinical trials are finished. Furthermore, lifecycle management approaches can result in submarine patents, which enable an unintentional extension of patent exclusivity.

A submarine patent is one that was filed before to the 1995 law change but issued years later as a result of a delay, such as an interference proceeding. The patent application remains hidden at the patent office because it was filed before the deadline to publish the application, and it unexpectedly appears submarine. As a result, a patent is given years after technology has advanced, and the patent is valid for 17 years from the date of issue, as it was under the prior statute's laws.

Furthermore, because product development is inherently inventive, subsequently found advances, such as a better purification process or method of application, might provide extra patent exclusivity in the form of a late issuing patent. As a result, the regulatory market and patent exclusivity can run concurrently or not.

State trade secret laws vary, but they all have one thing in common: the knowledge must be of economic benefit to the owner. The owner took and continues to take appropriate precautions to protect the information from public disclosure.

Available matter considered exclusive trade secrets by a biopharmaceutical company are often excluded from patent disclosures subject to public disclosure.

Trade secret confidentiality is unrestricted by law and can provide organizations with a competitive advantage. For example, the biologic manufacturer could maintain track of critical process controls utilized throughout manufacturing or downstream bioprocess steps in the reference product's creation. If the trade secret/ proprietary knowledge is kept hidden, the producer will gain a competitive advantage. Manufacturing process controls are developed and established for each product/process, and they are crucial in determining the quality and purity of biological medicinal products.

7.11 Broad coverage

Several patents may cover nucleic acid and amino acid sequences, expression vectors, cell-based expression systems, upstream and downstream methods for producing and purifying the drug substance, optimized formulations developed to stabilize the drug product, administration devices, general methods of use, and indications for a single biological product. Given the patentable subject matter scope, it is not uncommon to identify anywhere from 50 to more than 100 patent filings relevant to a single biological product. The patent arguments that may be applicable in the case of an antibody product are summarized in table 7.4.

Table 7.4. Possible patent claims for antibody products.

Antibody product	Possible patent claims
Amino acid sequence	Amino acid sequence of: Complete heavy and light chains. Heavy and light chain variable regions. CDR regions. Modifications made to the framework, CDR, or Fc regions.
Analytical methods	Assays developed to monitor the quality or purity of the product.
Culture conditions	Media components. Culture method/feed media. Optimized culture conditions.
Device	Device for administration and use thereof.
Diagnostic methods and kits	Methods and kits used to identify select patents which are more or less likely to respond to treatment.
Expression system	Host cells engineered to express the product.
Expression vector	Every individual element and combination of the vector elements express the sequence in a suitable host cell, including promoter, enhancer, other regulatory sequences, and selection marker.
Formulation	Pharmaceutical compositions comprising the drug product.
Methods of use	Broad mechanism-based methods of use. Disease-specific methods of use. Indication-specific treatment regimens corresponding to the product label.
Nucleic acid sequence	Nucleic acid sequences encoding any or all of the above-listed amino acid sequences.
Platform technology	Platform technologies and assays used to discover or optimize the structural and functional features of the product or processes used to manufacture or purify the product.
Purification	Chromatography methods claiming the use of resins alone or in series. Optimized conditions. Compositions having a defined level of purity or homogeneity.

7.12 The *Purple Book*

On 9 September 2014 the FDA released the first version of the biologic equivalent of the *Orange Book*. The *Purple Book* is formally known as *Lists of Approved Biological Products with Reference Product Exclusivity and Biopharmaceutical Interchangeability Evaluations*, and it contains a list of biological products, including biopharmaceutical and interchangeable biological products that have been licensed by the FDA under the Public Health Service Act (the PHS Act). Unlike the *Orange*

Book, however, it does not contain patents unique to the biological innovator product. The lists only include the date a biological product was approved under section 351(a) of the PHS Act and if the FDA reviewed the biological product for reference product exclusivity under section 351(k)(7) of the PHS Act. The FDA has determined whether a biological product approved under section 351(k) of the PHS Act is biopharmaceutical or interchangeable with a reference biological product, according to the *Purple Book* (an already-licensed FDA biological product). The reference product that exhibited biopharmaceutical or interchangeability will be classified under the biopharmaceutical and interchangeable biological products authorized under section 351(k) of the PHS Act. Separate listings for biological goods controlled by the Center for Drug Evaluation and Research (CDER) and the Center for Biologics Evaluation and Research (CBER) will be updated on a regular basis.

7.13 Patent term extension

Patents related to human 'drug products', including biological products, are among those in the United States that may be qualified for a regulatory delay extension. A patent that claims (i) a prescription product for human use, (ii) a technique of using the product, or (iii) a method of producing the product can have its term extended provided it meets six standards outlined in 35 U.S.C. 156. A fundamental component of the patent term extension regulations is that only one patent term extension is allowed per 'drug product'.

Another element of the extension is that the patent can be reinstated for a period of five years, subject to a constitutional limit on total patent duration. The product's total patent life with the patent extension cannot exceed 14 years from the date of approval in either circumstance. To put it another way, if the product's patent life after approval is 14 years or more, the patent will not be valid for a patent extension. Patent term extensions are not granted automatically; they must be requested within sixty days after the product's initial commercial marketing or use permission.

7.14 Patent term adjustment

7.14.1 Factors affecting a patent term

Although 20 years is the accepted modern standard for patent terms, not all countries hold to it yet, and other factors cause variation in terms. These generic variations include:

- Foreign legal sanctions and their effect, for example, a TRIPS Council decision, required Canada to extend the duration of some patents issued under 17 year term legislation to 20 years.
- Delays in obtaining a patent before it is granted (which can shorten or lengthen a term).
- Nonpayment of annuity payments, whether on purpose or by mistake, results in a premature lapse.
- Changes in the law can affect the duration of all pending or active proceedings.

- Only a subset of patents is affected by targeted legal changes (special provisions in the law for pharmaceutical patents).

7.14.2 Pharmaceutical patent variations

The pharmaceutical industry is the target of the most common limited amendments in patent terms, with certain countries allowing extensions for medical devices and agrochemicals. The underlying principle is to provide a way for a patent owner to be compensated for sales time lost due to mandatory registration requirements (testing safety and efficacy). It is acknowledged that these provisions prevent patent owners from infringing their patent property between the award of the patent and the approval of the marketing authorization, a penalty that no other industry faces. Different countries have implemented different mechanisms, but they all follow the same strategy to recover part of the lost exclusivity era.

To receive a term extension, there are essentially two mechanisms. The first is to expand the original patent's definition, which is the approach taken by the US and Japan. The European Union's solution is to establish an entirely new legal instrument that only comes into effect after the patent expires. A supplementary security certificate (SPC) is a common name for this new instrument.

In 1984 the United States enacted comprehensive term extension provisions (often referred to as the Hatch-Waxman Act), followed by Japan in 1987. Two laws from the European Union date from 1992 and 1996. Unfortunately for knowledge specialists, putting those provisions in place was not easy. Since France and Italy enacted national SPC legislation before the EU regulations took effect, certain goods could be covered by those laws rather than the EU regime.

Other nations, such as Australia, Bulgaria, Cyprus, Czech Republic, Estonia, Hungary, Iceland, Israel, Latvia, Mexico, Moldova, Norway, South Africa, South Korea, Switzerland, and Taiwan, have enacted or are considering patent term extension legislation. Most countries have opted for a variable-term extension, with a maximum extension duration ranging from two to seven years.

7.14.3 Annuity fees and term computation

Patent security is only maintained in most developed countries if such post-grant payments (annuity fees) are charged. They are due in the United States at 3.5, 7.5, and 11.5 years after the grant, while in Europe they are due on the third anniversary of filing, even though the patent is still pending. A public registry is often used to record payment notices. Annuity payments in many countries are based on a sliding scale that increases with patent age.

The national patent law plays a big role in determining how long a patent can last. For example, in the United Kingdom, 'a patent shall remain in effect until the end of a term of 20 years, starting on the date of applying'. This means that a UK patent issued on 1 April would usually expire on 31 March, 20 years later. In Germany, on the other hand, 'a patent shall be valid for 20 years from the date of filing of the application in respect of the invention'. While it may seem to be a minor difference, the computation of a day may be important in certain situations.

7.14.4 Information on aspects of the patent term

When dealing with legal status information sources, the first rule is to never depend on a single source. Not only can the actual information collected vary, but so can the time it takes for it to be included. Some references are less up to date than others. (The sources vary not because they are incorrect but because they have different coverage and updating policies.) Furthermore, some sources only include partial details, so users must choose carefully.

National registers, where they exist in searchable form, are authoritative but pay attention to the format. Inside the patent office, machine-readable documents may exist, but the public web copies may be incomplete. The INPADOC file, which is now accessible via the esp@cenet® website (http://ep.espacenet.com), remains the best single source for multinational data. As a final caveat, do not forget the possibility of related documents (such as US reissue or reexamination cases) or other legal instruments (such as SPCs) that are not always explicitly linked to their parents on many web implementations of these data sources.

7.14.5 USPTO web sources

The basic public legal status source of the US Patent and Trademark Office is the PAIR service (http://pair.uspto.gov). This contains some annuity payment data (although it may be incomplete) and some information on Hatch-Waxman extensions under 35 USC § 156. A specific part of the website (www.uspto.gov/web/offices/pac/dapp/opla/term/156.html) discusses these actions. Hatch-Waxman extensions are issued as a Certificate of Correction to the 'master' patent. They are available in facsimile form (as TIFF files) in the main patent search area of the website.

7.14.6 Other web registers

More and more patent offices are making legal status data available on the Internet. The web version of the EPO Register is free and available under the epoline® umbrella (www.epoline.org/register.html). However, this site contains no data on any SPC applications that may cite a European Patent as the 'basic patent' because the process of grant (or refusal) of an SPC is entirely a matter for individual national governments within Europe. There is no requirement that data be fed back to the EPO for inclusion in its register.

Other patent offices with websites that contain legal status information include those in the United Kingdom (www.patent.gov.uk), Germany (www.dpma.de/index.htm), The Netherlands (www.bie.nl), and Australia (www.ipaustralia.gov.au). Typically, such sites are free, but they may require you to register for a user ID before using them. A limited amount of English-language legal status information is available from the free Japan Patent Office website (www.jpo.go.jp). More comprehensive information is only on the fee-based Patolis-e-service.

7.14.7 Commercial online files

In addition to the web-based sources described above, many electronic files are available on commercial host systems. Most of them have a mixture of general legal status and some actions specific to the pharmaceutical industry.

7.14.7.1 Non-patent-office sources

It is vital to note that drug exclusivity is not totally regulated by patents. Other government agencies have the authority to grant or deny marketing exclusivity. The US Food and Drug Administration's Center for Drug Evaluation and Research, for example, maintains approval listings on its website (www.fda.gov/cder/orange/adp.htm). The record of medicine patent expiration can be found in the electronic edition of the so-called *Orange Book* (www.fda.gov/cder/ob/default.htm). Based on the same dataset, there are new versions or derivatives. Minesoft (London, UK, www.minesoft.com) has created an alternative to the *Orange Book*, and FOI Services Inc (Gaithersburg, MD, www.foiservices.com) publishes *Drugs under Patent*.

A second mechanism for extending the term of a patent exists in the United States. The second extension, known as a Patent Term Adjustment (PTA), was created by the American Inventors Protection Act of 1999 (AIA). The AIA authorizes the USPTO to meet such deadlines during the patent examination phase. The failure of the USPTO to meet one or more of the AIA's time clocks or deadlines (the AIA's time clock requires the USPTO to issue the first Office Action within 14 months of the filing date, to reply within four months of an appeal or board decision, and to issue patents within four months of the issue date) is the most common reason for obtaining a patent term. The AIA also mandates that the first patent application procedure be completed within three years of the actual filing (unless the file party's ongoing applications and appeals cause delays). If the deadlines are not fulfilled, patent applicants may be rewarded or compensated for the USPTO's administrative delays by getting a day-for-day extension of the patent term resulting from the extended examination procedure. Patentees can acquire term extensions for their patents that span anywhere from a day to several years. Any time spent responding to the patent application's USPTO conduct will be removed from the PTA given.

7.15 Freedom-to-operate (FTO) opinions

All patents and patent applications considered applicable to the manufacture, use, or sale of a device, in this case a biopharmaceutical product, are thoroughly examined in the FTO decisions. The opinion goes over each patent/application and determines if it is infringed, invalid, or likely to expire before the product is released. It can be used for a variety of purposes by the biopharmaceutical application. The more traditional job is to provide a well-thought-out opinion that prevents a court from imposing treble damages if a biologic product infringes on a patent. It does, however, provide a competitive environment for designing around any patents that might be infringed, as well as a list of patents that the inventor (or, as the legislation calls it, the reference product sponsor (RPS)) might claim throughout the patent exchange process.

The gene sequence responsible for expressing the substance is included in the composition of matter; these patents are difficult to avoid. Fortunately, most of these patents are about to expire, and although the dates of these patents' expiration vary widely, there is concrete data to evaluate them. It is not uncommon for two to four years to pass between major markets; thus, the date of composition matter will decide the manufacturing location for the first launch.

7.15.1 Submarine patents

Submarine patents may arise just as the first composition of matter patent is about to expire, thanks to a complicated scheme of cross-licensing and overlapping patent applications filed before 1995. Interferon-alpha and etanercept are two such examples, both of which would enjoy decades of exclusivity, notwithstanding the spirit of the patent law. They were able to take advantage of a flaw in the US patenting system; the rules have since been modified, and the patent period is now 20 years from the date of the first filing, rather than the 17 years exploited by the submarine patents. Outside of the United States, this hazard does not exist. It is impossible to get around these patents.

7.15.2 System expression patents

Patents are covering basic speech technology, such as the Cabilly patents; these are more generic patents. In this case, a biosimilar developer may obtain a license unless Genentech manufactures the product. This patent was due to expire in 2017. It is impossible to get around these patents.

7.15.3 Process patents of originator

Although much of the focus is on the composition of matter patents, process patents can be the most difficult. This threat has surfaced as large market cap products such as adalimumab and etanercept approach their expiration dates. Although many of these patents may be contested, and some will be invalidated, the level of detail with which they carve out a protected area is astounding. For example, in the case of etanercept, you must show that the amino acid composition during upstream processing does not fit the stated distribution; most upstream processes do not track amino acid composition anyway. The scope of these patents can include media selection, upstream conditions, buffer pH and structure, downstream purification columns and their order of use, and even a higher purity claim.

Given the large number of bioprocessing patents held by both the originator and third parties, defining a suitable manufacturing method has already proven difficult and technically challenging. Patents that stipulate that the composition of amino acids during upstream processing must be within ranges demonstrate the difficulties with this aspect; ironically, this is not a standard examination. Even the media's composition is always unknown to the producer. However, once such patents are published, it is the developer's responsibility to research this aspect.

7.15.4 Third-party process patents

Although getting around the originator process patents can be difficult, third-party patents that refer to a particular product or a class of products are also possible. In certain cases, a biosimilar developer must devise a manufacturing route to circumvent the originator patent, then conduct a thorough review of third-party patents.

7.15.5 Formulation composition

Early biological product pioneers overlooked the importance of these patents, but now we are seeing formulation patents designed to lift the bar for demonstrating similarity. While the aim of creating these patents was to keep biosimilar developers out of business, this has not worked out if the agencies want Q/Q items. Realizing this, the agencies would allow for alternative formulations.

7.15.6 Lifecycle formulation projections

As a patent on the composition of matter approaches expiration, a recent trend has been to change the formulation, for example, switching from a lyophilized formulation to a solution or adopting a high concentration subcutaneous formulation instead of an intravenous solution (as has happened with some anticancer drugs, where the time to administer is significantly reduced using subcutaneous formulations). The manufacturer intends to turn its patients to a more user-friendly formulation, which may be a difficult marketing task. However, once the new formulation has been sold, it becomes difficult for the biosimilar product manufacturer to obtain reference samples against which to test the product, assuming any intellectual property issues may be overcome.

7.15.7 Alternative offering

Although the original product might have been released with a restricted presentation, these may change over time, mostly to provide useful diversity,for example, the use of prefilled syringes or injectors instead of vials alters the marketing and delivery environment due to reimbursement complexities.

7.15.8 Indications and dosage

Although the early pioneers of biological drugs overlooked the value of patenting particular uses of these products, such as indications, this is now the norm, and several patents will cover nearly any large molecule as it approaches expiration, including specific doses, conditions of use, and even dosing schedules. A patent held by Abbvie for prescribing precisely 40 mg of adalimumab every other week is an example. These patents aimed to prevent biosimilar developers from producing a product that was like what was already on the market, with the expectation that regulatory authorities would not allow for different dosing or indications. Recognizing the risk, regulatory agencies have opened their doors to alternative suggestions. These, on the other hand, can be very difficult to solve. Ironically, several of these patents have recently emerged, shocking major biosimilar product developers who may have already invested heavily in clinical trials only to discover that they will be unable to market these drugs after the trials. A biosimilar developer does not need to investigate the existing intellectual property but rather what could be developed in the future.

7.15.9 Delivery devices

Since all these devices are unique to the product, even though they are distributed by software manufacturing firms selling those devices to many customers, the originator may have several patents on the delivery device. The device chosen can have a huge effect on the biosimilar's marketability.

7.15.10 Developing freedom-to-operate

A person tasked with determining when a biopharmaceutical firm has freedom-to-operate must use a multi-faceted strategy to find all relevant patents and patent applications. To find all relevant patents/applications, start by making a list of search terms, at least a list of the parties involved in the discovery and manufacture of the substance, as well as a list of alternative names for the biologic used during the development process, should be included.

Due to the complicated history of biologic products, which sometimes involves many parties, it is feasible that multiple parties hold patents covering a single product. This is because it is common for a university or a cutting-edge biotech company to find a molecular target or lead molecule, then determine whether to license, sell, or partner with a pharmaceutical company to develop the candidate molecule into a biologic product. Because the medication research process would continue, any or all of the parties involved could have patents on the product.

It can scan sequence databases for patents claiming the amino acid sequence utilized to express the biologic product after identifying the amino acid sequence. A public sequence database that can be used for such searches is the Basic Local Alignment Search Tool (http://blast.ncbi.nlm.nih.gov/Blast.cgi). Many commercial databases and search providers, such as GenomeQuest, will seek for the applicant as an alternative.

After determining the search criteria, a thorough search of a patent database is required. Although several publicly accessible databases exist, such as the USPTO and the EPO, many open-ended operators are not supported. As a result, finding information in them is more difficult than finding information in paid databases, which appear to have more powerful search capabilities. You can check for generic methods for making the biologic, such as media and conditions for cultivating the cell line and expressing and purifying proteins monoclonal antibodies, in addition to product-specific patents/applications. For monoclonal antibody expression, Genentech was issued a general process patent (enter 6331 415 at http://patft.

uspto.gov/netahtml/PTO/srchnum.htm). This patent, which is also known as Cabilly II, is for a submarine. After a protracted prosecution that included interference actions and re-examination, US Patent 6 331 415 was issued in 2001. Despite having a priority claim dating back to 1983, this patent expired on December 18, 2018, after having been in effect for 17 years. This patent's claim 1 is limited to a technique for producing an antibody molecule in a single host cell that comprises at least the heavy and light chains' variable domains.

Hundreds, if not thousands, of patents will be sifted through after the search criteria are entered into the chosen database. From the search results, choose the patents that are most relevant to the biopharmaceutical applicant's cell line, media, production technique, bioreactor technology, purification process, formulation, and tests. To determine if the biopharmaceutical applicant has market freedom, the patents that are the closest to these must be analyzed claim by claim. This study is the foundation for the FTO's final opinion.

The person preparing the opinion would need to ascertain the expiration dates for each patent as well as the total number of patents to be examined. As previously stated, this entails determining the claims' proper filing date, the amount of patent term modification awarded by the USPTO for the delay, the impact of any terminal disclaimers submitted by the patentee, the amount of patent term extension awarded for regulatory approval delays (if any), and whether maintenance fees were charged on time. If the expiration date is known, some patents may become irrelevant to the study since they may expire before the biopharmaceutical product's planned launch date.

The FTO opinion should examine each argument to determine whether current processes, products, or therapeutic indications violate the patent. The most recent collection of claims must be examined to determine the likelihood of infringement as the claims read at the time and to assess the likelihood that the patent will issue for patent applications rather than patents that can be discovered. Assume that the patent application will most likely be violated. In any event, this application would need to be included in a watch list so that the prosecution could be tracked during the development of a biopharmaceutical product.

If patents do not expire prior to launch and are likely to be found infringing, the biopharmaceutical applicant must devise a strategy for moving forward once the FTO opinion is issued. Let us say you cannot design around the patent claims because it would be too costly, time-consuming, or result in a non-biopharmaceutical product. In that situation, you have the option of challenging the patent or requesting a license from the patent owners. If the conclusion is to appeal the patent, an invalidity opinion should be written. This opinion can be utilized to write a post-grant appeal for a patent. Alternatively, the invalidity verdict might be used as a negotiating tool during the patent exchange procedure under the BPCIA legislation provisions.

After the USPTO has awarded a patent, there are three avenues to contest it. These include things like *ex parte* re-examination, *inter partes* review, and post-grant review. Each option has benefits and drawbacks. *Ex parte* re-examination is the least expensive technique because it does not require the challenger to be named. The challenger, however, is unable to participate in the process once the initial appeal is filed. The patent reviewing corps conducts the re-examination. This procedure takes around two years to complete.

Early on, a strategy for controlling the product's lifecycle is developed. As the product proceeds through numerous production and regulatory approvals, it demands a thorough and intense intellectual property strategy. Proposals for specific treatments and doses are filed as a result of the success of many clinical trials. Patents keep popping up, in particular when composition or gene sequence patents are close to expiry. The authors anticipate that the constraints on drug manufacture, formulation, and use will deter biopharmaceutical companies from entering the market. They are also betting that the FDA will no longer designate them as biopharmaceuticals because of the product design modifications.

7.16 Conclusion

Intellectual property is a major consideration in the formulation of biopharmaceutical products; whereas this appears to be a legal issue for most, our experience tells us that the added complexities in the technology related to biological drugs make it an issue for all scientists engaged in the field of biopharmaceuticals, where the process itself can be patented and given the high financial stakes, litigation is common among the biopharmaceutical companies. This chapter teaches every team member in the development, manufacturing, and marketing how to avoid litigation, one of the prime causes of the high cost of development and delay in market entry.

Glossary

Term	Description
7-AAD	7-amino-actinomycin D
AAV	adeno-associated virus
Absorption	Removing a particular antibody or antigen from a sample (from serum, for example) by adding the corresponding antigen or antibody to that sample.
ADA	amino deaminase
Adjuvant	A pharmacological or immunological agent is added to enhance the effect of other agents. For example, in the case of vaccines, an adjuvant may be added to boost the immune response to produce more antibodies and longer-lasting immunity, thereby reducing the dose of antigen required.
Adsorption	The binding of molecules to a surface is because of a chemical or physio electric interaction between the membrane surface or the chromatographic resin and the molecule. Nonspecific adherence of substances in solution or suspension to cells or other particulate matter.
Adventitious agents	Acquired, sporadic, accidental contaminants. Examples include bacteria, yeast, mold, mycoplasma, or viruses that can potentially contaminate prokaryote or eukaryote cells used in production. Potential sources of adventitious organisms include the serum used in cell culture media, persistently or latently infected cells, or the environment.
Aerobe	An aerobic organism is one that grows in the presence of oxygen. A strict aerobe grows only under such a condition.
Agrobacterium tumefaciens	A bacterial plant pathogen, commonly found in soil, that contains a plasmid used to introduce desired sections of DNA into plants
Allergen	A substance, usually a protein, that can stimulate an abnormal immune response in the body, or what is known as an allergic reaction. Allergens come from a variety of sources, including dust mite excretion, pollen and pet dander.
Alpha helix and beta strand	In a protein, certain domains may form specific structures such as alpha helix and beta-strand, which constitute the secondary

doi:10.1088/978-0-7503-3175-3ch8

structure of the protein. An alpha helix has the following features: every 3.6 residues make one turn, the distance between two turns is 0.54 nm, the C=O (or N–H) of one turn is hydrogen-bonded to N–H (or C=O) of the neighboring turn. In a beta-strand, the torsion angle of N–Cα–C–N in the backbone is about 120 degrees.

Amino acid composition analysis	It is used to determine the amino acid composition and the protein quantity. A two-step process involves the protein's complete hydrolysis (chemical or enzymatic) into its component amino acids followed by chromatographic separation and quantitation via HPLC. The entire amino acid composition of the peptide or protein should include accurate values for methionine, cysteine, and tryptophan. The amino acid composition presented should be the average of at least three separate hydrolysates of each lot number. Integral values for those amino acid residues generally found in low quantities, such as tryptophan and methionine, could be obtained and used to support purity arguments.
Amino acid sequencing	A partial sequencing (8–15 residues) of amino acids within a protein or polypeptide by either amino-terminal or carboxy-terminal sequencing. This method is used to obtain information about the primary structure of the protein, its homogeneity, and the presence or absence of polypeptide cleavages. The sequence data determined by HPLC analysis is presented in tabular form and should include the total yield for every amino acid at each sequential cleavage cycle. Full sequence is often done by sequencing the peptide fragments isolated from HPLC fractionation. Amino acid sequencing is required in BP/EP for listed monographs of therapeutic proteins.
Amino acids	An amino acid is defined as a molecule containing an amino group (NH_2), carboxyl (COOH), and R. It has the following general formula, R-CH(NH2)-COOH. The R group differs among various amino acids. In a protein, the R group is also called a side chain. There are over 300 naturally occurring amino acids on earth, but the number of different amino acids in proteins is only 20.
Anaerobe	An anaerobic organism grows in the absence of air or oxygen. Some anaerobic organisms are killed by brief exposure to oxygen, whereas oxygen may just retard or stop the growth of others.
Antibody	A protein (immunoglobulin) is produced by the immune system of an organism in response to exposure to a foreign molecule (antigen). It is characterized by its specific binding to a site of that molecule (antigenic determinant or epitope).
Antibody affinity	A measure of the binding strength between antibody and a simple hapten or antigen determinant. It depends on the closeness of stereochemical fit between antibody combining sites and antigen determinants, the size of the area of contact between them, and the distribution of charged and hydrophobic groups. It includes the concept of 'avidity,' which refers to the strength of the antigen-antibody bond after the formation of reversible complexes.
Antibody-drug conjugates	Antibody-drug conjugates (ADCs) are defined as antibodies to which other molecules are bound through a chemical linker. The term 'conjugate' differentiates these modified antibody forms from peptide fusions that may be genetically fused to N- or C-termini of

	either the light or heavy chains of the antibody. The general principle of ADCs is to use the antibody to target a particular cell population, thereby carrying the conjugated molecule to that targeted cell population so that it can exert most, if not all, of its pharmacological activity upon just that population. By tethering a small molecule to a biologic, antibody-drug conjugates are engineered to deliver small molecules to targeted locations using biologic monoclonal antibodies as honing mechanisms.
Antigen	A compound (protein, polysaccharide, microorganism, virus) foreign to the body induces the production of specific antibodies.
Antigen determinant	The specific part of a structure of an antigen will induce an immune response, i.e., it will fit the receptors on T and B lymphocytes and be able to react with the antibodies produced.
Antigenicity	Antigenicity is the capacity of a substance to function as an antigen—to trigger an immune response. Any agent, often a large molecule that stimulates the production of an antibody that will react specifically with it. Each antigen may contain more than one site capable of binding to a particular antibody. An immunogen can cause the production of several antibodies with different specificities. The antigenicity of therapeutic proteins is one of the significant issues in the comparability of generic therapeutic proteins.
Antiserum	Blood serum contains antibodies against a particular antigen or immunogen. This frequently means serum from an animal that has been inoculated with the antigen.
Apoptosis	Apoptosis describes the molecular and morphological changes that characterize controlled cellular self-destruction, often called 'programmed cell death.'
Artificial chromosome	Synthesized DNA in chromosomal form for use as an expression vector.
Ascites	Liquid accumulations in the peritoneal cavity. Monoclonal antibodies can be purified from the ascites of mice that carry a transplanted hybridoma.
Attenuated	Weakened (attenuated) viruses are often used as vaccines; they can no longer produce disease but stimulate a robust immune response similar to the natural virus. Examples include oral polio, measles, mumps, and rubella vaccines.
Avidity	The total binding strength between all available binding sites of an antibody molecule and the corresponding determinants present on the antigen.
B lymphocytes	Also known as B cells, B lymphocytes are a type of white blood cell responsible for producing antibodies that stick to receptors on the surface of germs and other foreign substances and alert the immune system to the presence of intruders. B cells develop from stem cells in the bone marrow.
Bacterial expression	The most common microbial source for recombinant protein production in *E. coli* because of its well-understood genetics.
Bacteriophage (phage)	A bacteriophage, or phage, is a virus that infects and replicates within a bacterium. Various bacteriophages include R17, lambda, f1, phi X 174, and M13.

Baculovirus	A class of insect viruses is used as vectors for recombinant protein expression in insects; baculovirus is non-infective to humans.
Base pair	Two bases on different strands of nucleic acid join together. For example, in DNA, cytosine (C) always pairs with guanine (G), and adenine (A) links to thymine (T). In RNA molecules, adenine joins to uracil (U).
Binding site	The part of the antibody molecule that will specifically bind antigen.
Bioactivity	A protein's ability to function correctly after it has been delivered to the active site of the body (*in vivo*).
Bioavailability	A measure of the actual rate and the total amount of drug reaches the target tissue after administration.
Bioburden	The level and type (e.g., objectionable or not) of micro-organisms present in raw materials, API starting materials, intermediates, or APIs. Bioburden should not be considered contamination unless the levels have been exceeded or defined objectionable organisms have been detected.
Biosensors	The robust recognition systems of biological chemicals (enzymes, antibodies, DNA) are coupled to microelectronics to enable rapid, accurate low-level detection of such substances like sugars and proteins (such as hormones) in body fluids, pollutants in water, and gases in the air.
Bispecific antibodies	A type of antibody that can bind to two different antigens simultaneously.
cDNA	complementary DNA
cDNA library	A library containing cloned complementary DNA fragments is inserted into a collection of host cells, each of which has been inserted into a DNA vector (e.g., a circular DNA plasmid) and replicated in a bacterium like *E. coli*. The bacteria provide a pool of cDNAs for use in different experiments. Each bacteria population contains a single inserted cDNA which can be extracted and amplified using PCR.
Cell-based therapies	The therapeutic indication can classify cell therapies they aim to address, e.g., neurological, cardiovascular, ophthalmological; by whether they comprise cells taken from and administered to the same individual (autologous) or derived from a donor (allogeneic); or most commonly by the cell types, often using the EU regulatory classification. The EU regulatory classification of cell-based therapies discriminates between minimally manipulated cells for homologous use (transplants or transfusions) and those regulated as medicines that are required to demonstrate quality, safety, and efficacy standards to obtain a marketing authorization before becoming commercially available (referred to as Advanced Therapy Medicinal Products).
Cell culture	Cells are taken from a living organism and grown under controlled conditions ('in culture'). Methods used to maintain cell lines or strains. The *in vitro* growth of cells isolated from multicellular organisms. These cells are usually of one type.
Cell differentiation	The process whereby descendants of a common parental cell achieve and maintain specialization of structure and function.

Cell fusion	The formation of a hybrid cell with nuclei and cytoplasm from different cells, produced by fusing two cells of the same or different species.
Cell harvesting	The process of concentrating (dewatering) the cell mass after fermentation. Cell slurries above 70% wet cell weight are achievable. The cells may also be washed to prepare them for further processing, such as freezing or lysing. Unlike clarification processing, with cell harvesting, the cells are the target material.
Cell lines	Cells that acquire the ability to multiply indefinitely *in vitro*. A cell line is established when cells from the first culture (taken from the organism) make subsequent cultures. Thanks to genetic or other manipulations, immortal cell lines can replicate indefinitely.
Cell substrate	Cells are used to manufacture a product.
Cell-specific productivity	productivity of a cell culture expressed in terms of the amount of protein produced per cell per unit time (pg/cell/day)
Cells, Type of	All cells are divided into two types: prokaryotic cells and eukaryotic cells. The primary component of living organisms is composed of cells that can multiply. All cells contain cytoplasm, plasma membrane, and DNA.
Cellular therapy	Cellular therapy products include cellular immunotherapies, cancer vaccines, and other types of autologous and allogeneic cells for specific therapeutic indications, including hematopoietic stem cells and adult and embryonic stem cells. FDA, cellular and gene therapy products https://www.fda.gov/biologicsbloodvaccines/cellulargenetherapyproducts/
Central dogma	The flow of genetic information is generally in the direction: DNA > RNA > Protein. This rule was dubbed the 'central dogma' because it was thought that the same principle would apply to all organisms. However, we now know that the flow of genetic information starts from RNA for RNA viruses.
CFU	colony-forming unit
cGMP	The minimum requirements by law for the manufacture, processing, packaging, holding, or distribution of material as established in Title 21 of the Code of Federal Regulations. Examples are Part 211 for Finished Pharmaceuticals, Part 606 for Blood and Blood Components, Part 820 for Medical Devices and Quality System Regulations (QCR).
Chaperones	Proteins that help other proteins fold correctly, either by preventing them from folding incorrectly or by catalyzing their correct formation, are used to maximize the usable protein produced by a variety of expression systems.
Chemokines	Chemokines are the cytokines that may activate or chemoattract leukocytes. Each chemokine contains 65 ~ 120 amino acids, with an 8 ~ 10 kD molecular weight. Their receptors belong to G-protein-coupled receptors. Since the entry of HIV into host cells requires chemokine receptors, their antagonists are being developed to treat AIDS.
Chemostat	A growth chamber keeps a bacterial culture at a specific volume and growth rate by limiting nutrient medium and removing spent culture.

Chemotaxis	Net-oriented movement in a concentration gradient of specific compounds. Various sugars and amino acids can serve as attractants, while some substances such as acid or alkali serve as repellants in microbial chemotaxis. White blood cells and macrophages demonstrate chemotactic movement in the presence of bacterial products, complement proteins, and antigen-activated T cells to contribute to the local inflammatory reaction and resistance to pathogens.
Chimeric antibody	An antibody made by combining genetic material from a nonhuman source, like a mouse, with genetic material from a human being.
Chimeric antigen receptors	(CARs, also known as chimeric immunoreceptors, chimeric T cell receptors or artificial T cell receptors) are engineered receptors that combine a new specificity with an immune cell to target cancer cells. Typically, these receptors graft the specificity of a monoclonal antibody onto a T cell. The receptors are called chimeric because they are fused of parts from different sources. CAR-T cell therapy is a treatment that uses such transformed cells for cancer therapy.
Chimeric transgene	A transgene contains sequences derived from two different genes from two other species.
Chromosome	A long and complex DNA chain contains genetic information (genes). Prokaryotes contain only a single chromosome; eukaryotes have more than one, consisting of a complex of DNA, RNA, and protein. The exact number of chromosomes is species-specific. Humans have 23 pairs.
Circular dichroism	With optical rotary dispersion, one of the optical spectrophotometric methods is used to determine the secondary structure and quantitate the specific structure forms (alpha- helix, B-pleated sheet, and random coil) within a protein. The resultant spectra are compared to the natural protein structure or the reference standard for the recombinant.
Cistron	The smallest unit of genetic material is responsible for synthesizing a specific polypeptide.
Clone	To duplicate precisely, whether a gene or a whole organism; or an organism that is a genetically identical copy of another organism. A cell line stems from a single ancestral cell and typically expresses all the same genes. If this is a B lymphocyte clone, they will generally produce identical antibodies, i.e., monoclonal antibodies.
Cloning vectors	Methods of transferring desired genes to organisms that will be used to express them. Cloning vectors are used to make recombinant organisms. Vector is an agent that can carry a DNA fragment into a host cell. If used for reproducing the DNA fragment, it is called a cloning vector.
Codon	A Group of three nucleotide bases in DNA or RNA determines the composition of one amino acid in 'building' a protein and can code for chain termination.
Cohesion termini	DNA molecule with single-stranded ends with exposed (cohesive) complementary bases.

Complementary DNA (cDNA)	A DNA copy of a messenger RNA (mRNA) molecule produced by reverse transcription. The term is used to reflect the fact that its sequence is a complement of the original mRNA sequence. Scientists often use cDNA deployed in gene cloning or as gene probes or create a cDNA library. Some viruses also use cDNA to convert their viral RNA into mRNA, allowing viral proteins to take over the host cell.
Complementary DNA (cDNA) library	The advantage of the cDNA library is that it contains only the coding region of a genome. Therefore, to prepare a cDNA library, the first step is to isolate the total mRNA from the cell type of interest. Because eukaryotic mRNAs consist of a poly-A tail, they can easily be separated. Then the reverse enzyme transcriptase is used to synthesize a DNA strand complementary to each mRNA molecule. After the single-stranded DNA molecules are converted into double-stranded DNA molecules by DNA polymerase, they are inserted into vectors and cloned.
Composite membrane	A membrane is made up of two or more layers that are usually chemically or structurally different.
Conjugated Proteins	The covalently bonded to prosthetic groups such as glycoprotein and metalloprotein.
Constitutive promoter	A DNA sequence that controls gene expression and is always available.
Cosmid (cosmid vector)	An artificially constructed plasmid vector containing a specific bacteriophage gene allows it to carry up to 45,000 base pairs of desired DNA. A vector similar to a plasmid, but it also contains the cohesive sites (*cos* site) of bacteriophage lambda to permit the insertion of large fragments of DNA and *in vitro* packaging into a phage. The cosmid vector combines the plasmid vector and the COS site, which allows the target DNA to be inserted into the lambda head.
CPP	Critical process parameters are key variables affecting the production process that strongly affect product quality and quantity. Therefore, they are monitored and controlled attributes to maintain product output and quality concerning critical quality attributes.
CQA	A critical quality attribute is a chemical, physical, biological, and microbiological attribute that can be defined, measured, and continually monitored to ensure final product outputs remain within acceptable quality limits.
Cryopreservation	Maintenance of frozen cells, usually in liquid nitrogen.
CSF	Colony-stimulating factor.
CTD	Common Technical Document is a set of specifications for application dossier to register medicines and designed to be used across Europe, Japan, and the United States.
Cytokines	A protein acts as a chemical messenger to stimulate cell migration, usually toward where the protein was released. Interleukins, lymphokines, and interferons are the most common.
Cytopathic effect	Morphological alterations of cell lines are produced when cells are infected with a virus. Examples of cytopathic effects include cell

	rounding and clumping, a fusion of cell membranes, enlargement or elongation of cells, or lysis of cells.
Cytoplasm	The protoplasm of a cell outside the nucleus (inside the nucleus is called nucleoplasm). Protoplasm is a semifluid, viscous, translucent mixture of water, proteins, lipids, carbohydrates, and inorganic salts found in all plant and animal cells.
Cytostat	Something that retards cellular activity and production. This can refer to cytostatic agents or machinery, such as those that freeze cells.
Cytotoxic	They are damaging to cells.
Dalton	The unit of molecular weight is equal to a hydrogen atom's weight.
Denaturation	It is the unfolding of a protein molecule into a generally bio-inactive form—the disruption of DNA duplex into two separate strands.
Developmentally regulated promoters	A DNA sequence that controls gene expression is available only at certain times or stages.
DNA (deoxyribonucleic acid)	The nucleic acid-based on deoxyribose (a sugar) and the nucleotides G, A, T, and C. Occurring in a corkscrew-ladder shape, it is the primary component of chromosomes, thus carrying inheritable characteristics of life. The basic biochemical component of the chromosomes and the support of heredity. DNA contains the sugar deoxyribose and is the nucleic acid in which genetic information is stored (apart from some viruses). A DNA molecule has two strands, held together by the hydrogen bonding between their bases. Due to the specific base pairing, DNA's two strands complement each other. If we know the sequence of one strand, we can deduce the sequence of another strand. For this reason, a DNA database needs to store only the sequence of one strand. By convention, the sequence in a DNA database refers to the 5' to 3' strand (left to right). In a DNA molecule, the two strands are not parallel but intertwined. Each strand looks like a helix.
DNA cloning	Production of many identical copies of a defined DNA fragment. DNA cloning is a technique for reproducing DNA fragments. It can be achieved by either cell-based or polymerase chain reaction (PCR)-based technique. In the cell-based approach, a vector is required to carry the DNA fragment of interest into the host cell.
DNA fingerprinting	Sequences of nucleic acids in specific areas (loci) on a DNA molecule are polymorphic, meaning that the genes in those locations may differ from person to person. DNA fragments can be cut from those sequences using restriction enzymes. Fragments from various samples can be analyzed to determine whether they are from the same person. Analyzing restriction fragment length polymorphism (RFLP) is called DNA typing or DNA fingerprinting.
DNA hybridization (dot blot) analysis	Detection of DNA to the nanogram level using hybridization of cellular DNA with specific DNA probes. Manifestation can be by 32P labeling, chemiluminescence, chromogenic, or avidin-biotin assays.
DNA library	Set of cloned DNA fragments representing the entire genome or the transcription of a particular tissue.

DNA polymerase	An enzyme catalyzes the synthesis of double-stranded DNA from single-stranded DNA.
DNA replication	DNA polymerases synthesize DNA molecules from deoxyribonucleoside triphosphate (dNTP). The chemical reaction is like the synthesis of RNA strands. Both DNA and RNA polymerases can extend nucleic acid strands only in the 5' to 3' direction. However, the two strands in a DNA molecule are antiparallel. Therefore, only one strand (leading strand) can be synthesized continuously by the DNA polymerase. The other strand (lagging strand) is synthesized segment by segment. DNA replication is triggered by the expression of all required proteins, such as DNA polymerase, DNA primase, and cyclin. In yeast, the transcription factor regulating the expression of these proteins is called MCB binding factor. In mammals, the corresponding transcription factor is E2F.
DNA sequencing	The Sanger method being used today was pioneered by Fred Sanger in the 1970s. It is also known as the dideoxy method because a small number of dideoxynucleosides are mixed with normal deoxynucleotides during sequencing. The dideoxynucleotides lack both 2' and 3' hydroxyl groups, while the deoxynucleotides lack only the 2' hydroxyl group. The 3' hydroxyl group is essential for forming the phosphodiester bond that connects two nucleotides. Therefore, the dideoxynucleotide will be the terminator of a polynucleotide chain since it lacks the vital 3' hydroxyl group. The Sanger method may sequence a DNA fragment containing up to 500 nucleotides. A strategy known as shotgun sequencing is commonly used for large-scale sequencing (such as the entire human genome). In this approach, the DNA molecule of interest is randomly chopped into numerous small pieces. After these small pieces are sequenced, their common overlaps assemble the whole sequence.
DNA synthesis	The formation of DNA by the sequential addition of nucleotide bases.
DNA vaccines	Recombinant DNA vectors encoding antigens administered for the prevention or treatment of disease. The host cells take up the DNA, express the antigen, and present it to the immune system like that which would occur during natural infection. This induces humoral and cellular immune responses against the encoded antigens. The vector is called naked DNA because there is no need for complex formulations or delivery agents; the plasmid is injected in saline or other buffers.
DNase	An enzyme that produces single-stranded nicks in DNA. DNase is used in nick translation.
Domain antibodies	The smallest known antigen-binding fragments of antibodies, ranging from 11 kDa to 15 kDa.
Downstream processing	It starts with a feed stream free of cells and cell debris; the purification sequences involve chromatography and membrane separations to achieve final product purity: capture, intermediary purification, or polishing comprising the purification part of the process.

Dynamic binding capacity	Dynamic capacity describes the amount of sample that binds to a gel packed in a column run under defined conditions. The dynamic capacity for any media is highly dependent on running conditions, sample preparation, and even the origin of the sample. In general, the lower the flow rates, the higher the dynamic capacity. As the flow rate approaches zero, the dynamic capacity approaches the available capacity. Dynamic binding capacities are determined by loading a sample containing a known concentration of the target molecule and monitoring the molecule in the column flow-through while applying the sample.
Edman degradation	A type of protein sequencing from the amino terminus.
Endogenous pyrogen assay	An *in vitro* assay based on the release of endogenous pyrogen produced by endotoxin from human monocytes. This assay appears to be more sensitive than the USP Rabbit Pyrogen Test but is much less sensitive than the LAL assay. However, it does have the advantage that it can detect all substances that cause a pyrogenic response from human monocytes.
Endogenous virus	A viral entity whose genome is part of the germline of the species of origin of the cell line and is covalently integrated into the genome of the animal from which the parental cell line was derived. This document intentionally introduced nonintegrated viruses such as EBV used to immortalize cell substrates or Bovine Papilloma Virus fit in this category.
Endonuclease	A restriction enzyme breaks up nucleic acid molecules at specific sites along their length. Microorganisms naturally produce such enzymes as a defense against foreign nucleic acids.
Endoplasmic reticulum	A highly specialized and complex network of branching, interconnecting tubules (surrounded by membranes) is found in the cytoplasm of most animal and plant cells. The rough endoplasmic reticulum is where ribosomes make proteins. It appears 'rough' because it is covered with ribosomes. The smooth endoplasmic reticulum is the site for the synthesis and metabolism of lipids, and it is involved in detoxifying chemicals such as drugs and pesticides.
Endotoxin	The outer cell wall of gram-negative bacteria. Poison in the form of a fat/sugar complex (lipopolysaccharide) forms a part of the cell wall of some types of bacteria. It is released only when the cell ruptures and can cause septic shock and tissue damage. Pharmaceuticals are tested routinely for endotoxins. A heat-stable lipopolysaccharide is associated with the outer membrane of certain gram-negative bacteria. It is not secreted and is released only when the cells are disrupted.
Enhancers	Enhancers are the positive regulatory elements located either upstream or downstream of the transcriptional initiation site. However, most of them are located upstream. In prokaryotes, enhancers are close to the promoter, but eukaryotic enhancers could be far from the promoter. An enhancer region may contain one or more elements recognized by transcriptional activators.
Enzymes	Proteins catalyze biochemical reactions by causing or speeding up reactions without being changed in the process themselves. Enzymes are the catalysts of biochemical reactions in the cell

	and include such examples as oxide-reductases, transferases, hydrolases, proteases, nucleases, phosphatases, lyases, isomerases, and ligases.
Epithelium (epithelial)	The layer(s) of cells between an organism or its tissues or organs and their environment (skin cells, inner linings of lungs or digestive organs, outer linings of kidneys, and so on).
Epitope	Part of an antigen molecule to which an antibody attaches itself.
Ethylene oxide (EtO) sterilization	A sterilization process is still typical for biomedical products, in which the product is subjected to steam and highly toxic ethylene oxide gas. However, because many materials change properties when exposed to moisture and EtO byproducts, products destined for this process must be specially engineered for EtO sterilization.
Eukaryote	An organism whose cells contain a membrane-bound nucleus. Sometimes spelled eucaryote.
Exonucleases	Enzymes catalyze the removal of nucleotides from the ends of a DNA molecule.
Express	To translate a cell's genetic information into a specific protein stored in its DNA (gene).
Expression construct	The expression vector contains the coding sequence of the recombinant protein and the elements necessary for its expression.
Expression system	Organisms were chosen to manufacture (by expression) a protein of interest through recombinant DNA technology.
Expression vector	A way of delivering foreign genes to a host, creating a recombinant organism that will express the desired protein.
FACS	Fluorescence-activated cell sorter.
FBS	Fetal bovine serum.
Fed-batch	A production process based upon feeding a growth-limiting nutrient to the culture.
Fermentation	An anaerobic bioprocess, fermentation is used in various industrial processes to manufacture products such as alcohols, acids, and cheese by the action of yeasts, molds, and bacteria.
Fermenter	A bioreactor is used to grow bacteria or yeasts in liquid culture. See fermentation or fermentation technology.
Flanking Control Regions	Noncoding nucleotide sequences adjacent to the 5' and 3' end of the coding sequence of the product contain essential elements that affect the transcription, translation, or stability of the coding sequence. These regions include, e.g., promoter, enhancer, and splicing sequences and do not have origins of replication and antibiotic resistance genes.
Floc	A fluffy aggregate that resembles a woolly cloud.
FMEA	Failure Mode Effect Analysis is an approach in which all items associated with risk are listed and evaluated concerning severity, likelihood, and detection.
Fully human antibody	Antibodies are produced by one of two very different routes. The first route is very similar to the murine hybridoma process. The significant difference is that the mice used to produce fully human monoclonal antibodies have been genetically altered to carry human antibody genes rather than mouse antibody genes. Thus, no part of the eventual therapeutic monoclonal antibody is mouse derived. The second route uses a technology called phage display

	to identify optimal CDRs. Phage display involves inserting a genetic library of CDRs into a type of virus that infects bacteria (bacteriophages). The phages then express the CDRs, allowing easy screening of the CDRs exhibiting the most robust antigen binding. Once the best CDRs are identified, they are grafted onto a human antibody scaffold. Fully human monoclonal antibodies generally show a lower incidence of ADAs than their humanized counterparts, but immune responses to fully human monoclonal antibodies persist and vary widely by product and indication. Newer transgenic mice that have a fuller complement of human antibody genes are also being used to develop a new generation of fully human monoclonal antibodies
Fungal fermentation	Fungal fermentation can manufacture many recombinant therapeutic proteins. In addition, yeasts offer certain advantages over bacteria as a recombinant expression system, such as their ability as eukaryotic organisms to perform certain post-translational modifications on the proteins they make.
Fusion of protoplast	Fusion of two cells whose walls have been eliminated, making it possible to redistribute the genetic heritage of micro-organisms.
Fusion partner	The gene for a protein that can be joined is a medically helpful protein to optimize production in bacterial fermentation expression systems. When making a small protein or peptide in *E. coli*, it is often necessary to produce the protein fused to a larger protein to get high levels of stable expression. To yield the desired protein or peptide, the resulting fusion protein must be cleaved (chemically or enzymatically broken). The nonproduct fusion partner is leftover and usually thrown away.
Gene	The basic unit of heredity plays a part in the expression of a specific characteristic. The expression of a gene is the mechanism by which the genetic information is transcribed and translated to obtain a protein. A gene is a part of the DNA molecule that directs the synthesis of a specific polypeptide chain. It is composed of many codons. When the gene is considered a unit of function in this way, the term cistron is often used. By definition, a gene includes the entire nucleic acid sequence necessary to express its product (peptide or RNA). Such sequence may be divided into the regulatory region and transcriptional region. The regulatory region could be near or far from the transcriptional region. The transcriptional region consists of exons and introns. Exons encode a peptide or functional RNA. Introns will be removed after transcription. 'Gene family' refers to a set of genes with homologous sequences.
Gene expression	An organism may contain many somatic cells, each with a distinct shape and function. However, they all have the same genome. The genes in a genome do not affect cellular functions until they are 'expressed.' Different types of cells express different sets of genes, thereby exhibiting various shapes and functions. The essential steps involved in the expression of protein genes are transcription, where a DNA strand is used as the template to synthesize an RNA strand, which is called the primary transcript; RNA processing that

	modifies the primary transcript to generate a mature mRNA (for protein genes) or a functional tRNA or rRNA; for RNA genes (tRNA and rRNA), the expression is complete after an available tRNA or rRNA is generated. However, protein genes require additional steps of nuclear transport where mRNA is transported from the nucleus to the cytoplasm for protein synthesis. Finally, protein synthesis in the cytoplasm where mRNA binds to ribosomes and synthesizes a polypeptide based on mRNA sequence.
Gene mapping	Also called linkage mapping, gene mapping describes the method used to identify the position of a gene and the distances between genes. Gene mapping is used to find genes responsible for rare, single-gene inherited disorders like cystic fibrosis and Duchenne muscular dystrophy. Scientists are also using the method to work out what genes are involved in developing more common disorders like asthma, heart disease, diabetes, cancer, and psychiatric conditions.
Gene regulatory elements	Transcriptional regulation is mediated by the interaction between transcription factors and their DNA binding sites, the *cis*-acting elements. In contrast, the sequences encoding transcription factors are trans-acting elements. The *cis*-acting elements may be divided into the following four types, i.e., promoters, enhancers, silencers, and response elements. The transcription region consists of exons and introns. The regulatory elements include a promoter, response, enhancer, and silencer. Downstream refers to the direction of transcription, and upstream is opposite to the transcription direction. The numbering of base pairs in the promoter region is as follows. The number increases along the direction of transcription, with '+1' assigned for the initiation site. There is no '0' position. The base pair upstream of +1 is numbered '-1', not '0'.
Gene therapy	Human gene therapy is the administration of genetic material to modify or manipulate the expression of a gene product or to alter the biological properties of living cells for therapeutic use
Gene transcription	The process by which cells read out the genetic instructions in their genes to RNAs, such as messenger RNA, is the first phase in the synthesis of proteins.
Gene transfer	The use of genetic or physical manipulation to introduce foreign genes into host cells to achieve desired characteristics in progeny.
Genetic code	Protein synthesis is based on mRNA sequence, which is made up of nucleotides, while proteins are made up of amino acids. There must be a specific relationship between the nucleotide sequence and the amino acid sequence. This relationship is called a genetic code.
Genetic engineering	It is a technique used to modify the genetic information in a living cell, reprogramming it for the desired purpose (such as the production of a substance it would not naturally produce). It is altering the genetic structure of an organism (adding foreign genes, removing native genes, or both) through technological means rather than traditional breeding.
Genetic mutation	A change that happens in the DNA sequence. Such a change can occur either when mistakes are made in copying DNA or result from environmental factors, such as ultraviolet radiation from the

sun or cigarette smoke. Some mutations are only carried in an individual organism. A change can be carried into the organism's offspring, where the mutation occurs in a germline cell.

Genetically modified organism (GMO)
Any organism that has had its genes altered through genetic engineering.

Genome
The total set of genes in an organism or species. The human genome consists of about 3 billion DNA bases coding for about 30,000 genes, packaged in 23 pairs of chromosomes. All the genes are carried by a cell. 'Genome' is the total genetic information of an organism. For most organisms, it is the complete DNA sTheuses, the genome is the complete RNA for RNA viruses sequence since their genetic information is encoded in RNA. The genomes of prominent organisms are given below:

Organism	Genome size (Mb)	Gene number
Hepatitis D virus	0.0017	1
Hepatitis B virus	0.0032	4
HIV-1	0.0092	9
Bacteriophage λ	0.0485	80
Escherichia coli	4.6392	4400
S. cerevisiae (yeast)	12.155	6300
C. elegans (nematode)	97	19000
D. melanogaster (fruit fly)	137	13600
Mus musculus (mouse)	3000	30000–70000
Homo sapiens (human)	3000	30000–70000

1 Mb = 1 million base pairs (for double-stranded DNA or RNA) or 1 million bases (for single-stranded DNA or RNA).

Genomic Library
The genomic library is normally made by λ phage vectors instead of plasmid vectors because the entire human genome is about three $\times 10^9$ bp long, while a plasmid or λ phage vector may carry up to 20 kb fragment. This would require 1.5×10^5 recombinant plasmids or λ phages. When plating *E. coli* colonies on a 3-inch petri dish, the maximum number to allow isolation of individual colonies is about 200 colonies per dish. Thus, at least 700 Petri dishes are required to construct a human genomic library. By contrast, as many as five $\times 10^4$ λ phage plaques can be screened on a typical petri dish. This requires only 30 Petri dishes to construct a human genomic library. Another advantage of the λ phage vector is that its transformation efficiency is about 1000 times higher than the plasmid vector. Preparation of the genomic library using λ phage vectors. It is basically the cloning of all DNA fragments representing the entire genome.

Genotype
The set of genes in DNA that is responsible for a particular trait.

Germ cell
The 'sex cells' in higher animals and plants carry only half of the organism's genetic material and can combine to develop into new living things.

Glycosylation	The addition of one or more oligosaccharide groups to a protein. The covalent attachment of sugars to an amino acid in the protein portion of a glycoprotein.
Golgi body	A cell organelle consisting of stacked membranes where posttranslational modifications of proteins are performed; also called Golgi apparatus.
GRAS	Generally Regarded As Safe, a designation by FDA that a chemical or substance added to food or drugs is considered safe.
Hapten	A low molecular weight substance that alone can react with its corresponding antibody. To be immunogenic, haptens are bonded to molecules having molecular weights greater than 5000. An example would be the hapten digoxin covalently bonded to bovine serum albumin, forming the digoxin- BSA immunogen.
High-affinity antibody	Antibodies with a high affinity for antigen. These antibodies are predominantly IgG and produced during a secondary response to antigen. A low concentration of the antigen can trigger cells producing a high-affinity antibody.
HLA	human leukocyte antigen.
Host	A cell whose metabolism is used to grow and reproduce a virus, plasmid, or another form of foreign DNA.
Host related impurities	Impurities related to the culturing of cells (e.g., cell debris, nucleic acids, host cell proteins, cell culture media components, endotoxins) or transgenic milk components.
Humanized antibody	Humanized monoclonal antibodies are an extension of the chimera strategy. All regions of the mouse antibody are replaced with human counterparts except for the complementarity-determining regions (CDRs, i.e., the amino acids that make direct contact with the antigen). A humanized antibody is a type of antibody that replaces the hypervariable regions of the murine antibody with hypervariable regions of a human antibody. Humanized Abs consist of murine-derived CDR to enable specificity and affinity to the antigen. Humanized is different from chimeric. It differs in the extent of the protein sequence (amino acids) retained in the variable region. e.g., humanized will carry about 5–6 nonhuman consensus residues of the murine antibody in the heavy chain variable region, whereas chimeric will have about 25 murine derived residues in the heavy and light variable chain regions.
Hybrid	An animal or plant that has been produced through crossbreeding two different species or plant varieties.
Hybrid cell	A cell is formed by combining two cells of different origins in which the two nuclei merge into one.
Hybridoma	An immortalized cell line (usually derived by fusing B-lymphocyte cells with myeloma tumor cells) secretes desirable antibodies.
Hybridoma technology	Fusion between an antibody-forming cell (lymphocyte) and a malignant myeloma cell ('immortal') will result in a continuously growing cell clone (hybridoma) that can produce antibodies of a single specificity. Hybrid cells are made by combining tumor cells and plasma cells; the combination of normal B-lymphocytes and myeloma cells is commonly used in cell-culture expressions systems to produce monoclonal antibodies.

Immortalize	To alter cells (either chemically or genetically) to reproduce indefinitely.
Immortalized cells	B cells combined with cancer lymphocytes
Immunoassay	A qualitative or quantitative assay technique based on the measure of interaction of high-affinity antibody with antigen is used to identify and quantify proteins.
Immunoblotting	A technique for transferring antibody/antigen from a gel to a nitrocellulose filter can be complexed with their complementary antigen/antibody.
Immunocytokines	Antibody-cytokine fusion proteins, with the potential to preferentially localize on tumor lesions and to activate anticancer immunity at the site of disease. Various tumor targets (e.g., cell membrane antigens and extracellular matrix components) and antibody formats (e.g., intact igg and antibody fragments) have been considered for immunocytokine development, and some products have advanced to clinical trials.
Immunodiffusion (double, Ouchterlony technique)	A technique in which an antigen and antibody are placed in two adjacent wells cut into a medium such as agar. As they diffuse through the medium, they form visible precipitation lines of antigen/antibody complexes at the point where the respective concentrations are at the optimum ratio for lattice formation.
Immunodiffusion (single)	An identity diffusion technique whereby the product (antigen) is placed in a well-cut into a medium such as agar containing its complementary antibody. The product diffuses into the medium, forming a ring-shaped precipitate whose density is a function of antigen concentration.
Immunospecificity	A performance characteristic is determined by conducting cross-reactivity studies with structurally similar substances present in the analyte matrix. Specificity studies are determined with each new lot of polyclonal antibodies used in the immunoassay. For monoclonal antibodies, each subsequent new lot is usually characterized by biochemical and biophysical techniques instead of comprehensive specificity studies.
Immunotechnology	Technology-based on applications of cells and molecules of the immune system.
Immunotoxins	Semi-synthetic conjugates various toxic molecules, including radioactive isotopes and bacterial or plant toxins, with specific immune substances such as immunoglobulins, monoclonal antibodies, and antigens. The antitumor or antiviral immune substance carries the toxin to the tumor or infected cell, where the toxin exerts its poisonous effect. Mesh, 1990 Broader term: antibodies.
In situ hybridization	Hybridization with an appropriate probe is carried out directly on a chromosome preparation or histological section.
In vitro cell age	A measure of the period between thawing of the MCB vial(s) and harvest of the production vessel measured by elapsed chronological time in culture, population doubling level of the cells, or passage level of the cells when sub cultivated by a defined procedure for dilution of the culture.
Inclusion bodies	Very high expression levels of heterologous proteins expressed in bacteria may lead to the formation of *inclusion bodies*. In such

	cases, the protein molecules clump together (aggregate) in the cytoplasm to create irregular organelle-like structures (about 1 μm in diameter).
Inducer	A chemical or conditional change that activates the expression leads to the production of the desired product—a small molecule that interacts with a regulator protein and triggers gene transcription.
Integration site	The site where one or more copies of the expression construct are integrated into the host cell genome.
Interspersed repeats	Interspersed repeats are repeated DNA sequences located at dispersed regions in a genome. They are also known as mobile elements or transposable elements. A stretch of DNA sequence may be copied to a different location through DNA recombination. After many generations, such a sequence (the repeat unit) could spread over various regions.
IMPD	Investigational medicinal product dossier (IMPD) is a document submitted to the European. Medicines Agency (EMA) in Europe to obtain permission to conduct clinical trials.
IND	An investigational New Drug (IND) application is submitted to the US Food and Drug Administration (FDA) to obtain permission to conduct clinical trials.
Kb	kilobase.
lac operon	The *lac operon* governs the production of enzymes for metabolizing lactose. In the absence of lactose, the repressor substance binds to the operator, inhibiting the production of three enzymes. Lactose, however, represses the repressor, allowing the enzymes to be produced. The *lac* operon of *E. coli* consists of three genes: *lacZ*, *lacY*, and *lacA*, encoding β-galactosidase, lactose permease, and thiogalactoside transacetylase, respectively. Lactose permease is located on the cell membrane, capable of pumping lactose into the cell. β-galactosidase can convert lactose into glucose and galactose. Thiogalactoside transacetylase is responsible for degrading small molecules.
lac repressor	In the absence of lactose, transcription of the *lac* operon is inhibited by the *lac* repressor. The lactose can bind to the *lac* repressor, preventing it from interacting with its DNA binding site. Hence, the *lac* operon is quickly transcribed in a medium containing lactose, producing the enzymes to generate glucose, the primary energy source for *E. coli*.
Lambda phages	The lambda phages are viruses that can infect bacteria. The significant advantage of the λ phage vector is its high transformation efficiency, about 1000 times more efficient than the plasmid vector. The λ phages are commonly used in DNA cloning. They have two life cycles: lytic and lysogenic. In the lytic cycle, λ phages replicate rapidly and eventually cause host cell lysis. The viral DNA circularizes and integrates into the host DNA in the lysogenic cycle. Then, λ phages may replicate with the host cell. Under certain conditions (e.g., ultraviolet irradiation of cells), the λ phages may transform from the lysogenic to the lytic cycle. This transformation is mainly controlled by cI (also known as λ a

	repressor) and Cro. An increase in cI proteins promotes the lysogenic cycle, whereas Cro proteins promote the lytic cycle.
Library, genome, or cDNA	Suppose you have known the partial sequence of a gene (e.g., from the sequence of a homologous gene) and want to determine its entire sequence, then you may use the technique described in this section. DNA library is a collection of cloned DNA fragments. There are two types of DNA libraries. The genomic library contains DNA fragments representing the entire genome of an organism. The cDNA library contains only complementary DNA molecules synthesized from mRNA molecules in a cell.
Ligase	An enzyme used to join DNA molecules. An enzyme that causes fragments of DNA or RNA to link together; used with restriction enzymes to create recombinant DNA.
Limulus Amoebocyte Lysate (LAL) test	A sensitive test for the presence of endotoxins using the ability of the endotoxin to cause a coagulation reaction in the blood of a horseshoe crab. The LAL test is easier, quicker, less costly, and much more sensitive than the rabbit test, but it can detect only endotoxins and not all types of pyrogens and must therefore be thoroughly validated before being used to replace the USP Rabbit Pyrogen test. Various forms of the LAL test include a gel clot test, a colorimetric test, a chromogenic test, and a turbidimetric test.
Locus	The site of a gene on a chromosome.
Lymphokines	Substances are released predominantly from T-lymphocytes after reaction with the specific antigen. Lymphokines are biologically highly active and cause chemotaxis and activation of macrophages and other cell-mediated immune reactions. Gamma-interferon is a lymphokine.
Lysosomes	Cell organelles containing enzymes, responsible for degrading proteins and other materials ingested by the cell.
Macrokinetics	Movement of whole cells and their media within a bioreactor.
Macrovoid	A generally undesirable open space in a membrane filter is appreciably larger than the average pore openings in a given filter. Macrovoids can lead to pinhole defects resulting in an unwanted passage that directly impacts final product yield. Macrovoids can also affect the overall membrane strength and thus the device's ability to maintain integrity under pressure.
Marker gene	A gene whose presence is easily detectable, which is inserted into a GMO along with the desired gene.
Master validation plan	The master validation plan (MVP) is a document that describes the validation activities required to assure that the utilities, facilities, production equipment, analytics, and process perform reproducibly for the manufacture of safe products with proven efficacy
Meiosis	Cell division in which the daughter cells have half the number of chromosomes as the parent cell.
Messenger RNA (mRNA)	RNA serves as the template for protein synthesis; it carries the transcribed genetic code from the DNA to the protein-synthesizing complex to direct protein synthesis.
Messenger RNA (mRNA)	A molecule in cells conveys genetic information from DNA to the ribosome, the site of protein synthesis.

Microtubules	Cellular organelles typical in microorganisms: thin tubes that make structures involved in the cellular movement.
Mitochondria (mitochondrion singular)	Animal-cell organelles reproduce using their DNA. They metabolize nutrients to provide the cell with energy and have once been symbiotic bacteria. Chloroplasts are their plant-cell equivalents, the cellular organelle responsible for oxidative metabolism and phosphorylation in eukaryotic cells, widely believed to have originated as a symbiotic bacterium.
Motif	The motif is a characteristic domain structure consisting of two or more α helices or β strands. Common examples include coiled-coil, helix-loop-helix, zinc finger, leucine zipper, etc. Many proteins also contain specific domains, such as the SH2 (Src homology 2) domain.
Multicellular organisms	Refers to organisms composed of more than one cell—often billions of them, arranged in various organs, tissues, and systems.
Multimerization of genes	One way of increasing the molar ratio and increasing the amount of peptide produced is to produce a fusion protein with multiple copies of the target peptide. An additional beneficial effect is often obtained by this strategy since gene multimerization has also been shown to increase the proteolytic stability of the produced peptides. When the gene multimerization strategy is employed to increase the production yield, subsequent processing of the gene product to obtain the native peptide is needed.
Multi specific antibodies	A classical antibody's two arms will bind specifically to one antigen. However, expertise in molecular design has led to novel antibody formats with additional features. One approach implies engineering multiple antigen-binding domains into a single antibody molecule. These so-called bi-or multi-specific antibodies combine two or more antigen-recognizing elements into a single molecule, bind to two or more targets.
Murine antibody	An antibody produced in mouse B cells.
Mutagen	An agent (chemicals, radiation) that causes mutations in DNA.
Mutagenesis	The induction of genetic mutation by physical or chemical means to obtain a characteristic desired by researchers.
Mutation	A change in the genetic material, either of a single base pair (point mutation) or in the number or structure of the chromosomes. A permanent change in DNA sequence or chromosomal structure.
Mycoplasma	parasitic microorganisms that infect mammals possess some bacteria and viruses characteristics.
Necrosis	The localized nonapoptotic death of cells and tissues.
NGNA	N-glycolylneuraminic acid; forms the terminal saccharide unit for an N-linked oligosaccharide structure. The presence of NGNA has been linked to possible immunogenicity.
Nick	A break in the sugar-phosphate backbone of a DNA or RNA strand.
Nick translation	*In vitro* method was used to introduce radioactively labeled nucleotides into DNA.
Nonendogenous virus	Viruses from external sources are present in the MCB. See also Virus.

Nonspecific model virus	A virus is used for characterization of viral clearance of the process when the purpose is to characterize the capacity of the manufacturing process to remove and/or inactivate viruses in general, i.e., to characterize the robustness of the purification process.
Northern blot (blotting)	The technique for transferring RNA fragments from an agarose gel to a nitrocellulose filter can be hybridized to a complementary DNA. Northern blotting is used for detecting RNA fragments instead of DNA fragments. The technique is called 'Northern' simply because it is similar to 'Southern,' not because it was invented by a person named 'Northern.' In the Southern blotting, DNA fragments are denatured with an alkaline solution. RNA fragments are treated with formaldehyde in the Northern blotting to ensure linear conformation.
Nuclear transfer	Moving a part or all of an organism's genetic information into an unfertilized egg (whose nucleus had previously been removed); can be used for cloning or to produce transgenic animals (if the genes put into the egg have been recombined with genes from other species).
Nuclear transport	After RNA molecules (mRNA, tRNA, and rRNA) are produced in the nucleus, they must be exported to the cytoplasm for protein synthesis. On the other hand, many proteins operating in the nucleus must be imported from the cytoplasm. The traffic through the nuclear envelope is mediated by a protein family, divided into exportins and importins. The binding of a molecule (a 'cargo') to exportins facilitates its export to the cytoplasm. Importins facilitate import into the nucleus.
Nucleic acid chain	In a nucleic acid chain, two nucleotides are linked by a phosphodiester bond, which may be formed by the condensation reaction similar to the peptide bond formation.
Nucleic acids	DNA or RNA: long, chainlike molecules composed of nucleotides.
Nucleotides	Molecules are composed of a nitrogen-rich base, phosphoric acid, and sugar. The bases can be adenine (A), cytosine (C), guanine (G), thymine (T), or uracil (U). A nucleotide comprises three parts: pentose, base, and phosphate group. In DNA or RNA, a pentose is associated with only one phosphate group, but a cellular free nucleotide (such as ATP) may contain more than one phosphate group. If all phosphate groups are removed, a nucleotide becomes a nucleoside. A free nucleotide may contain one, two, or three phosphate groups in cells. For example, the energy carrier ATP (adenosine triphosphate) has three phosphate groups; ADP (adenosine diphosphate) has two; AMP (adenosine monophosphate) has one. A nucleotide becomes a nucleoside such as adenosine if all phosphate groups are removed.
Nucleus	The largest organelle, a sphere that contains all the cell's genetic material and a nucleolus that builds ribosomes.
Oleophobic	Membranes that repel nonpolar fluids such as oil and lubricants.
Oligonucleotide	Commonly made in the laboratory, an oligonucleotide is a short sequence of DNA (usually 2-50 bases). Oligonucleotides are important in artificial gene synthesis, polymerase chain reactions,

DNA sequencing, library construction and can be used as molecular probes.

Oligonucleotide therapeutics — Most ON therapies act through antisense mechanisms and are directed against various RNA species, as exemplified by gapmers, steric block ons, antagomirs, small interfering RNAs (siRNAs), micro-RNA mimics, and splice switching ons.

Oncogene — When expressed as a protein, a gene can lead cells to become cancerous, usually by removing the normal constraints on its growth.

Operator gene — A gene that switches on an adjacent structural gene(s).

Operon — A segment of DNA containing adjacent genes, including structural genes, an operator gene, and a regulatory gene. Complete unit of bacterial gene expression consisting of a regulator gene(s), control elements (promoter and operator), and adjacent structural gene(s).

Organelle — A structurally discrete component that performs a specific function inside a eukaryotic cell.

Organism — A single, autonomous living thing. Bacteria and yeasts are organisms; mammalian and insect cells used in culture are not.

Overload — In preparative chromatography, the overload condition is defined as the mass of sample injected onto the column at which efficiency and resolution begin to be affected if the sample size is further increased.

p53(LocusLink) — Involved in the control of transcription, this is a tumor suppressor protein, also known as 'Guardian of the Genome.' It plays an important role in cell cycle control and apoptosis. Defective p53 could allow abnormal cells to proliferate, resulting in cancer. As many as 50% of all human tumors contain p53 mutants.

PAT — Process Analytical Technology is a mechanism to design, analyze, and control pharmaceutical manufacturing processes through the measurement of critical process parameters (CPP), which affect critical quality attributes (CQA).

PBPC — Peripheral blood progenitor cell.

PCR — Polymerase chain reaction, a method of duplicating genes exponentially. See Polymerase Chain reaction.

Peptide — A compound is made up of two or more amino acids linked together in a chain.

Peptide bond — The chemical bond between the carboxyl (-COOH) group of one amino acid and the amino (-NH_2) group of another.

Peptide mapping — A powerful technique involves the breakdown of proteins into peptides using particular enzymes. The enzymes cleave the proteins at predictable and reproducible amino acid sites, and the resultant peptides are separated via HPLC or electrophoresis. A sample peptide map is compared to a map done on a reference sample as a confirmational step in identity profiling. It is also used to confirm disulfide bonds, location of carbohydrate attachment, sequence analysis, and identification of impurities and protein degradation.

Peptides — The peptide is a chain of amino acids linked together by peptide bonds. Polypeptides usually refer to long peptides, whereas oligopeptides are short peptides (< 10 amino acids). Proteins are made up of polypeptides with more than 50 amino acids. The primary

	structure of a protein refers to its amino acid sequence. The amino acid is also called a residue—proteins consisting of fewer than 40 amino acids.
Periodic countercurrent chromatography (PCC)	a column is loaded until complete saturation is achieved, and the material breaking through is loaded onto a second one. After loading is completed, the column is disconnected, and material is eluted while loading of the second column is continued and breakthrough loaded on a third column. After regeneration of the first column, another cycle can start.
Phenotype	The physical characteristics of an organism that result from the interactions between genes, environment, disease, molecular mechanisms, and chance.
Plaque	Clear area in a plated bacterial culture due to lysis by a phage.
Plasmid	A circular molecule of DNA that can replicate autonomously of other replicons and is commonly dispensable to the cell; used in genetic engineering. Hereditary material that is not part of a chromosome. Plasmids are extrachromosomal, circular, and self-replicating and found in the cytoplasm of cells (naturally in bacteria and some yeasts). They can be used as vectors (along with viruses) for introducing up to 10,000 base pairs of foreign DNA into recipient cells.
Plasmid cloning	A process by which a plasmid is used to import recombinant DNA into a host cell for cloning. In DNA cloning, a DNA fragment that contains a gene of interest is inserted into a cloning vector or plasmid.
Plasmid insertion	Plasmids are similar to viruses but lack a protein coat and cannot move from cell to cell in the same fashion as a virus. Plasmid vectors are small circular molecules of double-stranded DNA derived from natural plasmids in bacterial cells. A piece of DNA can be inserted into a plasmid if both the circular plasmid and the source of DNA have recognition sites for the same restriction endonuclease. The plasmid and the foreign DNA are cut by this restriction endonuclease (EcoRI in this example; see Restriction enzymes also), producing intermediates with sticky and complementary ends. Those two intermediates recombine by base-pairing and are linked by the action of DNA ligase. A new plasmid containing the foreign DNA as an insert is obtained.
Plasmid plastid	Any of several types of cellular organelle found in plants and algae but not in animals or prokaryotes.
Polyclonal	Derived from different types of cells.
Polyhedrin	Protein some viruses use to protect themselves from ultraviolet light.
Polymerase	An enzyme that catalyzes the production of nucleic acid molecules.
Polymerase chain reaction (PCR)	*In vitro* technique for amplifying nucleic acid. The technique involves a series of repeated cycles of high-temperature denaturation, low-temperature oligonucleotide primer annealing, and intermediate temperature chain extension. As a result, a nucleic acid can be amplified a million-fold after 25–30 cycles. PCR is a cell-free method of DNA cloning. However, it is much faster and more sensitive than cell-based cloning.

Post-translational processing	Protein processing done by the Golgi bodies after proteins have been constructed by ribosomes. Enzymatic processing of a protein, such as adding carbohydrate moieties or removing a signal sequence to direct a protein through a cell or organelle membrane. The endoplasmic reticulum (ER) is either smooth or rough. These membrane-bound networks of branching, interconnected tubules are like tiny manufacturing plants inside the cells of most eukaryotes. The smooth ER synthesizes and metabolizes lipid molecules and helps detoxify cells. The rough ER is covered with ribosomes, which are the site of protein synthesis, where RNA from the cell nucleus is translated into amino acid sequences based on the genetic code. Bacteria have ribosomes, but they do not have ER. Yeasts have the same organelles as other eukaryotes (plants and animals), but they do not function in the same way. Protein synthesis does not stop at the amino acid chain in a eukaryotic cell. The Golgi apparatus perform complex post-translational modifications in a way that has barely begun to be understood by cellular biologists. These stacked-membrane structures put the finishing touch on glycoproteins and other complex polypeptides. It is not well understood how they differentiate one from the other or recognize molecules.
Primary and secondary antibodies	Two groups of antibodies are classified based on whether they bind to antigens or proteins directly or target another (primary) antibody that, in turn, is bound to an antigen or protein; a primary antibody can be very useful for the detection of biomarkers for diseases such as cancer, diabetes, Parkinson's and Alzheimer's disease and they are used for the study of the absorption, distribution, metabolism, and excretion and multi-drug resistance of therapeutic agents. Secondary antibodies can be conjugated to enzymes such as horseradish peroxidase or alkaline phosphatase (ap); or fluorescent dyes such as fluorescein isothiocyanate, rhodamine derivatives, Alexa fluor dyes; or other molecules to be used in various applications. In addition, secondary antibodies are used in many biochemical assays, includingElisaa, including many HIV tests, western blot, immunostaining, immunohistochemistry, immunocytochemistry.
Primer	Template strand of DNA used to generate a new double-strand of DNA.
Prions	Resembling viruses, these pathogens are composed only of protein, with no detectable nucleic acid.
Probes	A probe is a piece of DNA or RNA used to detect specific nucleic acid sequences by hybridization (binding of two nucleic acid chains by base pairing). They are radioactively labeled so that the hybridized nucleic acid can be identified by autoradiography. The size of probes ranges from a few nucleotides to hundreds of kilobases. Long probes are usually made by cloning. Initially, they may be double-stranded, but the active probes must be single-stranded. Short probes (oligonucleotide probes) can be made by chemical synthesis.
Production cells	Cell substrate is used to manufacture a product.

Prokaryote	An organism (e.g., bacterium, virus, blue-green algae) whose DNA is not enclosed within a nuclear membrane or whose cell contains neither a membrane-bound nucleus nor other membrane-bound organelles such as mitochondria and plastids. Includes the 'true bacteria' and the archaeans. Sometimes spelled procaryote.
Promoter	A region of DNA that regulates the level of function of other genes. (Source: US Department of Agriculture.)
Promoters	DNA sequence initiates transcription of a gene to produce mRNA, used in genetic engineering to direct cells to manufacture a protein of interest.
Protease	An enzyme that speeds the breakdown of proteins into amino acids.
Protein	Macromolecules whose structures are coded in an organism's DNA. Each is a chain of more than 40 amino acids folded back upon itself in a particular way. A polypeptide consisting of amino acids. In their biologically active states, proteins function as catalysts in metabolism and, to some extent, as structural elements of cells and tissues.
Protein sorting	Proteins are synthesized on ribosomes which are located mainly in the cytosol. Only a small number of ribosomes are in mitochondria and chloroplasts. Proteins synthesized on these ribosomes can be directly incorporated into the compartments within these organelles. However, most mitochondrial and chloroplast proteins are encoded by nuclear DNA and synthesized on cytosolic ribosomes. These and all other proteins synthesized in the cytosol must be transported to appropriate locations in the cell. This is made possible by the specific signal sequence in the newly synthesized peptide.
Protein synthesis	The process of protein synthesis goes through the following steps: *Initiation*: Peptide synthesis always starts from methionine (Met). Therefore, the initial aminoacyl-tRNA is Met-tRNA$_i$, where the subscript 'I' specifies 'initiation.' In bacteria, the methionine of the initial aminoacyl-tRNA has been modified by adding a formyl group (HCO) to its amino group. The modified methionine is called formylmethionine (fMet), which is unique for bacteria. Thus, fMet is an obvious foreign substance in eukaryotes. It can elicit a strong immune response. In humans, the immune response elicited by the peptide 'fMet-Leu-Phe' is about a thousand times greater than 'Met-Leu-Phe.'*Elongation*: A ribosome contains two major tRNA-binding sites: A and P sites. After the large subunit joins the initiation complex, the initial Met-tRNA$_i$ enters the P site and the newly arrived aminoacyl-tRNA is always placed at the A site ('A' for 'aminoacyl'). Then, methionine is transferred to the new aminoacyl-tRNA, forming a 'peptidyl-tRNA' where a peptide is attached to the tRNA. Subsequently, the empty tRNA at the P site is ejected from the ribosome, and the peptidyl-tRNA jumps to the P site ('P' for 'peptidyl'). During this translocation step, the ribosome also moves one codon down the mRNA chain. Similar steps are repeated in the subsequent cycles of elongation. *Termination*: Protein synthesis will terminate when the ribosome

	arrives at one of three stop codons. The termination process is assisted by special proteins called termination factors which recognize the stop codons. Their association stimulates the release of the peptidyl-tRNA from the ribosome. Subsequently, the released peptidyl-tRNA divides into tRNA and a newly synthesized peptide chain. The ribosome also divides into the large and small subunits, ready for synthesizing another peptide.
Protein transport	A protein destined for the nucleus and cytoplasm contains a specific sequence that can be recognized directly by importin/exportin or an adaptor protein.
Proteolytic	Capable of lysing (denaturing or breaking down) proteins.
Proteomics	Study of 'proteome,' which contains all proteins in a cell at a particular time. While all cells in an organism have the same genome, they usually have different proteomes.
QbD	quality by design (QbD) emphasizes that quality should be built into the process that produces the product, achieved via an understanding of the product and process by which it is developed and manufactured along with a knowledge of the risks involved in manufacturing the product and how best to mitigate those risks.
QTPP	quality target product profile (QTPP) is inclusive of the information described in the target protein profile, drug product profile, drug substance profile, and target protein profile
Radioimmunoassay (RIA)	A generic term for immunoassays has a radioactive label (tag) on either the antigen or antibody. Standard labels include I125 and H3, used for assay detection and quantitation. Classical RIAs are competitive binding assays where the antigen and tagged antigen compete for a limited fixed number of binding sites on the antibody. The antibody-bound tagged complex is inversely proportional to the concentration of the antigen.
RCA	Replication-competent adenovirus.
RCR	Replication-competent retrovirus.
RCV	Replication-competent virus.
rDNA technology	The rDNA technology generally relates to the manipulation of genetic materials and, more particularly, to recombinant procedures making possible the production of polypeptides possessing part or all of the primary structural conformation and one or more of the biological properties of naturally occurring proteins such as erythropoietin.
Recombinant	Containing genetic material from another organism. Genetically altered microorganisms are usually recombinant, whereas plants and animals so modified are called transgenic (see transgenics).
Recombinant DNA	Also known as gene cloning or splicing, recombinant DNA is a technique that produces identical copies (clones) of a gene. The procedure involves joining together DNA segments in a cell-free system (e.g., in a test tube outside living cells or organisms). The recombinant DNA molecule is then introduced into a cell, where it will replicate itself, either as an independent entity (autonomously) or as an integral part of a cellular chromosome.
Recombinant proteins production	Many proteins used for medical treatment or research are typically expressed at deficient concentrations. However, through

	recombinant DNA technology, many proteins can be produced. This involves cloning the gene encoding the desired protein into an 'expression vector' which must contain a promoter so that the protein can be expressed.
Recombination	The formation of new combinations and arrangements of genes during meiosis; recombination is achieved by crossing over, independent assortment, and segregation. NHLBI can be natural or synthetic. Narrower terms: genetic recombination, recombinant antibodies, recombinant proteins.
Reconciliation	A comparison between the theoretical quantity and the actual quantity.
Reference standard	Highly characterized physical specimens used in testing by pharmaceutical and related industries to help ensure medicines' identity, strength, quality, and purity.
Regeneration	It returns the packing in the column to its initial state after gradient elution; the mobile phase is passed through the column stepwise or in a gradient. The stationary phase is solvated to its original condition. In ion-exchange chromatography, regeneration involves replacing ions taken up in the exchange process with the original ions that occupied the exchange sites. Regeneration can also refer to bringing back any column to its original state (e.g., removing impurities with a strong solvent).
Regulatory gene	A gene controls the expression of one or more structural genes. It does this either directly by DNA binding or indirectly by activating a cellular protein.
Relevant genotypic and phenotypic markers	Those markers permit the identification of the cell line strain that should include the recombinant protein expression or the presence of the expression construct.
Relevant virus	The virus used in process evaluation studies is either the identified virus or of the same species as the known virus or likely to contaminate the cell-substrate or any other reagents or materials used in the production process.
Restriction enzyme	A bacterial enzyme that cuts DNA molecules at the location of sequences of base pairs. The role of these enzymes in bacteria is to 'restrict' the invasion of foreign DNA by cutting it into pieces. Hence, these enzymes are known as restriction enzymes. Restriction enzymes, also called restriction nucleases *(e.g., EcoRI from E. Coli)*, surrounds the DNA molecule at the point it seeks *(sequence GAATTC)*. It cuts one strand of the DNA double helix at one point and the second strand at a different, complementary point (between the G and the A base). The separated pieces have single-stranded 'sticky-ends,' which allow the complementary pieces to combine.
Restriction map	A linear arrangement of various restriction enzyme sites.
Restriction site	Base sequence recognized by an enzyme.
Retrovirus	RNA virus, which replicates via conversion into a DNA duplex.
Reverse transcriptase	An enzyme that catalyzes the synthesis of DNA from RNA.
Reverse vaccinology	Reverse vaccinology reduces the time required for identifying candidate vaccines and provides new solutions for those vaccines that have been difficult or impossible to develop. The basic idea

	behind reverse vaccinology is that an entire pathogenic genome can be screened using bioinformatics approaches to find genes. Some of the traits that the genes are monitored for that may indicate antigenicity include genes that code for proteins with extracellular localization, signal peptides, and B-cell epitopes. Next, those genes are filtered for desirable attributes that would make good vaccine targets such as outer membrane proteins. Once the candidates are identified, they are produced synthetically and are screened in animal models of the infection.
Ribonucleic acid (RNA)	A chemical substance made up of nucleotides compound of sugars, phosphates, and derivatives of the four bases adenine (A), guanine (G), cytosine (C), and uracil (U). RNAs function in cells as messengers of information from DNA translated into protein or as molecules that have certain structural or catalytic functions in the synthesis of proteins. RNA is also the carrier of genetic information for specific viruses. RNAs may be single or double-stranded.
Ribosome	Cell organelles that translate RNA to build proteins.
RNA	Ribonucleic acid.
RNA polymerase	An enzyme that catalyzes the synthesis of RNA in transcription.
RNA processing	RNA processing generates a mature mRNA (for protein genes) or a functional tRNA or rRNA from the primary transcript.
SDS PAGE (sodium dodecyl sulfate-polyacrylamide gel electrophoresis)	Electrophoretic separation of proteins based on their molecular weights. A uniform net negative charge is imposed on the molecules by adding SDS. Under these conditions, migration toward the anode through a gel matrix allows separation via size, not charge, with the smaller molecules migrating the longest distance. However, this technique is unreliable for sizes below an MW of ca. 8000. Proteins are observed via Coomassie blue or silver staining or further transferred to membranes for antigen/antibody specificity testing.
Sequence	The precise order of bases in a nucleic acid or amino acids in a protein.
Silencers	Silencers are the DNA elements that interact with repressors (proteins) to inhibit transcription. In prokaryotes, silencers are known as operators, found in many genes such as *lac* operon and *trp* operon. A DNA element may act either as an enhancer or silencer in a few cases, depending on the binding protein. For example, specific genes contain an element called E box (consensus CACGTG) which can bind either Max/Myc dimer or Max/Mad dimer. The Max/Myc dimer activates transcription, whereas the Max/Mad dimer suppresses transcription of these genes.
SMB	Simulated moving bed; chromatography method in which a countercurrent movement of the stationary and mobile phase is simulated by a complex interconnection of at least four chromatography columns.
Somatic cell	In higher organisms, a cell that (unlike germ cells).
Southern blot (blotting)	Technique for transferring DNA fragments from an agarose gel to a nitrocellulose filter can be hybridized to complementary DNA. Southern blotting is a technique for detecting specific DNA

	fragments in a complex mixture. The technique was invented in the mid-1970s by Edward Southern. It has been applied to detect Restriction Fragment Length Polymorphism (RFLP) and Variable Number of Tandem Repeat Polymorphism (VNTR). The latter is the basis of DNA fingerprinting.
Specific model virus	A virus is closely related to the known or suspected virus (same genus or family), having similar physical and chemical properties to those of the observed or suspected virus.
SPTFF	Single-pass tangential flow filtration (SPTFF): a tangential-flow filtration (TFF) without a recirculation loop, providing the performance of TFF with the simplicity of direct-flow filtration.
Strain	A population of cells all descended from a single cell. A group of organisms of the same species has distinctive characteristics but is not usually considered a separate breed or variety.
T-helper cells	T-lymphocytes have the specific capacity to help other cells, such as B-lymphocytes, make antibodies. T-helper cells are also required for the induction of other T-lymphocyte activities. A synonym is T inducer cell, T4 cell, or CD 4 lymphocyte.
T-suppressor cells	T-lymphocytes with a specific capacity to inhibit T-helper cell function.
Tailing	The phenomenon in which the normal Gaussian peak has an asymmetry factor > 1. The peak will have a skew in the trailing edge. Tailing is caused by sites on the packing that has stronger-than-normal retention for the solute.
Tandem repeats	Tandem repeats are an array of consecutive repeats of DNA sequence. They include three subclasses: satellites, minisatellites, and microsatellites. The name 'satellites' comes from their optical spectra. The size of a satellite DNA ranges from 100 kb to over 1 Mb. In humans, a well-known example is an alphoid DNA located at the centromere of all chromosomes. Its repeat unit is 171 bp, and the repetitive region accounts for 3–5% of the DNA in each chromosome. Other satellites have a shorter repeat unit. Most satellites in humans or other organisms are located at the centromere.
Target product profile (TPP)	outlines the desired 'profile' or characteristics of a target product aimed at a particular disease or disease. TPPs state intended use, target populations, and products' desired attributes, including safety and efficacy-related characteristics.
TCID50	Tissue culture infectious dose, 50%.
Telomerase	In eukaryotes, the chromosome ends are called telomeres which have at least two functions to protect chromosomes from fusing and solve the end-replication problem. In the absence of telomerase, the telomere will become shorter after each cell division. The cell may cease to divide and die when it reaches a certain length. Therefore, telomerase plays a critical role in the aging process.
Three-dimensional structure	The three-dimensional (3D) structure is also called the tertiary structure. If a protein molecule consists of more than one polypeptide, it also has a quaternary structure, which specifies the relative positions among the polypeptides (subunits) in a protein.

Topo isomers	During replication, the unwinding of DNA may cause the formation of tangling structures, such as supercoils or catenates. The major role of topoisomerases is to prevent DNA tangling.
Transcription	The first stage in gene expression uses genetic information transmitted from the chromosomes' DNA to messenger RNA. Next, transcription is a process in which one DNA strand is used as a template to synthesize complementary RNA. There are three classes of RNA polymerases in eukaryotes: I, II, and III. In prokaryotes, binding the polymerase's σ factor to the promoter can catalyze the unwinding of the DNA double helix. The most important σ factor is Sigma 70, whose structure has been determined by x-ray crystallography, but its complexity with DNA has not been solved. After the DNA strands have been separated at the promoter region, the core polymerase can then start to synthesize RNA based on the sequence of the DNA template strand.
Transduction	Transfer genes from one bacterium to another using a phage (virus).
Transfection	Permanently changing a cell using viral DNA.
Transformation	Permanent genetic change following incorporation of new DNA.
Transgene	A foreign gene is incorporated by transformation into the germline.
Transgenic organism	An organism has had its genes deliberately altered to give it specific characteristics it would not otherwise possess naturally.
Transgenics	The alteration of plant or animal DNA contains a gene from another organism. There are two types of cells in animals and plants, germline cells (the sperm and egg in animals, pollen and ovule in plants) and somatic cells (all other cells). The germ-line DNA is altered in transgenic animals and plants, so those alterations are passed on to offspring. Transgenic animals are used to produce therapeutics, study disease, or improve livestock strains. Transgenic plants have been created for increased resistance to disease and insects and make biopharmaceuticals.
Translation	The process of transferring information from DNA by RNA specifies the sequence of amino acids in a polypeptide (protein) chain.
Transmembrane pressure (TMP)	The force drives liquid flow through a cross-flow membrane. During filtration, the feed side of the membrane is under higher pressure than the permeate side. The pressure difference forces liquid through the membrane. TMP = {(feed pressure + retentate pressure) / 2} – permeate pressure.
Transplastomic	Transformation of plastids.
tRNA translation	The translation is a process by which the nucleotide sequence of mRNA is converted into the amino acid sequence of a peptide. It starts from the initiation codon and follows the mRNA sequence strictly with three nucleotides for one amino acid'.
Tryptic fragment analysis	Quantitating the resultant fragments caused by tryptic digestion.
Vector (molecular biology)	A vector in molecular biology is a vehicle that is used to carry foreign genetic material from one cell into another, where it can be replicated and expressed. Typically, a vector is a DNA sequence

	that consists of an insert (transgene) and a larger sequence that serves as the 'backbone' of the vector. The most common form of vector used in genetic engineering is plasmids.
Virus	The simplest form of life: RNA or DNA wrapped in a shell of protein, sometimes with a means of injecting that genetic material into a host organism (infection). Viruses cannot reproduce on their own but require the aid of a host.
Working cell bank (WCB)	The WCB is prepared from a homogeneous suspension of cells obtained from culturing the MCB under defined culture conditions. A working cell bank is created from the master cell bank by reviving the live cells: thawing and then culturing them on an agar medium. It may take two days to grow enough new cells to begin fermentation from a frozen bank. From a lyophilized bank, it can take longer. Refrigerated cultures need only a day or so.
Yeast artificial chromosome (YAC)	The yeast artificial chromosome (YAC) vector can carry a large DNA fragment (up to 2 Mb), but its transformation efficiency is very low. A vector constructed from the telomeric, centromeric, and replication origin sequences needed for replication in yeast cells used to clone pieces of DNA

Suggested reading

Preface

Arora R and CAB International 2010 *Medicinal Plant Biotechnology* (Cambridge, MA: CABI)

Atun R A and Sheridan D J 2007 *Innovation in the Biopharmaceutical Industry* (Hackensack NJ: World Scientific)

Bawa R, Szebeni J, Webster T J and Audette G F 2018 *Immune Aspects of Biopharmaceuticals and Nanomedicines* (Singapore: Pan Stanford)

Beck A 2013 *Glycosylation Engineering of Biopharmaceuticals: Methods and Protocols* (New York: Humana/Springer)

Behera B K 2021 *Biopharmaceuticals: Challenges and Opportunities* 1st edn (Boca Raton, FL: CRC)

Brooks G 1998 *Biotechnology in Healthcare: An Introduction to Biopharmaceuticals* (London: Pharmaceutical)

Burke M and Walmsley S 2001 *Biopharmaceuticals: A New Era of Discovery in the Biotechnology Revolution?* (Richmond, VA: PJB)

Castilho L R 2008 *Animal Cell Technology: From Biopharmaceuticals to Gene Therapy* (New York: Taylor and Francis)

Cavagnaro J A 2008 *Preclinical Safety Evaluation of Biopharmaceuticals: A Science-Based Approach to Facilitating Clinical Trials* (Hoboken, NJ: Wiley)

Crick F and Knablein J 2005 *Modern Biopharmaceuticals: Design, Development and Optimization* (Weinheim: Wiley)

Das T K 2014 *Biophysical Methods for Biotherapeutics: Discovery and Development Applications* (Hoboken, NJ: Wiley)

Dutton R L and Scharer J M 2007 *Advanced Technologies in Biopharmaceutical Processing* 1st edn (Ames, IA: Blackwell)

Geigert J 2019 *The Challenge of CMC Regulatory Compliance for Biopharmaceuticals* 3rd edn (Cham: Springer)

Grindley J N and Ogden J E 2000 *Understanding Biopharmaceuticals: Manufacturing and Regulatory Issues* (Denver, CO: Interpharm)

Hefferon K L 2010 *Biopharmaceuticals in Plants: Toward the Next Century of Medicine* (Boca Raton, FL: CRC/Taylor and Francis)

doi:10.1088/978-0-7503-3175-3ch9

Hill R G and Rang H P 2013 *Drug Discovery and Development: Technology in Transition* 2nd edn (Edinburgh: Churchill Livingstone/Elsevier)

Ho R J Y and Gibaldi M 2013 *Biotechnology and Biopharmaceuticals: Transforming Proteins and Genes into Drugs* 2nd edn (Hoboken, NJ: Wiley-Blackwell)

Houde D J and Berkowitz S A 2015 *Biophysical Characterization of Proteins in Developing Biopharmaceuticals* (Amsterdam: Elsevier)

Huxsoll J F 1994 *Quality Assurance for Biopharmaceuticals* (New York: Wiley)

Jameel F and Hershenson S 2010 *Formulation and Process Development Strategies for Manufacturing Biopharmaceuticals* (Oxford: Wiley)

Jørgensen L and Nielsen H M 2009 *Delivery Technologies for Biopharmaceuticals: Peptides, Proteins, Nucleic Acids and Vaccines* (Hoboken, NJ: Wiley)

Knäblein J 2013 *Modern Biopharmaceuticals: Recent Success Stories* (Weinheim: Wiley-Blackwell)

Komives C and Zhou W 2019 *Bioprocessing Technology for Production of Biopharmaceuticals and Bioproducts* 1st edn (Hoboken, NJ: Wiley)

Krause S O 2007 *Validation of Analytical Methods for Biopharmaceuticals: A Guide to Risk-Based Validation and Implementation Strategies* (Bethesda, MD: PDA)

Lee C-J 2006 *Clinical Trials of Drugs and Biopharmaceuticals* (Boca Raton, FL: Taylor and Francis)

Niazi S 2018 *Biosimilarity: The FDA Perspective* (Boca Raton, FL: CRC Press)

Niazi S 2015 *Biosimilars and Interchangeable Biologicals: Strategic Elements* (Boca Raton, FL: CRC)

Niazi S 2015 *Biosimilars and Interchangeable Biologics: Tactical Elements* (Boca Raton, FL: CRC)

Niazi S 2012 *Disposable Bioprocessing Systems* (Boca Raton, FL: CRC)

Niazi S 2003 *Filing Patents Online* (Boca Raton, FL: CRC)

Niazi S 2015 *Fundamentals of Modern Bioprocessing* ed S K Niazi and J L Brown (Boca Raton, FL: CRC)

Niazi S 2007 and 2014 *Handbook of Bioequivalence Testing* (New York: Informa Healthcare)

Niazi S 2005 *Handbook of Biogeneric Therapeutic Proteins: Manufacturing, Regulatory, Testing and Patent Issues* (Boca Raton, FL: CRC)

Niazi S 2004, 2009 and 2020 *Handbook of Pharmaceutical Manufacturing Formulations: Sterile Products* **vol 6** 2nd edn (New York: Informa Healthcare)

Niazi S 2009 *Handbook of Pharmaceutical Manufacturing Formulations: Compressed Solids* **vol 1** 2nd edn (New York: Informa Healthcare)

Niazi S 2004, 2009 and 2020 *Handbook of Pharmaceutical Manufacturing Formulations: Uncompressed Solids* **vol 2** 2nd edn (New York: Informa Healthcare)

Niazi S 2004, 2009 and 2020 *Handbook of Pharmaceutical Manufacturing Formulations: Liquid Products* **vol 3** 2nd edn (New York: Informa Healthcare)

Niazi S 2004, 2009 and 2020 *Handbook of Pharmaceutical Manufacturing Formulations: Semisolid Products* **vol 4** 2nd edn (New York: Informa Healthcare)

Niazi S 2004, 2009 and 2020 *Handbook of Pharmaceutical Manufacturing Formulations: Over the Counter Products* **vol 5** 2nd edn (New York: Informa Healthcare)

Niazi S 2006 and 2019 *Handbook of Preformulation: Chemical, Biological and Botanical Drugs* (New York: Informa Healthcare)

Niazi S 2004 and 2020 Pharmacokinetic and pharmacodynamic modeling in early drug development *The Process of New Drug Discovery and Development* 2nd edn, ed C G Smith and J T O'Donnell (New York: CRC)

Niazi S 1979 *Textbook of Biopharmaceutics and Clinical Pharmacokinetics* (New York: Wiley)
Prazeres D M F 2011 *Plasmid Biopharmaceuticals: Basics, Applications, and Manufacturing* (Oxford: Wiley-Blackwell)
Rathore A S and Mhatre R 2009 *Quality by Design for Biopharmaceuticals: Principles and Case Studies* (Oxford: Wiley)
Rathore A S and Sofer G K 2005 *Process Validation in Manufacturing of Biopharmaceuticals: Guidelines, Current Practices, and Industrial Case Studies* (Boca Raton, FL: Taylor and Francis)
Rathore A S and Sofer G K 2012 *Process Validation in Manufacturing of Biopharmaceuticals* 3rd edn (Boca Raton, FL: Taylor and Francis/CRC)
Rathore A S 2012 *Process Validation in Manufacturing of Biopharmaceuticals* 3rd edn (Hoboken: CRC)
Razinkov V I and Kleemann G R 2017 *High-Throughput Formulation Development of Biopharmaceuticals: Practical Guide to Methods and Applications* (Amsterdam: Elsevier/Woodhead)
Rehbinder E 2009 *Pharming: Promises and Risks of Biopharmaceuticals Derived From Genetically Modified Plants and Animals* (Berlin: Springer)
Reid E C T Q 2007 A tool-kit for in-process determination and control of structural and conformational authenticity of complex biopharmaceuticals *PhD Thesis* University of London
Rickwood S and Southworth A 1995 *New Technologies in Biopharmaceuticals* (London: Financial Times Pharmaceuticals and Healthcare)
Rojanasakul Y and Wu-Pong S 2008 *Biopharmaceutical Drug Design and Development* 2nd edn (Totowa, NJ: Humana)
Rosales Mendoza S 2016 *Algae-Based Biopharmaceuticals* (Cham: Springer)
Sandset T 2020 *'Ending AIDS' in the Age of Biopharmaceuticals: The Individual, the State and the Politics of Prevention* 1st edn (New York: Taylor & Francis)
Schmidt S R 2013 *Fusion Protein Technologies for Biopharmaceuticals: Applications and Challenges* (Hoboken, NJ: Wiley)
Segarra C 2013 Development of an integrated platform for the production of recombinant biopharmaceuticals *PhD Thesis* University of Sheffield
Takahashi M 2015 *Proc. 2015 Int. Conf. on Medicine and Biopharmaceuticals* (*China, 15–16 August*)
Tavakoli-Keshe R 2014 Quantifying the impact of the physical environment during processing and storage of biopharmaceuticals *PhD Thesis* University College London
Thomaz-Soccol V, Pandey A and Resende R R 2017 *Current Developments in Biotechnology and Bioengineering. Human and Animal Health Applications* (Amsterdam: Elsevier)
Tovey M G 2011 *Detection and Quantification of Antibodies to Biopharmaceuticals: Practical and Applied Considerations* (Hoboken, NJ: Wiley)
Walsh G and Murphy B 1999 *Biopharmaceuticals: An Industrial Perspective* (Dordrecht: Kluwer Academic)
Walsh G 2007 *Pharmaceutical Biotechnology: Concepts and Applications* (Chichester: Wiley)
Walsh G 2009 *Post-translational Modification of Protein Biopharmaceuticals* (Weinheim: Wiley)
Weert M and Møller E H 2008 *Immunogenicity of Biopharmaceuticals* (New York: Springer/Arlington, VA: AAPS Press)
Wittrup K D and Verdine G L 2012 *Protein Engineering for Therapeutics. Part B* 1st edn (San Diego, CA: Academic)

Yang H 2017 *Emerging Nonclinical Biostatistics in Biopharmaceutical Development and Manufacturing* (Boca Raton, FL: CRC/Taylor and Francis)

Chapter 1

Arora R and CAB International 2010 *Medicinal Plant Biotechnology* (Cambridge, MA: CABI)

Atun R A and Sheridan D J 2007 *Innovation in the Biopharmaceutical Industry* (Hackensack, NJ: World Scientific)

Bawa R, Szebeni J, Webster T J and Audette G F 2018 *Immune Aspects of Biopharmaceuticals and Nanomedicines* (Singapore: Pan Stanford)

Beck A 2013 *Glycosylation Engineering of Biopharmaceuticals: Methods and Protocols* (New York: Humana/Springer)

Behera B K 2021 *Biopharmaceuticals: Challenges and Opportunities* 1st edn (Boca Raton, FL: CRC)

Bharati S L and Chaurasia P K 2018 *Research Advancements in Pharmaceutical, Nutritional, and Industrial Enzymology* (Hershey, PA: Medical Information Science Reference)

1997 *BioDrugs: Clinical Immunotherapeutics, Biopharmaceuticals and Gene Therapy* vol 7 (Auckland: Adis International)

Biopharmaceuticals (Austin, TX: Landes Bioscience)

Bluemel J 2015 *The Nonhuman Primate in Nonclinical Drug Development and Safety Assessment* (Amsterdam: Elsevier/Academic)

Canadian Agency for Drugs and Technologies in Health 2017 Pharmacoeconomic review report. Abobotulinum toxin A (Dysport therapeutic): (Ipsen Biopharmaceuticals Canada, Inc) *CADTH Common Drug Review* 1st edn (Ottawa: Canadian Agency for Drugs and Technologies in Health)

Castilho A 2015 *Glyco-engineering: Methods and Protocols* (New York: Humana)

Center for Biologics Evaluation and Research 2010 *What Is a Biological Product?* (Silverspring, MD: US Food and Drug Administration)

Centers for Disease Control, Vaccines Acronyms and Abbreviations https://www.cdc.gov/vaccines/terms/vacc-abbrev.html

Costantino H R and Pikal M J 2004 *American Association of Pharmaceutical Scientists. Lyophilization of Biopharmaceuticals* (Arlington, VA: AAPS Press)

Das T K 2014 *Biophysical Methods for Biotherapeutics: Discovery and Development Applications* (Hoboken, NJ: Wiley)

Dutton R L and Scharer J M 2007 *Advanced Technologies in Biopharmaceutical Processing* 1st edn (Ames, IA: Blackwell)

EMA 2012 *Questions and Answers on Biosimilar Medicines (Similar Biological Medicinal Products)* (London: European Medicines Agency)

EMA 2016 *Advanced Therapy Medicinal Products* (London: European Medicines Agency)

Faye L C and Gomord V R 2009 *Recombinant Proteins from Plants: Methods and Protocols* (New York: Humana)

FDA, Cellular and Gene Therapy Guidances https://www.fda.gov/vaccines-blood-biologics/biologics-guidances/cellular-gene-therapy-guidances

Flynne W G 2008 *Biotechnology and Bioengineering* (New York: Nova Biomedical)

Foster L 2002 *Patenting in the Biopharmaceutical Industry—Comparing the US with Europe* https://web.archive.org/web/20060316164416/http://scientific.thomson.com/free/ipmatters/pii/8180019/

Franks F and Auffret T 2008 *Freeze-drying of Pharmaceuticals and Biopharmaceuticals: Principles and Practice* (Cambridge: RSC Publishing)

Geigert J 2019 *The Challenge of CMC Regulatory Compliance for Biopharmaceuticals* 3rd edn (Cham: Springer)

Grindley J N and Ogden J E 2000 *Understanding Biopharmaceuticals: Manufacturing and Regulatory Issues* (Denver, CO: Interpharm)

Günther C, Hauser A and Huss R 2016 *Advances in Pharmaceutical Cell Therapy: Principles of Cell-Based Biopharmaceuticals* (Hackensack, NJ: World Scientific)

Hefferon K L 2010 *Biopharmaceuticals in Plants: Toward the Next Century of Medicine* (Boca Raton, FL: CRC/Taylor and Francis)

HHS, Vaccines Glossary https://www.vaccines.gov/resources/glossary/index.html

Hill R G and Rang H P 2013 *Drug Discovery and Development: Technology in Transition* 2nd edn (Edinburgh: Churchill Livingstone/Elsevier)

Ho R J Y and Gibaldi M 2013 *Biotechnology and Biopharmaceuticals: Transforming Proteins and Genes into Drugs* 2nd edn (Hoboken, NJ: Wiley-Blackwell)

Houde D J and Berkowitz S A 2015 *Biophysical Characterization of Proteins in Developing Biopharmaceuticals* (Amsterdam: Elsevier)

Huxsoll J F 1994 *Quality Assurance for Biopharmaceuticals* (New York: Wiley)

International Union of Pure and Applied Chemistry 1997 *Compendium of Chemical Terminology: Recommendations* ed A D McNaught and A Wilkinson (Oxford: Blackwell Science) 'Gold Book' http://goldbook.iupac.org/

International Union of Pure and Applied Chemistry 2009 Glossary of terms related to pharmaceutics *Pure Appl. Chem.* **81** 971–999 https://www.iupac.org/publications/pac/81/5/0971/

Industrial Biotechnology Association and Paine Webber Inc 1990 *Biopharmaceuticals in Transition* (Woodlands, TX: Portfolio)

Jameel F and Hershenson S 2010 *Formulation and Process Development Strategies for Manufacturing Biopharmaceuticals* (Hoboken, NJ: Wiley)

Jørgensen L and Nielsen H M 2009 *Delivery Technologies for Biopharmaceuticals: Peptides, Proteins, Nucleic Acids and Vaccines* (Chichester: Wiley)

Kingham R, Klasa G and Carver K 2014 *Key Regulatory Guidelines for the Development of Biologics in the United States and Europe* (New York: Wiley) pp 75–88

Kiss B, Gottschalk U, Pohlscheidt M and Abraham E 2018 *New Bioprocessing Strategies: Development and Manufacturing of Recombinant Antibodies and Proteins* (Cham: Springer)

Knäblein J 2005 *Modern Biopharmaceuticals: Design, Development, and Optimization* (Weinheim: Wiley)

Knäblein J 2013 *Modern Biopharmaceuticals: Recent Success Stories* (Weinheim: Wiley-Blackwell)

Komives C and Zhou W 2019 *Bioprocessing Technology for Production of Biopharmaceuticals and Bioproducts* 1st edn (Hoboken, NJ: Wiley)

Lamanna W C, Holzmann J, Cohen H P, Guo X, Schweigler M, Stangler T, Seidl A and Schiestl M 2018 Maintaining consistent quality and clinical performance of biopharmaceuticals *Expert Opin. Biol. Ther.* **18** 369–79

Lazarus A H and Semple J W 2010 *Immunoglobulin Therapy* (Bethesda, MD: AABB)

Lill J R and Sandoval W N 2017 *Analytical Characterization of Biotherapeutics* (Hoboken, NJ: Wiley)

Liu C and Morrow J 2017 *Biosimilars of Monoclonal Antibodies: A Practical Guide to Manufacturing, Preclinical, and Clinical Development* (Hoboken, NJ: Wiley)

Mandenius C-F and Titchener-Hooker N J 2013 *Measurement, Monitoring, Modelling and Control of Bioprocesses* (Heidelberg: Springer)

Marks L 2018 *Engineering Health: How Biotechnology Changed Medicine* (London: The Royal Society of Chemistry)

Mire-Sluis A R (ed) 2005 *State of the Art Analytical Methods for the Characterization of Biological Products and Assessment of Comparability: Congress, Bethesda, MD, June 2003 (Developments in Biologicals)* **vol 122** (Basel: Karger)

Morbidelli M, Wolf M and Bielser J-M 2020 *Perfusion Cell Culture Processes for Biopharmaceuticals: Process Development, Design, and Scale-up* (New York: Cambridge University Press)

Nature *Biologics: Latest Research and Reviews* https://www.nature.com/subjects/biologics

Neves J D and Sarmento B 2014 *Mucosal Delivery of Biopharmaceuticals: Biology, Challenges and Strategies* (New York: Springer)

NFCR Center for Therapeutic Antibody Engineering, Glossary, National Foundation for Cancer Research, Dana Farber Cancer Institute http://research.dfci.harvard.edu/nfcr-ctae/research/tech_glossary.php

Niazi S and Brown J L 2016 *Fundamentals of Modern Bioprocessing* (Boca Raton, FL: CRC)

Niazi S 2019 *Biosimilarity: The FDA Perspective* (Boca Raton, FL: CRC)

Niazi S 2016 *Biosimilars and Interchangeable Biologics. Strategic Elements* (Boca Raton, FL: CRC/Taylor and Francis)

Niazi S 2016 *Biosimilars and Interchangeable Biologics. Tactical Elements* (Boca Raton, FL: Taylor and Francis/CRC)

Niazi S 2015 *Handbook of Bioequivalence Testing* 2nd edn (Boca Raton, FL: CRC/Taylor and Francis Group)

Niazi S 2006 *Handbook of Biogeneric Therapeutic Proteins: Regulatory, Manufacturing, Testing, and Patent Issues* (Boca Raton, FL: Taylor and Francis)

Niazi S 2019 *Handbook of Pharmaceutical Manufacturing Formulations* 3rd edn (Boca Raton, FL: CRC)

Niazi S 2019 *Handbook of Preformulation: Chemical, Biological, and Botanical Drugs* 2nd edn (Boca Raton, FL: CRC)

Niazi S 1979 *Textbook of Biopharmaceutics and Clinical Pharmacokinetics* (New York: Appleton-Century-Crofts)

Nick C 2012 The US Biosimilars Act: challenges facing regulatory approval *Pharm. Med.* **26** 145–52

Pandit N K 2007 *Introduction to the Pharmaceutical Sciences* 1st edn (Baltimore, MD: Lippincott Williams and Wilkins)

Pathak Y and Benita S 2012 *Antibody Mediated Drug Delivery Systems: Concepts, Technology, and Applications* (Hoboken, NJ: Wiley)

Pham P V 2018 *Stem Cell Drugs: A New Generation of Biopharmaceuticals* (Cham: Springer)

Prazeres D M F 2011 *Plasmid Biopharmaceuticals: Basics, Applications, and Manufacturing* (Hoboken, NJ: Wiley)

Primrose S B and Twyman R M 2004 *Genomics: Applications in Human Biology* (Malden, MA: Blackwell)

Pugsley M K and Curtis M J 2015 *Principles of Safety Pharmacology* (Heidelberg: Springer)

Rader R A 2008 (Re)defining biopharmaceutical *Nat. Biotechnol.* **26** 743–51

Rathore A S and Mhatre R 2009 *Quality by Design for Biopharmaceuticals: Principles and Case Studies* (Hoboken, NJ: Wiley)

Rathore A S and Sofer G K 2005 *Process Validation in Manufacturing of Biopharmaceuticals: Guidelines, Current Practices, and Industrial Case Studies* (Boca Raton, FL: Taylor and Francis)

Rathore A S and Sofer G K 2012 *Process Validation in Manufacturing of Biopharmaceuticals* 3rd edn (Boca Raton, FL: Taylor and Francis/CRC)

Razinkov V I and Kleemann G R 2017 *High-throughput Formulation Development of Biopharmaceuticals: Practical Guide to Methods and Applications* (Amsterdam: Elsevier/Woodhead)

Rehbinder E 2009 *Pharming: Promises and Risks of Biopharmaceuticals Derived from Genetically Modified Plants and Animals* (Berlin: Springer)

Rojanasakul Y and Wu-Pong S 2008 *Biopharmaceutical Drug Design and Development* 2nd edn (Totowa, NJ: Humana)

Schiestl M, Stangler T, Torella C, Cepeljnik T, Toll H and Grau R 2011 Acceptable changes in quality attributes of glycosylated biopharmaceuticals *Nat. Biotechnol.* **29** 310–12

Schmidt S R 2013 *Fusion Protein Technologies for Biopharmaceuticals: Applications and Challenges* (Hoboken, NJ: Wiley)

Shire S J 2010 *Current Trends in Monoclonal Antibody Development and Manufacturing* (New York: Springer/AAPS Press)

Thomaz-Soccol V, Pandey A and Resende R R 2017 *Current Developments in Biotechnology and Bioengineering. Human and Animal Health Applications* (Amsterdam: Elsevier)

Tovey M G 2011 *Detection and Quantification of Antibodies to Biopharmaceuticals: Practical and Applied Considerations* (Hoboken, NJ: Wiley)

Walsh G 2003 *Biopharmaceuticals: Biochemistry and Biotechnology* 2nd edn (New York: Wiley)

Walsh G and Murphy B 1999 *Biopharmaceuticals, an Industrial Perspective* (Dordrecht: Kluwer Academic)

Walsh G 2003 *Biopharmaceuticals: Biochemistry and Biotechnology* 2nd edn (Chichester: Wiley)

Wittrup K D and Verdine G L 2012 *Protein Engineering for Therapeutics. Part B* 1st edn (San Diego, CA: Academic)

Chapter 2

Alfaleh M A, Jones M L, Howard C B and Mahler S M 2017 Strategies for selecting membrane protein-specific antibodies using phage display with cell-based panning *Antibodies* **6** 10

Beck-Sickinger A G 2016 Directed evolution in drug and antibody development: From the Nobel Prize to broad clinical application *Internist (Berl.)* **60** 1014–20

Bolton G R and Mehta K K 2016 The role of more than 40 years of improvement in protein a chromatography in the growth of the therapeutic antibody industry *Biotechnol. Prog.* **32** 1193–202

Buyel J F, Twyman R M and Fischer R 2017 Very-large-scale production of antibodies in plants: the biologization of manufacturing *Biotechnol. Adv.* **35** 458–65

Chahar D S, Ravindran S and Pisal S S 2020 Monoclonal antibody purification and its progression to commercial scale *Biologicals* **63** 1–13

Chan S K and Lim T S 2017 Immune human antibody libraries for infectious diseases *Adv. Exp. Med. Biol.* **1053** 61–78

Chan S K, Rahumatullah A, Lai J Y and Lim T S 2017 Naïve human antibody libraries for infectious diseases *Adv. Exp. Med. Biol.* **1053** 35–59

Chen G, Sidhu S S and Nilvebrant J 2017 Synthetic antibodies in infectious disease *Adv. Exp. Med. Biol.* **1053** 79–98

Donini M and Marusic C 2019 Current state-of-the-art in plant-based antibody production systems *Biotechnol. Lett.* **41** 335–46

Drake P M, Szeto T H, Paul M J, Teh A Y and Ma J K 2017 Recombinant biologic products versus nutraceuticals from plants—a regulatory choice? *Br. J. Clin. Pharmacol.* **83** 82–7

Dumont J, Euwart D, Mei B, Estes S and Kshirsagar R 2016 Human cell lines for biopharmaceutical manufacturing: history, status, and future perspectives *Crit. Rev. Biotechnol.* **36** 1110–22

Elgundi Z, Reslan M, Cruz E, Sifniotis V and Kayser V 2017 The state-of-play and future antibody therapeutics *Adv. Drug Deliv. Rev.* **122** 2–19

Foltz I N, Gunasekaran K and King C T 2016 Discovery and bio-optimization of human antibody therapeutics using the XenoMouse® transgenic mouse platform *Immunol. Rev.* **270** 51–64

Frenzel A *et al* 2017 Designing human antibodies by phage display *Transfus. Med. Hemother.* **44** 312–18

Frenzel A, Schirrmann T and Hust M 2016 Phage display-derived human antibodies in clinical development and therapy *MAbs* **8** 1177–94

Gao W *et al* 2016 Recent advances in site-specific conjugations of antibody drug conjugates (ADCs) *Curr. Cancer Drug Targets* **16** 469–79

Gaughan C L 2016 The present state of the art in expression, production, and characterization of monoclonal antibodies *Mol. Divers* **20** 255–70

Ghosh C, Sarkar P, Issa R and Haldar J 2019 Alternatives to conventional antibiotics in the era of antimicrobial resistance *Trends Microbiol.* **27** 323–38

Glick B R and Patten C L 2017 *Molecular Biotechnology: Principles and Applications of Recombinant DNA* 5th edn (Washington, DC: ASM Press)

Gupta S K and Shukla P 2018 Glycosylation control technologies for recombinant therapeutic proteins *Appl. Microbiol. Biotechnol.* **102** 10457–68

Holtz B R *et al* 2015 Commercial-scale biotherapeutics manufacturing facility for plant-made pharmaceuticals *Plant Biotechnol. J.* **13** 1180–90

Igawa T 2017 Next generation antibody therapeutics using bispecific antibody technology *Yakugaku Zasshi* **137** 831–6

Ito Y 2017 Development of a rapid identification method for a variety of antibody candidates using high-throughput sequencing *Yakugaku Zasshi* **137** 823–30

Jafari B, Hamzeh-Mivehroud M, Morris M B and Dastmalchi S 2019 Exploitation of phage display to develop anticancer agents targeting fibroblast growth factor signaling pathways: new strategies to tackle an old challenge *Cytokine Growth Factor Rev.* **46** 54–65

Jin Y, Lei C, Hu D, Dimitrov D S and Ying T 2017 Human monoclonal antibodies as candidate therapeutics against emerging viruses *Front Med.* **11** 462–70

Joubert S, Dodelet V, Béliard R and Durocher Y 2019 Biomanufacturing of monoclonal antibodies *Med. Sci.* **35** 1153–9

Kelley B, Kiss R and Laird M 2018 A different perspective: how much innovation is needed for monoclonal antibody production using mammalian cell technology? *Adv. Biochem. Eng. Biotechnol.* **165** 443–62

Kennedy P J, Oliveira C, Granja P L and Sarmento B 2018 Monoclonal antibodies: technologies for early discovery and engineering *Crit. Rev. Biotechnol.* **38** 394–408

Kim M G, Kim D, Suh S K, Park Z, Choi M J and Oh Y K 2016 Current status and regulatory perspective of chimeric antigen receptor-modified T cell therapeutics *Arch. Pharm. Res.* **39** 437–52

Kumar R, Parray H A, Shrivastava T, Sinha S and Luthra K 2019 Phage display antibody libraries: a robust approach for generating recombinant human monoclonal antibodies *Int. J. Biol. Macromol.* **135** 907–18

Kunert R and Reinhart D 2016 Advances in recombinant antibody manufacturing *Appl. Microbiol. Biotechnol.* **100** 3451–61

Lai J Y and Lim T S 2020 Infectious disease antibodies for biomedical applications: a mini review of immune antibody phage library repertoire *Int. J. Biol. Macromol.* **163** 640–8

Lu R M *et al* 2020 Develop therapeutic antibodies for the treatment of diseases *J. Biomed. Sci.* **27** 1

Lushova A A, Biazrova M G, Prilipov A G, Sadykova G K, Kopylov T A and Filatov A V 2017 Next-generation techniques for discovering human monoclonal antibodies *Mol. Biol.* **51** 899–906

Marschall A L, Dübel S and Böldicke T 2016 Recent advances with ER targeted intrabodies *Adv. Exp. Med. Biol.* **917** 77–93

Mastrangeli R, Palinsky W and Bierau H 2019 Glycoengineered antibodies: towards the next-generation of immunotherapeutics *Glycobiology* **29** 199–210

Ministro J, Manuel A M and Goncalves J 2020 Therapeutic antibody engineering and selection strategies *Adv. Biochem. Eng. Biotechnol.* **171** 55–86

Nagano K 2016 Challenge to the development of molecular targeted therapy against a novel target candidate identified by antibody proteomics technology *Yakugaku Zasshi* **136** 145–9

Nicolini F *et al* 2020 Fully human antibodies for malignant pleural mesothelioma targeting *Cancers* **12** 915

Reader R H, Workman R G, Maddison B C and Gough K C 2019 Advances in the production and batch reformatting of phage antibody libraries *Mol. Biotechnol.* **61** 801–15

Shim H 2017 Antibody phage display *Adv. Exp. Med. Biol.* **1053** 21–34

Shukla A A, Wolfe L S, Mostafa S S and Norman C 2017 Evolving trends in mAb production processes *Bioeng. Transl. Med.* **2** 58–69

Somasundaram R, Choraria A and Antonysamy M 2020 An approach towards developing monoclonal IgY antibodies against SARS CoV-2 spike protein (S) using phage display method: a review *Int. Immunopharmacol.* **85** 106654

Ubah O and Palliyil S 2017 Monoclonal antibodies and antibody like fragments derived from immunised phage display libraries *Adv. Exp. Med. Biol.* **1053** 99–117

Unkauf T, Miethe S, Fühner V, Schirrmann T, Frenzel A and Hust M 2016 Generation of recombinant antibodies against toxins and viruses by phage display for diagnostics and therapy *Adv. Exp. Med. Biol.* **917** 55–76

Van Hoecke L and Roose K 2019 How mRNA therapeutics are entering the monoclonal antibody field *J. Transl. Med.* **17** 54

Viardot A and Bargou R 2018 Bispecific antibodies in haematological malignancies *Cancer Treat Rev.* **65** 87–95

Voigt A, Semenova T, Yamamoto J, Etienne V and Nguyen C Q 2018 Therapeutic antibody discovery in infectious diseases using single-cell analysis *Adv. Exp. Med. Biol.* **1068** 89–102

Wang W and Manmohan S 2014 *Biological Drug Products: Development and Strategies* (Hoboken, NJ: Wiley)

Wei B, Berning K, Quan C and Zhang Y T 2017 Glycation of antibodies: modification, methods and potential effects on biological functions *MAbs* **9** 586–94

Yusibov V, Kushnir N and Streatfield S J 2016 Antibody production in plants and green algae *Annu. Rev. Plant Biol.* **67** 669–701

Zhao A, Tohidkia M R, Siegel D L, Coukos G and Omidi Y 2016 Phage antibody display libraries: a powerful antibody discovery platform for immunotherapy *Crit. Rev. Biotechnol.* **36** 276–89

Chapter 3

Al-Rubeai M and Naciri M 2014 *Stem Cells and Cell Therapy* (*Cell Engineering*) (Berlin: Springer)

Alarcon J B, Waine G W and McManus D P 1999 DNA vaccines: technology and application as anti-parasite and anti-microbial agents *Adv. Parasitol.* **42** 343–410

André S, Seed B, Eberle J, Schraut W, Bültmann A and Haas J 1998 Increased immune response elicited by DNA vaccination with a synthetic gp120 sequence with optimized codon usage *J. Virol.* **72** 1497–503

Armbruster N, Jasny E and Petsch B 2019 Advances in RNA vaccines for preventive indications: a case study of a vaccine against rabies *Vaccines* **7** 132

Annals of the New York Academy of Sciences 2012 *Evolving Challenges in Promoting Cardiovascular Health* (New York: Wiley)

Barouch D H *et al* 1998 Augmentation and suppression of immune responses to an HIV-1 DNA vaccine by plasmid cytokine/Ig administration *J. Immunol.* **161** 1875–82

Benteyn D, Heirman C, Bonehill A, Thielemans K and Breckpot K 2014 mRNA-based dendritic cell vaccines *Exp. Rev. Vaccines* **14** 161–76

Berkhout B, Ertl H C J and Weinberg M S 2015 *Gene Therapy for HIV and Chronic Infections* (*Advances in Experimental Medicine and Biology*) (Berlin: Springer)

Boelens J J and Wynn R F 2013 *Stem Cell Therapy in Lysosomal Storage Diseases* (*Stem Cell Biology and Regenerative Medicine*) (Berlin: Springer)

Boulis N, O'Connor D and Donsante A 2017 *Molecular and Cellular Therapies for Motor Neuron Diseases* (New York: Elsevier/Academic)

CDC 2005 CDC and Fort Dodge Animal Health achieve first licensed DNA vaccine *CDC* https://www.cdc.gov/media/pressrel/r050718.htm

Carralot J P *et al* 2004 Polarization of immunity induced by direct injection of naked sequence-stabilized mRNA vaccines *Cell Mol. Life Sci.* **61** 2418–24

Casaroli-Marano R P and Zarbin M A 2014 Cell-based therapy for retinal degenerative disease *Developments in Ophthalmology* (Basel: Karger)

Cathomen T, Hirsch M and Porteus M H 2016 *Genome Editing: the Next Step in Gene Therapy* (*Advances in Experimental Medicine and Biology*) (Berlin: Springer)

Chase L G and Vemuri M C 2013 *Mesenchymal Stem Cell Therapy* (*Stem Cell Biology and Regenerative Medicine*) (New York: Humana)

Chen J, Lin L, Li N and She F 2012 Enhancement of *Helicobacter pylori* outer inflammatory protein DNA vaccine efficacy by co-delivery of interleukin-2 and B subunit heat-labile toxin gene encoded plasmids *Microbiol. Immunol.* **56** 85–92

Chen Y, Webster R G and Woodland D L 1998 Induction of CD8+ T cell responses to dominant and subdominant epitopes and protective immunity to Sendai virus infection by DNA vaccination *J. Immunol.* **160** 2425–32

Dzau V J and Liew C-C 2007 *Cardiovascular Genetics and Genomics for the Cardiologist* (London: Blackwell Futura)

Freese A 2005 *Principles of Molecular Neurosurgery* (*Progress in Neurological Surgery*) (Basel: Karger)

Fynan E F, Webster R G, Fuller D H, Haynes J R, Santoro J C and Robinson H L 1993 DNA vaccines: protective immunizations by parenteral, mucosal, and gene-gun inoculations *Proc. Natl Acad. Sci. USA* **90** 11478–82

Galli M C and Serabian M 2015 *Regulatory Aspects of Gene Therapy and Cell Therapy Products: a Global Perspective* (*Advances in Experimental Medicine and Biology*) (Berlin: Springer)

Gao Q *et al* 2019 Therapeutic potential of CRISPR/Cas9 gene editing in engineered T-cell therapy *Cancer Med.* **8** 4254–64

Greaves D R and Gordon S 2005 Thematic review series: the immune system and atherogenesis. Recent insights into the biology of macrophage scavenger receptors *J. Lipid Res.* **46** 11–20

Greenwell P and McCulley M 2007 *Molecular Therapeutics: 21st Century Medicine* (New York: Wiley)

Grier E V 2004 *Focus on Stem Cell Research* (New York: Nova Biomedical)

Grier E V 2006 *Stem Cell Therapy* (New York: Nova Biomedical)

Hakim N S 2010 *Pancreas, Islet, and Stem Cell Transplantation for Diabetes* 2nd edn (Oxford: Oxford University Press)

Harrison R P and Chauhan V M 2017 Enhancing cell and gene therapy manufacture through the application of advanced fluorescent optical sensors (review) *Biointerphases* **13** 01A301

Hostiuc S 2018 *Clinical Ethics at the Crossroads of Genetic and Reproductive Technologies* (New York: Elsevier/Academic)

Huebener N, Fest S, Strandsby A, Michalsky E, Preissner R, Zeng Y, Gaedicke G and Lode H N 2008 A rationally designed tyrosine hydroxylase DNA vaccine induces specific antineuroblastoma immunity *Mol. Cancer Therap.* **7** 2241–51

Jain K K 2013 *Applications of Biotechnology in Neurology* (Totowa, NJ: Humana)

Kanagavelu S K *et al* 2012 Soluble multi-trimeric TNF superfamily ligand adjuvants enhance immune responses to a HIV-1 Gag DNA vaccine *Vaccine* **30** 691–702

Klinman D M, Yamshchikov G and Ishigatsubo Y 1997 Contribution of CpG motifs to the immunogenicity of DNA vaccines *J. Immunol.* **158** 3635–9

Kreiter S *et al* 2011 FLT3 ligand enhances the cancer therapeutic potency of naked RNA vaccines *Cancer Res.* **71** 6132–42

Kuhn A N *et al* 2010 Phosphorothioate cap analogs increase stability and translational efficiency of RNA vaccines in immature dendritic cells and induce superior immune response *in vivo* *Gene Ther.* **17** 961–71

Kutzler M A and Weiner D B 2008 DNA vaccines: ready for prime time? *Nat. Rev. Genet.* **9** 776–88

Leri A, Anversa P and Frishman W H 2007 *Cardiovascular Regeneration and Stem Cell Therapy* (Maiden, MA: Blackwell Futura)

Lewis P J and Babiuk L A 1999 DNA vaccines: a review *Adv. Virus Res.* **54** 129–88

Loudon P T *et al* 2010 GM-CSF increases mucosal and systemic immunogenicity of an H1N1 influenza DNA vaccine administered into the epidermis of non-human primates *PLoS One* **5** e11021

Mancini-Bourgine M, Fontaine H, Bréchot C, Pol S and Michel M L 2006 Immunogenicity of a hepatitis B DNA vaccine administered to chronic HBV carriers *Vaccine* **24** 4482–9

Mejía Vázquez D M D C and Navarro S 2010 *New Approaches in the Treatment of Cancer* (*Cancer Etiology, Diagnosis and Treatments*) (Hauppauge, NY: Nova Science)

Morstyn G and Sheridan W 1996 *Cell Therapy: Stem Cell Transplantation, Gene Therapy, and Cellular Immunotherapy* (*Cancer: Clinical Science in Practice*) (Cambridge: Cambridge University Press)

Ng P and Brunetti-Pierri N 2017 *Therapeutic Applications of Adenoviruses* (*Gene and Cell Therapy Series*) (Boca Raton, FL: CRC)

Odé Z, Condori J, Peterson N, Zhou S and Krenciute G *et al* 2020 CRISPR-mediated non-viral site-specific gene integration and expression in T cells: protocol and application for T-cell therapy *Cancers* **12** 1704

Pardi N, Hogan M J, Porter F W and Weissman D 2018 mRNA vaccines—a new era in vaccinology *Nat. Rev. Drug Discov.* **17** 261–79

Pardi N and Weissman D 2016 Nucleoside modified mRNA vaccines for infectious diseases *Meth. Mol. Biol.* **1499** 109–21

Pascolo S 2006 Vaccination with messenger RNA *Meth. Mol. Med.* **127** 23–40

Perales M-A, Abutalib S A and Bollard C 2019 *Cell and Gene Therapies* (*Advances and Controversies in Hematopoietic Transplantation and Cell Therapy*) (Berlin: Springer)

Petite H 2005 *Engineered Bone* (Boca Raton, FL: CRC)

Polak J M 2010 *Cell Therapy for Lung Disease* (London: Imperial College Press)

Rees R C 2014 *Tumor Immunology and Immunotherapy* 1st edn (Oxford: Oxford University Press)

Reichmuth A M, Oberli M A, Jaklenec A, Langer R and Blankschtein D 2016 mRNA vaccine delivery using lipid nanoparticles *Therap. Deliv.* **7** 319–34

Robinson H L and Pertmer T M 2000 DNA vaccines for viral infections: basic studies and applications *Adv. Virus Res.* **55** 1–74

Sedegah M, Jones T R, Kaur M, Hedstrom R, Hobart P, Tine J A and Hoffman S L 1998 Boosting with recombinant vaccinia increases immunogenicity and protective efficacy of malaria DNA vaccine *Proc. Natl Acad. Sci. USA* **95** 7648–53

Segal B H 2018 *Management of Infections in the Immunocompromised Host* (Berlin: Springer)

Sideman S and Beyar R 2004 *Cardiac Engineering: From Genes and Cells to Structure and Function* (*Annals of the New York Academy of Sciences*) (New York: New York Academy of Sciences)

Sreebny L M and Vissink A 2010 *Dry Mouth: The Malevolent Symptom: A Clinical Guide* (Hoboken, NJ: Wiley-Blackwell)

Templeton N S 2004 *Gene and Cell Therapy: Therapeutic Mechanisms and Strategies* 2nd edn (New York: Marcel Dekker)

Terai S and Suda T 2016 *Gene Therapy and Cell Therapy through the Liver: Current Aspects and Future Prospects* (Berlin: Springer)

Tousoulis D 2018 *Coronary Artery Disease: From Biology to Clinical Practice* (New York: Elsevier/Academic Press)

van den Berg J H *et al* 2010 Shielding the cationic charge of nanoparticle-formulated dermal DNA vaccines is essential for antigen expression and immunogenicity *J. Control. Release* **141** 234–40

Verbeke R, Lentacker I, De Smedt S C and Dewitte H 2019 Three decades of messenger RNA vaccine development *Nano Today* **28** 100766

Viroj W 2009 *Cell, Gene, and Molecular Therapy: New Concepts* (New York: Nova Biomedical)

Vogel A B *et al* 2018 Self-amplifying RNA vaccines give equivalent protection against influenza to mRNA vaccines but at much lower doses *Mol. Ther.* **26** 446–55

Wang P J 2005 *New Arrhythmia Technologies* (Maiden, MA: Blackwell Futura)

Weiss W R, Ishii K J, Hedstrom R C, Sedegah M, Ichino M, Barnhart K, Klinman D M and Hoffman S L 1998 A plasmid encoding murine granulocyte-macrophage colony-stimulating factor increases protection conferred by a malaria DNA vaccine *J. Immunol.* **161** 2325–32

Whitehouse D and Rapley R 2012 *Molecular and Cellular Therapeutics* (Hoboken, NJ: Wiley-Blackwell)

Widera G *et al* 2000 Increased DNA vaccine delivery and immunogenicity by electroporation *in vivo J. Immunol.* **164** 4635–40

Xu H *et al* 2010 CD40-expressing plasmid induces anti-CD40 antibody and enhances immune responses to DNA vaccination *J. Gene Med.* **12** 97–106

Chapter 4

Abramovich R A, Bykov V A, Elagina I A, Papazova N A and Vorob'ev A N 2012 Scientific approaches to development of medicinal formulation based on biotechnological substance *Antibiot. Khimioter.* **57** 13–16

Agersø H, Møller-Pedersen J, Cappi S, Thomann P, Jesussek B and Senderovitz T 2002 Pharmacokinetics and pharmacodynamics of a new formulation of recombinant human growth hormone administered by ZomaJet 2 Vision, a new needle-free device, compared to subcutaneous administration using a conventional syringe *J. Clin. Pharmacol.* **42** 1262–8

Agrawal G, Wakte P and Shelke S 2017 Formulation optimization of human insulin loaded microspheres for controlled oral delivery using response surface methodology *Endocr. Metab. Immune. Disord. Drug Targets* **17** 149–65

Alebouyeh M *et al* 2016 Rapid formulation assessment of filgrastim therapeutics by a thermal stress test *Biologicals* **44** 150–6

Allison A C and Byars N E 1986 An adjuvant formulation that selectively elicits the formation of antibodies of protective isotypes and of cell-mediated immunity *J. Immunol. Meth.* **95** 157–68

Ameri M *et al* 2014 Human growth hormone delivery with a microneedle transdermal system: preclinical formulation, stability, delivery and PK of therapeutically relevant doses *Pharmaceutics* **6** 220–34

Anderson P M and Sorenson M A 1994 Effects of route and formulation on clinical pharmacokinetics of interleukin-2 *Clin. Pharmacokinet.* **27** 19–31

Andya J D *et al* 1999 The effect of formulation excipients on protein stability and aerosol performance of spray-dried powders of a recombinant humanized anti-IgE monoclonal antibody *Pharm. Res.* **16** 350–8

Anish C, Upadhyay A K, Sehgal D and Panda A K 2014 Influences of process and formulation parameters on powder flow properties and immunogenicity of spray dried polymer particles entrapping recombinant pneumococcal surface protein A *Int. J. Pharm.* **466** 198–210

Aubin Y, Hodgson D J, Thach W B, Gingras G and Sauvé S 2015 Monitoring effects of excipients, formulation parameters and mutations on the high order structure of filgrastim by NMR *Pharm. Res.* **32** 3365–75

Balasubramanian S V, Bruenn J and Straubinger R M 2000 Liposomes as formulation excipients for protein pharmaceuticals: a model protein study *Pharm. Res.* **17** 344–50

Bednarek E *et al* 2017 Structure and pharmaceutical formulation development of a new long-acting recombinant human insulin analog studied by NMR and MS *J. Pharm. Biomed. Anal.* **135** 126–32

Bei R *et al* 1998 The use of a cationic liposome formulation (DOTAP) mixed with a recombinant tumor-associated antigen to induce immune responses and protective immunity in mice *J. Immunother.* **21** 159–69

Bello-Rivero I *et al* 2018 HeberFERON, a new formulation of IFNs with improved pharmacodynamics: perspective for cancer treatment *Semin. Oncol.* **45** 27–33

Bellomi F *et al* 2007 Immunogenicity comparison of interferon beta-1a preparations using the BALB/c mouse model: assessment of a new formulation for use in multiple sclerosis *New Microbiol.* **30** 241–6

Bhambhani A, Kissmann J M, Joshi S B, Volkin D B, Kashi R S and Middaugh C R 2012 Formulation design and high-throughput excipient selection based on structural integrity and conformational stability of dilute and highly concentrated IgG1 monoclonal antibody solutions *J. Pharm. Sci.* **101** 1120–35

Bittner B *et al* 2012 Development of a subcutaneous formulation for trastuzumab-nonclinical and clinical bridging approach to the approved intravenous dosing regimen *Arzneimittelforschung* **62** 401–9

Bogard W C *et al* 1989 Practical considerations in the production, purification, and formulation of monoclonal antibodies for immunoscintigraphy and immunotherapy *Semin. Nucl. Med.* **19** 202–20

Boven K *et al* 2005 The increased incidence of pure red cell aplasia with an Eprex formulation in uncoated rubber stopper syringes *Kidney Int.* **67** 2346–53

Brückl L, Hahn R, Sergi M and Scheler S 2016 A systematic evaluation of mechanisms, material effects, and protein-dependent differences on friction-related protein particle formation in formulation and filling steps *Int. J. Pharm.* **511** 931–45

Bush L, Webb C, Bartlett L and Burnett B 1998 The formulation of recombinant factor IX: stability, robustness, and convenience *Semin. Hematol.* **35** 18–21

Bye J W, Platts L and Falconer R J 2014 Biopharmaceutical liquid formulation: a review of the science of protein stability and solubility in aqueous environments *Biotechnol. Lett.* **36** 869–75

Bysted B V, Scharling B, Møller T and Hansen B L 2007 A randomized, double-blind trial demonstrating bioequivalence of the current recombinant activated factor VII formulation and a new robust 25 degrees C stable formulation *Haemophilia* **13** 527–32

Chang B S and Hershenson S 2002 Practical approaches to protein formulation development *Pharm. Biotechnol.* **13** 1–25

Chang B S, Reeder G and Carpenter J F 1996 Development of a stable freeze-dried formulation of recombinant human interleukin-1 receptor antagonist *Pharm. Res.* **13** 243–9

Chen B L, Arakawa T, Hsu E, Narhi L O, Tressel T J and Chien S L 1994 Strategies to suppress aggregation of recombinant keratinocyte growth factor during liquid formulation development *J. Pharm. Sci.* **83** 1657–61

Chen F M *et al* 2007 Novel glycidyl methacrylated dextran (Dex-GMA)/gelatin hydrogel scaffolds containing microspheres loaded with bone morphogenetic proteins: formulation and characteristics *J. Control. Release* **118** 65–77

Chen S *et al* 2012 Investigation on formulation and preparation of adenovirus encoding human endostatin lyophilized powders *Int. J. Pharm.* **427** 145–52

Cun D, Wan F and Yang M 2015 Formulation strategies and particle engineering technologies for pulmonary delivery of biopharmaceuticals *Curr. Pharm. Des.* **21** 2599–610

Davio S R and Hageman M J 1993 Characterization and formulation considerations for recombinantly derived bovine somatotropin *Pharm. Biotechnol.* **5** 59–89

Dawson P J 1992 Effect of formulation and freeze-drying on the long-term stability of rDNA-derived cytokines *Dev. Biol. Stand.* **74** 273–82

Devrim B, Bozkir A and Canefe K 2011 Preparation and evaluation of PLGA microparticles as carrier for the pulmonary delivery of rhIL-2 I. Effects of some formulation parameters on microparticle characteristics *J. Microencapsul.* **28** 582–94

Di Minno G *et al* 2010 Longer-acting factor VIII to overcome limitations in haemophilia management: the PEGylated liposomes formulation issue *Haemophilia* **16** 2–6

Eng M *et al* 1997 Formulation development and primary degradation pathways for recombinant human nerve growth factor *Anal. Chem.* **69** 4184–90

Engler H *et al* 2003 Development of a formulation that enhances gene expression and efficacy following intraperitoneal administration in rabbits and mice *Mol. Ther.* **7** 558–64

Fabregas B 2008 New formulation of recombinant coagulation factor: simplicity, rapidity and manageability *Soins* **2008** 22–3

Fatouros A and Sjöström B 2000 Recombinant factor VIII SQ—the influence of formulation parameters on structure and surface adsorption *Int. J. Pharm.* **194** 69–79

Frokjaer S and Otzen D E 2005 Protein drug stability: a formulation challenge *Nat. Rev. Drug Discov.* **4** 298–306

García-García I *et al* 2016 Pharmacokinetic and pharmacodynamic characterization of a novel formulation containing co-formulated interferons alpha-2b and gamma in healthy male volunteers *BMC Pharmacol. Toxicol.* **17** 58

Giannos S A, Kraft E R, Zhao Z Y, Merkley K H and Cai J 2018 Formulation stabilization and disaggregation of bevacizumab, ranibizumab and aflibercept in dilute solutions *Pharm. Res.* **35** 78

Gibbons A, McElvaney N G and Cryan S A 2010 A dry powder formulation of liposome-encapsulated recombinant secretory leukocyte protease inhibitor (rSLPI) for inhalation: preparation and characterisation *AAPS Pharm. Sci. Tech.* **11** 1411–21

Gietz U, Arvinte T, Häner M, Aebi U and Merkle H P 2000 Formulation of sustained release aqueous Zn-hirudin suspensions *Eur. J. Pharm. Sci.* **11** 33–41

Gombotz W R, Pankey S C, Bouchard L S, Phan D H and MacKenzie A P 1996 Stability, characterization, formulation, and delivery system development for transforming growth factor-beta 1 *Pharm. Biotechnol.* **9** 219–45

Gourbatsi E, Povey J F and Smales C M 2018 The effect of formulation variables on protein stability and integrity of a model IgG4 monoclonal antibody and translation to formulation of a model ScFv *Biotechnol. Lett.* **40** 33–46

Govardhan C *et al* 2005 Novel long-acting crystal formulation of human growth hormone *Pharm. Res.* **22** 1461–70

Grumetto L, Prete A D, Ortosecco G, Borrelli A, Prete S D and Mancini A 2016 A gel formulation containing a new recombinant form of manganese superoxide dismutase: a clinical experience based on compassionate use-safety of a case report *Case Rep. Ophthalmol. Med.* **2016** 7240209

Guo P, Yu C, Wang Q, Zhang R, Meng X and Feng Y 2018 Liposome lipid-based formulation has the least influence on rAAV transduction compared to other transfection agents *Mol. Ther. Methods Clin. Dev.* **9** 367–75

Gurny R and Friess W 2011 Unmet needs in protein formulation science *Eur. J. Pharm. Biopharm.* **78** 183

Hahn S K, Kim J S and Shimobouji T 2007 Injectable hyaluronic acid microhydrogels for controlled release formulation of erythropoietin *J. Biomed. Mater. Res.* A **80** 916–24

Hahn S K, Kim S J, Kim M J and Kim D H 2004 Characterization and *in vivo* study of sustained-release formulation of human growth hormone using sodium hyaluronate *Pharm. Res.* **21** 1374–81

Hamizi S *et al* 2013 Subcutaneous trastuzumab: development of a new formulation for treatment of HER2-positive early breast cancer *Oncol. Targets Ther.* **6** 89–94

Hancock G E, Heers K M and Smith J D 2000 QS-21 synergizes with recombinant interleukin-12 to create a potent adjuvant formulation for the fusion protein of respiratory syncytial virus *Viral. Immunol.* **13** 503–9

Heller M C, Carpenter J F and Randolph T W 1999 Protein formulation and lyophilization cycle design: prevention of damage due to freeze-concentration induced phase separation *Biotechnol. Bioeng.* **63** 166–74

Herman A C, Boone T C and Lu H S 1996 Characterization, formulation, and stability of Neupogen (Filgrastim), a recombinant human granulocyte-colony stimulating factor *Pharm. Biotechnol.* **9** 303–28

Hofer C, Göbel R, Deering P, Lehmer A and Breul J 1999 Formulation of interleukin-2 and interferon-alpha containing ultradeformable carriers for potential transdermal application *Anticancer Res.* **19** 1505–7

Hora M S *et al* 1992 Development of a lyophilized formulation of interleukin-2 *Dev. Biol. Stand.* **74** 295–303

Hughes H P *et al* 1994 A slow release formulation for recombinant bovine interferon alpha I-1 *Antiviral Res.* **23** 33–44

Huyghebaert N, Vermeire A, Rottiers P, Remaut E and Remon J P 2005 Development of an enteric-coated, layered multi-particulate formulation for ileal delivery of viable recombinant *Lactococcus lactis Eur. J. Pharm. Biopharm.* **61** 134–41

Iglesias E *et al* 2006 Influence of aluminum-based adjuvant on the immune response to multiantigenic formulation *Viral Immunol.* **19** 712–21

Iyer L K *et al* 2019 Pulse proteolysis: an orthogonal tool for protein formulation screening *J. Pharm. Sci.* **108** 842–50

Jaber A, Driebergen R, Giovannoni G, Schellekens H, Simsarian J and Antonelli M 2007 The Rebif new formulation story: it's not trials and error *Drugs R D* **8** 335–48

Jacobsen L V, Rolan P, Christensen M S, Knudsen K M and Rasmussen M H 2000 Bioequivalence between ready-to-use recombinant human growth hormone (rhGH) in liquid formulation and rhGH for reconstitution *Growth Horm. IGF Res.* **10** 93–8

Kamat M S, Tolman G L and Brown J M 1996 Formulation development of an antifibrin monoclonal antibody radiopharmaceutical *Pharm. Biotechnol.* **9** 343–64

Kapoor R and Shome D 2018 Intradermal injections of a hair growth factor formulation for enhancement of human hair regrowth-safety and efficacy evaluation in a first-in-man pilot clinical study *J. Cosmet. Laser Ther.* **20** 369–79

Karsdal M A *et al* 2011 Lessons learned from the development of oral calcitonin: the first tablet formulation of a protein in phase III clinical trials *J. Clin. Pharmacol.* **51** 460–71

Kim S J, Hahn S K, Kim M J, Kim D H and Lee Y P 2005 Development of a novel sustained release formulation of recombinant human growth hormone using sodium hyaluronate microparticles *J. Control. Release* **104** 323–35

Kim S J and Kim C W 2015 Characterization of recombinant human growth hormone variants from sodium hyaluronate-based sustained release formulation of rhGH under heat stress *Anal. Biochem.* **485** 59–65

Kim S J and Kim C W 2016 Development and characterization of sodium hyaluronate microparticle-based sustained release formulation of recombinant human growth hormone prepared by spray-drying *J. Pharm. Sci.* **105** 613–22

Kim S Y, Lee S J and Lim S J 2014 Formulation and *in vitro* and *in vivo* evaluation of a cationic emulsion as a vehicle for improving adenoviral gene transfer *Int. J. Pharm.* **475** 49–59

Kraiem H *et al* 2017 Two-dimensional isoelectric focusing OFFGEL, micro-fluidic lab-on-chip electrophoresis and FTIR for assessment of long-term stability of rhG-CSF formulation *IEEE Trans. Nanobiosci.* **16** 694–702

Krasner A, Pohl R, Simms P, Pichotta P, Hauser R and De Souza E 2012 A review of a family of ultra-rapid-acting insulins: formulation development *J. Diabetes Sci. Technol.* **6** 786–96

Kumar P S, Ramakrishna S, Saini T R and Diwan P V 2006 Influence of microencapsulation method and peptide loading on formulation of poly(lactide-co-glycolide) insulin nanoparticles *Pharmazie* **61** 613–7

Kumar V, Sharma V K and Kalonia D S 2009 *In situ* precipitation and vacuum drying of interferon alpha-2a: development of a single-step process for obtaining dry, stable protein formulation *Int. J. Pharm.* **366** 88–98

López M, González L R, Reyes N, Sotolongo J and Pujol V 2004 Stabilization of a freeze-dried recombinant streptokinase formulation without serum albumin *J. Clin. Pharm. Ther.* **29** 367–73

Lv B H, Tan W, Zhu C C, Shang X and Zhang L 2018 Properties of a stable and sustained-release formulation of recombinant human parathyroid hormone (rhPTH) with chitosan and silk fibroin microparticles *Med. Sci. Monit.* **24** 7532–40

Lynch J M, Barbano D M, Bauman D E, Hartnell G F and Nemeth M A 1992 Effect of a prolonged-release formulation of N-methionyl bovine somatotropin (sometribove) on milk fat *J. Dairy Sci.* **75** 1794–1809

Macdougall I C 2004 Pure red cell aplasia with anti-erythropoietin antibodies occurs more commonly with one formulation of epoetin alfa than another *Curr. Med. Res. Opin.* **20** 83–6

Mach H and Arvinte T 2011 Addressing new analytical challenges in protein formulation development *Eur. J. Pharm. Biopharm.* **78** 196–207

Maeda H *et al* 2003 Design of long-acting formulation of protein drugs with a double-layer structure and its application to rhG-CSF *J. Control. Release* **91** 281–97

Mahjoubi N *et al* 2015 Preventing aggregation of recombinant interferon beta-1b in solution by additives: approach to an albumin-free formulation *Adv. Pharm. Bull.* **5** 497–505

Malek-Sabet N, Masoumian M R, Zeinali M, Khalilzadeh R and Mousaabadi J M 2013 Production, purification, and chemical stability of recombinant human interferon-γ in low oxygen tension condition: a formulation approach *Prep. Biochem. Biotechnol.* **43** 586–600

Malyala P and Singh M 2010 Micro/nanoparticle adjuvants: preparation and formulation with antigens *Methods Mol. Biol.* **626** 91–101

Mandal A, Pal D, Agrahari V, Trinh H M, Joseph M and Mitra A K 2018 Ocular delivery of proteins and peptides: challenges and novel formulation approaches *Adv. Drug Deliv. Rev.* **126** 67–95

Mattern M, Winter G, Kohnert U and Lee G 1999 Formulation of proteins in vacuum-dried glasses. II. Process and storage stability in sugar-free amino acid systems *Pharm. Dev. Technol.* **4** 199–208

Maurice T *et al* 2013 Intranasal formulation of erythropoietin (EPO) showed potent protective activity against amyloid toxicity in the $A\beta_{25-35}$ non-transgenic mouse model of Alzheimer's disease *J. Psychopharmacol.* **27** 1044–57

Mitragotri S, Burke P A and Langer R 2014 Overcoming the challenges in administering biopharmaceuticals: formulation and delivery strategies *Nat. Rev. Drug Discov.* **13** 655–72

Mizoguchi M *et al* 2015 Novel oral formulation approach for poorly water-soluble drug using lipocalin-type prostaglandin D synthase *Eur. J. Pharm. Sci.* **74** 77–85

Mizoguchi M, Nakatsuji M, Takano J, Ishibashi O, Wada K and Inui T 2016 Development of pH-independent drug release formulation using lipocalin-type prostaglandin D synthase *J. Pharm. Sci.* **105** 2735–42

Mordenti J *et al* 1999 Intraocular pharmacokinetics and safety of a humanized monoclonal antibody in rabbits after intravitreal administration of a solution or a PLGA microsphere formulation *Toxicol. Sci.* **52** 101–6

Murányi A, Bartoš P, Tichý E, Lazová J, Pšenková J and Žabka M 2015 Development of gel-forming lyophilized formulation with recombinant human thrombin *Drug Dev. Ind. Pharm.* **41** 1566–73

Naito M *et al* 2013 CD40L-Tri, a novel formulation of recombinant human CD40L that effectively activates B cells *Cancer Immunol. Immunother.* **62** 347–57

Naughton C A, Duppong L M, Forbes K D and Sehgal I 2003 Stability of multidose, preserved formulation epoetin alfa in syringes for three and six weeks *Am. J. Health Syst. Pharm.* **60** 464–8

Nguyen T H and Ward C 1993 Stability characterization and formulation development of alteplase, a recombinant tissue plasminogen activator *Pharm. Biotechnol.* **5** 91–134

Norbury L J *et al* 2018 Intranasal delivery of a formulation containing stage-specific recombinant proteins of *Fasciola hepatica* cathepsin L5 and cathepsin B2 triggers an anti-fecundity effect and an adjuvant-mediated reduction in fluke burden in sheep *Vet. Parasitol.* **258** 14–23

Osterberg T, Fatouros A and Mikaelsson M 1997 Development of freeze-dried albumin-free formulation of recombinant factor VIII SQ *Pharm. Res.* **14** 892–8

Osterberg T, Fatouros A, Neidhardt E, Warne N and Mikaelsson M 2001 B-domain deleted recombinant factor VIII formulation and stability *Semin. Hematol.* **38** 40–3

Page C, Dawson P, Woollacott D, Thorpe R and Mire-Sluis A 2000 Development of a lyophilization formulation that preserves the biological activity of the platelet-inducing cytokine interleukin-11 at low concentrations *J. Pharm. Pharmacol.* **52** 19–26

Panjwani N, Hodgson D J, Sauvé S and Aubin Y 2olo Assessment of the effects of pH, formulation and deformulation on the conformation of interferon alpha-2 by NMR *J. Pharm. Sci.* **99** 3334–42

Park K 2012 Significance of handling, formulation and storage conditions on the stability and bioactivity of rhBMP-2 *J. Control. Release* **162** 654

Patro S Y, Freund E and Chang B S 2002 Protein formulation and fill-finish operations *Biotechnol. Annu. Rev.* **8** 55–84

Pellequer Y, Ollivon M and Barratt G 2004 Formulation of liposomes associated with recombinant interleukin-2: effect on interleukin-2 activity *Biomed. Pharmacother.* **58** 162–7

Pereira P, Kelly S M, Cooper A, Mardon H J, Gellert P R and van der Walle C F 2007 Solution formulation and lyophilisation of a recombinant fibronectin fragment *Eur. J. Pharm. Biopharm.* **67** 309–19

Peters E E, Ameri M, Wang X, Maa Y F and Daddona P E 2012 Erythropoietin-coated ZP-microneedle transdermal system: preclinical formulation, stability, and delivery *Pharm. Res.* **29** 1618–26

Piedmonte D M and Treuheit M J 2008 Formulation of Neulasta (pegfilgrastim) *Adv. Drug Deliv. Rev.* **60** 50–8

Rasmussen T, Tantipolphan R, van de Weert M and Jiskoot W 2010 The molecular chaperone alpha-crystallin as an excipient in an insulin formulation *Pharm. Res.* **27** 1337–47

Remmele R L, Nightlinger N S, Srinivasan S and Gombotz W R 1998 Interleukin-1 receptor (IL-1R) liquid formulation development using differential scanning calorimetry *Pharm. Res.* **15** 200–8

Richard J and Prang N 2010 The formulation and immunogenicity of therapeutic proteins: product quality as a key factor *IDrugs* **13** 550–8

Ruiz L *et al* 2006 Long-term stabilization of a new freeze-dried and albumin-free formulation of recombinant human interferon alpha 2b *PDA J. Pharm. Sci. Technol.* **60** 72–8

Ruiz L, Rodriguez I, Baez R and Aldana R 2007 Stability of an extemporaneously prepared recombinant human interferon alfa-2b eye drop formulation *Am. J. Health Syst. Pharm.* **64** 1716–9

Santana H, García G, Vega M, Beldarraín A and Páez R 2015 Stability studies of a freeze-dried recombinant human epidermal growth factor formulation for wound healing *PDA J. Pharm. Sci. Technol.* **69** 399–416

Senet P, Mons B, Aractangi S and Tilleul P 2002 Evaluation of the stability and efficacy of rhGM-CSF as a topical agent in a gel formulation *J. Wound Care* **11** 132–4

Shire S J 1996 Stability characterization and formulation development of recombinant human deoxyribonuclease I (Pulmozyme, (dornase alpha)) *Pharm. Biotechnol.* **9** 393–426

Singh M, Shirley B, Bajwa K, Samara E, Hora M and O'Hagan D 2001 Controlled release of recombinant insulin-like growth factor from a novel formulation of polylactide-co-glycolide microparticles *J. Control Release* **70** 21–8

Sønderby P, Bukrinski J T, Hebditch M, Peters G H J, Curtis R A and Harris P 2018 Self-interaction of human serum albumin: a formulation perspective *ACS Omega* **3** 16105–17

Steiner S *et al* 2008 A novel insulin formulation with a more rapid onset of action *Diabetologia* **51** 1602–06

Stote R, Marbury T, Shi L, Miller M and Strange P 2010 Comparison pharmacokinetics of two concentrations (0.7% and 1.0%) of Nasulin, an ultra-rapid-acting intranasal insulin formulation *J. Diabetes Sci. Technol.* **4** 603–9

Stote R, Miller M, Marbury T, Shi L and Strange P 2011 Enhanced absorption of Nasulin™, an ultrarapid-acting intranasal insulin formulation, using single nostril administration in normal subjects *J. Diabetes Sci. Technol.* **5** 113–9

Takeshita A *et al* 2000 Efficacy of a new formulation of lenograstim (recombinant glycosylated human granulocyte colony-stimulating factor) containing gelatin for the treatment of neutropenia after consolidation chemotherapy in patients with acute myeloid leukemia *Int. J. Hematol.* **71** 136–43

Thompson C A 2001 New interferon alfa formulation licensed for treatment of hepatitis C *Am. J. Health Syst. Pharm.* **58** 452

Vemuri S, Yu C T and Roosdorp N 1993 Formulation and stability of recombinant alpha 1-antitrypsin *Pharm. Biotechnol.* **5** 263–86

Volkin D B and Middaugh C R 1996 The characterization, stabilization, and formulation of acidic fibroblast growth factor *Pharm. Biotechnol.* **9** 181–217

Wang S, Zhang X, Wu G, Tian Z and Qian F 2017 Optimization of high-concentration endostatin formulation: harmonization of excipients' contributions on colloidal and conformational stabilities *Int. J. Pharm.* **530** 173–86

Wei Y *et al* 2012 A novel sustained-release formulation of recombinant human growth hormone and its pharmacokinetic, pharmacodynamic and safety profiles *Mol. Pharm.* **9** 2039–48

Wright J F, Qu G, Tang C and Sommer J M 2003 Recombinant adeno-associated virus: formulation challenges and strategies for a gene therapy vector *Curr. Opin. Drug Discov. Devel.* **6** 174–8

Yatuv R *et al* 2008 Enhancement of factor VIIa haemostatic efficacy by formulation with PEGylated liposomes *Haemophilia* **14** 476–83

Chapter 5

Behera B K 2021 *Biopharmaceuticals: Challenges and Opportunities* 1st edn (Boca Raton, FL: CRC)

CBER 2021 Administrative processing of original biologics license applications (BLA) and new drug applications (NDA) *Report* SOPP 8401 https://www.fda.gov/media/85659/download

Dutton R L and Scharer J M 2007 *Advanced Technologies in Biopharmaceutical Processing* 1st edn (Ames, IA: Blackwell)

FDA 2002 Science and the regulation of biological products https://www.fda.gov/about-fda/histories-product-regulation/science-and-regulation-biological-products

FDA, CMC and GMP guidances https://www.fda.gov/vaccines-blood-biologics/general-biologics-guidances/cmc-and-gmp-guidances

Geigert J 2019 *The Challenge of CMC Regulatory Compliance for Biopharmaceuticals* 3rd edn (Cham: Springer)

Grindley J N and Ogden J E 2000 *Understanding Biopharmaceuticals: Manufacturing and Regulatory Issues* (Denver, CO: Interpharm)

Hill R G and Rang H P 2013 *Drug Discovery and Development: Technology in Transition* 2nd edn (Edinburgh: Churchill Livingstone/Elsevier)

Mackler B F 1992 *Evolving Regulatory Concerns in Safety Evaluation of Biopharmaceuticals* (Potomac, MD: BioConferences International)

Rathore A S and Sofer G K 2012 *Process Validation in Manufacturing of Biopharmaceuticals* 3rd edn (Boca Raton, FL: Taylor and Francis/CRC)

Thomaz-Soccol V, Pandey A and Resende R R 2017 *Current Developments in Biotechnology and Bioengineering: Human and Animal Health Applications* (Amsterdam: Elsevier)

Chapter 6

Australian Government, Department of Health 2018 How are biosimilar medicines assessed and monitored? https://www1.health.gov.au/internet/main/publishing.nsf/Content/biosimilar-hp-how-are-biosimilar -medicines-assessed-and-monitored

Avery R L *et al* 2017 Systematic pharmacokinetics and pharmacodynamics of intravitreal aflibercept, bevacizumab, and ranibizumab *Retina* **37** 1847–58

Brian K 2009 Industrialization of mAb production technology: the bioprocessing industry at a crossroads *mAbs* **1** 443–52

Brennan Z 2018 AbbVie sees 80% discounts in Nordic market with new Humira biosimilars *RAPS Regulatory Focus* 2 November https://www.raps.org/news-and-articles/news-articles/2018/11/abbvie-sees-80-discounts-in-nordic-market-with-ne

BIO Citizen Petition, Follow-on Therapeutic Proteins, 23 April 2003 http://www.fda.gov/OHRMS/DOCKETS/DOCKETS/03p0176/03p-0176-cp00001-01-vol11.pdf

Colburn W A 1983 A time-dependent volume of distribution term used to describe linear concentration–time profiles *J. Pharmacokin. Biopharm.* **11** 389–400

Cauchi R 2018 State laws and legislation related to biologic medications and substitution of biosimilars *NCSL* 22 October http://www.ncsl.org/research/health/state-laws-and-legislation-related-to-biologic-medications-and-substitution-of-biosimilars.aspx

Dudzinski D M 2005 Reflections on historical, scientific, and legal issues relevant to designing approval pathways for generic versions of recombinant protein-based therapeutics and monoclonal antibodies *Food Drug Law J.* **60** 143–260

Durkin E 2016 WHO unveils final biological naming plan that differs from FDA's *InsideHealthPolicy.com FDA Week* 29 January

Davio K 2018 FDA's Leah Christl provides an update on the biosimilar action plan *AJMC The Center for Biosimilars* https://www.centerforbiosimilars.com/conferences/dia-2018/fdas-leah-christl-provides-an-update-on-the-biosimilar-action-plan

DiMasi J A, Hansen R W and Grabowski H G 2003 The price of innovation: new estimates of drug development costs *J. Health Econ.* **22** 151–85

Dong X, Tsong Y, Shen M and Zhong J 2015 Using tolerance intervals for assessment of pharmaceutical quality *J. Biopharm. Stat.* **25** 317–27

EMA 2015 Benepali EPAR EMA/CHMP/819219/2015 http://www.ema.europa.eu

EMA 2017 Cyltezo EPAR EMA/CHMP/750187/2017 http://www.epa.europa.eu

EMA 2016 Efgratin EPAR. EMA/18691/2017; Withdrawal assessment report http://www.ema.europa.eu

EMA 2017 Biosimilar medicines: Overview https://www.ema.europa.eu/en/human-regulatory/overview/biosimilar-medicines-overview

EMA 2009 Filgrastim Hexal EPAR EMEA/CHMP/651324/2008 https://www.ema.europa.eu/en/documents/assessment-report/filgrastim-hexal-epar-public-assessment-report_en.pdf

EMA 2018 Filgrastim, EMEA EMEA/CHMP/BMWP/31329/2005 Rev 1 http://www.ema.europa.eu

EMA Flixabi: EPAR EMA http://www.ema.europa.eu

EMA 2017 Guideline on the evaluation of anticancer medicinal products in man; EMA/CHMP/205/95 Rev5 https://www.ema.europa.eu/en/documents/scientific-guideline/guideline-evaluation-anticancer-medicinal-products-man-revision-5_en.pdf

EMA 2017 Herzuma EPAR EMA/44005/2018 https://www.ema.europa.eu/en/medicines/human/EPAR/herzuma

EMA Hyrimoz EPAR EMA/CHMP/404076/2018 https://www.ema.europa.eu/en/medicines/human/EPAR/hyrimoz

EMA Kanjinti EPAR EMA/CHMP/261937/2018 https://www.ema.europa.eu/en/documents/assessment-report/kanjinti-epar-public-assessment-report_en.pdf;

EMA 2017 Mvasi EPAR EMA/798844/2017 https://www.ema.europa.eu/en/medicines/human/EPAR/mvasi

EMA 2017 Ontruzant EPAR EMA/CHMP/9855/2018 https://www.ema.europa.eu/en/medicines/human/EPAR/ontruzant

EMA 2010 Prolia EPAR EMA/21672/2010 https://www.ema.europa.eu/en/documents/assessment-report/prolia-epar-public-assessment-report_en.pdf

EMA 2017 Rixathon EPAR EMA/303207/2017 https://www.ema.europa.eu/en/documents/assessment-report/rixathon-epar-public-assessment-report_en.pdf

EMA Soliris. EPAR 2007 https://www.ema.europa.eu/en/medicines/human/EPAR/soliris

EMA 2016 Terrosa EPAR EMA/84371/2017 https://www.ema.europa.eu/en/documents/assessment-report/terrosa-epar-public-assessment-report_en.pdf

EMA 2016 Truxima EPAR EMA/CHMP/75695/2017 https://www.ema.europa.eu/en/documents/assessment-report/truxima-epar-public-assessment-report_en.pdf

EMA 2017 Ziextenzo (pegfilgrastim), EPAR EMA/47326/2017, Withdrawal of the marketing authorisation application

FDA 2020 Drug Approval Package: ABRILADA https://www.accessdata.fda.gov/drugsatfda_-docs/nda/2019/761118Orig1s000TOC.cfm

FDA 2016 Drug Approval Package: Amjevita (adalimumab-atto) https://www.accessdata.fda.gov/drugsatfda_docs/nda/2016/761024_toc.cfm

FDA 2020 Drug Approval Package: AVSOLA https://www.accessdata.fda.gov/drugsatfda_docs/nda/2019/761086Orig1s000TOC.cfm

FDA 2018 Drug Approval Package: Cyltezo (adalimumab-adbm) https://www.accessdata.fda.gov/drugsatfda_docs/nda/2017/761058Orig1s000TOC.cfm

FDA 2016 Drug Approval Package: Erelzi (etanercept-szzs) https://www.accessdata.fda.gov/drugsatfda_docs/nda/2016/761042Orig1_toc.cfm

FDA 2019 Drug Approval Package: Eticovo https://www.accessdata.fda.gov/drugsatfda_docs/nda/2019/761066Orig1s000TOC.cfm

FDA 2018 Biosimilars action plan: balancing innovation and competition https://www.fda.gov/media/114574/download

FDA *Purple Book: Lists of Licensed Biological Products with Reference Product Exclusivity and Biosimilarity or Interchangeability Evaluations* https://www.fda.gov/drugs/developmentapprovalprocess/howdrugsaredevelopedandapproved/approvalapplications/therapeuticbiologicapplications/biosimilars/ucm411418.htm

FDA Significant dates in US food and drug law history http://www.fda.gov/AboutFDA/WhatWeDo/History/Milestones/ucm128305.htm

FDA 2016 Biosimilar User Fee Act (BsUFA II) Reauthorization Public Meeting, 20 October, slide 31 https://www.fda.gov/downloads/ForIndustry/UserFees/BiosimilarUserFeeActBsUFA/UCM526071.pdf

FDA 2019 Considerations in demonstrating interchangeability with the reference product, guidance for industry https://www.fda.gov/downloads/Drugs/GuidanceComplianceRegulatoryInformation/Guidances/UCM537135.pdf

FDA 2016 Implementation of the biologics price competition and innovation Act of 2009 http://www.fda.gov/Drugs/GuidanceComplianceRegulatoryInformation/ucm215089.htm

FDA 2018 Interpretation of the 'deemed to be a license' provision of the Biologics Price Competition and Innovation Act of 2009, guidance for industry https://www.fda.gov/downloads/Drugs/GuidanceComplianceRegulatoryInformation/Guidances/UCM490264.pdf

FDA 2018 Labeling for proposed biosimilar products: guidance for industry http://www.fda.gov/downloads/Drugs/GuidanceComplianceRegulatoryInformation/Guidances/UCM493439.pdf

FDA 2017 Nonproprietary naming of biological products: guidance for industry https://www.fda.gov/downloads/Drugs/GuidanceComplianceRegulatoryInformation/Guidances/UCM459987.pdf

FDA *Orange Book: Approved Drug Products with Therapeutic Equivalence Evaluations* https://www.accessdata.fda.gov/Scripts/cder/ob/index.cfm

FDA 2018 Drug Approval Package: Fulphila https://www.accessdata.fda.gov/drugsatfda_docs/nda/2018/761075Orig1s000TOC.cfm

FDA 2019 Drug Approval Package: HADLIMA https://www.accessdata.fda.gov/drugsatfda_-docs/nda/2019/761059Orig1s000TOC.cfm

FDA 2019 Drug Approval Package: Herzuma https://www.accessdata.fda.gov/drugsatfda_docs/nda/2018/761091Orig1s000TOC.cfm

FDA 2019 Drug Approval Package: Hyrimoz https://www.accessdata.fda.gov/drugsatfda_docs/nda/2018/761071Orig1s000TOC.cfm

FDA 2016 Inflectra (infliximab-dyyb) for Injection https://www.accessdata.fda.gov/drugsatfda_-docs/nda/2016/125544Orig1s000TOC.cfm

FDA 2018 Drug Approval Package: Ixifi (infliximab-qbtx) https://www.accessdata.fda.gov/drugsatfda_docs/nda/2017/761072Orig1s000TOC.cfm

Franklin J 2018 Proposed biosimilar and interchangeable products: brief update from FDA *FDLI Annual Conference (3 May)* slide 3 https://www.fdli.org/wp-content/uploads/2018/05/Biosimilars-New-Developments-and-Updates-3.pdf

FDA 2019 Drug Approval Package: Kanjinti https://www.accessdata.fda.gov/drugsatfda_docs/nda/2019/761073Orig1s000TOC.cfm

FDA 2018 Drug Approval Package: Mvasi (bevacizumab-awwb) https://www.accessdata.fda.gov/drugsatfda_docs/nda/2017/761028Orig1s000TOC.cfm

FDA 2019 Drug Approval Package: Nivestym (filgrastim-aafi) https://www.accessdata.fda.gov/drugsatfda_docs/nda/2018/761080Orig1s000TOC.cfm

FDA 2018 Drug Approval Package: OGIVRI (Trastuzumab-dkst) https://www.accessdata.fda.gov/drugsatfda_docs/nda/2017/761074Orig1s000TOC.cfm

FDA 2019 Drug Approval Package: Ontruzant (trastuzumab-dttb) https://www.accessdata.fda.gov/drugsatfda_docs/nda/2019/761100Orig1s000TOC.cfm

FDA 2018 Drug Approval Package: Renflexis (infliximab-abda) https://www.accessdata.fda.gov/drugsatfda_docs/nda/2017/761054Orig1s000TOC.cfm

FDA 2018 Drug Approval Package: Retacrit (epoetin alfa-epbx) https://www.accessdata.fda.gov/drugsatfda_docs/nda/2018/125545Orig1s000TOC.cfm

FDA 2019 Drug Approval Package: RUXIENCE https://www.accessdata.fda.gov/drugsatfda_-docs/nda/2019/761103Orig1s000TOC.cfm

FDA 2019 Drug Approval Package: Trazimera (trastuzumab-qyyp) https://www.accessdata.fda.gov/drugsatfda_docs/nda/2019/761081Orig1s000TOC.cfm

FDA 2019 Drug Approval Package: TRUXIMA (rituximab-abbs) https://www.accessdata.fda.gov/drugsatfda_docs/nda/2018/761088Orig1s000TOC.cfm

FDA 2019 Drug Approval Package: UDENYCA https://www.accessdata.fda.gov/drugsatfda_-docs/nda/2018/761039Orig1s000TOC.cfm

FDA 2015 Zarxio (filgrastim-sndz) https://www.accessdata.fda.gov/drugsatfda_docs/nda/2015/125553Orig1s000TOC.cfm

FDA 2019 Drug Approval Package: Zirabev https://www.accessdata.fda.gov/drugsatfda_docs/nda/2019/761099Orig1s000TOC.cfm

FDA 2019 Drug Approval Package: ZIEXTENZO https://www.accessdata.fda.gov/drugsatfda_-docs/nda/2019/761045Orig1s000TOC.cfm

Government of Canada 2019 Biosimilar biologic drugs https://www.canada.ca/en/health-canada/services/drugs-health-products/biologics-radiopharmaceuticals-genetic-therapies/biosimilar-biologic-drugs.html

Gotham D, Barber M and Hill A 2018 Production costs and potential prices for biosimilars of human insulin and insulin analogues *BMJ Global Health* **3** e000850

Goetze A M *et al* 2011 High-mannose glycans on the Fc region of therapeutic IgG antibodies increase serum clearance in humans *Glycobiology* **21** 949–59

IMS Health 2016 The impact of biosimilar competition, p 4 https://www.medicinesforeurope.com/wp-content/uploads/2016/08/IMS-Impact-of-Biosimilar-Competition-2016.pdf

Indian proposed biosimilar guidance http://nib.gov.in/NIB-DBT2016.pdf

Initiative for Medicines, Access, and Knowledge (I-MAK), Overpatented, overpriced: how excessive pharmaceutical patenting is extending monopolies and driving up drug prices, p 7 https://www.i-mak.org/wp-content/uploads/2018/08/I-MAK-Overpatented-Overpriced-Report.pdf

IQVIA Institute 2018 Medicine use and spending in the US: A review of 2017 and Outlook to 2022 *Report* April, p 11

Japanese Ministry of Health, Labour and Welfare 2013 Proposed Biosimilar Guidance Guideline for the Quality, Safety, and Efficacy Assurance of Follow-on Biologics https://www.pmda.go.jp/files/000153851.pdf

Janet Woodcock, Deputy Commissioner, Chief Medical Officer, FDA, testimony before the House Committee on Oversight and Government Reform, March 26, 2007 http://www.fda.gov/newsevents/testimony/ucm154070.htm

Kesselheim A S and Avorn J 2016 The high cost of prescription drugs in the United States: origins and prospects for reform *JAMA* **316** 858–71

Kang P and Saif M 2007 Infusion-related and hypersensitivity reactions of monoclonal antibodies used to treat colorectal cancer—identification, prevention, and management *J. Support. Oncol.* **5** 451–7

Kaitin K I, Bryant N R and Lasagna L 1993 The role of the research-based pharmaceutical industry in medical progress in the United States *J. Clin. Pharmacol.* **33** 412–7

Kozlowski S, Woodcock J, Midthun K and Sherman R B 2011 Developing the nation's biosimilars program *N. Engl. J. Med.* **365** 385–8

Keown A 2018 FDA withdraws draft guidance on biosimilar development *BioSpace* https://www.biospace.com/article/fda-withdraws-draft-guidance-on-biosimilar-development/

Kneller R 2010 The importance of new companies for drug discovery: origins of a decade of new drugs *Nat. Rev. Drug Discov.* **9** 867–82

National Medical Products Association, Chinese Proposed biosimilar guidance http://subsites.chinadaily.com.cn/nmpa/index.html

Niazi S 1976 Volume of distribution as a function of time *J. Pharm. Sci.* **65** 452–4

Niazi S and Brown J L 2016 *Fundamentals of Modern Bioprocessing* (Boca Raton, FL: CRC)

Niazi S 2019 *Biosimilarity: The FDA Perspective* (Boca Raton, FL: CRC)

Niazi S 2016 *Biosimilars and Interchangeable Biologics: Strategic Elements* (Boca Raton, FL: CRC/Taylor and Francis)

Niazi S 2016 *Biosimilars and Interchangeable Biologics: Tactical Elements* (Boca Raton, FL: Taylor and Francis/CRC)

Niazi S 2006 *Handbook of Biogeneric Therapeutic Proteins: Regulatory, Manufacturing, Testing, and Patent Issues* (London: Taylor and Francis)

Niazi S 2019 *Handbook of Preformulation: Chemical, Biological, and Botanical Drugs* 2nd edn (Boca Raton, FL: CRC)

Niazi S 1979 *Textbook of Biopharmaceutics and Clinical Pharmacokinetics* (New York: Appleton-Century-Crofts)

Niazi S 1976 Volume of distribution and tissue level errors in instantaneous intravenous input assumptions *J. Pharm. Sci.* **65** 1541–3

Niazi S 1976 Volume of distribution as a function of time *J. Pharm. Sci.* **65** 452–4

Niazi S 2019 Critical analysis of draft FDA guidance on the development of therapeutic protein biosimilars *The Center for Biosimilars* https://www.centerforbiosimilars.com/contributor/sarfaraz-niazi/2019/06/a-critical-analysis-of-the-fda-draft-guidance-on-development-of-therapeutic-protein-biosimilars-comparative-analytical-assessment-and-other-qualityrelated-considerations

Niazi S 2019 A critical analysis of the final guidance on demonstrating interchangeability of a biosimilar with its reference product *The Center for Biosimilars* https://www.centerforbiosimilars.com/contributor/sarfaraz-niazi/2019/05/a-critical-analysis-of-the-final-guidance-on-demonstrating-interchangeability-of-a-biosimilar-with-its-reference-product

Niazi S 2019 A critical analysis of the FDA draft guidance on development of therapeutic protein biosimilars: comparative analytical assessment and other quality-related considerations *The Center for Biosimilars* https://www.centerforbiosimilars.com/contributor/sarfaraz-niazi/2019/06/a-critical-analysis-of-the-fda-draft-guidance-on-development-of-therapeutic-protein-biosimilars-comparative-analytical-assessment-and-other-qualityrelated-considerations

Niazi S 2019 FDA allows waiver of clinical trials for insulin biosimilars as recommended in Niazi citizen petition *BioSpace* https://www.biospace.com/article/releases/fda-allows-waiver-of-clinical-trials-for-insulin-biosimilars-as-recommended-in-niazi-citizen-petition/

Roskos L, Lum F, Lockbaum F, Schwab G and Yang B 2006 Pharmacokinetic/pharmacodynamic modeling of pegfilgrastim in healthy subjects *Clin. Pharmacol.* **46** 747–57

Rup B *et al* 2015 Standardizing terms, definitions and concepts for describing and interpreting unwanted immunogenicity of biopharmaceuticals: recommendations of the Innovative Medicines Initiative ABIRISK consortium *Clin. Exp. Immunol.* **181** 385–400

Shankar G *et al* 2014 Assessment and reporting of the clinical immunogenicity of therapeutic proteins and peptides—harmonized terminology and tactical recommendations *AAPS J.* **16** 658–73

Stevens A J *et al* 2011 The role of public-sector research in the discovery of drugs and vaccines *New England J. Med.* **364** 535–41

Silverman E 2018 AbbVie offers a ‘huge’ discount on Humira to fend off European rivals *STAT+ Pharmalot* 1 November

Silverman E 2018 Behind the patent thicket: tactics AbbVie allegedly used to thwart proposed biosimilar versions of Humira *STAT* 7 November https://www.statnews.com/pharmalot/2018/11/07/abbvie-biosimilars-humira-patents/

The Insulin Amendments, PL 77–366, codified at 21 USC 356, were repealed by PL 105–115, the Food and Drug Administration Modernization Act (FDAMA)

Törnqvist E, Annas A, Granath B, Jalkesten E, Cotgreave I and Öberg M 2014 Strategic focus on 3R principles reveals major reductions in the use of animals in pharmaceutical toxicity testing *PLoS One* **9** e101638

Whalen J 2015 Why the US pays more than other countries for drugs *The Wall Street Journal* 1 December

Witkovsky V 2014 On the exact two-sided tolerance intervals for univariate normal distribution and linear regression *Aust. J. Statist.* **43** 279–92

Wesolowski C A, Wesolowski M J, Babyn P S and Wanasundara S N 2016 Time varying apparent volume of distribution and drug half-lives following intravenous bolus injections *PLoS One* **11** e0158798

World Health Organization 2010 Guidelines on the approval of biotherapeutic products https://www.who.int/biologicals/areas/biological_therapeutics/BIOTHERAPEUTICS_FOR_WEB_22APRIL2010.pdf

Chapter 7

Adams S 2002 Information sources on post-grant actions to pharmaceutical patents *J. Chem. Inf. Comput. Sci.* **42** 467–72

Bollok M, Resina D, Valero F and Ferrer P 2009 Recent patents on the *Pichia pastoris* expression system: expanding the toolbox for recombinant protein production *Recent Pat. Biotechnol.* **3** 192–201

Cockburn I and Long G 2015 The importance of patents to innovation: updated cross-industry comparisons with biopharmaceuticals *Expert Opin. Ther. Pat.* **25** 739–42

German Patent Law of 16 December 1980 as amended by Laws of 16 July and 6 August 1998, section 16(1)

Murhammer D W 1991 Review and patents and literature. The use of insect cell cultures for recombinant protein synthesis: engineering aspects *Appl. Biochem. Biotechnol.* **31** 283–310

Niazi S 2003 *Filing Patents Online* (Boca Raton FL: CRC)

Picanco-Castro V, de Freitas M C, Bomfim A S and de Sousa Russo E M 2014 Patents in therapeutic recombinant protein production using mammalian cells *Recent Pat. Biotechnol.* **8** 165–71

United Kingdom Public and General Acts, 35 and 36 Eliz. 2 ch. 37, Patents Act 1977, s.25(1)

World Trade Organization 1994 Agreement on trade-related aspects of intellectual property rights, Article 33 www.wto.org/english/docs_e/legal_e/27-trips_01_e.htm

www.ingramcontent.com/pod-product-compliance
Ingram Content Group UK Ltd.
Pitfield, Milton Keynes, MK11 3LW, UK
UKHW051042120926
8556IPUK00030B/805

9 780750 331739